Handbook of Coffee Processing By-Products

Handbook of Coffee Processing By-Products
Sustainable Applications

Edited by

Charis M. Galanakis
Research & Innovation Department
Galanakis Laboratories
Chania, Greece

ACADEMIC PRESS
An imprint of Elsevier

Academic Press is an imprint of Elsevier
125 London Wall, London EC2Y 5AS, United Kingdom
525 B Street, Suite 1800, San Diego, CA 92101-4495, United States
50 Hampshire Street, 5th Floor, Cambridge, MA 02139, United States
The Boulevard, Langford Lane, Kidlington, Oxford OX5 1GB, United Kingdom

Notices
Knowledge and best practice in this field are constantly changing. As new research and experience broad-
en our understanding, changes in research methods, professional practices, or medical treatment may
become necessary.

Practitioners and researchers must always rely on their own experience and knowledge in evaluating and
using any information, methods, compounds, or experiments described herein. In using such information
or methods they should be mindful of their own safety and the safety of others, including parties for whom
they have a professional responsibility.

To the fullest extent of the law, neither the Publisher nor the authors, contributors, or editors, assume any
liability for any injury and/or damage to persons or property as a matter of products liability, negligence
or otherwise, or from any use or operation of any methods, products, instructions, or ideas contained in
the material herein.

Library of Congress Cataloging-in-Publication Data
A catalog record for this book is available from the Library of Congress

British Library Cataloguing-in-Publication Data
A catalogue record for this book is available from the British Library

ISBN: 978-0-12-811290-8

For information on all Academic Press publications visit our website at
https://www.elsevier.com/books-and-journals

Working together
to grow libraries in
developing countries

www.elsevier.com • www.bookaid.org

Publisher: Nikki Levy
Acquisition Editor: Megan R. Ball
Editorial Project Manager: Jaclyn A. Truesdell
Production Project Manager: Lisa Jones
Designer: Greg Harris

Contents

List of Contributors

Rita C. Alves
REQUIMTE/LAQV, University of Porto, Porto, Portugal

Rafaela P. Andrade
University of Lavras (UFLA), Lavras, Minas Gerais, Brazil

Valentina Aristizábal-Marulanda
National University of Colombia, Manizales, Colombia

Ana Belščak-Cvitanović
University of Zagreb, Zagreb, Croatia

Carlos A. Cardona Alzate
National University of Colombia, Manizales, Colombia

Yéssica Chacón-Perez
National University of Colombia, Manizales, Colombia

Marcelo M.R. de Melo
University of Aveiro, Aveiro, Portugal

Maria D. del Castillo
Institute of Food Science Research (UAM–CSIC), Madrid, Spain

Eustáquio S. Dias
University of Lavras (UFLA), Lavras, Minas Gerais, Brazil

Jorge Domínguez
University of Vigo, Vigo, Spain

Benjamin M. Dorsey
Illinois State University, Normal, IL, United States

Whasley F. Duarte
University of Lavras (UFLA), Lavras, Minas Gerais, Brazil

Beatriz Fernandez-Gomez
Institute of Food Science Research (UAM–CSIC), Madrid, Spain

Suzana E. Hikichi
University of Lavras (UFLA), Lavras, Minas Gerais, Brazil

Amaia Iriondo-DeHond
Institute of Food Science Research (UAM–CSIC), Madrid, Spain

Maite Iriondo-DeHond
Madrid Institute for Research and Rural Development, Agriculture, and Food, Alcalá de Henares, Spain

Mejdi Jeguirim
Institute of Materials Science of Mulhouse, Mulhouse, France

Marjorie A. Jones
Illinois State University, Normal, IL, United States

Draženka Komes
University of Zagreb, Zagreb, Croatia

Madona Labaki
Lebanese University, Fanar, Jdeidet, Lebanon

Lionel Limousy
Institute of Materials Science of Mulhouse, Mulhouse, France

Nuria Martinez-Saez
Institute of Food Science Research (UAM–CSIC), Madrid, Spain

Maria Antónia Nunes
REQUIMTE/LAQV, University of Porto, Porto, Portugal

M. Beatriz P.P. Oliveira
REQUIMTE/LAQV, University of Porto, Porto, Portugal

Inês Portugal
University of Aveiro, Aveiro, Portugal

Francisca Rodrigues
REQUIMTE/LAQV, University of Porto, Porto, Portugal

Juan C. Sanchez-Hernandez
University of Castilla-La Mancha, Toledo, Spain

Carlos M. Silva
University of Aveiro, Aveiro, Portugal

Armando J.D. Silvestre
University of Aveiro, Aveiro, Portugal

Wen-Tien Tsai
National Pingtung University of Science and Technology, Pingtung, Taiwan

Ana F. Vinha
REQUIMTE/LAQV, University of Porto; University Fernando Pessoa, Porto, Portugal

Jin-Rong Zhou
Beth Israel Deaconess Medical Center, Boston, MA, United States

Preface

Coffee is one of the most consumed beverages all over the world and the second largest commodity in trading volume, being surpassed only by oil. Currently, more than 70 countries produce coffee. Global coffee production was estimated to be 145 million 60 kg bags in 2015–16, and consumption rounded the 152.1 million. On the other hand, while billions of cups of coffee are consumed worldwide each day, the bulk of the coffee plant biomass is considered to be waste. Following the several steps of production, the coffee industry generates huge amounts of residues, such as coffee silverskin, spent coffee grounds, coffee pulp, coffee husks, cut coffee stems, and wastewater. For instance, in Brazil the production of coffee in the years 2008–13 averaged 2.9 million tons, generating about 1.4 million tons of waste each year. These residues constitute a source of severe contamination of water bodies and lands around production units, and a serious environmental problem for all the producing countries. The problem is now getting even worse, taking into consideration the recent concerns that global warming will severely affect coffee cultivation if not make it disappear.

The current state-of-the-art handling of coffee processing residues includes management practices that either degrade the substrate or lead to diminution of their pollution load without advancing high-added-value ingredients like antioxidants. These practices have a negative ecologic and economic impact on the coffee industry, and cannot be continued within the sustainability and bioeconomy frame of the years to come. The urgent need for sustainability within the coffee industry has turned the interest of researchers to investigate the handling of coffee by-products with another perspective (e.g., by adapting more profitable options). Therefore, modern coffee industries are driven to develop valorization strategies that allow not only the recovery of high-added-value ingredients, but also their recycling through the generation of new products that find applications in diverse biotechnological fields, such as pharmaceutical, food, or cosmetics industries.

Following these considerations, there is a need for a new guide covering the latest developments in this particular direction. The current book aims to indicate the alternative sustainable solutions of upgrading coffee processing residues, as well as denoting their industrial potential as a source for the recovery of bioactive compounds and their reutilization in the previously noted sectors. It fills the existing gap in the current literature by providing a reference for all the involved partners active in the field trying to optimize the performance of coffee-processing industries and reduce their environmental impact. This is conducted by denoting advantages, disadvantages, and real potentiality of relevant processes, as well as highlighting success stories that are already applied in some countries. The ultimate goal is to support the scientific community, professionals, and producers that aspire to develop real high-scale commercial applications.

The book consists of 12 chapters. Chapter 1 discusses the state of the art in the field of coffee processing by-products by describing the steps involved in coffee processing from the field to the cup, the respective generation of by-products along the chain, and their characteristics. In addition, it provides an overview of the methods proposed for the sustainable management of these by-products, as well as legislative frameworks and policy recommendations. Chapter 2 explores the healthy components (e.g., caffeine, chlorogenic acids, trigonelline, and diterpenes) of coffee and coffee processing by-products, and gives some background on antioxidants (what they are and how to study them) and how these relate to health. Chapter 3 deals with the industrial valorization of relevant residues within the integrated concept of biorefinery, taking into account production scale, design, and technical and economic issues. Chapter 4 explores the extraction and formulation of bioactive compounds from coffee processing by-products using conventional extraction techniques and approaches. An overview of patented recovery methodologies and potential applications resulting from the use of the recovered bioactives is provided, too. Chapter 5 discusses the recovery of target compounds using emerging technologies (e.g., supercritical fluids, subcritical water, ultrasound, and microwave-assisted extraction).

The rest of the chapters deal with the sustainable applications of recovered ingredients in different sectors. In particular, Chapter 6 summarizes applications of coffee by-products in food products due to their biological, nutritional, and technological functions. Chapter 7 concentrates on the potential applications of bioactive compounds from coffee processing by-products as active ingredients for skin care products. Their potential UV protective action, emollient capacity, and antiwrinkle and antimicrobial activity are critically reviewed and discussed. Chapters 8 and 9 deal with the biotechnological (e.g., as a substrate for the cultivation of microorganisms) and environmental (e.g., generation of activated carbons) applications of coffee processing by-products, respectively. Chapter 10 explores the potential of utilizing exhausted coffee residues as a precursor for the production of biochar and their application for agricultural purposes. Chapter 11 describes the possibilities of using coffee processing by-products for energy applications (e.g., biofuels, biodiesel, and bioethanol). In particular, the recovery of energy from biomass through thermochemical processes (e.g., gasification, combustion, hydrothermal treatment) and biochemical processes is presented. Finally, Chapter 12 provides an overview of the main current applications of spent coffee grounds and discusses the results obtained during their processing through vermicomposting on a pilot scale.

Conclusively, the book provides a handbook for agricultural, chemical, and environmental engineers, as well as food scientists and technologists who work in the coffee-processing industry and are seeking to improve their by-products management by actively utilizing them in effective applications. It addresses professionals, researchers, specialists, and new product developers working at the edge of the food and environmental sectors. It could be used as a textbook for ancillary reading at the graduate and postgraduate levels, and in multidisciplinary courses dealing with agricultural science, food technology, and environmental, bioresource, and chemical engineering. Along these lines, it could become a target reference for libraries and institutes dealing with coffee production all around the world (e.g., Brazil, Vietnam, Colombia, Indonesia, USA).

I would like to take this opportunity to thank all the contributors of this book for their fruitful collaboration and high-quality work in bringing together different topics and sustainable applications in an integral and comprehensive text. Their acceptance of my invitation to participate in this book, as well as their dedication to editorial guidelines and the book's concept is highly appreciated. In addition, I would also like to thank the acquisitions editor, Megan Ball, and the book manager, Jackie Truesdell, for their assistance during editing, as well as Lisa Jones and all the team of Elsevier during the production process. I would also like to acknowledge the support and expertise of the Food Waste Recovery Group of the ISEKI Food Association that provided us with tools and insights in the field. The ability of the group to support the food and beverage industries in order to recover food waste and improve sustainability is remarkable.

Last but not least, a message for the readers: in such big collaborative project, it is impossible to avoid minor errors or gaps. Therefore, if you find any mistakes or have any objections regarding the content of the book, please do not hesitate to contact me. Instructive comments are and always will be welcome.

Charis M. Galanakis
Research & Innovation Department, Galanakis Laboratories, Chania, Greece
e-mail: cgalanakis@chemlab.gr
Food Waste Recovery Group, ISEKI Food Association, Vienna, Austria
e-mail: foodwasterecoverygroup@gmail.com

State of the art in coffee processing by-products

1

**Rita C. Alves*, Francisca Rodrigues*, Maria Antónia Nunes*,
Ana F. Vinha*,**, M. Beatriz P.P. Oliveira***

*REQUIMTE/LAQV, University of Porto, Porto, Portugal;

**University Fernando Pessoa, Porto, Portugal

ABSTRACT

This chapter describes the steps involved in coffee processing from the field to the cup and the respective generation of by-products along the chain. The chemical composition of coffee husks, pulp, immature, and defective beans, coffee silverskin, and spent coffee grounds is detailed and methods for the sustainable management of these by-products are addressed, as well as legislative frameworks and policy recommendations. Although coffee by-products have a high potential of application in different fields, more integrated strategies with the involvement of coffee producers, industries, academic institutions, governmental and nongovernmental organizations are still needed to convert coffee by-products into really profitable substrates.

Keywords: coffee production; processing; waste; by-products; valorization; innovation; frameworks

1.1 INTRODUCTION

Coffee is one of the most popular beverages all over the world. Behind each hot and tasteful cup of coffee, which can be presented by so many different ways, a real journey is hidden. The genus *Coffea*, which belongs to the Rubiaceae family, embraces two of the more important plant species of the international coffee trade: *Coffea arabica* L. and *Coffea canephora* Pierre, widely known as Arabica and Robusta. *C. arabica* L., considering the different varieties and cultivated forms, originates about 65%–70% of the world coffee production. Its origin remounts to the mountains of Ethiopia (Yemen, AD 850) and it is an autogamic plant (self-fertile) (Alves et al., 2011; Ferrão, 2009). Its cultivation is carried out in regions of moderate temperature from tropical and subtropical areas. Some of the *C. arabica* varieties with higher commercial interest are the *typica* Cramer, the *bourbon* (B. Rodr.) Choussy, the *caturra* K.M.C., the *culumnaris* Ottotandr. ex Cramer, the *mokka* Hort. ex Cramer, and the *xanthocarpa* (Caminhoá) Froehner (Ferrão, 2009).

Handbook of Coffee Processing By-Products. http://dx.doi.org/10.1016/B978-0-12-811290-8.00001-3

Over the years, as the coffee market achieved great importance, an outstanding scientific and technical investment was performed. The preparation of new cultivars, ecologically well adapted, more productive, resistant to pests and diseases, giving origin to a commercial product of high quality, has been one of the fields in which success has been achieved. The selection and improvement of coffee have been fundamental tools in this process, which implied the use of hybridizations and crossings to assemble as many desirable characteristics as possible. Along this process, coffee plants that were not originally interesting due to the quality of the produced beverage were used to induce advantageous characteristics. Thus, besides the natural *C. arabica* varieties with their typical chromossomal and genetic composition, several others also have been emerging as a result of the genetic improvement (e.g., cultivar Catimor, cultivar Sarchimor) or, even, by natural and spontaneous crossing along the time (e.g., cultivar Mundo Novo, cultivar Bourbon-amarelo). Although not pure, their behavior in culture and their final product presents characteristics similar to those of natural Arabicas (Ferrão, 2009).

C. canephora Pierre, in turn, is indigenous from Equatorial African lowland forests from Guinea to Uganda and its cultivation was extended to Asia and South America. It is an allogamic species (self-sterile) that represents about 10%–25% of the worldwide coffee production. The organoleptic characteristics of these coffees are considered inferior to those of Arabica, but they contain higher levels of caffeine and total soluble solids. Besides, they present higher resistance to diseases, particularly to the coffee leaf rust (*Helimeia vastatrix*) and coffee berry disease (*Colletotrichum kahawae*). Also, their roasted seeds produce a "neutral" brew that easily accept the Arabica flavor, and because it is cheaper, *C. canephora* sp. have been assuming increased interest in international markets. Their applications are essentially to increase the body of the beverages (e.g., espresso coffee) and to produce instant coffee (Alves et al., 2011; Ferrão, 2009; Illy and Viani, 2005). We can cite as examples of varieties of this coffee species the *laurentii* De Wild, the *kouillensis* Pierre ex De Wild, the *ugandae* Cramer, and the *welwitschii* Chev (Ferrão, 2009). A summary of the main differences between Arabica and Robusta coffees is depicted in Table 1.1.

Besides the referenced main species—*C. arabica* L. and *C. canephora* Pierre—others can be listed, as *C. liberica* or *C. stenophyla*, but they present low economical importance compared to the first two, since from all the commercial coffees that

Table 1.1 Characteristics of Arabica and Robusta Coffees

Arabica Coffee	Robusta Coffee
• Superior cup quality	• Smaller bean size
• More appreciated organoleptic characteristics	• Usually cheaper
• Lower total soluble solids content	• Double caffeine content
• More vulnerable to pests and diseases	• Higher yield of extractable solids
	• More resistant to pests and diseases

currently circulate in the international market, about 98% correspond to Arabica and Robusta coffees (Ferrão, 2009).

Currently, more than 70 countries produce coffee. In 2015–16, the global coffee production was about 145 million of 60 kg bags, while the consumption rounded the 152.1 million. Brazil is the world's largest producer of coffee (~43 million 60 kg bags in 2015), followed by Vietnam (27.5 million 60 kg bags). Colombia and Indonesia are in third and fourth place, respectively (International Coffee Organization, 2016).

Along the several steps of coffee production (from the small producers to the big companies of coffee processing and roasting) a huge amount of residues is generated. For instance, in Brazil the production of coffee from 2008 to 2013 averaged 2.9 million tons, being generated about 1.4 million tons of wastes each year (Oliveira and Franca, 2015). Considering all the producing countries, coffee wastes and by-products constitute a source of severe contamination and a serious environmental problem. It is very important that coffee industries make an effort to valorize the by-products that result from coffee processing in order to increase the sustainability of the process. Simultaneously to an environmentally friendly approach, this can be seen as an opportunity to increase economical incomes and create new jobs. In fact, the different types of by-products are rich in valuable chemical compounds with potential applications in diverse biotechnological fields, such as pharmaceutical, food, or cosmetic ones.

The next subsections of this chapter detail the different steps of coffee beans processing (from the field to the fork) in order to show the variety of by-products generated in each phase and, subsequently, to highlight the suggested strategies to recover and use those by-products for innovative and useful applications.

1.2 COFFEE PROCESSING

1.2.1 THE POSTHARVESTING PROCESSING

The postharvesting processing aims to separate the seed from the remaining parts of the coffee fruit and guarantee a good preservation of the final product. Moreover, the technique has to be adequate in order to protect coffee from the acquisition of undesirable characteristics during all this process (Ferrão, 2009).

The coffee fruit has five layers of protective material that need to be removed in order to reveal the bean inside. From outside to inside, it is composed by:

1. the skin (epicarp or exocarp), a monocellular layer covered with a waxy substance; when ripe it can be red, yellow, or pink, according to the coffee variety;
2. the pulp (mesocarp), composed by a fleshy pulp and, in ripe fruits, a slimy pectinaceous layer of mucilage;
3. the parchment (endocarp), a thin polysaccharide covering;
4. the silverskin (or chaff), a thin tegument that directly coats the seed; and
5. two seeds with elliptical form (Farah and Santos, 2015; Instaurator, 2008).

The high quality of a commercial coffee can only be achieved when all (or almost all) the fruits are harvested in a perfect stage of maturation. However, this highly increases the costs of the process so, in a normal harvest, perfectly mature fruits (that should be the great majority) are usually mixed with some fruits that are excessively mature or, instead, immature (Ferrão, 2009). According to local conditions of the producer country, coffee processing can be performed by different methods (Alves et al., 2011). Each one has its own advantages and disadvantages; therefore it is not possible to select one as the best. It is possible to obtain commercial coffees of good quality using all the processes, if well conducted, and the technique selection depends a lot on the local possibilities (e.g., water availability) (Ferrão, 2009).

In the dry processing method, the cherries are dried and then mechanically dehusked. This process is used for most Brazilian, Ethiopian, and Haitian Arabica coffees, and for Robusta coffee in most parts of the world (Alves et al., 2011). In general, in this technique, excessively mature and immature beans are not usually separated from those perfectly mature and, thus, they will all compose the final batch. The fruits are harvested, and disposed in thin layers (5–10 cm) as quickly as possible. This, together with an adequate mixing along the process, should avoid pulp decomposition (due to its high content in water and sugars) that could originate the much-unappreciated "fermented beans" or "black beans." The drying process can be performed under the sun in yards (natural drying) or in mechanical dryers. The latest are recommended in regions where rain is frequent during the fruit-drying period or to finish the natural drying. During this process, the coffee beans detach from the parchment (endocarp) and after 3–4 weeks, depending on the drying conditions, the fruits are ready to be dehusked (moisture <12%). Nevertheless, if the dried fruits could rest (e.g., in silos) for some months before dehusking, the quality of the final product can be improved. During dehusking, the pericarp (skin, pulp, and parchment) is removed from the beans (Ferrão, 2009).

The wet method is a more sophisticated procedure compared to the previous one. It is based on depulp of the fruits, followed by fermentation. Although this process demands water in abundance and specific technical equipment, it generally allows the obtention of higher quality coffees with higher economical value. It is mainly used for Arabica coffees and coffees of higher quality (Alves et al., 2011; Ferrão, 2009). In order to be wet-processed, the fruits should be in a perfect state of maturation; therefore a careful selection of the cherries is needed, often with manual harvesting or with machinery that allows the separation of the mature beans. In this case, selection and washing tanks in which the coffee will be processed are unevenly disposed in a way that the materials can be separated by gravity. The well-mature beans present a slightly higher density than water and tend to deposit. The green and excessively mature fruits usually float. Therefore, it is possible to separate them and only the mature ones will proceed to the subsequent step: the depulping phase. This operation intends to remove the epicarp and the mesocarp of the fruit. At the end of depulping, the seed is still involved in the endocarp (Fig. 1.1). The gelatinous layer

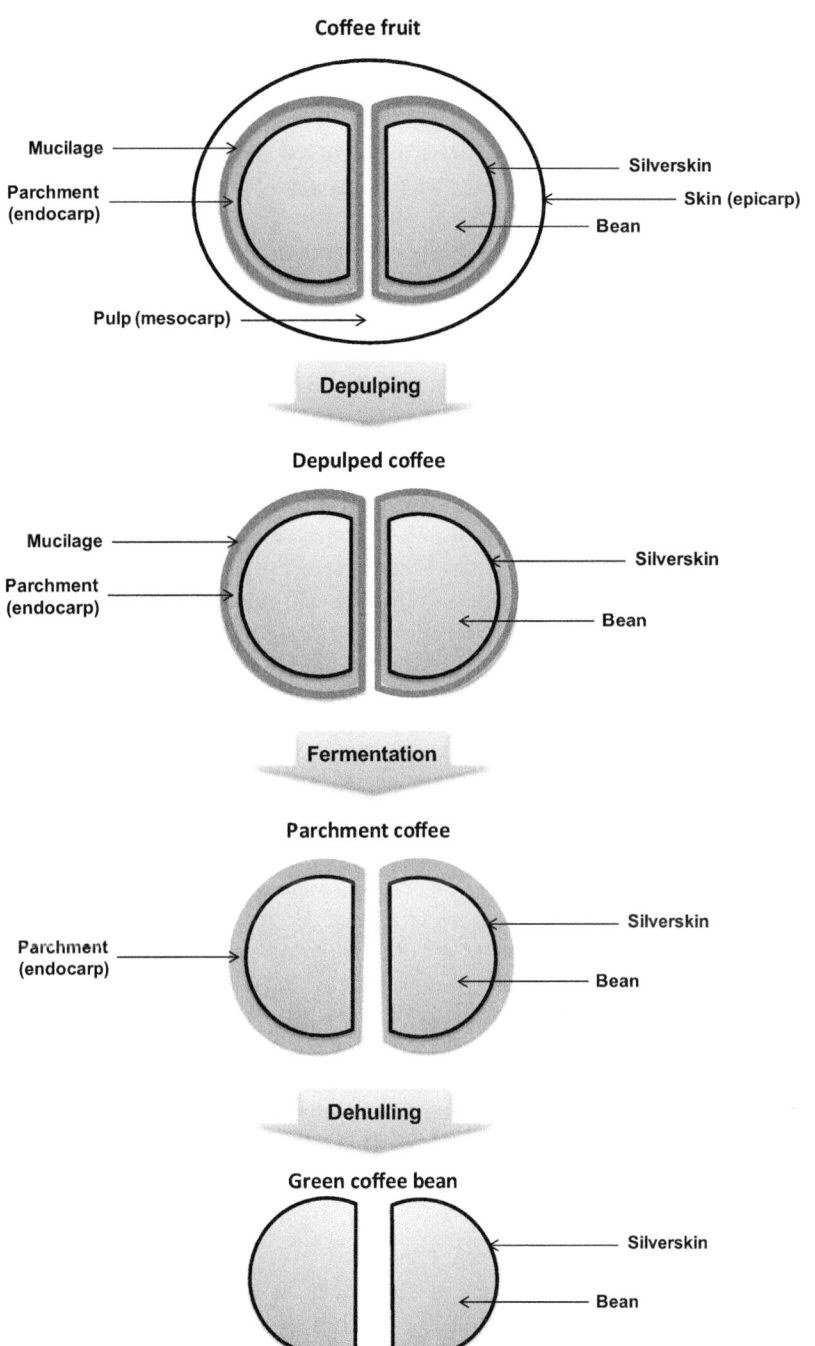

FIGURE 1.1 Coffee Processing Steps in the Wet Postharvesting Method

(mucilage) that coats the endocarp facilitates depulping by reducing the number of broken seeds and the strength to be applied. However, its tendency to retain water and slippery characteristics can impair the following phases. This mucilage, composed mainly by pectins, can be eliminated by fermentation, a process that involves a complex group of chemical and biological reactions. In this phase, the coffee stays in rest to allow the enzymatic and other processes to occur naturally, causing the degradation of mucilage. During this process, the temperature often increases, due to the alcoholic fermentation of the pulp sugar remaining, which is favorable to the enzymatic action of pectinases. The ideal time of fermentation is between 24 and 72 h. Otherwise, the color of the beans can be affected (fermented beans). Besides the natural action of pectinases, commercial enzymes or chemical agents can be also added to increase the efficiency of the process. However, it is still not a very common procedure (Ferrão, 2009).

In a third process, called "semidry" or "semiwashed," concepts of both dry and wet methods are combined. This method consists of washing and selecting the fruits in flotation tanks, followed by depulping, but excluding the fermentation step (Farah and Santos, 2015). Then, the depulped coffee, which contains the mucilage remains, can be directly dried. This process has been used in Central Africa and Brazil, producing the "natural depulped coffee". Both, wet- and semiwashed methods require an additional step for parchment removal, an inner membrane that stays adherent to the beans (Alves et al., 2011).

In addition, in the last years, the mucilage elimination by friction (mechanical action), instead of water, has been gaining supporters. Machines and equipments based on this principle have been appearing in the market, allowing the use of a method that simultaneously saves costs and water. In several regions where the dry-method was a tradition due to the water scarcity, these equipments were, indeed, very well accepted. These machines receive directly the fruits from the washing and calibration and, in sequence, remove the pulp, the mucilage, and wash the parchment coffee that is then ready to be dried and processed (Ferrão, 2009).

As previously referenced, along all of these types of processing, several by-products are generated. Just as an example, values described for washed Colombia coffee show that each 100 kg of mature fruits are constituted by 39 kg of pulp, 22 kg of mucilage, and 39 kg of parchment coffee (Ferrão, 2009). Therefore, considering the millions of bags produced in just 1 year, it can be highlighted that the amount of generated residues is extremely high. Fig. 1.2 summarizes the described postharvesting techniques and the main by-products generated in each one.

1.2.2 THE COFFEE ROAST

The roasting of raw beans is usually carried out in the consumer countries due to the friability and flavor characteristics of roasted beans, which would not resist the necessary movements of international circulation (Ferrão, 2009).

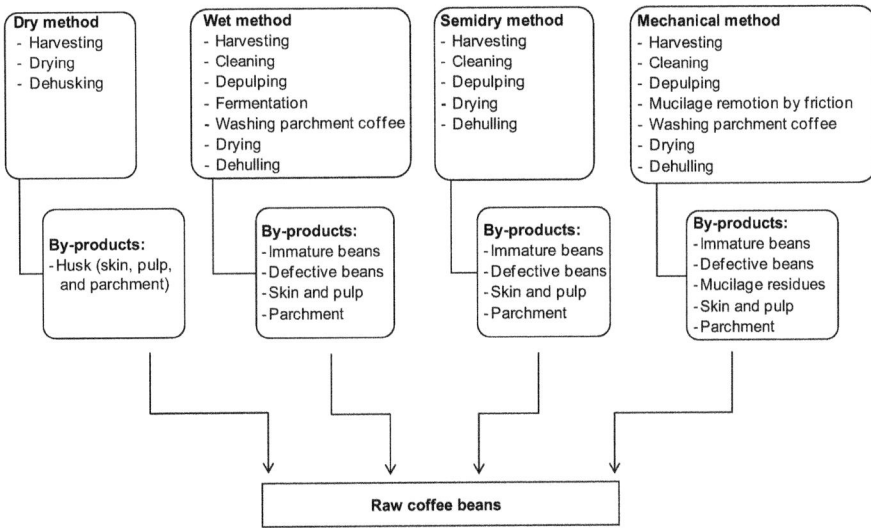

FIGURE 1.2 The Main Processes of Coffee Fruit Postharvesting Processing and the Respective Generated By-Products

After reception and confirmation of their quality, the beans are stored till roasting. When the bags are opened, coffee is usually subjected to a new cleaning step to remove any defective or immature beans, small stones, or metal pieces, for instance, by using a system of sieves and a metal detector, respectively.

Different types of roast can be employed, based on several points of view, namely mechanical, thermal, and operational. Summing up, the modern technologies present as basic principle the passage of a forced convective flow of hot gases through a moving bed of coffee beans. The beans movement can be created by rotation or by the flow of roasting gases (Illy and Viani, 2005).

The chemical composition of green beans is already very complex. During the roast, several physical and chemical reactions occur (with the formation and/or degradation of several chemical compounds), and consequently, the organoleptic characteristics of the beans are completely modified. At this stage, the silverskin (the thin tegument that coats the bean) is detached and can be separated from the final product by air flow, representing the main by-product of coffee roast industry (Fig. 1.3). Industrial data revealed that the roasting of about 4 tons of coffee produces about 30 kg of silverskin. Considering the millions of tons of coffee beans roasted annually all over the world, a huge amount of this by-product is produced. Since silverskin is not usually employed to prepare coffee beverages, it is discarded and often used as firelighters or dispatched to landfills (Costa et al., 2014). After roast, more cleaning steps can be performed. Then, the roasted beans can be directly packed or ground. In this last case, ground coffee can be sold as it is or used to prepare capsules or instant powder.

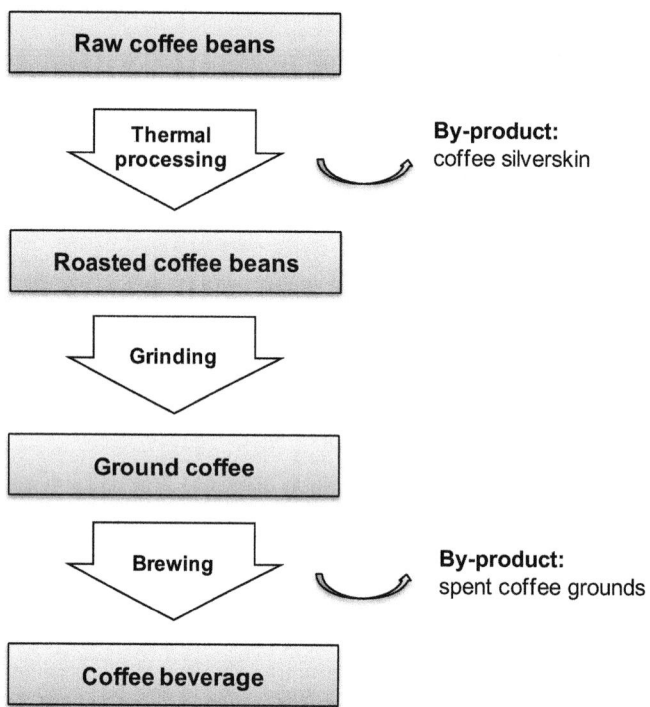

FIGURE 1.3 Raw Coffee Beans Processing and Generated By-Products

1.2.3 **THE COFFEE BEVERAGE**

The coffee beverage can be prepared by many different ways, namely by decoction (e.g., boiled coffee, Turkish coffee, percolator coffee), infusion (e.g., filtered and napoletana coffees), and pressure (e.g., press-pot, mocha, and espresso coffees) (Petracco, 2001). The composition of the final beverage will depend not only on the brew method (e.g., coffee grinding degree, powder/water ratio, water temperature, time of extraction) but also on the coffee species used to prepare the commercial blend. Moreover, different studies have been showing differences in the extractability of compounds according to coffee roast conditions (Alves et al., 2007; Alves et al., 2010a,b). By this way, the composition of the residue that remains after coffee beverage preparation will vary, too. However, a main conclusion can be achieved: the chemical compounds of coffee are not all extracted along brewing and the residue is still rich in different chemicals with important bioactivities.

In the same perspective, and considering the industrial preparation of instant/soluble coffee, several methods can be employed for its preparation, which in general are based on the production of a dried soluble portion of coffee aqueous extracts by percolation, concentration, and dehydration (e.g., evaporation, freeze-drying, spray-drying) (Alves et al., 2011). In this case, coffees with a higher roast degree (preferably Robusta) are usually employed in order to increase the yield of extraction. Besides, the spent coffee from the instant coffee industries is more exhaustively extracted, compared to the "richer", but more disperse spent coffee resulting from beverage preparation at coffee shops, restaurants, and homes. Soluble coffees are consumed all over the world, representing more than 70% of coffee consumed in Great Britain, Ireland, and Australia, >50% in Japan, >40% in USA and Canada, ~20% in Spain, and ~5% in Portugal. In this last case, it represents a consumption of more than 100 ton of imported soluble coffee per year (Ferrão, 2009). This type of coffee is appreciated essentially for its easiness of preparation for consumption.

Depending on the conditions of coffee extraction different weight percentages of spent coffee can be achieved. This worthless residue is normally simply discarded into dustbins, and finally, sent to landfills (Low et al., 2015).

1.3 COFFEE BY-PRODUCTS COMPOSITION AND POTENTIAL APPLICATIONS

Coffee processing by-products include those derived from postharvesting processing, coffee roasting, and coffee consumption, namely, immature/defective beans, husks, skin and pulp, parchment, silverskin, and spent coffee. Along the years, an increase in these wastes has been observed, directly related to the rise of coffee consumption in the world. Being these by-products a great source of bioactive compounds, their valorization and use can be of interest for different industries and fields, such as food, pharmaceutical, or cosmetic ones. Although all of the coffee by-products could be used for new purposes, a lot still needs to be done by both industries and researchers to achieve a real economic feasibility. The dispersion of the residues, the high perishability of the majority, together with the high cost for their separation, collection, and transportation to industrial facilities where they could be treated and transformed, makes this process a real challenge.

1.3.1 COFFEE HUSKS/PULP

Coffee husks are the main solid residues obtained during coffee dry processing. They are composed of the dried skin, the pulp, the mucilage, and the parchment, all together in a single fraction (Esquivel and Jiménez, 2012). On a dry-weight basis, husks represent about 12% of the cherry, and for each ton of harvested coffee fruit, about 0.18 tons of coffee husks are produced (Murthy and Naidu, 2012).

In contrast, the wet-processing method allows the recovery of the skin and pulp in one fraction, which is generally called coffee pulp (40%–50% of the fresh weight of coffee berries) (Dias et al., 2015; Esquivel and Jiménez, 2012; Hernandez et al., 2009). Coffee pulp is produced in large amounts: generally, about 1 ton is obtained from every 2 tons of produced coffee (Murthy and Naidu, 2012; Roussos et al., 1995). In 2008, only in Mexico, the coffee pulp production was estimated in 707 million dry tons (Torres-Mancera et al., 2013).

Several studies have been carried out to assess the chemical composition of coffee husks and coffee pulp. Franca and Oliveira (2009) compiled data from different authors, describing that husks are composed (in dry weight) by protein (8%–11%), lipids (0.5%–3%), minerals (3%–7%), carbohydrates (58%–85%), reducing sugars (14%), caffeine (~1%), and tannins (~5%). According to Bekalo and Reinhardt (2009), coffee husks contain 24.5% of cellulose, 29.7% of hemicelluloses, 23.7% of lignin, and 6.2% of ash. Coffee husks are also known for their high content in secondary metabolites, such as caffeine and polyphenols (Esquivel and Jiménez, 2012; Murthy and Naidu, 2012; Pandey et al., 2000). Mullen et al. (2013) described 5-O-caffeoylquinic acid as the major phenolic present in Arabica and Robusta coffee husks from Mexico and India (0.2–1.9 mg/g). Other compounds were also identified although in minor amounts (in the µg/g range), namely quercetin-3-O-rutinoside, quercetin-3-O-glucoside, quercetin-3-O-galactoside, (+)-catechin, and (−)-epicatechin, and procyanidin dimers, trimers, and tetramers. In general, the phenolic profile of the husks varied widely according to the geographical origin and species. For example, total procyanidin contents varied from 1.3 µg/g (Chinese Robusta husks) and 534 µg/g (Indian Robusta husks) and total flavonols ranged from 5 µg/g (Indian Robusta husks) to 261 µg/g (Mexican Arabica husks) (Mullen et al., 2013).

When compared to husks (1.2%), coffee pulp presented a higher total phenolic content (1.5%) (Murthy et al., 2012). Ramirez-Coronel et al. (2004) being identified four major classes of phenolic compounds in Arabica coffee pulp, namely flavan-3-ols, hydroxycinnamic acids, flavonols, and anthocyanidins. Also, in fresh coffee pulp, Ramirez-Martinez (1988) reported the presence of 5-caffeoylquinic acid (identified as the major phenolic), epicatechin, 3,4-dicaffeoylquinic acid, 3,5-dicaffeoylquinic acid, 4,5-dicaffeoylquinic acid, catechin, rutin, protocatechuic acid, and ferulic acid. Other phenolics have been found by other authors, such as 5-feruloylquinic acid (Clifford and Ramirez-Martinez, 1991), cyanidin-3-rutinoside, and cyanidin-3-glucoside (Esquivel et al., 2010). The soluble and bound hydroxycinnamates in Arabica coffee pulp from seven cultivars were evaluated by Rodríguez-Durán et al. (2014), considering three ripening stages. Chlorogenic acid was the main phenolic acid (94%−98%) in the soluble fraction, whereas caffeic acid was the most abundant hydroxycinnamate found in the bound fraction (72%−88%). Ferulic and p-coumaric acids were detected, too, although in small amounts. Considering the ripening stage, the maximum content of total hydroxycinnamates in pulp was at the semiripe stage. However, its concentration decreased at the ripe stage in six of the seven studied cultivars. In addition, coffee pulp is very rich in fiber (~61%) and has a high content of protein and sugars (~12% and 14%, respectively), minerals (especially potassium), tannins, and caffeine (Murthy et al., 2012). It is also described that yellow coffee varieties have higher proanthocyanidin content than the red ones (De Colmenares et al., 1994). This group of compounds largely contributes to organoleptic features, such as bitterness and astringency (Ramirez-Coronel et al., 2004).

The disposal of coffee husks/pulp represents an environmental burden, especially due to their chemical composition, namely their content in caffeine and tannins (Giannetti et al., 2011; Salmones et al., 2005). For instance, tannins are considered antinutritional factors and are the reason why the amount of coffee pulp in animal feed should be limited to 10% (Pandey et al., 2000).

Nevertheless, other approaches have been suggested to use coffee husks/pulp namely as fertilizers, for composting or vermicomposting, as biosorbents, for bioethanol production or caffeine extraction (Bonilla-Hermosa et al., 2014; Gurram et al., 2016; Hughes et al., 2014; Murthy et al., 2012; Mussatto et al. 2011a; Pandey et al., 2000; Shemekite et al., 2014; Tello et al., 2011). Their content in tannins and fermentable sugars makes them ideal substrates for bioprocesses, too. Considering that gallic acid can be produced by microbial hydrolysis of tannic acid by tannase, secreted by microorganisms, coffee pulp was suggested as a potential raw material to produce gallic acid by microbial transformation (Bhoite et al., 2013).

The extraction and recovery of bioactive compounds, as phenolic acids and related compounds, is one of the most promising applications to valorize these byproducts (Murthy and Naidu, 2012). The production of value-added products, such as enzymes, organic acids, flavor and aroma compounds has also been studied (Pandey et al., 2000; Rodríguez-Durán et al., 2014).

Besides, strategies that detoxify coffee husks and pulp from their phytotoxic compounds and antinutritional factors, or at least that degrade them to a plausibly safe level, led to different possibilities of use, namely as biofertilizer, feed, or even as a substrate for the production of edible mushrooms.

For instance, Brand et al. (2000) tested biological detoxification of coffee husk by filamentous fungi (Rhizopus, *Phanerochaete*, and *Aspergillus* spp.) using a solid-state fermentation system in which coffee husk was used as the sole source of carbon and nitrogen source. *R. arrizus* LPB-79 showed great results on the degradation of caffeine and tannins (87% and 65%, respectively), which were obtained in 6 days (pH = 6.0; moisture: 60%). With *P. chrysosporium* BK, maximum degradation rates of 70.8% and 45% for caffeine and tannins, respectively, were obtained in 14 days (pH = 5.5; moisture: 65%). An *Aspergillus* strain, isolated from the coffee husk, showed the best biomass formation on coffee husk extract-agar medium. Optimization assays were carried out using a factorial design and surface response experiments with *Aspergillus* sp. The best detoxification rates achieved were 92% for caffeine and 65% for tannins. These results showed good prospects of using these fungal strains, in particular *Aspergillus* sp., for the detoxification of coffee husk.

1.3.2 IMMATURE AND DEFECTIVE COFFEE BEANS

The presence of defective and immature beans results from problems during harvesting and preprocessing operations (Franca and Oliveira, 2008). According to Ramalakshmi et al. (2007), the defective coffee beans represent about 15%–20%

of coffee production on a weight basis. The most important defects are black, sour, or brown, immature, bored or insect-damaged, and broken beans. The immature beans, which result from immature fruits, normally contribute to beverage astringency. Both black and sour defects are associated with bean fermentation and decrease significantly the quality of the beverage. Immature-black beans usually fall on the ground while immature, remaining in contact with the soil where they suffer fermentation (Franca et al., 2005; Mazzafera, 1999). In the case of green coffee, it is possible to differentiate defective and nondefective (healthy) beans by color, size, acidity levels, sucrose levels, and the presence of histamine. However, in the case of roasted coffee, only an evaluation of the volatile profile will effectively provide the means for differentiation. The macro differences could be observed based on the volume of immature coffee beans, which is lower. Likewise, defective beans attain a lighter roasting degree than nondefective ones under the same roasting conditions (Franca and Oliveira, 2008; Vasconcelos et al., 2007). Mazzafera (1999) described that nondefective beans are heavier and present higher moisture contents compared to immature and black beans. Protein and oil levels are higher in nondefective beans, but free amino acids (especially asparagine) and soluble phenol contents are higher in defective beans. The amount of 5-caffeoylquinic acid was approximately 35% higher in immature coffees, which is in accordance with data published by Farah et al. (2005) that also reported significantly higher levels of all chlorogenic acids in immature beans.

The total mineral content of defective coffee beans is higher (~6% dry basis) comparing to nondefective ones (~5% dry basis). Potassium is the predominant mineral, followed by calcium and magnesium (Oliveira et al., 2006; Vasconcelos et al., 2007). Lower sugar and sucrose contents were observed in immature beans comparatively to healthy ones, showing that the amount of sugar is primarily associated with the developmental stage of the fruit (Mazzafera, 1999; Vasconcelos et al., 2007). In addition, higher caffeine levels have been associated to the presence of defective or low quality coffees (Franca et al., 2005; Mazzafera, 1999). Trigonelline, together with caffeine, receives considerable attention in coffee chemistry research, because this alkaloid is responsible for aroma compounds production. During roasting, trigonelline is partially degraded to produce two important compounds—pyridines and nicotinic acid (vitamin B3). According to Franca et al. (2005), trigonelline levels were about 1% in nondefective, immature and sour coffee beans, but lower values (~0.8%) were found for black-immature beans. In addition, significantly lower levels of 5-caffeoylquinic acid were detected.

Regarding lipids, the fatty acid composition of oils from defective beans was not significantly different from healthy mature coffee beans (Oliveira et al., 2006). Oliveira et al. (2006) reported that linoleic and palmitic acids are the predominant fatty acids, with averages of 44% and 34%, respectively, while miristic and palmitoleic acids are present in trace amounts.

According to literature, in Brazil, the defective beans are separated from the nondefective ones prior to commercialization in the international market and dumped in the Brazilian internal market, thus depreciating the quality of the roasted coffee consumed in that country (Franca et al., 2005). The extraction of the oil (Oliveira et al., 2006) or their bioactive compounds, such as chlorogenic acid or caffeine, for potential applications in the food and pharmaceutical sectors, can be considered an alternative use for those low-grade coffee beans.

1.3.3 SILVERSKIN

Coffee silverskin is a thin layer that is directly in contact with the coffee bean. Strongly adherent, it is only detached during roasting, because it does not expand like the bean during the thermal processing. It is the main by-product of coffee-roasting industries, which have to collect it mandatorily. Coffee silverskin is currently used as direct fuel (such as firelighters), for composting and soil fertilization (Costa et al., 2014; Mussatto et al., 2011a). Compared to the other coffee by-products, it is a relatively stable product due to the lower moisture content (~7%) (Borrelli et al., 2004) acquired during the roast, and it could be easily gathered for further processing. However, the produced amount is lower compared to other coffee by-products, since silverskin represents a minor fraction of coffee production. Even so, based on data kindly provided by an industrial coffee roaster, which revealed that for 120 tons of roasted coffee about 1 ton of silverskin is produced, and considering the millions of coffee bags produced around the world every year (see Section 1.1), this by-product presents a huge potential for being used and valorized.

Silverskin is rich in protein (19%) and dietary fiber (62%), especially the soluble one (86% of total dietary fiber) (Borrelli et al., 2004). It contains 18% of cellulose and 13% of hemicellulose, being this last composed by xylose (4.7%), arabinose (2.0%), galactose (3.8%), and mannose (2.6%) (Carneiro et al., 2009). Napolitano et al. (2007) reported fat contents varying from 1.6% to 3.3% and caffeine levels ranging from 0.8% to 1.4%. Both were dependent on the geographical origin of the samples. In terms of fat composition, the detected profile varied according to the method used for lipids extraction (Toschi et al., 2014). With a classic Soxhlet extraction with n-hexane, triacylglycerols were found to be the major components of lipids (48%), followed by free fatty acids (21%), esterified sterols (15%), free sterols (13%), and diacylglycerols (4%). In what concerns to the fatty acids profile, C18:2n-6 and C16:0 were the major ones (29% and 28%, respectively), followed by C22:0 and C20:0 (11% each) (Toschi et al., 2014).

According to Ballesteros et al. (2014), silverskin is also an interesting source of minerals (5% of ash), containing mainly potassium (21,100 mg/kg of dry silverskin), calcium (9,400 mg/kg), magnesium (3,100 mg/kg), sulfur (2,800 mg/kg), phosphorous (1,200 mg/kg), and iron (843 mg/kg), among others. Nevertheless, the authors also reported the presence of aluminum (470 mg/kg). Borrelli et al. (2004) found that ochratoxin A levels were below to 4 µg/kg, the maximum level suggested by the Istituto Superiore Sanità, Italy. However, Toschi et al. (2014) found 5- to 9-fold higher levels: from 17.8 to 36.1 µg/kg, which should be more explored due to safety issues.

Silverskin antioxidant activity, due to melanoidins (formed through Maillard reactions during roasting) and phenolic compounds, has also been shown in different studies (Ballesteros et al. 2014; Borrelli et al., 2004; Costa et al., 2014), highlighting the potential of this by-product as an alternative functional ingredient for the food industry. In fact, the amount of water-soluble melanoidins of this by-product is about 4.5%, comparable to that observed in coffee brews (Borrelli et al., 2004). In what concerns total phenolics content, Costa et al. (2014) described 302.5 ± 7.1 mg of gallic acid equivalents (GAE) per liter of extract, which was prepared using 1 g of silverskin in 50 ml of ethanol:water (1:1). The authors also found tannins (0.43 ± 0.06 mg tannic acid equivalents/L) and flavonoids (83.0 ± 1.4 mg epicatechin equivalents/L). In another study, Toschi et al. (2014) reported 0.39–0.73 g gallic acid equivalents per 100 g of silverskin. Using HPLC, Narita and Inouye (2012) were able to detect 1.1 mg of 5-caffeoylquinic acid per gram of silverskin.

It was found that, in association with its antioxidant activity, silverskin aqueous extracts have in vitro antiglycative properties, protecting against the formation of advanced glycation end-products and trapping of carbonyl reactive species, such as methylglyoxal (Mesías et al., 2014). Moreover, silverskin showed to be efficient as prebiotic, namely for bifidobacteria. Instead, *Lactobacillus* spp. and coliforms showed a limited aptitude to use silverskin for their growth, while *Bacteroides* spp. and clostridia growth was inhibited (Borrelli et al., 2004).

Based on its promising health benefits, Martinez-Saez et al. (2014) used coffee silverskin to prepare a novel antioxidant beverage for body weight control and Mussatto et al. (2011a) suggested that the incorporation of silverskin in flakes, breads, biscuits, and snacks can be an interesting approach. Moreover, Pourfarzad et al. (2013) used coffee silverskin after subjected to a treatment with alkaline hydrogen peroxide to give higher quality, shelf life, sensory, and image properties to Barbari flat bread, simultaneously reducing the caloric density and increasing dietary fiber content of the product.

In another perspective, Mussatto et al. (2011a) suggested that the chemical composition of silverskin opens up possibilities for innovative applications. For instance, cellulose and hemicelluloses can be converted to polysaccharides, oligosaccharides, and monosaccharides using acid treatments or enzymes. Subsequently, those sugars can be used to produce added-value compounds (e.g., glucose into ethanol or butanol, manose into mannitol) (Mussatto et al., 2011a).

Finally, silverskin has also been a focus of study for the cosmetic field. Rodrigues et al. (2015) demonstrated with in vitro and in vivo assays that silverskin extracts are not irritants and can be regarded as safe for topical application. In another study, Rodrigues et al. (2016) reported for the first time the successful use of silverskin as a cosmetic active ingredient with similar results to hyaluronic acid in the improvement of skin hydration and firmness.

1.3.4 SPENT COFFEE GROUNDS

Spent coffee grounds are the main by-product of the coffee brewing process and are obtained by both domestic brew preparation (at coffee shops, restaurants, homes) or during the industrial preparation of instant coffee. They consist of a dark brown solid residue with high moisture, being those from the first origin richer in chemical compounds compared to spent coffee from instant coffee industries. This is understandable since for soluble coffee production, extraction is maximized in order to obtain the highest yields.

Ballesteros et al. (2014) analyzed the nutritional composition of spent coffee grounds derived from mixtures of Arabica and Robusta coffee varieties, provided by a coffee roaster industry, highlighting the richness of this by-product in polysaccharides, lignin, and protein. The ash content is a minority, representing 1.3% w/w, and being composed by a variety of elements, including potassium, calcium, magnesium, sulfur, phosphorus, iron, manganese, boron, copper, and others. Potassium is the most predominant element. These minerals are micronutrients that could have different body functions, such as hormonal and enzymatic activities, electrolyte balance, and normal growth.

The fat content of fresh spent coffee grounds is about 2% w/w (Ballesteros et al., 2014; Jiménez-Zamora et al., 2015). In turn, different authors reported that dried spent coffee contains between 13% and 18% of oil (Al-Hamamre et al., 2012; Kulkarni and Dalai, 2006; Petrik et al., 2014). According to Couto et al. (2009), palmitic (C16:0) and linoleic (C18:2) acids are the major fatty acids and comprise about 35% each of the total fatty acid content of the extracted oil. Coffee oil is also rich in vitamin E, namely α- and β-tocopherols (no other vitamers are present), and according to Alves et al. (2010b) only a small percentage (approximately 1%) of the total tocopherol amount in the coffee cake is extracted during the preparation of classic espresso coffee. Although the use of servings or capsules may increase 5-fold the amount of vitamin E extracted into the brew, 95% of the original tocopherols still remains in the coffee cake. This makes spent coffee grounds a very rich source of this liposoluble antioxidant vitamin. This goes in accordance to Gross et al. (1997) that stated that lipids and several hydrophobic compounds are mainly retained in the spent coffee.

Regarding protein, the reported contents in spent coffee grounds range between 14% and 17.5% w/w (Ballesteros et al., 2014; Mussatto et al., 2011b; Ravindranath et al., 1972). The suitability of the proteins in spent coffee grounds is considered poor due to its thermal history and contents of phenolics and melanoidins (Monente et al., 2015). According to Lago et al. (2001), glutamic acid and leucine are the predominant amino acids, but in lower amounts than in coffee beans.

The content of total dietary fiber is about 62% w/w, being the insoluble portion predominant: 5 times higher than the soluble fraction (Ballesteros et al., 2014; Jiménez-Zamora et al., 2015). Hemicellulose (constituted by mannose, galactose, and arabinose) and cellulose are the most representative polysaccharides (Ballesteros et al., 2014; Mussatto et al., 2011b). In terms of sugars, the spent coffee is composed by ~37% mannose, ~32% galactose, ~24% glucose, and ~7% arabinose. Xylose is not present (Ballesteros et al., 2014). Lignin, a macromolecule composed by a great variety of functional groups, is also present in significant amounts (~24% w/w) (Ballesteros et al., 2014; Stewart, 2008).

Besides, in addition to its interesting nutritional profile, spent coffee also contains a wide range of components formed through the Maillard reactions during the roast, such as melanoidins (Borrelli et al., 2004). In fact, based on its chemical composition, authors have been suggesting the potential interest of spent coffee for different industries, such as cosmetic, nutraceutical, or even pharmaceutical (Esquivel and Jiménez, 2012; Mussatto et al., 2011a).

Several studies determined the bioactivity of spent coffee ground extracts, using different solvents and methods (Bravo et al., 2012; Panusa et al., 2013; Ramalakshmi et al., 2009). Ramalakshmi et al. (2009) compared the DPPH scavenging activity and the oxygen radical absorbance capacity (ORAC) of methanolic extracts of spent coffee grounds and low-grade green coffee beans. No significant differences in radical-scavenging activity between samples were observed, with results ranging between 82% and 92%, respectively, for coffee beans and spent coffee. Bravo et al. (2012) reported that the antioxidant capacities of the aqueous spent coffee extracts depends on the coffee brew preparation, ranging between 46.0% and 102.3% for filter, 59.2% and 85.6% for espresso, and almost 42% for plunger, in comparison to their respective coffee brews.

Most of this antioxidant activity seems to be due to the phenolic compounds present in this by-product. Panusa et al. (2013) extracted and analyzed spent coffee grounds in order to evaluate the recovery of relevant natural antioxidants. The results showed total phenolics ranging between 17 and 35 mg of GAE/g dry sample. Other studies reported slightly lower values: 16–19 mg GAE/g (Mussatto et al., 2011c; Zuorro and Lavecchia, 2012). Chlorogenic and caffeic acids are the most relevant phenolic components in this by-product (Maydata, 2002). Such compounds can play an important role in health due to their antioxidant properties. Nevertheless, their contents are mainly dependent on coffee species and beans maturity. Among phenolics, the main compounds are caffeoylquinic acids, feruloylquinic acids, p-coumaroylquinic acids and mixed diesters of caffeic and ferulic acids with quinic acid (Esquivel and Jiménez, 2012; Farah and Donangelo, 2006). According to Ramalakshmi et al. (2009) the major antioxidant compound present in spent coffee is chlorogenic acid (5-caffeoylquinic acid): almost 6%. Bravo et al. (2012) also demonstrated that spent coffee had relevant amounts of total caffeoylquinic acids (6.22–13.24 mg/g), mainly dicaffeoylquinic acids (3.31–5.79 mg/g), which were 4- to 7-fold higher than in their respective coffee brews. Panusa et al. (2013) reported that the total content of chlorogenic acid and derivatives varied between 1.65 and 6.09 mg 5-caffeoylquinic acid equivalents/gram dry basis. More recently, Monente et al. (2015) assessed the total phenolic compounds in spent coffee ground extracts and reported that free and bound caffeoylquinic, dicaffeoylquinic, caffeic, ferulic, p-coumaric, sinapic, and 4-hydroxybenzoic acids were detected. According to the same study, phenolic compounds with one or more caffeic acid molecules were approximately 54% linked to macromolecules, such as melanoidins.

Besides chlorogenic acid and its derivatives, methylxanthines are present in spent coffee and caffeine is the major one recovered, representing 1%–2% of dry matter (Esquivel and Jiménez, 2012). According to Bravo et al. (2012), the caffeine content ranged from 3.59 to 8.09 mg/g of spent coffee, depending on the preparation of coffee brews with the most common coffeemakers (filter, espresso, plunger, and mocha). In another study, different levels of caffeine were detected in extracts of spent coffee from capsules and coffee bars, varying between 0.96 and 0.97 mg/g and 5.99 and 11.50 mg/g, respectively (Panusa et al., 2013).

Table 1.2 summarizes some of the bioactive compounds most commonly found in spent coffee grounds. These chemical components are not all completely

Table 1.2 Chemical Structures of Some of the Most Commonly Found Compounds in Spent Coffee Grounds

Name	Chemical Structure	References
Caffeine		Bravo et al. (2012); Panusa et al. (2013); Ramalakshmi et al. (2009)
Caffeoylquinic acids		Bravo et al. (2012); Panusa et al. (2013); Monente et al. (2015)
Feruloylquinic acids		Bravo et al. (2012); Monente et al. (2015); Panusa et al. (2013)
p-Coumaroylquinic acids		Panusa et al. (2013)
4-Hydroxybenzoic acids		Monente et al. (2015)

extracted during beverage preparation, and a considerable amount of different chemical compounds are speculated to remain in the spent coffee grounds, which are most often thrown away as a waste. To date, some potential applications have been proposed to use spent coffee, for example, for the production of fuel for industrial boilers (Silva et al., 1998), as animal feed (Givens and Barber, 1986), as substrate for fungus growth (Machado et al., 2012), as raw material to produce fuel ethanol (Machado et al., 2012; Rocha et al., 2014), as adsorbent for the removal of heavy metals (Yeung et al., 2014), or for preparation of a distilled beverage with coffee aroma (Sampaio et al., 2013). In spite of these possible applications, spent coffee grounds are still underutilized as a valuable material for industrial processes. Nevertheless, several studies describing its bioactivity, sugars, and even amino acids contents have also been performed aiming to find alternatives for the use of this residue (Ballesteros et al., 2014; Bravo et al., 2012; Campos-Vega et al., 2015; Lago et al., 2001; Monente et al., 2015; Mussatto, 2015; Mussatto et al., 2011a,b,c).

1.4 LEGISLATIVE FRAMEWORKS AND POLICY RECOMMENDATIONS

Sustainability is a dynamic process in which long-term environmental, social, and economical requirements should be fulfilled by an integrated way without compromising the capacity of future generations to meet their own needs (World Commission on Environment and Development, 1987).

As mentioned before, the wastes produced along the coffee chain in both producing and importer countries are undoubtedly a source of contamination and a serious environmental problem. For that reason, several efforts have been made to investigate and develop processes for their valorization and use. Nevertheless, integrated strategies are still necessary always having in mind the sustainability of the coffee chain.

International Coffee Organization (2005) spread a copy of a document about the potential uses of coffee wastes and by-products that was prepared by a team working on the reformulation of a project entitled "Use of coffee by-products and alternative uses for low-grade coffee" submitted by Costa Rica and approved by the Council in 2003. The development of a full-scale project was coordinated with the International Center for Science and High Technology, United Nations Industrial Development Organization (ICS-UNIDO). In that document, Rathinavelu and Graziosi summarize several of the different possibilities of coffee wastes/by-products applications, namely in the production of feed, beverages, vinegar, biogas, caffeine, pectin, pectic enzymes, protein, and compost. This could be seen as a way to inspirit ICO members and community to give attention to such practices, in addition to publicize the relevant work that was being developed.

Currently, in the European Union, Directive 2008/98/EC on Waste (Waste Framework Directive) is the core legislative act that regulates waste management, strengthening the actions that must be taken in order to prevent wastes and introducing an approach that takes into account the whole life-cycle of products. One of the aims of this directive was to repeal and replace Directive 2006/12/EC, because key concepts, such as the definitions of waste, recovery, and disposal needed to be clarified and measures for waste prevention had to be strengthened. Moreover, the introduction of an approach that takes into account the whole life-cycle of products and not only the waste phase was a very relevant issue. One of the main focuses of Directive 2008/98/EC is, therefore, the reduction of the environmental impacts of waste generation and waste management by increasing the waste economic value: the recovery of wastes is encouraged in order to conserve natural resources. Thus, the main aim of waste policies should be to minimize the negative effects of the generation and management of waste on both human health and environment. Waste policies should also aim to reduce the use of resources and incentive the application of the waste hierarchy. This concept lays down a priority order of what represents the best environmental option in waste legislation and policies, except when technical feasibility, economic viability or environmental protection are not possible to be achieved. In general, the following priority order in waste prevention and management should be applied:

1. prevention;
2. preparing for reuse;
3. recycling;
4. other recovery (e.g., energy recovery); and
5. disposal.

In this context, Member States shall take measures to encourage the options that deliver the best overall environmental outcome, which may require specific waste streams departing from the hierarchy whenever justified. One of the main aims of this Directive is, indeed, to help the European Union to be closer to a "recycling society" (European Union, 2008).

The polluter-pays principle is a guiding principle at European and international levels, according to which the waste producers and the waste holders should manage the wastes in order to guarantee the protection of the environment and human health. The introduction of extended producer responsibility in Directive 2008/98/EC is the basis to support the design and production of goods taking into account the efficient use of resources during their whole life-cycle, including their repair, reuse, disassembly and recycling without compromising the free circulation of goods on the internal market.

Still, according to this Directive, a substance that results from a production process not primarily aimed to produce that item can be considered a by-product and not a waste, only if this is consistent with the protection of the environment and human health, and under environmental licenses or general environmental rules. Therefore, the following conditions have to be met:

1. further use of the substance is certain;
2. the substance can be used directly without any further processing other than normal industrial practice;
3. the substance is produced as an integral part of a production process; and
4. further use is lawful.

Considering this, Member States shall take the necessary measures to guarantee that waste management is performed without endangering human health or environment (especially water, air, soil, plants, or animals), without causing a nuisance (through noise or odors), and without adversely affect the countryside or places of special interest.

The European Coffee Federation (ECF) was founded in 1981 to represent the interests of both the coffee trade and roasting industry. Their members have been working together and in partnership with stakeholders in the coffee chain to guarantee that coffee products are manufactured in a responsible way. ECF is a founding partner in the Sustainable Coffee Program that brings together coffee industries, trade, and export partners, governmental and nongovernmental organizations, among others, aiming to bring sustainable coffee production to scale to meet increasing requirements, to improve farmer livelihoods, and sustain natural resources. In 2014, a consortium of leading coffee companies and stakeholders was selected to run a pilot for the European Commission's Product Environmental Footprint (PEF) project. Among other objectives, the coffee PEF pilot aimed to engage with European Union policy developments, to help developing future ones, and to create methodologies to evaluate and improve the environmental performance of coffee based beverages (European Coffee Federation, 2014).

The general concept of the circular economy, which is based on the maintenance of natural resources in the economy as long as possible, while simultaneously their economic value and technical properties are preserved, is highly applicable to the coffee chain. Among other things, a chain of value should involve cross-sector collaborations that create business opportunities for developing innovative products or processes that emerge from the use of resources that were previously considered wastes (Talmon-Gross et al., 2016). Indeed, circular economy is inspired in industrial ecology that regards the flow of material and energy through industrial systems in a way similar to the natural ecosystem: in nature, nothing is wasted and nutrients are recycled in a closed loop. In accordance, circular economy proponents aim to make use of all wastes as input for further value creation (Pike, 2016).

The transition to a circular economy is a great challenge, but the opportunities that rise from all coffee by-products produced along the chain can represent a clear picture of this model. As described along this chapter, such products are rich in nutrients and several bioactive compounds, being added-value substrates. It is thus essential to develop strategies to collect those by-products and create new opportunities of business that can offer a more efficient and profitable use of the different resources in the circular economy. This will certainly have crucial implications not only in small businesses, but also in the global economy. Moreover, consumers are increasingly concerned about the social and environmental impacts of the products they consume, including coffee, being often motivated to also become part of the solution (Pike, 2016; Talmon-Gross et al., 2016). Recently, some initiatives and companies have been created based on the use of coffee by-products to create new products. An example is the firm bio-bean (http://www.bio-bean.com/), from London, that collect spent coffee grounds from coffee shops and convert them into biofuels (e.g., barbecue coals and biomass pellets), being the first company in the world to industrialize this process. The possibility of selling the pellets back to coffee shops to be used, for instance, on coffee roast, can create a true circular economy: the waste becomes the input power for the production activities that created it. Besides, coffee shops give this firm free-of-charge material (in both directions) instead of paying disposal fees to send spent coffee grounds to landfills (Pike, 2016).

Although successful attempts, such as the case of bio-bean, have been created, to date, a lot still needs to be performed to recover all the potential that all coffee by-products present. Governmental organizations can play here a crucial role in promoting changes in the economic behavior, encouraging the adoption of circular economic practices, and promoting public policies for waste reduction (Pike, 2016). However, an interconnected system composed by producers, distributors, and also consumers, together with governments and organizations, entrepreneurs, and researchers is mandatory in order to achieve a true circular economy.

1.5 CONCLUSIONS

In general, coffee by-products represent a relevant environmental burden due to their production in high amounts and their richness in phytotoxic and/or antinutrient compounds (e.g., caffeine, tannins, and polyphenols) that can limit their use directly for soil and feed applications. However, they can be a good source to extract such compounds, which in turn, can be used for food, cosmetic, or pharmaceutical fields. Furthermore, coffee by-products are very rich in polysaccharides and contain high protein and mineral contents, what makes them a product with high biotechnological value (Mussatto et al., 2011a). Several studies have been performed by industries and academic institutions in order to valorize and use the different coffee by-products produced along the coffee chain (in both producer and consumer countries) in distinct perspectives. Nevertheless, the great majority is still dispatched for landfields or used for energy production. Although National and European legislation has been created for waste management in general, specific recommendations and policies about coffee waste/by-products are still lacking in both producing and consumer countries. Integrated strategies in which industry, academic institutions, governmental and nongovernmental organizations are involved together are crucial to transform coffee by-products in really profitable substrates.

ACKNOWLEDGMENTS

Francisca Rodrigues and M. Antónia Nunes are thankful for the research grant from project UID/QUI/500006/2013.

The authors thank the financial support to the project Operação NORTE-01-0145-FEDER-000011—Qualidade e Segurança Alimentar uma abordagem (nano)tecnológica. This work was also supported by the project UID/QUI/50006/2013—POCI/01/0145/FED-

ER/007265 with financial support from FCT/MEC through national funds and cofinanced by FEDER. Francisca Rodrigues and M. Antónia Nunes is thankful for the research grants from project UID/QUI/50006/2013.

REFERENCES

Al-Hamamre, Z., Foerster, S., Hartmann, F., Kröger, M., Kaltschmitt, M., 2012. Oil extracted from spent coffee grounds as a renewable source for fatty acid methyl ester manufacturing. Fuel 96, 70–76.

Alves, R.C., Casal, S., Oliveira, M.B.P.P., 2007. Factors influencing the norharman and harman contents in espresso coffee. J. Agr. Food Chem. 55 (5), 1832–1838.

Alves, R.C., Soares, C., Casal, S., Fernandes, J.O., Oliveira, M.B.P.P., 2010a. Acrylamide in espresso coffee: influence of species, roast degree and brew length. Food Chem. 119 (3), 929–934.

Alves, R.C., Casal, S., Oliveira, M.B.P.P., 2010b. Tocopherols in coffee brews: influence of coffee species, roast degree and brewing procedure. J. Food Comp. Anal. 23 (8), 802–808.

Alves, R.C., Oliveira, M.B.P.P., Casal, S., 2011. Coffee authenticity. In: Oliveira, M.B.P.P., Mafra, I., Amaral, J.S. (Eds.), Current Topics on Food Authentication. Transworld Research Network, Kerala, India.

Ballesteros, L.F., Teixeira, J.A., Mussatto, S.I., 2014. Chemical, functional, and structural properties of spent coffee grounds and coffee silverskin. Food Bioproc. Technol. 7 (12), 3493–3503.

Bekalo, S.A., Reinhardt, H.W., 2009. Fibers of coffee husk and hulls for the production of particleboard. Mater. Struct. 43 (8), 1049–1060.

Bhoite, R.N., Navya, P.N., Murthy, P.S., 2013. Statistical optimization of bioprocess parameters for enhanced gallic acid production from coffee pulp tannins by *Penicillium verrucosum*. Prep. Biochem. Biotechnol. 43 (4), 350–363.

Bonilla-Hermosa, V.A., Duarte, W.F., Schwan, R.F., 2014. Utilization of coffee by-products obtained from semi-washed process for production of value-added compounds. Biores. Technol. 166, 142–150.

Borrelli, R.C., Esposito, F., Napolitano, A., Ritieni, A., Fogliano, V., 2004. Characterization of a new potential functional ingredient: coffee silverskin. J. Agr. Food Chem. 52 (5), 1338–1343.

Brand, D., Pandey, A., Roussos, S., Soccol, C.R., 2000. Biological detoxification of coffee husk by filamentous fungi using a solid-state fermentation system. Enzyme Microb. Technol. 26 (1–2), 127–133.

Bravo, J., Juániz, I., Monente, C., Caemmerer, B., Kroh, L.W., De Peña, M.P., Cid, C., 2012. Evaluation of spent coffee obtained from the most common coffeemakers as a source of hydrophilic bioactive compounds. J. Agr. Food Chem. 60 (51), 12565–12573.

Campos-Vega, R., Loarca-Piña, G., Vergara-Castañeda, H.A., Oomah, B.D., 2015. Spent coffee grounds: a review on current research and future prospects. Trends Food Sci. Technol. 45 (1), 24–36.

Carneiro, L.M., Silva, J.P.A., Mussatto, S.I., Roberto, I.C., Teixeira, J.A., 2009. Determination of total carbohydrates content in coffee industry residues. In: Book of abstracts of the 8th International Meeting of the Portuguese Carbohydrate Group. Braga, Portugal: GLUPOR.

Clifford, M.N., Ramirez-Martinez, J.R., 1991. Phenols and caffeine in wet-processed coffee beans and coffee pulp. Food Chem. 40 (1), 35–42.

Costa, A.S.G., Alves, R.C., Vinha, A.F., Barreira, S.V.P., Nunes, M.A., Cunha, L.M., Oliveira, M.B.P.P., 2014. Optimization of antioxidants extraction from coffee silverskin, a roasting by-product, having in view a sustainable process. Ind. Crops Prod. 53, 350–357.

Couto, R.M., Fernandes, J., da Silva, M.D.R.G., Simões, P.C., 2009. Supercritical fluid extraction of lipids from spent coffee grounds. J. Supercrit. Fluids 51 (2), 159–166.

De Colmenares, N.G., Ramírez-Martínez, J.R., Aldana, J.O., Clifford, M.N., 1994. Analysis of proanthocyanidins in coffee pulp. J. Sci. Food Agr. 65 (2), 157–162.

Dias, M., Melo, M.M., Schwan, R.F., Silva, C.F., 2015. A new alternative use for coffee pulp from semi-dry process to β-glucosidase production by *Bacillus subtilis*. Lett. Appl. Microbiol. 61 (6), 588–595.

Esquivel, P., Jiménez, V.M., 2012. Functional properties of coffee and coffee by-products. Food Res. Int. 46 (2), 488–495.

Esquivel, P., Kramer, M., Carle, R., Jiménez, V., 2010. Anthocyanin profiles and caffeine contents of wet-processed coffee (*Coffea arabica*) husks by HPLC-DAD-MS/MS. Elsevier, The Netherlands.

European Coffee Federation, 2014. Coffee sustainability. Available from: http://www.ecf-coffee.org/about-coffee/coffee-sustainability

European Union, 2008. Directive 2008/98/EC of the European Parliament and of the Council of 19 November 2008 on waste and repealing certain directives. Official J. Eur. Union L 312, 3–30.

Farah, A., Donangelo, C.M., 2006. Phenolic compounds in coffee. Braz. J. Plant Physiol. 18 (1), 23–36.

Farah, A., Santos, T.F., 2015. The coffee plant and beans: an introduction. In: Preedy, V.R. (Ed.), Coffee and Health and Disease Prevention. Academic Press, London, United Kingdom.

Farah, A., de Paulis, T., Trugo, L.C., Martin, P.R., 2005. Effect of roasting on the formation of chlorogenic acid lactones in coffee. J. Agr. Food Chem. 53 (5), 1505–1513.

Ferrão, J.E.M., 2009. O Café, A Bebida Negra Dos Sonhos Claros. Chaves Ferreira-Publicações S. A, Lisboa, Portugal.

Franca, A., Oliveira, L., 2008. Chemistry of defective coffee beans. In: Koeffer, E.N. (Ed.), Progress in Food Chemistry. Nova Science Publishers Inc, New York.

Franca, A.S., Oliveira, L.S., 2009. Coffee processing solid wastes: current uses and future perspectives. In: Ashworth, G.S., Azevedo, P. (Eds.), Agricultural Wastes. Nova Science Publishers, Inc, New York.

Franca, A.S., Oliveira, L.S., Mendonça, J.C.F., Silva, X.A., 2005. Physical and chemical attributes of defective crude and roasted coffee beans. Food Chem. 90 (1–2), 89–94.

Giannetti, B.F., Ogura, Y., Bonilla, S.H., Almeida, C.M.V.B., 2011. Accounting emergy flows to determine the best production model of a coffee plantation. Energy Pol. 39 (11), 7399–7407.

Givens, D.I., Barber, W.P., 1986. In vivo evaluation of spent coffee grounds as a ruminant feed. Agr. Wastes 18 (1), 69–72.

Gross, G., Jaccaud, E., Huggett, A.C., 1997. Analysis of the content of the diterpenes cafestol and kahweol in coffee brews. Food Chem. Toxicol. 35 (6), 547–554.

Gurram, R., Al-Shannag, M., Knapp, S., Das, T., Singsaas, E., Alkasrawi, M., 2016. Technical possibilities of bioethanol production from coffee pulp: a renewable feedstock. Clean Technol. Environ. Pol. 18 (1), 269–278.

Hernandez, C.E., Chen, H.-H., Chang, C.-I., Huang, T.-C., 2009. Direct lipase-catalyzed lipophilization of chlorogenic acid from coffee pulp in supercritical carbon dioxide. Ind. Crops Prod. 30 (3), 359–365.

Hughes, S.R., López-Núñez, J.C., Jones, M.A., Moser, B.R., Cox, E.J., Lindquist, M., Galindo-Leva, L.Á., Riaño-Herrera, N.M., Rodriguez-Valencia, N., Gast, F., Cedeño, D.L.,

Tasaki, K., Brown, R.C., Darzins, A., Brunner, L., 2014. Sustainable conversion of coffee and other crop wastes to biofuels and bioproducts using coupled biochemical and thermo-chemical processes in a multi-stage biorefinery concept. Appl. Microbiol. Biotechnol. 98 (20), 8413–8431.

Illy, A., Viani, R., 2005. Espresso coffee: the science of quality, second ed. Elsevier Academic Press, London.

Instaurator, 2008. The Espresso Quest. In: Everage, L., Oak, E. (Eds.), Loowedge Publishing, Copacabana, Australia.

International Coffee Organization, 2005. Potential alternative use of coffee wastes and by-products by Rajkumar Rathinavelu and Giorgio Graziosi. Available from: http://www.ico.org/documents/ed1967e.pdf

International Coffee Organization, 2016. Total production by exporting countries. Available from: http://www.ico.org/prices/po-production.pdf

Jiménez-Zamora, A., Pastoriza, S., Rufián-Henares, J.A., 2015. Revalorization of coffee by-products, prebiotic, antimicrobial and antioxidant properties. LWT Food Sci. Technol. 61 (1), 12–18.

Kulkarni, M.G., Dalai, A.K., 2006. Waste cooking oil: an economical source for biodiesel: a review. Ind. Eng. Chem. Res. 45 (9), 2901–2913.

Lago, R.C.A., Antoniassi, R., Freitas, S.C., 2001. Centesimal composition and amino acids of raw, roasted and spent ground of soluble coffee. Embrapa Café, Brasília, Brasil.

Low, J.H., Rahman, W.A.W.A., Jamaluddin, J., 2015. The influence of extraction parameters on spent coffee grounds as a renewable tannin resource. J. Clean. Prod. 101, 222–228.

Machado, E.M.S., Rodriguez-Jasso, R.M., Teixeira, J.A., Mussatto, S.I., 2012. Growth of fungal strains on coffee industry residues with removal of polyphenolic compounds. Biochem. Eng. J. 60, 87–90.

Martinez-Saez, N., Ullate, M., Martin-Cabrejas, M.A., Martorell, P., Genovés, S., Ramon, D., del Castillo, M.D., 2014. A novel antioxidant beverage for body weight control based on coffee silverskin. Food Chem. 150, 227–234.

Maydata, A.G., 2002. Café, antioxidantes y protección a la salud. Medisan 64, 72–81.

Mazzafera, P., 1999. Chemical composition of defective coffee beans. Food Chem. 64 (4), 547–554.

Mesías, M., Navarro, M., Martínez-Saez, N., Ullate, M., del Castillo, M.D., Morales, F.J., 2014. Antiglycative and carbonyl trapping properties of the water soluble fraction of coffee silverskin. Food Res. Int. 62, 1120–1126.

Monente, C., Ludwig, I.A., Irigoyen, A., De Peña, M.-P., Cid, C., 2015. Assessment of total (free and bound) phenolic compounds in spent coffee extracts. J. Agr. Food Chem. 63 (17), 4327–4334.

Mullen, W., Nemzer, B., Stalmach, A., Ali, S., Combet, E., 2013. Polyphenolic and hydroxy-cinnamate contents of whole coffee fruits from China, India, and Mexico. J. Agr. Food Chem. 61 (22), 5298–5309.

Murthy, P.S., Naidu, M.M., 2012. Sustainable management of coffee industry byproducts and value addition: a review. Res. Conserv. Recycl. 66, 45–58.

Murthy, P.S., Manjunatha, M., Sulochannama, G., Naidu, M.M., 2012. Extraction, character-ization and bioactivity of coffee anthocyanins. Eur. J. Biol. Sci. 4 (1), 13–19.

Mussatto, S.I., 2015. Generating biomedical polyphenolic compounds from spent coffee or silverskin. In: Preedy, V.R. (Ed.), Coffee in Health and Disease Prevention. Elsevier, The Netherlands.

Mussatto, S.I., Machado, E.M.S., Martins, S., Teixeira, J.A., 2011a. Production, composition, and application of coffee and its industrial residues. Food Bioproc. Technol. 4, 661–672.

Mussatto, S.I., Carneiro, L.M., Silva, J.P.A., Roberto, I.C., Teixeira, J.A., 2011b. A study on chemical constituents and sugars extraction from spent coffee grounds. Carbohy. Polym. 83 (2), 368–374.

Mussatto, S.I., Ballesteros, L.F., Martins, S., Teixeira, J.A., 2011c. Extraction of antioxidant phenolic compounds from spent coffee grounds. Sep. Puri. Technol. 83, 173–179.

Napolitano, A., Fogliano, V., Tafuri, A., Ritieni, A., 2007. Natural occurrence of ochratoxin A and antioxidant activities of green and roasted coffees and corresponding byproducts. J. Agr. Food Chem. 55 (25), 10499–10504.

Narita, Y., Inouye, K., 2012. High antioxidant activity of coffee silverskin extracts obtained by the treatment of coffee silverskin with subcritical water. Food Chem. 135 (3), 943–949.

Oliveira, L.S., Franca, A.S., 2015. An overview of the potential uses for coffee husks. In: Preedy, V.R. (Ed.), Coffee in Health and Diseases Prevention. Academic Press, New York.

Oliveira, L.S., Franca, A.S., Mendonça, J.C.F., Barros-Júnior, M.C., 2006. Proximate composition and fatty acids profile of green and roasted defective coffee beans. LWT Food Sci. Technol. 39 (3), 235–239.

Pandey, A., Soccol, C.R., Mitchell, D., 2000. Biotechnological potential of coffee pulp and coffee husk for bioprocesses. Biochem. Eng. J. 6 (2), 153–162.

Panusa, A., Zuorro, A., Lavecchia, R., Marrosu, G., Petrucci, R., 2013. Recovery of natural antioxidants from spent coffee grounds. J. Agr. Food Chem. 61 (17), 4162–4168.

Petracco, M., 2001. Technology IV—beverage preparation: brewing trends for the new millennium. In: Clarke, R.J., Vitzthum, O.G. (Eds.), Coffee: Recent Developments. World Agriculture Series, Cornwall, UK.

Petrik, S., Obruča, S., Benešová, P., Márová, I., 2014. Bioconversion of spent coffee grounds into carotenoids and other valuable metabolites by selected red yeast strains. Biochem. Eng. J. 90, 307–315.

Pike, D., 2016. What Goes Around… How Coffee Waste is Fueling a Circular Economy. Available from: http://www.roastmagazine.com/resources/Articles/Roast_MayJune16_WhatGoesAround.pdf

Pourfarzad, A., Mahdavian Mehr, H., Sedaghat, N., 2013. Coffee silverskin as a source of dietary fiber in bread-making: optimization of chemical treatment using response surface methodology. LWT Food Sci. Technol. 50 (2), 599–606.

Ramalakshmi, K., Kubra, I.R., Rao, L.J., 2007. Physicochemical characteristics of green coffee: comparison of graded and defective beans. J. Food Sci. 72 (5), S333–S337.

Ramalakshmi, K., Rao, L.J.M., Takano-Ishikawa, Y., Goto, M., 2009. Bioactivities of low-grade green coffee and spent coffee in different in vitro model systems. Food Chem. 115 (1), 79–85.

Ramirez-Coronel, M.A., Marnet, N., Kolli, V.S.K., Roussos, S., Guyot, S., Augur, C., 2004. Characterization and estimation of proanthocyanidins and other phenolics in coffee pulp (*Coffea arabica*) by thiolysis−high-performance liquid chromatography. J. Agr. Food Chem. 52 (5), 1344–1349.

Ramirez-Martinez, J.R., 1988. Phenolic compounds in coffee pulp: quantitative determination by HPLC. J. Sci. Food Agr. 43 (2), 135–144.

Ravindranath, R., Khan, R.Y.A., Obi Reddy, T., Thirumala Rao, S.D., Reddy, B.R., 1972. Composition and characteristics of Indian coffee bean, spent grounds and oil. J. Sci. Food Agr. 23 (3), 307–310.

Rocha, M.V.P., de Matos, L.J.B.L., Lima, L.P.D., Figueiredo, P.M.D.S., Lucena, I.L., Fernandes, F.A.N., Gonçalves, L.R.B., 2014. Ultrasound-assisted production of biodiesel and ethanol from spent coffee grounds. Biores. Technol. 167, 343–348.

Rodrigues, F., Pereira, C., Pimentel, F.B., Alves, R.C., Ferreira, M., Sarmento, B., Amaral, M.H., Oliveira, M.B.P.P., 2015. Are coffee silverskin extracts safe for topical use? An in vitro and in vivo approach. Ind. Crop Prod. 63, 167–174.

Rodrigues, F., Matias, R., Ferreira, M., Amaral, M.H., Oliveira, M.B.P.P., 2016. In vitro and in vivo comparative study of cosmetic ingredients coffee silverskin and hyaluronic acid. Exper. Dermatol. 25 (7), 572–574.

Rodríguez-Durán, L.V., Ramírez-Coronel, M.A., Aranda-Delgado, E., Nampoothiri, K.M., Favela-Torres, E., Aguilar, C.N., Saucedo-Castañeda, G., 2014. Soluble and bound hydroxycinnamates in coffee pulp (*Coffea arabica*) from seven cultivars at three ripening stages. J. Agr. Food Chem. 62 (31), 7869–7876.

Roussos, S., de los Angeles Aquiáhuatl, M., del Refugio Trejo-Hernández, M., Gaime Perraud, I., Favela, E., Ramakrishna, M., Raimbault, M., Viniegra-González, G., 1995. Biotechnological management of coffee pulp - isolation, screening, characterization, selection of caffeine-degrading fungi and natural microflora present in coffee pulp and husk. Appl. Microbiol. Biotechnol. 42 (5), 756–762.

Salmones, D., Mata, G., Waliszewski, K.N., 2005. Comparative culturing of *Pleurotus* spp. on coffee pulp and wheat straw: biomass production and substrate biodegradation. Biores. Technol. 96 (5), 537–544.

Sampaio, A., Dragone, G., Vilanova, M., Oliveira, J.M., Teixeira, J.A., Mussatto, S.I., 2013. Production, chemical characterization, and sensory profile of a novel spirit elaborated from spent coffee ground. LWT Food Sci. Technol. 54 (2), 557–563.

Shemekite, F., Gómez-Brandón, M., Franke-Whittle, I.H., Praehauser, B., Insam, H., Assefa, F., 2014. Coffee husk composting:a investigation of the process using molecular and non-molecular tools. Waste Manage. 34 (3), 642–652.

Silva, M.A., Nebra, S.A., Machado Silva, M.J., Sanchez, C.G., 1998. The use of biomass residues in the Brazilian soluble coffee industry. Biomass Bioener. 14 (5–6), 457–467.

Stewart, D., 2008. Lignin as a base material for materials applications: chemistry, application and economics. Ind. Crops Prod. 27 (2), 202–207.

Talmon-Gross, L., Miedzinski, M., Technopolis Group, 2016. Framework conditions to support emerging industries and clusters in the area of circular economy: from recycling to product-service systems. European Cluster Observatory. Available from: http://ec.europa.eu/DocsRoom/documents/16266/attachments/1/translations/en/renditions/native

Tello, J., Viguera, M., Calvo, L., 2011. Extraction of caffeine from Robusta coffee (*Coffea canephora* var. Robusta) husks using supercritical carbon dioxide. J. Supercrit. Fluids 59, 53–60.

Torres-Mancera, M.T., Baqueiro-Peña, I., Figueroa-Montero, A., Rodríguez-Serrano, G., González-Zamora, E., Favela-Torres, E., Saucedo-Castañeda, G., 2013. Biotransformation and improved enzymatic extraction of chlorogenic acid from coffee pulp by filamentous fungi. Biotechnol. Prog. 29 (2), 337–345.

Toschi, T.G., Cardenia, V., Bonaga, G., Mandrioli, M., Rodriguez-Estrada, M.T., 2014. Coffee silverskin: characterization, possible uses, and safety aspects. J. Agr. Food Chem. 62 (44), 10836–10844.

Vasconcelos, A.L.S., Franca, A.S., Glória, M.B.A., Mendonça, J.C.F., 2007. A comparative study of chemical attributes and levels of amines in defective green and roasted coffee beans. Food Chem. 101 (1), 26–32.

World Commission on Environment and Development, 1987. Our common future. Available from: http://www.un-documents.net/our-common-future.pdf

Yeung, P.-T., Chung, P.-Y., Tsang, H.-C., Cheuk-On Tang, J., Yin-Ming Cheng, G., Gambari, R., Chui, C.-H., Lam, K.-H., 2014. Preparation and characterization of bio-safe activated charcoal derived from coffee waste residue and its application for removal of lead and copper ions. RSC Adv. 4, 38839–38847.

Zuorro, A., Lavecchia, R., 2012. Spent coffee grounds as a valuable source of phenolic compounds and bioenergy. J. Clean. Prod. 34, 49–56.

Healthy components of coffee processing by-products

2

Benjamin M. Dorsey, Marjorie A. Jones

Illinois State University, Normal, IL, United States

ABSTRACT

Harvesting and processing of coffee beans generates large amounts of biomass that is typically discharged to the environment causing ecological problems to the respective coffee-producing countries. On the other hand, coffee processing by-products have high amounts of antioxidant compounds, such as caffeine, chlorogenic acids, trigonelline, and diterpenes that could be recovered and used as natural food preservatives, additives in cosmetics, and as nutritional supplements. In addition, these compounds have not been subjected to the roasting process that the coffee beans undergo, thereby their levels and quality remains relatively unchanged. This chapter explores the antioxidant components of coffee and coffee by-products, gives some background on antioxidants (what they are and how to study them) and how these relate to health.

Keywords: coffee by-products; antioxidants; reactive oxygen species

2.1 INTRODUCTION

While billions of coffee cups are consumed worldwide each day, the bulk of the coffee plant biomass is considered to be waste. Currently there are only a few economically viable uses of this biowaste, and thus such materials may be considered an ecological disaster especially for the countries that produce the major supplies of coffee, such as Brazil, Vietnam, Indonesia, and Colombia (Hughes et al., 2014). As reported by Index mundi (2014), these four countries produced together more than 103 million 60 kg bags of green coffee annually. Estimates have been reported that for every 1 million 60 kg bags of dried coffee beans, some 218,400 tons of coffee cherry pulp and mucilage wastes are generated (Veenstra, 1995). This sheer mass of coffee processing by-products (coffee pulp, coffee husks, silver skin, and spent coffee grounds) is largely untapped at present and represents a large resource that can be used for value-added products. Coffee pulp is an abundant coffee cherry fruit by-product because it is approximately 28% of the coffee fruit on a dry weight basis. Coffee husks are about 12% of the dry weight of the coffee fruit while the coffee bean is some 50%–55% of the dry weight of the coffee fruit (http://www.feedipedia.org/node/549). Research is being directed toward these waste streams being used for the production of biofuels (Hughes et al., 2014). However, other uses of coffee processing by-product wastes are now being discovered. This chapter explores the antioxidant components of coffee and coffee by-products, gives some background on antioxidants (what they are and how to study them) and how these relate to health. Also the potential pharmaceutical applications for coffee by-products are explored.

2.1.1 SPENT COFFEE GROUNDS

In addition to the coffee processing by-products, more than 6 million tons of spent coffee grounds (grounds that remain after brewing the beverage) are generated each year. To a limited extent such wastes are being recycled for a variety of home uses, such as cleaning a dirty grill, exterminating fleas from pets and snails from gardens, and use as a skin exfoliate (http://www.movoto.com/blog/opinions/coffee-grounds/; http://www.dailymail.co.uk/femail/food/article-3043176/The-20-surprising-uses-coffee-grounds.html). Chavan et al. (2016) have recently reported the novel use of spent coffee grounds formulated with polysiloxane to make a sponge-like material that traps lead and mercury from water. Such spent coffee wastes are generally not considered economically useful wastes, especially for the harvesting of antioxidant molecules.

2.2 BACKGROUND ON ANTIOXIDANTS

2.2.1 THE CHEMICAL BASIS OF OXIDATION AND REDUCTION: MOVEMENT OF ELECTRONS

Oxidation and reduction reactions are linked (coupled) chemical reactions in which movement of electrons occur. One famous memory device to help understand such reactions is "LEO goes GER". The LEO indicates loss of electrons from a system (oxidation) and the GER indicates gain of electrons to a system (reduction). Such reactions are "coupled" with one atom or molecule giving up (losing) an electron while another atom or molecule accepting (gaining) this electron. Such coupled reactions are called "redox" reactions. An oxidant is a chemical species that oxidizes something else as it becomes reduced. In Eq. (2.1), Y is the oxidant. The reductant reduces something else as it becomes oxidized and X is the reductant in Eq. (2.1). In this equation, the hydrogen is the "source" of the electron:

$$Y^{\cdot} + X : H \rightarrow X^{\cdot} + H : Y \qquad (2.1)$$

Antioxidants can also be oxidized by generally forming a more stable radical or chemical species thereby protecting other molecules in cells, such as proteins, lipids, and DNA from oxidative damage. Normally, oxidation and reduction reactions occur with high frequency in cells and such reactions are an important part of energy metabolism and biosynthesis reactions. Such reactions are highly controlled in normal cells.

2.2.2 OXIDANTS AND ANTIOXIDANTS AND WHY WE NEED THEM

In vivo, antioxidants are small or large molecules that have the capacity to reduce or prevent chemical oxidation caused by reactive oxygen species (ROS) and reactive nitrogen species (RNS). These include radicals, such as dioxygen ($O_2\cdot\cdot$), superoxide anion (O_2^-), hydroxyl (OH·), singlet oxygen (1O2), peroxyl (ROO·), alkoxyl (RO·), and nitric oxide (NO·) radicals (Kohen and Nyska, 2002). Radicals can also occur from exposure to UV light (from the sun) or radioactivity. Oxygen radicals contain at least one unpaired electron, in their molecular orbitals, whose spin allows the acceptance or donation of an electron. Nonradical oxygen derivatives include hydrogen peroxide (H_2O_2), organic peroxide (ROOH), hypochlorous acid (HOCl), ozone (O_3), aldehydes (HCOR), and peroxynitrite (ONOOH). These nonradical oxygen derivatives are strong biological oxidizing agents (Kohen and Nyska, 2002). Oxidative stress is the general term used to indicate an imbalance in normal levels of oxidized and reduced molecules in cells. Oxidative stress is correlated with aging, cancers, and a number of neurological problems. Oxidative damage can occur to lipids (lipid peroxidation), proteins (mainly by OH·, RO·, and nitrogen-based free radicals), and DNA (mainly by OH· or UV light). Some metal ions that can lead to the formation of the types of radicals are also considered sources of biological oxidants. For example, Fe^{3+} in the presence of H_2O_2 will lead to the formation of both OH· and O_2^-, which can then damage lipids, proteins, and DNA. The well-known Fenton reaction (Eqs. (2.2) and (2.3)) involves different oxidation states of iron reacting with H_2O_2 to generate other ROS, especially the hydroxyl radical (·OH), which is one of the most reactive chemical species known, with only fluorine being more reactive (Bishop, 1968).

$$Fe^{2+} + H_2O_2 \rightarrow Fe^{3+} + OH^- + \dot{}OH \tag{2.2}$$

$$Fe^{3+} + H_2O_2 \rightarrow Fe^{2+} + \dot{}OOH + H^+ \tag{2.3}$$

2.2.2.1 Important biological roles of antioxidants

Antioxidants can be found exogenously or endogenously. Exogenous antioxidants are found naturally or as additions to food and drink and include: phytochemicals, vitamins (vitamin A, beta carotenes, vitamin C, and vitamin E), synthetic antioxidants, such as butylated hydroxytoluene (BHT), butylated hydroxyanisole (BHA), propyl-, octyl-, and dodecyl-gallates, *tert*-butyl hydroquinone (TBHQ), and others. Some of the most common exogenous antioxidant molecules involved are found in fruits, vegetables, and other food sources. These include vitamin C, vitamin E, beta carotenes, and vitamin A. These are found in low or very low amounts in coffee beans and coffee processing by-products. However, coffee and coffee processing by-products contain important classes of antioxidants involving purines, phenols, flavonoids, alkaloids, and isoprene derivatives. Murthy and Naidu (2012) reported that functional compounds, especially antioxidants, can be recovered from coffee processing by-products. Endogenous antioxidants are produced by the biological system itself, and include enzymes and small molecules: superoxide dismutase (SOD), catalase (CAT), and glutathione peroxidase (GPx) are important enzymes; small molecules include the tripeptide glutathione (GSH), alpha lipoic acid (ALA), coenzyme Q10 (CoQ10), urate, and others (Bouayed and Bihn, 2010; Wayner et al., 1987). Endogenous and exogenous antioxidants work synergistically to maintain or reestablish the homeostatic oxidative levels of the various molecules involved in controlling the redox cellular environment. An example of this is the regeneration of vitamin E by GSH to prevent lipid peroxidation (Bouayed and Bihn, 2010).

2.2.2.2 Dualistic activities of antioxidants

A certain amount of oxidation is required for normal cell function. Resting cells are normally in a reductive state while active aerobic cells are in an oxidative state (Kohen and Nyska, 2002). While defending against infection and during the induction of mitogenic responses, the cellular environment becomes increasingly more oxidative (Kohen and Nyska, 2002). Antioxidants appear to have dualistic behaviors as a function of concentration and environment, both in vitro and in vivo. Antioxidants and redox active vitamins taken at doses comparable to what are found in food can have beneficial effects, while those same compounds taken at high doses, like those found in concentrated supplements, can act as prooxidants (Bouayed and Bihn, 2010). Morgan et al. (1999) reported that phenolic iron complexes could have both prooxidants and antioxidant properties. Prooxidants can generate ROS and/or inhibit antioxidant systems. This can thus contribute to cellular oxidative stress. Phenolic compounds, especially polyphenols, such as catechin, can act as both antioxidants and prooxidants depending on concentration of metal ions, pH, as well as the phenolic structure. This indicates that phenolic compounds may be antioxidants for some cell types and conditions while being a prooxidants for other cell types (Yordi et al., 2012). Interestingly enough, an antioxidant that has a more negative Standard Reduction Potential than Fe^{3+} could serve as substrate for the Fenton reaction, and thus the production of hydroxyl radicals (Lemire et al., 2013). The benefit of ingesting the food itself, as opposed to a concentrated form of it, appears to be that the foods contain a diverse spectrum of phytochemicals, metal ions, and vitamins at concentrations where they behave as antioxidants, rather than as prooxidants (Bouayed and Bihn, 2010).

Antioxidants react with free radicals through three potential mechanisms: hydrogen atom transfer (HAT), single electron transfer (SET), or the combination of both HAT and SET (Haung et al., 2005; Liang and Kitts, 2014). HAT is defined as the concerted movement of an electron and a proton, as a hydrogen atom, via a single kinetic step as shown in Eq. (2.4). The antioxidant loses a hydrogen to the oxidant, therefore becoming a radical itself while simultaneously terminating the oxidant's propagation. The thermodynamic barrier to the HAT reaction is the bond dissociation enthalpy of the antioxidant, that is, the species donating a hydrogen atom to the oxidant. The smaller the bond dissociation enthalpy, the easier it is for the antioxidant to transfer a hydrogen atom (Mader et al., 2007).

$$H : A + X \rightarrow \cdot A + H : X \tag{2.4}$$

SET is defined as a single electron transfer from the antioxidant to the oxidant as shown in Eq. (2.5). The antioxidant becomes a radical cation, and the oxidant becomes an anion intermediate, that is then terminated by addition of H^+ to the anion Eq. (2.6); Liang and Kitts, 2014).

$$A : + X \cdot \rightarrow A \cdot^+ + X :^- \tag{2.5}$$

$$X :^- + H^+ \rightarrow HX \tag{2.6}$$

The thermodynamic barrier determining the antioxidants activity is the ionization potential of the antioxidant itself (Liang and Kitts, 2014). The smaller the value of the ionization potential, measured as electron volts (eV), the smaller the energetic barrier is to the antioxidant losing an electron (Liang and Kitts, 2014). Distinguishing between HAT and SET reactions is difficult, as these reactions can take place simultaneously, but assays do exist that allow only HAT or SET mechanisms to occur so that quantitative evaluations can be done to assess the reactivity (Liang and Kitts, 2014).

2.2.3 HOW WE CAN MEASURE ANTIOXIDANT AMOUNTS

A number of assays, consisting of varied methods, exist to measure specific types of antioxidant mechanisms. There are many variables to consider when selecting an assay for measuring antioxidant capacity, including temperature, solubility of the reagents and oxidants, and their relevance to in vivo conditions, pH stability of reagents and oxidant, concentration of reagents and oxidant, how the mechanism of antioxidant in the assay couples to the oxidant, and the stability of the reagents and products measured (Liang and Kitts, 2014). What follows are some of the standard methods for analyzing antioxidant capacity of different antioxidants, as well as some of the limitations of the assays.

2.2.3.1 DPPH (2,2-diphenyl-1-picrylhydrazyl) assay

2,2-Diphenyl-1-picrylhydrazyl, or DPPH, is an organic free radical that possesses an electron that is delocalized across the pi electron system (Blios, 1958; Kedare and Singh, 2011). This molecule absorbs light at 519 nm, in ethanol. When an antioxidant is added to a solution containing DPPH, the absorption of DPPH, at 519 nm, decreases as the DPPH radical is terminated by the addition of antioxidant. The DPPH radical can be terminated by both HAT and SET mechanisms; thus this assay is unable to differentiate between the two mechanisms (Prior et al., 2005). Data from DPPH assays are typically reported as kinetic rates (moles DPPH consumed/

time/mole antioxidant) (Xie and Schaich, 2014). It should be noted that depending on the type of antioxidant mechanism (HAT or SET), the data collected and used to quantify antioxidants via the DPPH method might not be reliable. Xie and Schaich (2014) report that significant differences in the rates at which antioxidants quench the DPPH radical can be due to a kinetic barrier for the reaction, especially in the case of HAT. They report that HAT reactions occur far more slowly than do SET reactions. The authors' data suggest that the rate of radical quenching is dependent on solvent, pH, the concentration of antioxidant present, as well as the mechanism by which the antioxidant functions (Xie and Schaich, 2014). Comparing antioxidants under different assay conditions may not be reliable via the DPPH method (Xie and Schaich, 2014).

2.2.3.2 ABTS (2,2-azino-bis-3-ethylbenzothiazoline-6-sulphonic acid) assay

Using 2,2-azino-*bis*-3-ethylbenzothiazoline-6-sulphonic acid, or ABTS, a radical cation can be generated. This radical cation absorbs light at 734 nm, in water (Liang and Kitts, 2014). When antioxidants are added to a solution containing ABTS radical cation, the absorption decreases as the ABTS radical cation is terminated (Liang and Kitts, 2014). This assay measures HAT antioxidants only, but can be utilized for aqueous and lipophilic systems (Pellegrini et al., 1999). The results of the ABTS assay are comparable to the results of the DPPH assay, as both radicals react with the same stoichiometry with the water-soluble vitamin E analog, Trolox. The stoichiometry is 2:1 of the radical to Trolox. The Trolox equivalent antioxidant capacity (reported in Trolox equivalent units) is a measurement of antioxidant strength based on Trolox (Pellegrini et al., 1999). This serves as a reasonable way to compare results between assays (Antonio, 1998; Gil et al., 2000). The results of the ABTS assay are typically reported in a dose-dependent manner (moles ABTS radical cation consumed/moles antioxidant per a given time interval) (Pellegrini et al., 1999).

2.2.3.3 Thiobarbituric acid reactive substances: assay for lipid peroxidation

Thiobarbituric acid reactive substances (TBARS) are formed as a by-product of lipid oxidative damage (i.e., as degradation products of fats) and can be detected by the TBARS assay using thiobarbituric acid (TBA) as a reagent. This is an indirect measure of ROS. The TBA reacts with malondialdehyde (MDA), which is one of the several low-molecular weight end products formed from the decomposition of some primary and secondary lipid peroxidation products. However, not all peroxidation reactions generate MDA, therefore TBA data may be misleading. Also MDA is neither the sole end product of fatty peroxide formation and decomposition, nor a substance generated exclusively through lipid peroxidation. These and other considerations from the extensive literature on MDA, TBA reactivity, and oxidative lipid degradation support the conclusion that MDA determination and the TBA test can offer, at best, a limited view of the complex process of lipid peroxidation (Janero, 1990).

2.2.3.4 Two superoxide scavenging assays

Hydroethidine, or HE, reacts with the superoxide radical to produce ethidium. Ethidium is a red color compound that can be measured by fluorimetry (excitation: 520 nm; emission: 610 nm). This assay is typically used in vitro (Munzel et al., 2002). The 1,3-diphenylisobenzofuran, or DPBF, molecule incorporates into phospholipid liposomes and becomes fluorescent (excitation: 410 nm; emission: 455 or 477 nm). When this fluorescent compound interacts with superoxide, it loses fluorescence (Ohyashiki et al., 1999).

Nonspecific oxidation of the probe from nonsuperoxide sources (such as cytochrome c in the electron transport chain of aerobes) can affect the validity of the measurements in the HE assay. Using large concentrations of HE can produce increases in fluorescence not associated with oxidation of HE by superoxide. It is further reported that in cells the HE method might not be tightly coupled to superoxide because of the superoxide dismutase enzyme that converts superoxide to peroxide (Benov et al., 1998; Tarpey et al., 2004). The fluorescent DBPF liposomal compound can be nonspecifically quenched by O_2 specific generating systems (Bystryak et al., 1995; Fukuzawa et al., 1997; Reddi et al., 1991; Wozniak et al., 1991) resulting in overestimation of antioxidant capacity.

While here we have reviewed only some of the most commonly used methods to detect amounts of antioxidants, we note that each method has its limitations and data should be carefully evaluated relative to appropriate standards and limits of detection. Yet using one or more of these established assays allows for experimental studies of the redox abilities of molecules in our diet, in our vitamin supplements, and in other ingested materials, such as those extracted from coffee processing by-products.

2.2.4 THE TYPES OF ANTIOXIDANTS FOUND IN COFFEE (UNDER DIFFERENT ROASTING CONDITIONS) AND COFFEE PROCESSING BY-PRODUCTS

Coffee beans and coffee processing by-products contain a very large variety of compounds, many of which have potential roles as antioxidants. Mez-Ruiz et al. (2007) reported in vitro antioxidant activity of a variety of compounds, as well as the metabolites of these compounds, in coffee. Some of these important classes of molecules are summarized in Table 2.1.

2.2.4.1 Caffeine (1,3,7-trimethylxanthine)

Structurally, caffeine is a purine derivative that is an odorless white powder with a bitter taste, and with a low water solubility (21.6 g/L) (https://pubchem.ncbi.nlm.nih.gov/compound/caffeine#section=Melting-Point). Caffeine is found in a wide variety of plants, generally in low amounts, and appears to have a protective role in these plants. Caffeine has an oral LD50 of 192 mg/kg body weight in adults and 320 mg/kg body weight in child (http://www.sciencelab.com/msds.

Table 2.1 Important Classes of Molecules in Coffee and Coffee Processing by-Products

Compound	Some Biological Effects	Structure
Caffeine	Antioxidant by increasing the glutathione production, activates glutathione reductase and superoxide dismutase, adenosine receptor agonist, decrease cyclic nucleotide phosphodiesterase thereby increasing cyclic adenosine monophosphate, suppress inflammatory, and immunocompetent cells	
Chlorogenic or caffeic acids	Antioxidant, protects against DNA and lipid peroxidation damage, modulates glucose metabolism, inhibits fatty acid synthase, and HMG-CoA reductase, antiinflammatory behavior	
Maillard reaction products (MRP) from coffee	Inhibits nuclear factor kappa beta, increase nuclear translation of NRf2 and NF-κβ	 A general MRP, 6-acetyl-2,3,4,5-tetrahydropyridine

Background on antioxidants   **35**

Table 2.1 Important Classes of Molecules in Coffee and Coffee Processing by-Products (*cont.*)

Compound	Some Biological Effects	Structure
Methylglyoxal	Inducer of advanced glycation end products, increase glyoxylase system	
Trigonelline	Source of niacin, increases glutathione, antiinflammatory, increases serum lipids	
Kahweol	Antioxidant, increase glutathione, antiinflammatory, increases serum lipids	
Cafestol	Antioxidant, increase glutathione, antiinflammatory, increases serum lipids	

php?msdsId=9927475). However, coffee, cocoa, tea, and yerba mate plants have commercially useful levels of this important compound. (https://pubchem.ncbi. nlm.nih.gov/compound/caffeine#section=3D-Conformer). In an analysis of various types of commercially available coffee, the caffeine content in regular coffee ranged from 10.9 mg caffeine/g coffee bean to 16.5 mg caffeine/g coffee bean (Fujioka and Shibamoto, 2007). Caffeine content is reported to vary by the type of coffee bean tested. *Arabica* beans are reported to have 10–12 mg caffeine/g coffee bean. *Robusta* beans contain 19–21 mg caffeine/g coffee bean (Casal et al., 2000; Fox et al., 2013).

2.2.4.2 Caffeine as an antioxidant

As an antioxidant, caffeine, or its metabolite 1-methyluric acid shows activity in the protection of DNA degradation against hydroxyl radicals, protection of LDL-cholesterol against oxidation, and protection against Fenton reaction products. To combat hydroxyl radicals, caffeine functions by several different mechanisms; caffeine inhibits the production of hydroxyl radicals produced from L-DOPA and Cu^{2+} relative to controls, and caffeine quenches radicals via the HAT reaction and radical adduct formation (Leon-Carmona and Galano, 2011). 1-Methyluric acid is reported, using the ABTS assay, to be particularly effective against the oxidation of LDL-cholesterol (Gomez-Ruiz et al., 2007). Caffeine is also reported to protect against radiation exposure by slowing the oxygen-dependent pathways of radiation damage (Devasagayam and Kesavan, 1996). Adding to its utility, caffeine has positive effects on the antioxidant systems in rat brains by reducing lipid peroxidation of brain membranes, increasing the concentration of reduced glutathione, and by increasing the activities of glutathione reductase and superoxide dismutase (Abreu et al., 2011). As a direct antioxidant and via enzymatic systems that act as antioxidants, caffeine proves useful against oxidative damage. Caffeine has been reported to react with (quench) hydroxyl radicals generated by the Fenton reaction (Carmona and Galano, 2011). Caffeine does this through a bimolecular reaction (Brezova et al., 2009; Shi et al., 1991). The rate constant for this reaction is 2.6×10^{-9} $M^{-1}*s^{-1}$ (Parras et al., 2007). Caffeine has also been reported to have antioxidant activity against the superoxide radical (Kumar et al., 2001). Furthermore, crude caffeine has been shown to be a potent hydrophilic antioxidant [145 µmol Trolox equivalent (TE)/g] and a potent lipophilic antioxidant (66 µmol TE/g) (Chu et al., 2012b). The main caffeine metabolites in humans, 1-methylxanthine and 1-methylurate, are reported to be potent antioxidants in vitro and in vivo (Gomez-Ruiz et al., 2007). These compounds also showed strong antioxidant behavior, greater than that of caffeine (Lee, 2000). The author of this study showed that caffeine and its metabolites 1-methylxanthine and 1-methylurate were protective against the oxidation of low-density lipoprotein, through their antioxidant activity (Lee, 2000). Thus caffeine and its metabolites confer protection against oxidation under a variety of experimental conditions. Caffeine or its metabolites have been reported to have many direct and indirect effects on a variety of important physiological systems ranging from neurological, digestive, cardiac, reproductive, and immunological interactions. This is likely through its behavior as a nonspecific adenosine receptor antagonist. Depending on dose, the effects on one or more of these systems can be complex. Caffeine is a nonspecific antagonist of the adenosine receptors (A1, A2a, A2b, and A3), and is known to have effects on sleep, arousal, learning, cognition, and memory (Benowitz, 1990). It is a central nervous system stimulant, causing the secretion of norepinephrine, dopamine, and serotonin (Lieberman et al., 1987; Sawyer et al., 1982). Caffeine affects the cardiovascular system, increasing blood pressure, by inducing peripheral vasoconstriction of blood vessels (Pincomb et al., 1985; Robertson et al., 1990). In addition, it affects the respiratory tract by increasing respiratory rate (Murat et al., 1981). Caffeine also has metabolic effects, including increasing circulating epinephrine, free fatty acids, cortisol, metabolic rate, and blood glucose levels (Lung et al., 1981; Robertson et al., 1990). Furthermore, chronic coffee consumption may lead to increases in serum cholesterol (Davis et al., 1988; Williams et al., 1985). At the molecular level, caffeine and other methylxanthines have been shown to inhibit the cyclic nucleotide phosphodiesterase enzymes (PDE; now understood to be a large super family of enzymes), which thereby increased the levels of 3'-5'-cyclic AMP (Victoria et al., 2006). This metabolite activates protein kinase A, which in turn phosphorylates a number of other enzymes thereby affecting their activity (Manni et al., 2008). This common mechanism then helps explain the wide variety of physiological effects associated with caffeine and other methylxanthines, such as theophylline (Boswell-Smith et al., 2006).

2.2.4.3 Caffeine as an immune modulator

Caffeine can have a pronounced effect on cyclic AMP, and cyclic AMP is a potent immune modulator with the capacity to suppress inflammatory and immunocompetent cells. Thus, caffeine or its metabolites can be expected to influence both innate and adaptive immune responses (Horrigan et al., 2004). A review done by Horrigan et al. (2006) indicates that a large number of studies have shown caffeine to have immune-modulating effects on the following: innate and adaptive immunity, cytokine production, free radical production, lymphocyte proliferation antibody production, leukocyte counts, immune organ weights, leukocyte chemotaxis, natural killer cell function, histamine release, myeloperoxidase production, in vivo hypersensitivity reactions, and also immune cell apoptosis. Caffeine has also been shown to suppress tumor necrosis factor alpha by the cyclic AMP/protein kinase A pathway (Horrigan et al., 2004). Others have shown that responses by the lectins phytohemagglutinin (PHA) and concanavalin A (Con A) were lowered by 33% during coffee consumption compared to no coffee consumption, and this may be related to caffeine content (Melamed et al., 1990). These results indicate that caffeine is an immune active substance with therapeutically accessible dosages to the average coffee consumer. As a therapeutic, caffeine is known to be an analgesic adjuvant, and can be found in over-the-counter (OTC) products used in the treatment of tension headaches. The combination of caffeine with an analgesic substance, especially the nonsteroidal antiinflammatories, is reported to be significantly superior than just analgesic alone (Migliardi et al., 1994). Interestingly enough, the addition of caffeine to aspirin for the treatment of patients with sore throats was superior to no treatment and treatment with only aspirin (Bernard et al., 1991). Thus, caffeine enjoys extensive use in OTC cold products. Adenosine is reported to be involved in the response to tissue damage, such that its release attenuates the immune response (Hasko and Cronstein, 2004). In an ex vivo experiment, administration of caffeine (400 or 600 mg/day) increased A_{2A} adenosine receptor density in human neutrophils, and increased the affinity of these receptors for the [^{3}H]-ZM241385 ligand, in binding assays (Varani et al., 2005). The analgesic effect of caffeine may be caused by its synergistic effect with adenosine via the stimulation of A_{2A} receptor production and increased adenosine binding (Hasko and Cronstein, 2004). It should be noted that the immunomodulatory effects of caffeine would likely not be the same between users of different ages due to differential immune system responses. Furthermore, investigation leading to the comparison of acute and chronic coffee consumption is needed to determine the practicality of potential therapies involving coffee (Horrigan et al., 2006). Caffeine is a cheap adenosine receptor antagonist and may play a role in treating the neurological component of several medical indications. Adenosine receptors play a role in the inflammatory response after spinal cord injury, and the blockade of the A1 receptor by caffeine has been shown to mediate neuroprotection post spinal cord injury, likely by the previously mentioned mechanism (Palacios et al., 2012; Song et al., 2009; Stone et al., 2009). It has been shown to be a downregulator of NO production and other neuroinflammatory responses involved with pathogenesis of the nervous system (Salvemini et al., 2013; Tsutsui et al., 2004; Yaday et al., 2012).

2.2.4.4 Caffeine and its relationship in reducing some diseases

Caffeine has been shown to have significant beneficial effects for Parkinson's disease patients that lead to the improvement of motor activity. These effects were mediated via neuroprotective effects and neurorestoration via tropic proteins (TGF-beta, glial cell derived neurotropic factor, neuturin, and bone morphogenic proteins) (Chen and Chern, 2011; Postuma et al., 2012; Rosim et al., 2011). It may also slow neurodegeneration (Li et al., 2008; Morelli et al., 2012; Sonsalla et al., 2012). In addition, caffeine appears to have utility in Alzheimer's disease prevention and reversal of Alzheimer's disease symptoms via inhibition and decreasing formation of amyloid-β plaque formations (Arendash et al., 2009; Cao et al., 2009; Chu et al., 2012a; Cupino and Zabel, 2013; Gahr et al., 2013). There are several proposed mechanisms for the attenuation of amyloid-β plaque formation. The first is that long-term consumption of caffeine increases the production of cerebral spinal fluid. This increase is associated with increased expression of the sodium–potassium ATPase and increased cerebral blood flow. One of the proposed hypotheses of Alzheimer's disease is that there is a decrease in the turnover of cerebral spinal fluid and thus less clearance of amyloid-β plaques (Wostyn et al., 2011). By increasing the turnover of cerebral spinal fluid, it is thought that Alzheimer's disease can be attenuated (Wostyn et al., 2011). Another mechanism of caffeine's Alzheimer's disease benefits appears to be mediated through the decrease in the expression of presilin 1 and β-secretase, proteins associated with Alzheimer's disease onset or amyloid β-plaque formation (Arendash et al., 2006). It is clear that caffeine has utility as a therapeutic in a wide range of applications. Perhaps caffeine derivatives can perform more selectively in some of the aforementioned applications and thus harvesting these compounds from coffee processing by-products should be undertaken.

2.2.4.5 Chlorogenic acids and caffeic acid

Chlorogenic acid and derivatives are members of the phenolic class of compounds and are composed of quinic acid and caffeic acid (a hydroxycinnamic acid). The monoester of chlorogenic acid has a molecular weight of 354.309 g/mol (https://pubchem.ncbi.nlm.nih.gov/compound/1794427#section=Chemical-and-Physical-Properties).

These compounds are classified by the number and location of the esterification position of the hydroxycinnamic acids to the quinic acid ring (Wijewickreme and Kitts, 1998). These compounds are found in a very wide variety of plants and have a variety of roles, such as precursors to ferulic acid, as well as lignin production and as antioxidants. Chlorogenic acids are formed preferentially with the esterification occurring at C5, C3, or C4. Furthermore, monoesters of quinic acid with hydroxycinnamic acids, di-esters of quinic acid with caffeic acid, and dichlorogenic acids such as dicaffeoylquinic acids, feruloylquinic acids, and p-coumaroylquinic acids all exist in either green or roasted coffee beans. It should be noted that chlorogenic acids can undergo a multitude of chemical reactions during the roasting process, including isomerization, epimerization, lactonization, degradation to low molecular weight compounds, and incorporation into melanoids (dark pigments). The actual reactions that take place vary by the roasting conditions, and will be different for different bean compositions (Perrone and Farah, 2009; Tareke et al., 2002). Chlorogenic acids comprise 5%–10% by weight of raw coffee beans and some 50% is lost during the roasting process in a medium roast (Fujioka and Shibamoto, 2007). Chlorogenic acids show antioxidant activity against DNA damaging radicals and linoleic acid oxidation suggesting these acids are valuable sources of cellular protectant in the diet. A study of commercially available coffee brews were assayed for three caffeolylquinic acids (3-caffeoylquinic acid, 4-caffeoylquinic acid, and 5-caffeoylquinic acid), three feruloylquinic acids (3-feruloylquinic acid, 4-feruloylquinic acid, and 5-feruloylquinic acid), and three dicaffeoylquinic acids (3,5-dicaffeoylquinic acid, 3,4-dicaffeoylquinic acid, and 4,5-dicaffeoylquinic acid). This study showed that caffeinated coffee contained slightly greater total amounts of chlorogenic acids (5.26–17.1 mg chlorogenic acids/g coffee) than did decaffeinated coffee (2.10–16.1 mg chlorogenic acids/g coffee) (Fujioka and Shibamoto, 2007). The antioxidant behavior of these compounds has been evaluated on multiple occasions. Ohnishi et al. (1994) reported that chlorogenic acids could protect against linoleic acid peroxidation. Chlorogenic acids also show antioxidant behavior is important in the protection against DNA damage, and the authors report that the position of esterification does not affect the antioxidant behavior (Xu et al., 2012). Using the DPPH assay it was discovered that both chlorogenic acids and caffeic acid inhibited the oxidation of linoleic acid. These authors also report that 3,5-dicaffeoylquinic acid was a stronger antioxidant than caffeic acids (Ohnishi et al., 1994). Caffeic acid, a hydroxycinnamic acid, has greater antioxidant activity than does chlorogenic acids in an in vitro study utilizing Caco-2 cells. The authors report that the uptake of chlorogenic acids was much smaller than was the uptake of caffeic acid by these cells, and this may account for the observed difference in antioxidant activity (Sato et al., 2011). Furthermore, these results speak to the importance of the radical quenching mechanism of the antioxidant in question, and the method of measuring antioxidant behavior. Comparisons of the antioxidant and free radical scavenging activities of caffeic acid (CA), caffeic acid phenethyl ester (CAPE), ferulic acid (FA), ferulic acid phenethyl ester (FAPE), rosmarinic acid (RA), and chlorogenic acid (CHA) to R-tocopherol (vitamin E) and BHT were made by several methods, including the rancimat test (which measures oxidation stability of lipids using either lard or corn oil), and the DPPH assay. The authors report that the type of lipids being used in the assay, and the method of antioxidant assessment affected which antioxidants performed best in the assay (Chen and Ho, 1997). The individual addition of these compounds to either lard (with more saturated fatty acids) or corn oil (with more unsaturated fatty acids) did slow the onset of lipid oxidation. The onset of lard oxidation took longer than did the oxidation of corn oil (Chen and Ho, 1997). Thus, chlorogenic acids show potential use as antioxidant food additives and extraction from coffee processing by-products could prove to be an important source of these additives.

2.2.4.6 Chlorogenic acids modulate glucose and lipid metabolism

Chlorogenic acids also play a role in affecting glucose and lipid metabolism, and thus may be potentially a very useful therapeutic for diabetic patients. Chlorogenic acids resulted in a reduction in the plasma glucose peak in an oral glucose tolerance test (Bassoli et al., 2008). They also decreased early fasting glucose and insulin responses in overweight men performing oral glucose tolerance tests (van Dijk et al., 2009). Chlorogenic acids are further reported to improve glucose tolerance and insulin resistance, in obese Zuker rats, and thus show potential for being an antidiabetic agent (Hsu et al., 2000). They could serve to lower overall blood glucose through several mechanisms. One such mechanism involves increasing the uptake of glucose in skeletal muscle via the activation of the AMPK pathway (Ong et al., 2013). They inhibit glucose-6-phosphatase (G-6-Pase) translocase and reduce the sodium gradient used to drive glucose transporter I in the intestine (Arion et al., 1997). They also inhibit 40% of the G-6-Pase activity in microsomal preparations from hepatocytes, suggesting that chlorogenic acids can decrease hepatic glucose output (Arion et al., 1997; Herling et al., 2002; Prabhakar and Doble, 2009; Shin et al., 2013). Furthermore, chlorogenic acid inhibits two key enzymes, α-amylase and α-glucosidase, linked to type 2 diabetes, and this, too, is proposed as one of the mechanisms by which coffee or coffee by-products mediate blood sugar (Oboh et al., 2015). In chlorogenic acid-treated rats, fasting plasma cholesterol and triglycerides, as well as liver triglycerides, showed significant decreases (de Sotillo and Hadley, 2002). Chlorogenic acids may further have beneficial effects on body weight as they are reported to be potent inhibitors of fatty acid synthase, 3-hydroxy-3-methylglutaryl CoA reductase, and acetyl CoA cholesterol transferase (Cho et al., 2010).

2.2.4.7 Chlorogenic acids and their antiinflammatory activities

Chlorogenic acids also exhibit antiinflammatory behavior (Liang and Kitts, 2016). Because inflammation is a response to cellular injury or stress from exogenous or endogenous offenses, it is thought that controlling inflammation is required to return the offended tissue to normal tissue homeostasis (Elenkov et al., 2005; Medzhitov, 2008; Medzhitov and Janeway, 1997; Nathan, 2006). Chlorogenic acids downregulate IL-8, producing an antiinflammatory effect for TNF-α and peroxide induced inflammation, in vitro (Shin et al., 2015). Phenolic compounds, such as chlorogenic acids, decrease the activity of cyclooxygenase, and thus decrease the production of prostaglandins and leukotrienes (Mitjavila and Moreno, 2012). It is known that the nuclear factor kappa B pathway plays a key role in the release of adhesion molecules, cytokines, and chemokines that are proinflammatory (Lawrence, 2009). Other authors report in vitro, attenuation in the activation of nuclear factor kappa B and JNK/AP1 signaling pathways (Shan et al., 2009). Chlorogenic acids are reported to decrease activity of the nuclear factor kappa B pathway and inhibit neutrophil infiltration in animal studies, through the oral administration of 5-CQA, one specific chlorogenic acid (Stalmach et al., 2014). Chlorogenic acid has also been shown to enhance the activity of human lymphocyte proliferation, and the secretion of IFN-gamma in a lymphocyte-transformed BrdU immunoassay (Chiang et al., 2003). In a rat model of lipopolysaccharide (LPS)-induced arthritis, oral administration of 40 mg/kg of 5-CQA suppressed pro-inflammatory cytokines (TNF-α and IL-1β) (Chauhan et al., 2012). This led to faster healing of tissue damage caused by inflammation. Oral administration of 5-CQA (50 mg/kg/d) decreased MDA and NO levels, increased reduced GSH, and accelerated wound healing (Bagdas et al., 2015). There are also data available indicating that 5-CQA can decrease the amount of inflammation produced by IL-1β, TNF-α, and IL-6 caused by xenobiotic-induced liver inflammation by carbon tetrachloride (Shi et al., 2013). Thus, chlorogenic acids operate as antiinflammatory compounds, in vitro, via multiple mechanisms.

2.2.4.8 Maillard reaction products

The Maillard reaction is a general reaction that spontaneously occurs between the amino group of amino acids and proteins and the carbonyl group of reducing sugars, such as the monosaccharide glucose and the rate of the reaction increases with temperature [see Eq. (2.7) for general reaction model]. It occurs in many foods during the cooking process.

$$\text{Reducing sugar} + \text{amine} + \text{heat} \rightarrow N\text{-substituted glycosylamine (Schiff base)} + \text{water} \quad (2.7)$$

Especially important in the coffee industry, this reaction is responsible for the production of hundreds of chemical species that add to the taste and aroma of coffee. At temperatures 150–200°C, carbonyl groups in sugars, and amino groups in proteins react to form Schiff bases that directly possess aroma and flavor compounds, or that further react to produce substances with aroma and flavor properties. The darker the roasting, the more Maillard reaction products (MRPs) are produced (Rover, 2007). The Maillard reaction also produces products that are potentially carcinogenic, such as acrylamide, and may serve as the main dietary source of such toxins (Perrone and Farah, 2009; Tareke et al., 2002). This reaction also is implicated in advanced glycation products (AGEs) in which increased blood sugar (glucose) reacts with proteins in blood and tissues leading to crosslinked and damaged proteins (Grandhee and Monnier, 1991). Maillard reaction products have been reported to be antioxidants with strong activity against radicals (Manzocco et al., 2000). MRPs found following coffee roasting have notable antioxidant behavior against aldehydic, copper produced radicals, and oxygen-based radical systems. As antioxidants against aldehydic radicals, the pyrrole-containing MRPs exhibit the greatest activity, inhibiting hexanal oxidation by nearly 100% at a concentration of 50 µg/mL (Yanagimoto et al., 2002). MRPs that contain the amino-hexose-reductone structure, have high potential to inhibit the oxidation of LDL-cholesterol (Dittrich et al., 2003). The degree of roasting, and how it correlates to antioxidant activity is, however, still under debate. Some reports indicate that more roasting equates to more antioxidant behavior (Manzocco et al., 2000) while other reports indicate that less roasting is best for the most antioxidant behavior (del Castillo et al., 2002). While the Maillard reaction does occur internally especially under conditions of high blood glucose levels, we also get the products of this reaction from outside sources. The current estimate of human consumption of Maillard reaction products, or AGEs is thought to be 75 mg/day. These AGEs come from both food and drink (Henle, 2003). It is reported that only 10%–30% of AGEs are digested and absorbed, meaning the remaining 70%–90% reach the colon and have the opportunity to modulate bacterial gut growth (Virella et al., 2003; Vytasek et al., 2010). There are three proposed mechanisms by which AGEs function pathophysiologically: AGE modification of proteins results in structure/function changes, AGEs can catalyze ROS formation, and AGEs can bind to the receptor for AGEs (RAGE) and initiate a series of downfield pathogenic mediators. AGE-RAGE interaction is the primary candidate for AGE-induced inflammatory processes (Kellow and Coughlan, 2015). AGE binding to RAGE causes numerous cellular events, whose main cellular signature is inflammation and cell migration, including NADPH activation, sustained nuclear factor kappa B activation, transcription of proinflammatory cytokines (IL-6, TNF-α), C-reactive protein, MCP-1, thrombin, vascular endothelial growth factor, adhesion molecules (VCAM-1, ICAM-1), transcription factors, and extracellular signal regulating. Because AGEs are able to modify protein structure, these denatured proteins may be identified as antigens by B cells, producing an autoimmune response (Virella et al., 2003; Vytasek et al., 2010).

2.2.4.9 Maillard reaction products in coffee and immune modulating effects

An extensive body of research using animal models have implicated Maillard reaction products in a variety of physiological and biochemical effects (Maslowski and Mackay, 2011). Maillard reaction products from coffee have been shown to have a number of immune-modulating effects. As the roast of coffee becomes darker, the amount of Maillard reaction products increases. It was shown that as the roast of coffee became darker, nuclear factor kappa B was inhibited by more than 80%, while an electrophilic response element (EpRE) activity (which helps regulate gene expression of genes involved in oxidative stress) was increased by more than 2500%, in vitro (Meng et al., 2013). In a luciferase nuclear factor kappa B transgenic mouse model, a single dose of dark-roasted coffee led to attenuation of nuclear factor kappa B activation by 63%. This is of importance because nuclear factor kappa B is often considered to be involved in proinflammatory cell signaling (Paur et al., 2010). In vitro models utilizing Caco-2 (heterogeneous human epithelial colorectal adenocarcinoma cells) or macrophages show that the stimulation of these cell types by Maillard reaction products increased the nuclear translocation of Nrf2 by 500% after a 2 h exposure. After a 24 h exposure, Nrf2 nuclear transport was increased by 5000% (Sauer et al., 2011). Stimulation of macrophages, in vitro, with coffee extract in phosphate buffered saline increased nuclear factor kappa B translocation 13-fold (Muscat et al., 2007). When Maillard reaction products were added to the Caco-2 and macrophage cell types in the presence of catalase, the Nrf2 nuclear translocation was significantly suppressed. This suggests that Maillard reaction products mediate their effect through the production of H_2O_2 because catalase degrades H_2O_2 (Sauer et al., 2011). Maillard reaction products have also been shown to induce nuclear translocation of NF-κB in vitro (macrophages and Caco-2 cell lines) and ex vivo (primary human intestinal microvascular endothelial cells) up to 500% (Sauer et al., 2011). This is important because chronic activation of nuclear factor kappa B is associated with inflammatory bowel diseases (Sauer et al., 2011). It has also been shown that Maillard reaction products modulate gut microbiota, both in humans and in rats (Seiquer et al., 2014). Maillard reaction products (also referred to as AGEs) have been implicated in several degenerative diseases because of their capacity to promote gene products involved with the inflammatory process (Kellow and Coughlan, 2015). AGEs may potentially function through several mechanisms to increase inflammation: suppression of protective metabolism and the incretins (a group of metabolic hormones that lead to a decrease in blood glucose and an increase in blood insulin), as well as affecting immune-mediated cell signaling, and by changing the gut microbiota and metabolite sensors (Kellow and Coughlan, 2015). When placed on a 2-week diet, high in AGEs, the bacterial population in the colon of humans was altered, thus showing that AGEs are able to modify gut microbiota populations in relatively short time periods (Seiquer et al., 2014).

2.2.4.10 Methylglyoxal as an inducer of AGEs

Methylglyoxal is a toxic metabolite that arises from nonenzymatic phosphate elimination from glyceraldehyde phosphate and dihydroxyacetone phosphate, two intermediates of glycolysis, as well as from heating a variety of foods, especially those with simple sugars. Brewed coffee is reported by Hayashi and Shibamato (1985) to be an important dietary source of methylglyoxal (75.6 µg per average serving). Methylglyoxal is a potent inducer of AGEs, and has been shown to decrease the capacity of dendritic cells to stimulate the proliferation of allogenic T-cells. Methylglyoxal may have a role in pathogenesis by this mechanism (Cai et al., 2014). Humans have detoxification systems that are activated by the ingestion of AGEs. The glyoxalase system catalyzes the metabolism of numerous dicarbonyls, including methylglyoxal. This system is composed of two enzymes, (glyoxylase 1 and glyoxylase 2), and the tripeptide glutathione (GSH) is a cofactor (Thornalley, 2003). When operating normally, the glyoxylase system metabolizes methylglyoxal, and cellular inflammation is kept under control. When oral methylglyoxal was fed to mice with no glyoxylase 1, increases in AGE deposits and amyloid-β were measured. Both of these contributed to deficits in cognitive and motor function (Cai et al., 2014). The administration of methylglyoxal induces a number of physiological effects in rats, including early activation of markers for arterial sclerosis

(Yamawaki et al., 2008), leukocyte recruitment, activation of the nuclear factor kappa B pathway, upregulation of endothelial adhesion molecules (Su et al., 2012), inhibition of blood vessel contractility (Mukohoda et al., 2009), induction of hypertension (Vasdev et al., 1998), and with chronic intake of methylglyoxal, inhibition of the serine/threonine protein kinase pathway was triggered (Crisostomo et al., 2013). The Nrf2 pathway is a major mechanism in cellular defense against oxidative and electrophilic stress. It functions through the activation of the Nrf2-antioxidant response-signaling pathway. This pathway controls gene expression of proteins involved in deactivation/metabolism of oxidants and electrophilic agents, such as glutathione reductase, glutathione S-transferase, glutathione peroxidases, glyoxylase 1, aldehyde reductases, and aldehyde dehydrogenases (Cai et al., 2014; Nguyen et al., 2015; Xue et al., 2012). It has been shown that AGEs can activate glyoxylase 1, through the Nrf2 pathway, providing protein and DNA protection (Xue et al., 2012). It may be possible for AGEs to affect metabolite sensing G-protein coupled receptors, GPR41 and GPR43. Because AGEs affect the bacterial population in the colon, the short chain fatty acid producing bacteria may be affected. This can cause differences in the activation of GPRC41 and GPRC43, leading to differential activation of the inflammasome (Macia et al., 2012; Maslowski and Mackay, 2011). These results indicate that AGEs are immune active compounds with complex activities. Because coffee processing by-products need not be subjected to the same heating process as brewing the coffee beverage, the formation of methylglyoxal is less likely to occur to the same extent.

2.2.4.11 Trigonelline, kahweol, and cafestol

The alkaloid trigonelline is a niacin derivative that is also found in coffee beans. It has a molecular weight of 137.1360 g/mol (https://pubchem.ncbi.nlm.nih.gov/compound/5570#section=Chemical-and-Physical-Properties). The pyridinium ring nitrogen of niacin is methylated to produce trigonelline. This compound is a degradation product of mammalian niacin metabolism (https://pubchem.ncbi.nlm.nih.gov/compound/Trigonelline). During coffee bean roasting, trigonelline is converted to niacin, and serves as a source of vitamin B_3 for its users (Trugo, 2003). It has been shown that coffee roasting can increase the niacin content of coffee beans by up to 1000% (Mazzafera, 1990). Trigonelline also contributes to the bitter taste of coffee, and serves as the precursor of many volatile coffee roasting products, such as pyrroles and pyridines (Flament et al., 1968). The antioxidant activity of trigonelline has not been evaluated in specific cell models, but in diabetic rats has been shown to reduce oxidative stress by upregulation of antioxidant enzyme activity with a decrease in lipid peroxidation (Zhou et al., 2013). Trigonelline seems to be especially effective against the oxidative damage seen in diabetic model systems. In diabetic rats, where prior to intervention the animals showed increased amounts of enzymes or molecules associated with oxidative stress, trigonelline works as an antioxidant by normalizing the activities of superoxide dismutase, catalase, glutathione, and inducible NO synthase (iNOS) to control levels (Zhou et al., 2013). In another diabetic rat study, trigonelline had the effect of improving kidney function compared to the untreated diabetic group. The improvements were as follows: improvements in glomerular filtration rate, improvement in the activity of antioxidant membrane bound and nonmembrane bound proteins, and decreased levels of TNF-α and hydroxyproline in kidney tissues. These results indicate a potential role for trigonelline as a kidney protectant, in part via its effects on antioxidant enzymes (Ghule et al., 2012). An in vitro study shows that trigonelline attenuates oxidative stress generated by copper ascorbate in mitochondria isolated from goat tissues. These tissues include heart, liver, brain, lung, and kidney. Trigonelline functions in a dose-dependent manner in the protection of these tissues with maximum radical quenching observed at a trigonelline concentration of 0.8 mg/mL (Dutta and Bandyopadhyay, 2014). In the same copper-ascorbate oxidative stress model in mitochondria, incubation of goat heart, liver, brain, kidney, and lung with trigonelline (0.8 mg/mL), samples incubated with trigonelline showed significantly less NO production in a dose-dependent manner (as percent of the no trigonelline control; heart, 66.40%; liver, 37.72%; brain, 50.53%; lung, 63.10%; kidney, 40.55%) (Ghule et al., 2012). Mitochondrial intactness was also improved by trigonelline incubation, in this same model of mitochondrial oxidation compared to no trigonelline. Furthermore, trigonelline administration decreased markers of mitochondria lipid peroxidation (Ghule et al., 2012). As antioxidants, the diterpenes kahweol and cafestol also function in a dose-dependent manner to increase (up to 2.4-fold) the activity of γ-glutamylcysteine synthetase. This is associated with an increase in GSH of up to 3-fold. At the largest dose of kahweol and cafestol, the liver mRNAs of the heavy and light subunits of γ-glutamylcysteine, doubled. In the kidney, lungs, and colon, γ-glutamylcysteine synthetase activity, as well as GSH concentrations increased. Kahweol and cafestol increased GSH levels through the induction of γ-glutamylcysteine synthetase, by affecting the rate-limiting enzyme in GSH synthesis (Huber et al., 2002). Trigonelline shows therapeutic potential in several areas, including microbial diseases, diabetes, skeletal muscle stimulants, obesity, and cancer treatment. Trigonelline has been shown to have antibacterial activity with a reported IC50 of 2.2 mg/mL (Almeida et al., 2006). For the treatment of diabetes, trigonelline significantly reduced blood glucose, total cholesterol, and triglycerides in diabetic rats. Trigonelline further aided diabetic rats by normalizing values of pancreas-to-body weight ratio, insulin level, insulin sensitivity index, insulin content in the pancreas, monoaldehyde, NO, superoxide dismutase, catalase, and glutathione to near control values (Zhou et al., 2011). In yet another model of diabetic rats, trigonelline treatment produced glucose normalizing effects, HbA_{1c} (a result of the nonezymatic attachment of glucose of the *N*-terminal amino acid of hemoglobin and is thus a marker of diabetes), normalization of insulin levels, and normalization of insulin sensitivity in the diabetic rat population to those values near the control, nondiabetic rats (Zhou and Zhou, 2012). A patent has been filed for the use of trigonelline as an anabolic, and anticatabolic agent of skeletal muscle. The patent claims that trigonelline, without adding calories to the diet, is responsible for preventing muscle loss, supporting healthy muscle function, delaying sarcopenia, preventing muscle loss due to illness or surgery, maintaining muscle strength, increasing muscle strength, promoting myoblast differentiation, promoting muscle growth, and promoting muscle recovery and repair at a dose range of 5–5000 mg trigonelline/day (http://www.

google.ch/patents/US8323707). Trigonelline has also been shown to suppress lipid droplet accumulation in a dose-dependent manner (Ilavenil et al., 2013). Trigonelline does this by downregulating peroxisome proliferator activated receptor and CCAAT element binding protein(C/EBP-α) mRNA expression. These downregulations led to other genes being downregulated. These genes include adiponectin, leptin, resistin, adapogenin, and adipocyte fatty acid binding protein (Ilavenil et al., 2013). Trigonelline is also reported to have detrimental effects on pancreatic cancer cells, making them more susceptible to apoptosis. It does this through the inhibition of Nrf2's cytoprotective apoptotic genes. Furthermore, the addition of trigonelline to anticancer drug treatment in mice increased the antitumor responses in said mice (Arlt et al., 2013). Trigonelline shows utility in a number of important areas, and as such, deserves more attention as a potential therapeutic especially if it can be economically extracted from coffee processing by-products.

2.2.4.12 Diterpenes as antiinflammatory molecules

Diterpenes are isoprene derivatives that are widely found in living things and are important precursors for a number of important molecules, such as sterols, retinol, and phytol, and have important antiinflammatory properties. Two important diterpenes in coffee are kahweol and cafestol, which are reported to have anticarcinogen activity (Cavin et al., 2002). Structurally, these molecules are similar, as both are pentacyclic diterpene alcohols. Their base ring structure is that of the kaurane skeleton (Kolling-Speer and Speer, 2005). The effects of the diterpenes kahweol and cafestol are most frequently reported to be antiinflammatory and/or anticarcinogenic. These diterpenes affect inflammation by mediating iNOS and cyclooxygenase-2 (COX-2) expression (Kim et al., 2004a,b). Kahweol also exhibits neuroprotective effects on dopaminergic neurons by inducing heme oxygenase-1 via the PI3K and p38/Nrf2 pathways (Hwang and Jeong, 2008). These diterpenes also have effects on cholesterol metabolism, blood pressure, and metabolic syndrome (Bonita et al., 2007). In cell cultures, kahweol, and cafestol inhibit inflammatory responses in macrophages (Kim et al., 2004a,b). Kahweol also has the capacity to prevent the degradation of IκB proteins, perhaps explaining some of its antiinflammatory and anticarcinogenic effects (Cavin et al., 2002; Kim et al., 2004a,b). Kahweol and cafestol both suppress the protein levels of specific protein 1(SP1) a protein involved with cell differentiation, cell growth, apoptosis, and immune response effects (Lee et al., 2012). These diterpenes and their derivatives are of interest in a variety of applications ranging from their effects on serum cholesterol, cell signaling, anticarcinogenic effects, to their liver protective effects. However, they are not without controversy. High intakes have been associated with increased serum low-density lipoprotein cholesterol and increased homocysteine levels, both of which are associated with increased risk of coronary heart disease (Olthof et al., 2001). Kahweol and cafestol have organ protective effects as well. Using a carbon tetrachloride induced liver damage model, pretreatment with kahweol and cafestol prior to carbon tetrachloride administration significantly attenuated the increases in serum levels of liver damage enzyme markers (aspartate aminotransferase and alanine aminotransferase) (Lee et al., 2007). These compounds also reduced oxidative stress, as indicated by increased reduced glutathione content and decreases in measurements of lipid peroxidation in the liver. These effects were dose-dependent (Lee et al., 2007). Kahweol and cafestol also reduced the frequency of liver lesions caused by carbon tetrachloride, as indicated by histological examinations (Lee et al., 2007). Kahweol and cafestol also affected cytochrome P450 2E1 (CYP2E1) enzyme activity, decreasing it. CYP2E1 is the major isozyme involved in the bioactivation of carbon tetrachloride. Kahweol and cafestol function as antioxidants in mouse liver homogenate, and on superoxide radical scavenging activity. Like trigonelline, kahweol, and cafestol appear to function as antioxidants, indirectly through their impact on individual enzymes (Lee et al., 2007). The coffee diterpenes kahweol and cafestol show activity in the areas of stimulating iNOS expression in macrophages, protection against aflatoxin, protection against H_2O_2 induced oxidative stress and DNA damage, as well as being anticarcinogenic and chemoprotective agents. Serum cholesterol, the production of prostaglandin E2 in macrophages, the production of COX-2 in macrophages, and the enhancement of phase 2 detoxification enzymes and hepatic GSH levels have all been affected by diterpenes. Kahweol possess antiinflammatory activity through the inhibition of iNOS protein and iNOS mRNA in LPS-activated RAW 264.7 macrophages. Kahweol does this by blocking the LPS-induced activation of nuclear factor kappa B (Kim et al., 2004b). With a dose-dependency, kahweol and cafestol act as DNA protective agents in rat livers against aflatoxin B1 binding. A 50% reduction in aflatoxin B1 binding to DNA was measured in kahweol and cafestol treated rats versus control rats. The authors indicate two mechanisms for chemoprotective effects against aflatoxin B1. The first is a decrease in CYP2C11 and CYP3A2, which are cytochrome P450 enzymes responsible for activation of xenobiotics in rat liver. Second, there was a strong induction in the expression of glutathione-S-transferase subunit GST Yc2. This protein is known to detoxify aflatoxin B1 (Cavin et al., 1998). Kahweol and cafestol exhibit DNA protective effects against peroxide induced DNA damage in NIH3T3 cells treated with either compound. Measurements of cytotoxicity, lipid peroxidation, and ROS production induced by H_2O_2 were made, and results indicate a reduction in these metrics in a dose-dependent manner. Kahweol and cafestol were also protective against H_2O_2-induced oxidative DNA damage. Kahweol and cafestol also exhibit protective effects against the superoxide anion generated from the xanthine/xanthine oxidase system (Lee and Jeong, 2007). Kahweol and cafestol exhibit reduction in genotoxicity against several carcinogens (aflatoxin B_1, benzo[a]pyrene, 7,12-dimethylbenz[a]anthracene, and 2-amino-1-methyl-6-phenylimidazo[4,5-[b]pyridine]) by varied mechanisms (Cavin et al., 2002). These mechanisms include induction of phase 2 detoxification enzymes (glutathione-S-transferases, glucuronosyl S-transferases), an increased expression of cellular antioxidant defense proteins (γ-glutamyl cysteine synthetase and heme oxygenase-1), and an inhibition of the expression and/or activity of phase 1 cytochrome P450s that are involved in carcinogen activation from procarcinogens (CYP2C11, CYP3A2) (Cavin et al., 2002). Others have shown similar effects by kahweol and cafestol on the induction of hepatic glutathione-S-transferase, but also report that hepatic N-acyltransferase induction by 2-amino-1-methyl-6-phenylimidazo-[4,5-b]pyridine, a known carcinogen, was inhibited by 80% (Huber et al., 2004).

2.2.4.13 Diterpenes effects on blood lipids

Less encouraging in terms of pharmaceutical usefulness, when cafestol was fed (61–64 mg/day) to 10 healthy male volunteers, measurable effects on mean and total serum cholesterol (increase by 31 ± 5 mg/dL), LDL-cholesterol (increase 22 ± 5 mg/dL), fasting triglycerides (increase 58 ± 11 mg/dL), and alanine aminotransferase (increase 18 ± 2 U/L) were observed after 28 days (Urgert et al., 1997). When kahweol and cafestol were taken together (61–63 mg/day) additive increases were observed: total cholesterol (additional increase of 9 ± 6 mg/dL), LDL cholesterol (additional increase of 9 ± 6 mg/dL), triglycerides (additional increase of 8 ± 9 mg/dL), and alanine aminotransferase (additional increase of 35 ± 11 U/L) (Urgert et al., 1997). These data indicate that cafestol has a much larger effect on blood lipid biomarkers than does kahweol, but a kahweol-only group would be required to be certain of this result (Urgert et al., 1997). These coffee diterpenes activate a number of potentially useful processes in different model systems. Their further investigation as activators of cytochrome and antioxidant enzymes in humans is warranted.

2.2.5 USEFUL MATERIALS IN DIFFERENT COFFEE BY-PRODUCTS

As shown in Table 2.2, coffee processing by-products contain a number of molecules of major interest as antioxidants and antiinflammatory compounds. These are reviewed in the next sections.

2.2.5.1 Husks

The main by-product from the dry method of coffee bean processing is the coffee husk, which is composed of the dried skin, pulp, and parchment (Esquivel and Jimenez, 2012). For each ton of harvested coffee fruit, some 0.18 ton of coffee husk can be generated. Such husk material is rich in carbohydrates (both simple and complex), protein, with a small amount of lipid. The minor chemical constituents (some 2%–10% by weight), with potential antioxidant properties, reported in coffee husks are caffeine, chlorogenic acids, and polyphenols (Jaiswal et al., 2012; Nonthakaew et al., 2015; Volz et al., 2014). Thus, husks are an untapped source of useful materials. Some of the numerous potential uses of caffeine, chlorogenic acids, and polyphenols have been mentioned previously. Some

Table 2.2 Major Molecules of Interest Found in Coffee Processing By-Products

By-Product	Molecules of Major Interest
Husks	Caffeine, chlorogenic acids, polyphenols
Coffee pulp	Chlorogenic acids and caffeic acid
Coffee silver skin	Caffeine and chlorogenic acids

interesting polyphenols include flavonoids, which are the most common group of polyphenolics in the human diet and in plants (Spencer, 2008). Because of the presence of these compounds in coffee husks, coffee husks may serve as valuable source for dietary antioxidant supplements. Coffee husks may also serve as a valuable food additive for the purposes of adding fiber to food products (Esquivel and Jimenez, 2012; Murthy and Naidu, 2012). The total amounts of these compounds found in coffee husks are: caffeine (1.0–1.3 g/100 g coffee husk), chlorogenic acids (2.5 g/100 g coffee husks), and polyphenols (0.8–1.2 g/100 g coffee husk) (Bondesson, 2015). While the total chlorogenic acid amount is on the same order of magnitude between coffee husk types, the chlorogenic acid profile of the husks varies (Mullen et al., 2013). There are eight reported chlorogenic acids that are found in coffee husks taken from either Arabica or Robusta coffee fruits. These chlorogenic acids include 3-O-caffeoylquinic acid, 5-O-caffeoylquinic acid, 4-O-caffeoylquinic acid, 4-O-feruloylquinic acid, 5-O-feruloylquinic acid, 3,4-O-dicaffeoylquinic acid, 3,5-O-dicaffeoylquinic acid, and 4,5-O-dicaffeoylquinic acid (Mullen et al., 2013). The total polyphenol content and polyphenol profile varies between the Arabica or Robusta coffee fruits, and the region of the globe in which the coffee fruits are grown (Mullen et al., 2013). The procyanidin (a flavonoid) content of tested coffees is reported from largest relative content to smallest relative content, with the largest content being set to 100% (India Robusta, 100%; India Arabica, 47%; Mexico Arabica, 22%; China Arabica, 19%) (Mullen et al., 2013). The flavanol content is reported from largest relative content to smallest relative content with the largest content being set to 100% (Mexico Arabica, 100%; India Arabica, 13%; Mexico Arabica, 8%; India Robusta, 7%; China Arabica, 6%; China Robusta, 2%) (Mullen et al., 2013). These data show that coffee husks from fruits grown under different conditions have different profiles, and thus some will be more useful as sources for functional molecules to add to food products. These are then large sources of potentially useful molecules of economic value for coffee growing countries that are currently not being utilized. This is especially of interest since the dry method of husk processing does not use high temperatures and thus the antioxidant capacities will remain high relative to brewed coffee.

2.2.6 COFFEE PULP AND SILVER SKIN

2.2.6.1 Coffee pulp

Coffee pulp is generally considered to be a waste product from the coffee industry but a waste product that should be further evaluated for potential uses (Rathinavelu and Graziosi, 2005). Coffee pulp is reported to function as a substrate for microbial processes and have antioxidant behavior. Serving as a substrate for microbial processes, coffee pulp was used as a solid-state substrate in a fermentation utilizing *Aspergillus niger* (Pandey et al., 2000; Peñaloza et al., 1985). This fermentation was utilized to produce amino acids from the coffee pulp substrate, and served as a more useful chicken feed than did the original processed coffee pulp (Peñaloza et al., 1985). Coffee pulp has also been successfully used in addition to fermentations used for bioethanol production, and was shown to have no negative effects on yeast cell fermentation ability (Menezes et al., 2013). Coffee pulp may also serve as a useful replacement for yellow corn maize in the diets of farmed catfish (*Clarius isheriensis*) (Fagbenro and Arowosoge, 1991). Coffee pulp also shows utility as an antioxidant, and is reported to contain four different major classes of polyphenols, including flavan-3-ols, hydroxycinnamic acids, flavanols, and anthocyanidins (Ramirez-Coronel et al., 2004). Hydroxycinnamic acids are present in the soluble coffee pulp fraction, (68%–97%). Chlorogenic acid was the main phenolic acid in the soluble fraction (94%–98%). Caffeic acid was the most abundant hydroxycinnamic acid in the bound fraction (72%–88%) (Rodríguez-Duran et al., 2014). The activities of anthocyanidins as inhibitors of glucosidase and amylase enzyme activity have also been measured with reported IC50 values of 0.22 and 0.43 mg/mL, respectively (Murthy et al., 2012).

2.2.6.2 Coffee silver skin

Coffee silver skin (CSS) is reported to function as a prebiotic, as an antioxidant, and as a source of fermentable sugar. Coffee silver skin is also known to contain both caffeine and chlorogenic acids (Bresciani et al., 2014). Coffee silver skin fiber is reported as being 98% of the total carbohydrate content in coffee silver skin (Borrelli et al., 2004). As a prebiotic or as a source of fermentable sugar, coffee silver skin has been shown to support the growth of probiotic intestinal *Bifidobacteria* (Borrelli et al., 2004). Coffee silver skin can be used as a substrate to produce α-amylase using *Neurospora crassa* CFR 308 (Murthy et al., 2009). Coffee silver skin has also been used as a substrate in combination with *Aspergillus japonicas* under solid-state fermentation conditions to produce fructooligosaccharrides (Mussatto and Teixeira, 2010). As an antioxidant, coffee silver skin shows activity in the ABTS (Andlauder and Furst, 1998), and DPPH assays, suggesting antioxidant behavior potentially by both HAT and SET mechanisms and, therefore, may be considered for use in food products as an antioxidant (Bresciani et al., 2014; Costa et al., 2014; Narita and Inou, 2012). Chlorogenic acids are reported as being 0.93% by weight of the total mass of coffee silver skin (CSS) (Bresciani et al., 2014; Costa et al., 2014). As literally tons of coffees are produced, coffee silver skin is gaining more recognition as a useful by-product of the coffee production process. The total antioxidant capacity of CSS has been reported to be 139 mmol Fe^{2+}/kg, a value similar to other food sources of antioxidants like dark chocolate (Bresciani et al., 2014). However, CSS also contains undesirable compounds, such as ochratoxin A, which is reported to be a toxin and carcinogen so that more studies need to be done to avoid these potential hazards (Toschi et al., 2014). Both coffee pulp and coffee silver skin present the opportunity to be value added products, serving to aid biological fermentations and acting as antioxidants. However, the methods used in processing of these materials will influence the extracted compounds and their functionality especially as antioxidants (Vignoli et al., 2011).

2.2.7 COMPOSITION SIMILARITIES BETWEEN COFFEE AND COFFEE PROCESSING BY-PRODUCTS

As reported by Farah and Donangelo (2006), the coffee bean, as well as coffee processing by-products appear to contain the same general classes of secondary metabolites, such as caffeine and a variety of phenolic compounds, such as hydroxy-cinnamic acids and chlorogenic acids, although at different proportions. All of these molecules have antioxidant properties, as well as other potential roles especially in regulating the immune system. These same secondary metabolites in general make the coffee processing by-products less useful as animal feeds especially at high levels in the feeds.

2.3 CONCLUSIONS

In this chapter, the chemical basis of oxidation and reduction, the need for exogenous antioxidants in biological systems, and their synergism with endogenous antioxidants, how to measure the presence and activity of antioxidants, how coffee and coffee by-products serve as a source of exogenous antioxidants, and how coffee and coffee by-product constituents also appear to function as immune modulating substances are discussed. With such a multitude of potential functions, coffee processing by-products have great potential to add value in the discussed areas. This may be especially true because many of the useful antioxidant compounds in coffee by-products are not lost in the roasting process that the coffee bean undergoes.

REFERENCES

Abreu, R.V., Silva-Oliveira, E.M., Dutra Moraes, M.F., Pereira, G.S., Moraes-Santos, T., 2011. Chronic coffee and caffeine ingestion effects on the cognitive function and antioxidant system of rat brains. Pharm. Biochem. Behav. 99 (4), 659–664.

Almeida, A.A.P., Farah, A., Silva, D.A.M., Nunan, E.A., Gloria, B.A., 2006. Antibacterial activity of coffee extracts and selected coffee chemical compounds against *Enterobacteria*. Agric. Food Chem. 54, 8738–8743.

Andlauder, W., Furst, P., 1998. Antioxidant power of phytochemicals with special reference to cereals. Cereal Food World. 43, 356–360.

Antonio, C., 1998. An end-point method for estimation of the total antioxidant activity in plant material. Phytochem. Anal. 9, 196–202.

Arendash, G.W., Schleif, W., Rezai-Zadeh, K., Jackson, E.K., Zacharia, L.C., Cracchiolo, J.R., Shippy, D., Tan, J., 2006. Caffeine protects Alzheimer's mice against cognitive impairment and reduces brain β-amyloid production. Neuroscience 142, 941–952.

Arendash, G.W., Mori, T., Cao, C., Mamcarz, M., Runfeldt, M., Dickson, A., et al., 2009. Caffeine reverses cognitive impairment and decreases brain amyloid-beta levels in aged Alzheimer's disease mice. J. Alzheimer's Dis. 17 (3), 661–680.

Arion, W.J., Canfield, W.K., Ramos, F.C., et al., 1997. Chlorogenic acid and hydroxynitro-benzaldehyde: new inhibitors of hepatic glucose 6-phosphatase. Arch. Biochem. Biophys. 339 (2), 315–322.

Arlt, A., Sebens, S., Krebs, S., Geismann, C., Grossmann, M., Kruse, M.-L., Schreiber, S., Schäfer, H., 2013. Inhibition of the Nrf2 transcription factor by the alkaloid trigonelline renders pancreatic cancer cells more susceptible to apoptosis through decreased proteasomal gene expression and proteasome activity. Oncogene 32, 4825–4835.

Bagdas, D., Etoz, B.C., Gul, Z., Ziyanok, S., Inan, S., Turacozen, O., Gul, N.Y., Topal, A., Cinkilic, N., Tas, S., et al., 2015. In vivo systemic chlorogenic acid therapy under diabetic conditions: wound healing effects and cytotoxicity/genotoxicity profile. Food Chem. Toxicol. 81, 54–61.

Bassoli, B.K., Cassolla, P., Borba-Murad, G.R., et al., 2008. Chlorogenic acid reduces the plasma glucose peak in the oral glucose tolerance test: effects on hepatic glucose release and glycaemia. Cell Biochem. Funct. 26 (3), 320–328.

Benov, L., Sztejnberg, L., Fridovich, I., 1998. Critical evaluation of the use of hydroethidine as a measure of superoxide anion radical. Free Radic. Biol. Med. 25, 826–831.

Benowitz, N.L., 1990. Clinical pharmacology of caffeine. Ann. Rev. 41, 277–288.

Bernard, P.S., John, M.F., Alberta, C.L., William, R.T., Robert, I.B., 1991. Caffeine as an analgesic adjuvant: a double-blind study comparing aspirin with caffeine to aspirin and placebo in patients with sore throat. JAMA 151 (4), 733–737.

Bishop, 1968. Hydrogen peroxide catalytic oxidation of refractory organics in municipal waste waters. Ind. Eng. Chem. 7, 1110–1117.

Blios, M.S., 1958. Antioxidant determinations by the use of a stable free radical. Nature 181, 1199–1200.

Bondesson, E., 2015. A nutritional analysis on the by-product coffee husk and its potential utilization in food production. BS thesis, Swedish University of Agricultural Sciences, Uppsala, Sweden.

Bonita, J.S., Mandarano, M., Shuta, D., Vinson, J., 2007. Coffee and cardiovascular disease: in vitro, cellular, animal, and human studies. Pharmacol. Res. 55, 187–198.

Borrelli, R.C., Esposito, F., Napolitano, A., Fogliano, V., 2004. Characterization of a new potential functional ingredient: coffee silverskin. J Agric. Food Chem. 52, 1338–1343.

Boswell-Smith, V., Spina, D., Page, C.P., 2006. Phosphodiesterase inhibitors. Br. J. Pharmacol. 147 (Suppl 1), S252–S257.

Bouayed, J., Bihn, T., 2010. Exogenous antioxidants: double-edged swords in cellular redox state health beneficial effects at physiologic doses versus deleterious effects at high doses. Oxid. Med. Cell Longev. 3 (4), 228–237.

Bresciani, L., Calani, L., Bruni, R., Brighenti, F., Del Rio, D., 2014. Phenolic composition, caffeine content and antioxidant capacity of coffee silverskin. Food Res. Int. 61, 196–201.

Brezova, V., Slebodova, A., Stasko, A., 2009. Coffee as a source of antioxidants: an EPR study. Food Chem. 114, 859–868.

Bystryak, S., Goldiner, I., Niv, A., Nasser, A.M., Goldstein, L., 1995. A homogeneous immuno-fluorescence assay based on dye-sensitized photobleaching. Anal. Biochem. 225, 127–134.

Cai, W., Uribarri, J., Zhu, L., et al., 2014. Oral glycotoxins are a modifiable cause of dementia and the metabolic syndrome in mice and humans. Proc. Natl. Acad. Sci. USA 111, 4940–4945.

Cao, C., Cirrito, J.R., Lin, X., Wang, L., Verges, D.K., Dickson, A., et al., 2009. Caffeine suppresses amyloid-beta levels in plasma and brain of Alzheimer's disease transgenic mice. J. Alzheimer's Dis. 17 (3), 681–697.

Carmona, L., Galano, A., 2011. Is caffeine a good scavenger of oxygenated free radicals? J. Phys. Chem. 115, 4538–4546.

Casal, S., Beatriz, O., Ferreira, M.A., 2000. HPLC/diode-array applied to the thermal degradation of trigonelline, nicotinic acid and caffeine in coffee. Food Chem. 68, 481–485.

Cavin, C., Holzhauser, D., Constable, A., Huggett, A.C., Schilter, B., 1998. The coffee-specific diterpenes cafestol and kahweol protect against aflatoxin B1-induced genotoxicity through a dual mechanism. Carcinogenesis 19 (8), 1369–1375.

Cavin, C., Holzhaeuser, D., Scharf, G., Constable, A., Huber, W.W., Schilter, B., 2002. Cafestol and kahweol, two coffee specific diterpenes with anticarcinogenic activity. Food Chem. Toxicol. 40 (8), 1155–1163.

Chauhan, P.S., Satti, N.K., Sharma, P., Sharma, V.K., Suri, K.A., Bani, S., 2012. Differential effects of chlorogenic acid on various immunological parameters relevant to rheumatoid arthritis. Phytother. Res. 26, 1156–1165.

Chavan, A.A., Liakos, I., Bayer, I.S., Pinto, J., Lauciello, S., Athanassiou, A., Fragouli, D., 2016. Spent coffee bioelastomeric composite foams for the removal of Pb2+ and Hg2+ from water. ACS Sustain. Chem. Eng 4 (10), 5495–5502.

Chen, J.F., Chern, Y., 2011. Impacts of methylxanthines and adenosine receptors on neurodegeneration: human and experimental studies. Hand. Exp. Pharmacol. 200, 267–310.

Chen, J.H., Ho, C.-T., 1997. Antioxidant activities of caffeic acid and its related hydroxycinnamic acid compounds. J. Agric. Food Chem. 45, 2374–2378.

Chiang, L.C., Ng, L.T., Chiang, M.Y., Lin, C.C., 2003. Immunomodulatory activities of flavonoids, monoterpenoids, triterpenoids, iridoid glycosides and phenolic compounds of Plantago species. Planta. Med. 69 (7), 600–604.

Cho, A.-S., Jeon, S.-M., Kim, M.-J., et al., 2010. Chlorogenic acid exhibits anti-obesity property and improves lipid metabolism in high-fat diet-induced-obese mice. Food Chem. Toxicol. 48 (3), 937–943.

Chu, Y.F., Chang, W.H., Black, R.M., Liu, J.R., Sompol, P., Chen, Y., et al., 2012a. Crude caffeine reduces memory impairment and amyloid β (1–42) levels in an Alzheimer's mouse model. Food Chem. 135 (3), 2095 2102.

Chu, Y., Chen, Y., Brown, P.H., Lyle, B.J., Black, R.M., Cheng, I.H., Ou, B., Prior, R.L., 2012b. Bioactivities of crude caffeine: antioxidant activity, cyclooxygenase-2 inhibition, and enhanced glucose uptake. Food Chem. 131 (2), 564–568.

Costa, A.S.G., Alves, R.C., Vinha, A.F., Barreira, S.F.P., Nunes, M.A., Cunha, L.M., Oliveira, M.B.P.P., 2014. Optimization of antioxidants extraction from coffee silverskin, a roasting by-product, having in view a sustainable process. Ind. Crops Prod. 53, 350–357.

Crisostomo, J., Matafome, P., Santosd-Silva, D., et al., 2013. Methylglyoxal chronic administration promotes diabetes-like cardiac ischemia disease in Wistar normal rats. Nutr. Metab. Cardiovasc. 23, 1223–1230.

Cupino, T.L., Zabel, M.K., 2013. Alzheimer's silent partner: cerebral amyloid angiopathy. Transl. Stroke Res. 5 (3), 330–337.

Davis, B.R., Curb, J.D., Borhani, N., Prineas, R.J., Molteni, A., 1988. Coffee consumption and serum cholesterol in the hypertension detection and follow-up program. Am. J. Epidemial. 128, 124–136.

de Sotillo, R., Hadley, M., 2002. Chlorogenic acid modifies plasma and liver concentrations of: cholesterol, triacylglycerol, and minerals in (fa/fa) Zucker rats. J. of Nutr. Biochem. 13 (12), 717–726.

del Castillo, M.D., Ames, J.M., Gordon, M.H., 2002. Effect of roasting on the antioxidant activity of coffee brews. J. Agric. Food Chem. 50 (13), 3698–3703.

Devasagayam, T.P., Kesavan, P.C., 1996. Radioprotective and antioxidant action of caffeine: mechanistic considerations. Indian J. Exp. Biol. 34 (4), 291–297.

Dittrich, R., El-Massry, F., Kunz, K., Rinaldi, F., Piech, C.C., Beckmann, M.W., Pischetsrieder, M., 2003. Maillard reaction products inhibit oxidation of human low-density lipoproteins in vitro. J. Agric. Food Chem. 51, 3900–3904.

Dutta, M., Bandyopadhyay, D., 2014. Trigonelline [99%] protects against copper ascorbate induced oxidative damage to aortic mitochondria in vitro: involvement of antioxidant mechanism(s). Int. J. Pharm. Sci. Rev. Res. 8 (11), 1694–1718.

Elenkov, I.J., Iezzoni, D.G., Daly, A., Harris, A.G., Chrousos, G.P., 2005. Cytokine dysregulation, inflammation and well-being. Neuroimmunomodulation 12, 255–269.

Esquivel, Jimenez, 2012. Functional properties of coffee and coffee by-products. Food Res. Int. 46, 488–495.

Fagbenro, O.A., Arowosoge, I.A., 1991. Growth response and nutrient digestibility by *Clarias isheriensis* (Sydenham, 1980) fed varying levels of dietary coffee pulp as replacement for maize in low-cost diets. Biores. Technol. 37 (3), 253–258.

Farah, A., Donangelo, C.M., 2006. Phenolic compounds in coffee. Braz. J. Plant Physiol. 18, 223–236.

Flament, I., Gautschi, F., Winter, M., Willhalm, B., Stoll, M., 1968. 3rd International Colloquium on the Chemistry of Coffee. Trieste, pp. 197–215.

Fox, G.P., Wu, A., Yiran, L., Force, L., 2013. Variation in caffeine concentration in single coffee beans. J. Agric. Food Chem. 61, 10772–10778.

Fujioka, K., Shibamoto, T., 2007. Chlorogenic acid and caffeine contents in various commercial brewed coffees. Food Chem. 106, 217–221.

Fukuzawa, K., Matsuura, K., Tokumura, A., Suzuki, A., Terao, J., 1997. Kinetics and dynamics of singlet oxygen scavenging by alpha-tocopherol in phospholipid model membranes. Free Radic. Biol. Med. 22, 923–930.

Gahr, M., Nowak, D.A., Connemann, B.J., Schönfeldt-Lecuona, C., 2013. Cerebral amyloidal angiopathy: a disease with implications for neurology and psychiatry. Brain Res. 26 (1519), 19–30.

Ghule, A.E., Jadhav, S.S., Bodhankar, S.L., 2012. Trigonelline ameliorates diabetic hypertensive nephropathy by suppression of oxidative stress in kidney and reduction in renal cell apoptosis and fibrosis in streptozotocin induced neonatal diabetic (nSTZ) rats. Int. Immunopharmacol. 14 (4), 740–748.

Gil, M.I., Tomas-Barberan, F.A., Hess-Pierce, B., Holcroft, D.M., Kader, A.A., 2000. Antioxidant activity of pomegranate juice and its relationship with phenolic composition and processing. J. Agric. Food Chem. 48, 4581–4589.

Gomez-Ruiz, J.A., Leake, D.S., Ames, J.M., 2007. In vitro antioxidant activity of coffee compounds and their metabolites. J. Agric. Food Chem. 55, 6962–6969.

Grandhee, S.K., Monnier, V.M., 1991. Mechanism of formation of the Maillard protein cross-link pentosidine: glucose, fructose, and ascorbate as pentosidine precursors. J. Biol. Chem. 266 (18), 11649–11653, PMID 1904866.

Hasko, G., Cronstein, N., 2004. Adenosine: an endogenous regulator of innate immunity. TRENDS Immunol. 25 (1), 33–39.

Haung, D., Ou, B., Prior, R.L., 2005. The chemistry behind antioxidant capacity assays. J. Agric. Food Chem. 53, 1841–1856.

Hayashi, L., Shibamato, L., 1985. Analysis of methylglyoxal in foods and beverages. J. Agric. Food Chem. 33, 109–1093.

Henle, T., 2003. AGEs in foods: do they play a role in uremia? Kidney Int. Suppl. 63, S145–S147.

Herling, A.W., Schwab, D., Burger, H.-J., et al., 2002. Prolonged blood glucose reduction in mrp-2 deficient rats (GY/TR-) by the glucose-6-phosphate translocase inhibitor S 3025. Biochim. Biophys. Acta 1569 (1–3), 105–110.

Horrigan, L.A., Kelly, J.P., Connor, T.J., 2004. Caffeine suppresses TNF-α production via activation of the cyclic AMP/protein kinase A pathway. Int. Immunopharmacol. 4 (10–11), 1409–1417.

Horrigan, L.A., Kelly, J.P., Connor, T.J., 2006. Immunomodulatory effects of caffeine: friend or foe? Pharmacol. Ther. 111 (3), 877–889.

Hsu, F.-L., Chen, Y.-C., Cheng, J.-T., 2000. Caffeic acid as active principle from the fruit of *Xanthium strumarium* to lower plasma glucose in diabetic rats. Planta. Med. 66 (3), 228–230.

Huber, W.W., Scharf, G., Rossmanith, W., Prustomersky, S., Grasl-Kraupp, B., Peter, B., Turesky, R.J., Schulte-Hermann, R., 2002. The coffee components kahweol and cafestol induce γ-glutamylcysteine synthetase, the rate limiting enzyme of chemoprotective glutathione synthesis, in several organs of the rat. Arch Toxicol. 75 (11), 685–694.

Huber, W.W., Teitel, C.H., Coles, B.F., King, R.S., Wiese, F.W., Kaderlik, K.R., Casciano, D.A., Shaddock, J.G., Mulder, G.J., Ilett, K.F., Kadlubar, F.F., 2004. Potential chemoprotective effects of the coffee components kahweol and cafestol palmitates via modification of hepatic *N*-acetyltransferase and glutathione *S*-transferase activities. Environ. Mol. Mutagen. 44 (4), 265–276.

Hughes, S.R., López-Núñez, J.C., Jones, M.A., Moser, B.R., Cox, E.J., Lindquist, M., Galindo-Leva, L.A., Riaño-Herrera, N.M., Rodriguez-Valencia, N., Gast, N.F., Cedeño, D.L., Tasaki, K., Brown, R.C., Darzins, A., Brunner, L., 2014. Sustainable conversion of coffee and other crop waste to biofuels and bioproducts using coupled biochemical and thermochemical processes in a multi-stage biorefinery concept. Appl. Microbiol. Biotechnol 98 (20), 8412–8431.

Hwang, Y.P., Jeong, H.G., 2008. The coffee diterpene kahweol induces heme oxygenase-1 via the PI3K and p38/Nrf2 pathway to protect human dopaminergic neurons from 6-hydroxy-dopamine-derived oxidative stress. FEBS Lett. 6 (45).

Ilavenil, S., Arasu, M.V., Lee, J.C., Kim, D.H., Roh, S.G., Park, H.S., Choi, G.J., Mayakrishnan, V., Chi, K.C., 2013. Trigonelline attenuates the adipocyte differentiation and lipid accumulation in 3T3-L1 cells. Phytomedicine 21 (5), 758–765.

Index mundi, 2014. US Department of Agriculture. Available from: http://www.indexmundi.com/agriculture/?country=co&commodity=green-coffee&graph=production

Jaiswal, R., Matei, M.F., Golon, A., Witt, M., Kuhnert, N., 2012. Understanding the fate of chlorogenic acids in coffee roasting using mass spectrometry based targeted and non-targeted analytical strategies. Food Funct. 3, 976–984.

Janero, 1990. Malondialdehyde and thiobarbituric acid-reactivity as diagnostic indices of lipid peroxidation and peroxidative tissue injury. Free Radic. Biol. Med. 9 (6), 515–540, PMID 2079232.

Kedare, S.B., Singh, R.P., 2011. Genesis and development of DPPH method of antioxidant assay. J. Food Sci. Technol. 48, 412–422.

Kellow, N.J., Coughlan, M.T., 2015. Effect of diet-derived advanced glycation end products on inflammation. Nutr. Rev. 73 (11), 737–759.

Kim, J.Y., Jung, K.S., Jeong, H.G., 2004a. Suppressive effects of the kahweol and cafestol on cyclooxygenase-2 expression in macrophages. FEBS Lett. 569 (1–3), 321–326.

Kim, J.Y., Jung, K.S., Lee, K.J., Na, H.K., Chun, H.K., Kho, Y.H., Jeong, H.G., 2004b. The coffee diterpene kahweol suppress the inducible nitric oxide synthase expression in macrophages. Cancer Lett. 213 (2), 147–154.

Kohen, R., Nyska, A., 2002. Oxidation of biological systems: oxidative stress phenomena, antioxidants, redox reactions, and methods for their quantitation. Toxicol. Pathol. 30 (6), 620–650.

Kolling-Speer, L., Speer, L., 2005. The raw seed composition. In: Illy, A., Viani, R. (Eds.), Coffee: Emerging Health Effects and Disease Prevention. Elsevier Academic Press, Italy, pp. 148–178.

Kumar, S.S., Devasagayam, T.P., Jayashree, B., Kesavan, P.C., 2001. Mechanism of protection against radiation-induced DNA damage in plasmid pBR322 by caffeine. Int. J. Radiat. Biol. 77, 617–623.

Lawrence, T., 2009. The nuclear factor NF-kappaB pathway in inflammation. Cold Spring Harb. Pers. Biol. 1, a001651.

Lee, C., 2000. Antioxidant ability of caffeine and its metabolites based on the study of oxygen radical absorbing capacity and inhibition of LDL peroxidation. Clin. Chim. Acta 295 (1–2), 141–154.

Lee, K.J., Jeong, H.G., 2007. Protective effects of kahweol and cafestol against hydrogen peroxide-induced oxidative stress and DNA damage. Toxicol. Lett. 173 (2), 80–87.

Lee, K.J., Choi, K.J., Jeong, H.G., 2007. Hepatoprotective and antioxidant effects of the coffee diterpenes kahweol and cafestol on carbon tetrachloride-induced liver damage in mice. Food Chem. Toxicol. 45 (11), 2118–2125.

Lee, K.A., Chae, J.I., Shim, J.H., 2012. Natural diterpenes from coffee, cafestol and kahweol induce apoptosis through regulation of specificity protein 1 expression in human malignant pleural mesothelioma. J. Biomed. Sci. 19 (1), 60.

Lemire, J.A., Harrison, J.J., Turner, R.J., 2013. Box 3: the Fenton reaction, free radical chemistry and metal poisoning. Nature Rev. Microbiol. 11, 371–384.

Leon-Carmona, J.R., Galano, A., 2011. Is caffeine a good scavenger of oxygenated free radicals? J. Phys. Chem. B 115, 4538–4546.

Li, W., Dai, S., An, J., Li, P., Chen, X., Xiong, R., et al., 2008. Chronic but not acute treatment with caffeine attenuates traumatic brain injury in the mouse cortical impact model. Neuroscience 151 (4), 1198–1207.

Liang, N., Kitts, D.D., 2014. Antioxidant property of coffee components: assessment of methods that define mechanisms of action. Molecules 19, 19180–19208.

Liang, N., Kitts, D.D., 2016. Role of chlorogenic acids in controlling oxidative and inflammatory stress conditions. Nutrients 8 (1), 16.

Lieberman, H.R., Wurtman, R.J., Emde, G.G., Roberts, C., Coviella, I.L.G., 1987. The effects of low doses of caffeine on human performance and mood. Psychopharmacology 92, 308–312.

Lung, R.T., Shetty, P.S., James, W.P.T., Barrand, M.A., Callingham, B.A., 1981. Caffeine: its effect on catecholamines and metabolism in lean and obese humans. Clin. Sci. 60, 527–535.

Macia, L., Thorburn, A.N., Binge, L.C., et al., 2012. Microbiota influences on epithelial integrity and immune function as a basis for inflammatory disease. Immunol. Rev. 245, 164–176.

Mader, E.A., Davison, E.R., Mayer, J.M., 2007. Large ground-state entropy changes for hydrogen atom transfer reactions of iron complexes. J. Am. Chem. Soc. 129, 5153–5166.

Manni, S., Mauban, J.H., Ward, C.W., Bond, M., 2008. Phosphorylation of the cAMP-dependent protein kinase (PKA) regulatory subunit modulates PKA-AKAP interaction, substrate phosphorylation, and calcium signaling in cardiac cells. J. Biol. Chem. 283 (35), 24145–24154.

Manzocco, L., Calligaris, S., Mastrocola, D., Nicoli, M.C., Lerici, C.R., 2000. Review of non-enzymatic browning and antioxidant capacity in processed foods. Trends Food Sci. Technol. 11 (9–10), 340–346.

Maslowski, K.M., Mackay, C.R., 2011. Diet, gut microbiota and immune responses. Immunology 12, 5–9.

Mazzafera, P., 1990. Trigonelline in coffee. Phyto. Chem. 30 (7), 2309–2310.

Medzhitov, R., 2008. Origin and physiological roles of inflammation. Nature 454, 428–435.

Medzhitov, R., Janeway, C.A., 1997. Innate immunity: the virtues of a nonclonal system of recognition. Cell 91, 295–298.

Melamed, L., Kark, J.D., Spirer, Z., 1990. Coffee and the immune system. Int. J. Immunopharmacol. 1 (1), 129–134.

Menezes, E.G.T., de Carmo, J.R., Menezes, A.G.T., Alves, J.G.L.F., Pimenta, C.J., Queiroz, F.Q., 2013. Use of different extracts of coffee pulp for the production of bioethanol. Appl. Biochem. Biotechnol. 169 (2), 673–687.

Meng, S., Cao, J., Feng, Q., Peng, J., Hu, Y., 2013. Roles of chlorogenic acid on regulating glucose and lipids metabolism: a review. Evid. Based Compl. Alt. Med. 2013 (2013), 1–11.

Mez-Ruiz, et al., 2007. In vitro antioxidant activity of coffee compounds and their metabolites. Food Chem. 55 (17), 6962–6969.

Migliardi, J.R., Armellino, J.J., Friedman, Michael, Gillings, D.B., Beaver, W.T., 1994. Caffeine as an analgesic adjuvant in tension headache. Clin. Trials Ther. 56 (5), 576–586.

Mitjavila, M.T., Moreno, J.J., 2012. The effects of polyphenols on oxidative stress and the arachidonic acid cascade: implications for the prevention/treatment of high prevalence diseases. Biochem. Pharmacol. 84, 1113–1122.

Morelli, M., Blandini, F., Simola, N., Hauser, R.A., 2012. A(2A) receptor antagonism and dyskinesia in Parkinson's disease. Park. Dis. 2012, 489853.

Morgan, J.F., Klucas, R.V., Grayer, R.J., et al., 1999. Complexes of iron with phenolic compounds from soybean nodules and other legume tissues: prooxidant and antioxidant properties. Free Radic. Biol. Med. 22 (5), 861–870.

Mukohoda, M., Yamawaki, H., Nomura, H., et al., 2009. Methylglyoxal inhibits smooth muscle contraction in isolated blood vessels. J. Pharmacol. Sci. 109, 305–310.

Mullen, W., Nemzer, B., Stalmach, A., Ali, S., Combet, E., 2013. Polyphenolic and hydroxycinnamate contents of whole coffee fruits from China, India, and Mexico. J. Agric. Food Chem. 61, 5298–5309.

Munzel, T., Afanas'ev, I.B., Kleschyov, A.L., Harrison, D.G., 2002. Detection of superoxide in vascular tissue. Arterioscler. Thromb. Vasc. Biol. 22, 1761–1768.

Murat, L., Moriette, G., Blin, M.C., Couchard, M., Flouvat, B., et al., 1981. The efficacy of caffeine in the treatment of recurrent idiopathic apnea in premature infants. J. Pediatr. 99, 984–989.

Murthy, P.S., Naidu, M.M., 2012. Recovery of phenolic antioxidants and functional compounds from coffee industry by-products. Food Bioproc. Technol. 5, 897–903.

Murthy, P.S., Naidu, M.M., Srinivas, P., 2009. Production of α-amylase under solid-state fermentation utilizing coffee waste. J. Chem. Technol. Biotechnol. 84, 1246–1249.

Murthy, P.S., Manjunatha, M.R., Sulochannama, G., Naidu, M., 2012. Extraction, characterization and bioactivity of coffee anthocyanins. Eur. J. Biol. 4 (1), 13–19.

Muscat, S., Pelka, J., Hegele, J., Weigle, B., Münch, G., Pischetsrieder, M., 2007. Coffee and Maillard products activate NF-κB in macrophages via H_2O_2 production. Mol. Nutr. Food Res. 51 (5), 525–535.

Mussatto, S.I., Teixeira, J.A., 2010. Increase in the fructooligosaccharides yield and productively by solid-state fermentation with *Aspergillus japonicas* using agro-industrial residues as support and nutrient source. Biochem. Eng. J. 53, 154–157.

Narita, Y., Inou, K., 2012. High antioxidant activity of coffee silver skin extracts obtained by the treatment of coffee silver skin with subcritical water. Food Chem. 135, 943–949.

Nathan, C., 2006. Neutrophils and immunity: challenges and opportunities. Nat. Rev. Immunol. 6, 173–182.

Nguyen, T., Nioi, P., Pickett, C.B., 2015. The Nrf2-antioxidant response element signaling pathway and its activation by oxidative stress. J. Biol. Chem. 284, 13291–13295.

Nonthakaew, A., Matan, N., Aewsiri, T., Matan, N., 2015. Caffeine in foods and its antimicrobial activity. Int. Food Res. J. 22, 9–14.

Oboh, G., Agunloye, O.M., Adefegha, S.A., Akinyemi, A.J., Ademiluyi, A.O., 2015. Caffeic and chlorogenic acids inhibit key enzymes linked to type 2 diabetes (in vitro): a comparative study. J. Basic Clin. Physiol. Pharmacol. 26 (2), 165–170.

Ohnishi, M., Morishita, H., Iwahashi, H., Toda, S., Shirataki, Y., Kimura, M., Kido, R., 1994. Inhibitory effects of chlorogenic acids on linoleic acid peroxidation and haemolysis. Inter. Plant Biochem. 36 (3), 579–583.

Ohyashiki, T., Nunomura, M., Katoh, T., 1999. Detection of superoxide anion radical in phospholipid liposomal membrane by fluorescence quenching method using 1,3-diphenylisobenzofuran. Biochim. Biophys. Acta 1421, 131–139.

Olthof, M.R., Hollman, P.C., Zock, P.L., Katan, M.B., 2001. Consumption of high doses of chlorogenic acid, present in coffee, or of black tea increases total homocysteine concentrations in humans. Am. J. Clin. Nur. 73, 532–538.

Ong, J.K.W., Hsu, A., Tan, B.K., 2013. Anti-diabetic and anti-lipidemic effects of chlorogenic acid are mediated by ampk activation. Biochem. Pharmacol. 85 (9), 1341–1351.

Palacios, N., Gao, X., McCullough, M.L., Schwarzschild, M.A., Shah, R., Gapstur, S., et al., 2012. Caffeine and risk of Parkinson's disease in a large cohort of men and women. Mov. Disord. 27 (10), 1276–1282.

Pandey, A., Soccol, C.R., Nigam, P., Brand, D., Mohan, R., Roussos, S., 2000. Biotechnological potential of coffee pulp and coffee husk for bioprocesses. Biochem. Eng. J. 6 (2), 153–162.

Parras, P., Martinez-Tome, M., Jimenez, A.M., Murcia, M.A., 2007. Antioxidant capacity of coffees of several origins brewed following three different procedures. Food Chem. 102, 582–592.

Paur, I., Balstad, T.R., Blomhoff, R., 2010. Degree of roasting is the main determinant of the effects of coffee on NF-κB and EpRE. Free Radi. Biol. Med. 48 (9), 1218–1227.

Pellegrini, R., Pannala, N., Yang, A., Rice-Evans, M., 1999. Antioxidant activity applying an improved ABTS radical cation decolorization assay. Free Rad. Biol. Med. 26, 1231–1237.

Peñaloza, W., Molina, M.R., Brenes, R.G., Bressani, R., 1985. Solid-state fermentation: an alternative to improve the nutritive value of coffee pulp. Appl. Environ. Microbiol. 49 (2), 388–393.

Perrone, D., Farah, A., 2009. Application of mass spectrometry on the analysis of coffee components. In: Lang, J.K. (Ed.), Handbook on Mass Spectrometry: Instrumentation, Data Analysis, and Application. Nova Science Publishers, Hauppauge, New York, pp. 465–498.

Pincomb, G.A., Lovallo, W.R., Passey, R.B., Whitsett, T.L., Silverstein, S.M., et al., 1985. Effects of caffeine on vascular resistance, cardiac output and myocardial contractility in young men. Am. J. Cardial. 56, 119–122.

Postuma, R.B., Lang, A.E., Munhoz, R.P., Charland, K., Pelletier, A., Moscovich, M., et al., 2012. Caffeine for treatment of Parkinson disease: a randomized controlled trial. Neurology 79 (7), 651–658.

Prabhakar, P.K., Doble, B., 2009. Synergistic effect of phytochemicals in combination with hypoglycemic drugs on glucose uptake in myotubes. Phytomedicine 16 (12), 1119–1126.

Prior, R.L., Wu, X., Schiach, K., 2005. Standardized methods for the determination of antioxidant capacity and phenolics in foods and dietary supplements. J. Agric. Food Chem. 53, 4290–4302.

Ramirez-Coronel, M.A., Marnet, N., Kolli, V.S.K., Roussos, S., Guyot, S., Augur, C., 2004. Characterization and estimation of proanthocyanidins and other phenolics in coffee pulp (*Coffea arabica*) by thiolysis–high-performance liquid chromatography. J. Agric. Food Chem. 52 (5), 1344–1349.

Rathinavelu, R., Graziosi, G., 2005. Potential alternative use of coffee wastes and by-products. Int. Coffee Org., Available from:http://www.ico.org/documents/ed1967e.pdf.

Reddi, E., Valduga, G., Rodgers, M.A., Jori, G., 1991. Studies on the mechanism of the hematoporphyrin-sensitized photooxidation of 1,3-diphenylisobenzofuran in ethanol and unilamellar liposomes". Photochem. Photobiol. 54, 633–637.

Robertson, D., Frolich, J.C., Carr, R., 1990. Annu. Rev. Med. 41, 277–288.

Rodríguez-Duran, L.V., Ramírez-Coronel, R.A., Aranda-Delgado, E., Favela-Torres, K.M.N.E., Aguilar, C.N., Saucedo-Castañeda, G., 2014. Soluble and bound hydroxycinnamates in coffee pulp (*Coffea arabica*) from seven cultivars at three ripening stages. J. Agric. Food Chem. 62, 7869–7876.

Rosim, F.E., Persike, D.S., Nehlig, A., Amorim, R.P., de Oliveira, D.M., Fernandes, M.J., 2011. Differential neuroprotection by A(1) receptor activation and A(2A) receptor inhibition following pilocarpine-induced status epilepticus. Epilep. Behav. 22 (2), 207–213.

Rover, 2007. Available from: http://legacy.sweetmarias.com/coffee_chemistry/tweaking_coffees_flavor_chem.html

Salvemini, D., Kim, S.F., Mollace, V., 2013. Reciprocal regulation of the nitric oxide and cyclooxygenase pathway in pathophysiology: relevance and clinical implications. Am. J. Physiol. Regul. Integr. Comp. Physiol. 304 (7), R473–R487.

Sato, Y., Itagaki, S., Kurokawa, T., Ogura, J., Kobayashi, M., Hirano, T., Sugawara, M., Iseki, K., 2011. In vitro and in vivo antioxidant properties of chlorogenic acid and caffeic acid. Int. J. Pharm. 403 (1–2), 136–138.

Sauer, T., Raithel, M., Kressel, J., Muscat, S., Münch, G., Pischetsrieder, M., 2011. Nuclear translocation of NF-κB in intact human gut tissue upon stimulation with coffee and roasting products. Food Funct. 2, 529–540.

Sawyer, D.A., Julia, H.L., Turin, A.C., 1982. Caffeine and human behavior: arousal, anxiety, and performance effects. J. Behav. Med. 5, 415–439.

Seiquer, I., Rubio, L.A., Peinado, M.J., Delgado-Andrade, C., Navarro, M.P., 2014. Maillard reaction products modulate gut microbiota composition in adolescents. Mol. Nutr. Food Res. 58 (7), 1552–1560.

Shan, J., Fu, J., Zhao, Z., Kong, X., Huang, H., Luo, L., Yin, Z., 2009. Chlorogenic acid inhibits lipopolysaccharide-induced cyclooxygenase-2 expression in RAW264.7 cells through suppressing NF-κB and JNK/AP-1 activation. Int. Immunopharmacol. 9, 1042–1048.

Shi, X., Dalal, N.S., Jain, A.C., 1991. Antioxidant behavior of caffeine: efficient scavenging of hydroxyl radicals. Food Chem. 29, 1–6.

Shi, H.T., Dong, L., Jiang, J., Zhao, J.H., Zhao, G., Dang, X.Y., Lu, X.L., Jia, M., 2013. Chlorogenic acid reduces liver inflammation and fibrosis through inhibition of toll-like receptor 4 signaling pathway. Toxicology 303, 107–114.

Shin, J.Y., Sohn, J., Park, K.H., 2013. Chlorogenic acid decreases retinal vascular permeability in diabetic rat model. J. Korean Med. Sci. 28 (4), 608–613.

Shin, H.S., Satsu, H., Bae, M.J., Zhao, Z.H., Ogiwara, H., Totsuka, M., Shimizu, M., 2015. Anti-inflammatory effect of chlorogenic acid on the IL-8 production in Caco-2 cells and the dextran sulphate sodium-induced colitis symptoms in C57BL/6 mice. Food Chem. 168, 167–175.

Song, L., Kong, M., Ma, Y., Ba, M., Liu, Z., 2009. Inhibitory effect of 8-(3-chlorostyryl) caffeine on levodopa-induced motor fluctuation is associated with intracellular signaling pathway in 6-OHDA-lesioned rats. Brain Res. 1276, 171–179.

Sonsalla, P.K., Wong, L.Y., Harris, S.L., Richardson, J.R., Khobahy, I., Li, W., et al., 2012. Delayed caffeine treatment prevents nigral dopamine neuron loss in a progressive rat model of Parkinson's disease. Exp. Neurol. 234 (2), 482–487.

Spencer, J.P., 2008. Flavonoids: modulators of brain function? Br. Nutr. 99, ES60–ES77.

Stalmach, A., Williamson, G., Crozier, A., 2014. Impact of dose on the bioavailability of coffee chlorogenic acids in humans. Food Funct. 5, 1727–1737.

Stone, T.W., Ceruti, S., Abbracchio, M.P., 2009. Adenosine receptors and neurological disease: neuroprotection and neurodegeneration. Hand. Exp. Pharmacol. 193, 535–587.

Su, Y., Lei, X., Wu, L., et al., 2012. The role of endothelial cell adhesion molecules P-selectin, E-selectin and intercellular adhesion molecule-1 in leukocyte recruitment induced by exogenous methylglyoxal. Immunology 137, 65–79.

Tareke, E., Rydberg, P., Karlsson, P., Eriksson, S., Tornqvist, M., 2002. Analysis of acrylaminde, a carcinogen formed in heated foodstuffs. J. Agric. Food Chem. 50, 4998–5006.

Tarpey, M.M., Wink, D.A., Grisham, M.B., 2004. Methods for detection of reactive metabolites of oxygen and nitrogen: in vitro and in vivo considerations. Am. J. Physiol. Regul. Integr. Comp. Physiol. 286, R431–R444.

Thornalley, P.J., 2003. Gloxylase-1: structure, function and a critical role in the enzymatic defense against glycation. Biochem. Soc. Trans. 31, 1343–1348.

Toschi, T.G., Cardenia, V., Bonaga, G., Mandrioli, M., Rodriguez-Estrada, M.T., 2014. Coffee silver skin: characterization, possible uses, and safety aspects. J. Agric. Food Chem. 62 (44), 10836–10844.

Trugo, L.C., 2003. Coffee analysis. Encycl. Food Sci. Nutr. 2, 498.

Tsutsui, S., Schnermann, J., Noorbakhsh, F., Henry, S., Yong, V.W., Winston, B.W., et al., 2004. A1 adenosine receptor upregulation and activation attenuates neuroinflammation and demyelination in a model of multiple sclerosis. J. Neurosci. 24 (6), 1521–1529.

Urgert, R., Essed, N., van der Weg, G., Kosmeijer-Schuil, T.G., Katan, M.B., 1997. Separate effects of the coffee diterpenes cafestol and kahweol on serum lipids and liver aminotransferases. Am. J. Clin. Nutr. 65 (2), 519–524.

van Dijk, A.E., Olthof, M.R., Meeuse, J.C., Seebus, E., Heine, R.J., van Dam, R.M., 2009. Acute effects of decaffeinated coffee and the major coffee components chlorogenic acid and trigonelline on glucose tolerance. Diabetes Care 32 (6), 1023–1025.

Varani, K., Portaluppi, F., Gessi, S., Merighi, S., Vincenzi, F., Cattabriga, E., Dalpiaz, A., Bortolotti, F., Belardinelli, L., Borea, P.A., 2005. Caffeine intake induces an alteration in human neutrophil A_{2A} adenosine receptors. Cell. Mol. Life Sci. 62, 2350–2358.

Vasdev, S., Ford, C.A., Longerich, L., et al., 1998. Role of aldehydes in fructose induced hypertension. Mol. Cell Biochem. 181, 1–9.

Veenstra, S., 1995. Recovery of biogas from landfill sites. HE Delft. In: Curso taller Internacional sobre tratamiento anaerobio de aguas residuals. Santiago de Cali, Colombia. Memorias.

Victoria, B.-S., Domenico, S., Clive, P.P., 2006. Phosphodiesterase inhibitors. Br. J. Pharmacol. 147 (Suppl 1), S252–S257.

Vignoli, J.A., Bassoli, D.G., Benassi, M.T., 2011. Antioxidant activity, polyphenols, caffeine and melanoids in soluble coffee: the influence of processing conditions and raw material. Food Chem. 124, 863–868.

Virella, G., Thorpe, S.R., Alderson, N.L., et al., 2003. Autoimmune response to advanced glycosylation end-products of human LDL. J. Lipid Res. 44, 487–493.

Volz, R.K., McGhie, T.K., Kumar, S., 2014. Variation and genetic parameters of fruit color and polyphenolic composition in an apple seedling population segregating for red leaf. Tree Genet. Genomes 10, 953–964.

Vytasek, R., Sedova, L., Vilim, V., 2010. Increased concentration of two different advanced glycation end-products detected by enzyme immunoassays with new monoclonal antibodies in sera of patients with theumatoid arthritis. BMC Muscoskelet. Disord. 11, 83.

Wayner, D.D., Burton, G.W., Ingold, K.U., Barclay, L.R., Locke, S.J., 1987. The relative contributions of vitamin E, urate, ascorbate and proteins to the total peroxyl radical-trapping antioxidant activity of human blood plasma. Biochem. Biophys. Acta 924, 408–419.

Wijewickreme, A.N., Kitts, D.D., 1998. Modulation of metal-induced genotoxicity by Maillard reaction products isolated from coffee. Food Chem. 36, 543–553.

Williams, P.T., Wood, P.D., Vranizan, K.M., Albers, J.J., Garay, S.C., et al., 1985. Coffee intake and elevated cholesterol and apolipoprotein B levels in men. J. Am. Med. Assoc. 253, 1407–1411.

Wostyn, P., Van Dam, D., Audenaert, K., De Deyn, P.P., 2011. Increased cerebral spinal fluid production as a possible mechanism underlying caffeine's protective effect against Alzheimer's disease. Int. J. Alzheimer's Dis. 2011, 617420.

Wozniak, M., Tanfani, F., Bertoli, E., Zolese, G., Antosiewicz, J., 1991. A new fluorescence method to detect singlet oxygen inside phospholipid model membranes. Biochim. Biophys. Acta 1082, 94–100.

Xie, J., Schaich, K.M., 2014. Re-evaluation of the 2,2-diphenyl-1-picrylhydrazyl free radical (DPPH) assay for antioxidant activity. J. Agric. Food Chem. 62, 4251–4260.

Xu, J.G., Hu, Q.P., Liu, Y., 2012. Antioxidant and DNA-protective activities of chlorogenic acid isomers. J. Agric. Food Chem. 60 (16), 11625–11630.

Xue, M., Rabbani, N., Morniji, H., et al., 2012. Transcriptional control of glyoxylase 1 by Nrf2 provides a stress-responsive defense against dicarbonyls glycation. Biochem. J. 443, 213–222.

Yaday, S., Gupta, S.P., Srivastava, G., Srivastava, P.K., Singh, M.P., 2012. Role of secondary mediators in caffeine-mediated neuroprotection in maneb- and paraquat-induced Parkinson's disease phenotype in the mouse. Neurochem. Res. 37 (4), 875–884.

Yamawaki, H., Saito, K., Okada, M., Hara, Y., 2008. Methylglyoxal mediates vascular inflammation via JNK and p38 in human endothelial cells. Am. J. Physiol. Cell Physiol. 295, C1510–C1517.

Yanagimoto, K., Lee, K.-G., Ochi, H., Shibamoto, T., 2002. Antioxidative activity of heterocyclic compounds found in coffee volatiles produced by Maillard reaction. J. Agric. Food Chem. 50, 5480–5484.

Yordi, E.G., Perez, E.M., Matos, M.J. Villares, E.U., 2012. Antioxidant and pro-oxidant effects of polyphenolic compounds and structure-activity relationship evidence. Intech.

Zhou, J.Y., Zhou, S.-W., 2012. Protection of trigonelline on experimental diabetic peripheral neuropathy. Evid. Based Compl. Alter. Med. 2012, 1–8.

Zhou, J., Zhou, S., Zeng, S., 2011. Experimental diabetes treated with trigonelline: effect on β cell and pancreatic oxidative parameters. Fundam. Clin. Pharmacol. 27 (3), 279–287.

Zhou, J., Zhou, S., Zeng, S., 2013. Experimental diabetes treated with trigonelline: effect on beta cell and pancreatic oxidative parameters. Fundam. Clin. Pharmacol. 27, 279–287.

FURTHER READING

Azam, S., Hadi, N., Khan, N.U., Hadi, S.M., 2003. Antioxidant and prooxidant properties of caffeine, theobromine and xanthine. Med. Sci. Monit. 9 (9), BR325–BR330.

Sauer, T., Raithel, M., Kressel, J., Münch, G., Pischetsrieder, M., 2013. Activation of the transcription factor Nrf2 in macrophages, Caco-2 cells and intact human gut tissue by maillard reaction products and coffee. Amino Acids 44 (6), 1427–1439.

Slow, S., Miller, W.E., McGregor, D.O., Lee, M.B., Lever, M., George, P.M., Chambers, S.T., 2004. Trigonelline is not responsible for the acute increase in plasma homocysteine following ingestion of instant coffee. Eur. J. Clin. Nutr. 58, 1253–1256.

Tang, B., Zhang, L., Geng, Y., 2005. Determination of the antioxidant capacity of different food natural products with a new developed flow injection spectrofluorimetry detection hydroxyl radicals. Talanta 65, 769–775.

Yang, X.F., Guo, X.Q., 2001. Investigation of the anthracene–nitroxide hybrid molecule as a probe for hydroxyl radicals. Analyst 126, 1800–1804.

The biorefinery concept for the industrial valorization of coffee processing by-products

3

Valentina Aristizábal-Marulanda, Yéssica Chacón-Perez, Carlos A. Cardona Alzate

National University of Colombia, Manizales, Colombia

ABSTRACT

During the cultivation and processing of coffee different residues can be obtained, such as, coffee cut-stems, coffee silverskin, spent coffee grounds, coffee pulp, and coffee husk. In the last years, some research has demonstrated that this type of biomass can be a good source of bioproducts through its transformation using stand-alone processes or biorefineries. In this sense, this chapter shows a description of coffee grain and an overview of the world production, consumption, and processing stages to obtain the by-products. In order to know the potential and identify the important compounds in the coffee processing, the physicochemical composition of by-products is analyzed. Thus, an analysis of possibilities of integral valorization of by-products as raw materials to obtain antioxidants, biofuels, bioenergy, bioproducts, and biofertilizers is performed. Finally, the biorefinery concept is considered to generate biochemicals and added-value products from the abovementioned residues, taking into account the design of biorefineries, the production scale, and a technical and economic assessment.

Keywords: coffee silverskin; spent coffee grounds; coffee pulp; coffee cut-stems; biorefinery; bioproducts

3.1 COFFEE

Ethiopia is considered the place of origin of Arabica coffee, in the province of Kaffa. Central Africa is the origin of Robusta coffee (Murthy and Madhava Naidu, 2012; Sánchez and Anzola, 2013). Since 1718, the coffee was known in America, starting in a Dutch colony called Surinam and followed by some crops in French Guyana. In 1730, the British colonizers carried the coffee to Jamaica and, then it was extended to Central and South America (Sánchez and Anzola, 2013).

Handbook of Coffee Processing By-Products. http://dx.doi.org/10.1016/B978-0-12-811290-8.00003-7

The coffee tree is a shrub that grows in tropical regions of the world and belongs to the family *Rubiaceae*. This plant contains 500 genus and 8000 species. The most known genus is *Coffea* L. and it has 10 civilized species. At least, 50 species are native (Federacion Nacional, 2010). Commercially, two species are explored, *Coffea arabica* (Arabica) and *Coffea canephora* (Robusta). The first specie has availability around 75% of world's production and grows in high altitudes from 600 to 2000 m. The second specie provides 25% of world's production, grows in altitudes below 600 m and is more resistant to plagues (Mussatto et al., 2011a; Sánchez and Anzola, 2013).

Countries as Brazil, Vietman, Colombia, and Indonesia are the largest world producers of coffee as is indicated in Table 3.1. These countries take up the first, second,

Table 3.1 Annual Total Productions by All Exporting Countries (in Thousands of 60 kg Bags)

Year	2014	2015
Total	142,278	143,306
Arabicas	84,515	83,268
Colombian Milds	14,571	14,845
Other Milds	25,768	25,755
Brazilian Naturals	44,176	42,668
Robusta	57,763	60,038
Africa	16,055	17,449
Côte d'Ivoire	1,750	1,800
Ethiopia	6,625	6,700
Uganda	3,744	4,755
Others	3,936	4,194
Asia and Oceania	45,667	47,805
India	5,450	5,833
Indonesia	11,418	12,317
Vietnam	26,500	27,500
Others	2,299	2,155
Mexico and Central America	17,269	16,753
Guatemala	3,310	3,400
Honduras	5,400	5,750
Mexico	3,591	2,800
Others	4,968	4,803
South America	63,287	61,298
Brazil	45,639	43,235
Colombia	13,333	13,500
Peru	2,883	3,300
Others	1,432	1,263

From ICO, International Coffee Organization, 2015. Available from: http://www.ico.org/

third, and fourth place, respectively, in the world coffee production. In 2015, South America produced 42.7%, Asia and Oceania 33%, Africa 12.1%, and Mexico and Central America 11.7% of the total production (Table 3.1).

3.2 COFFEE PROCESSING

Soluble coffee or roasted and ground coffee processes use the green coffee beans as raw material. Green coffee comes from the successive removal of pericarp from the ripe coffee beans, which are called coffee cherries (Fig. 3.1). In this way, four process steps can be identified in the agro-production chain of coffee, as shown in Fig. 3.2. First process step takes place in the agricultural sector where the coffee cherries are produced and collected. This implies the planting of seeds that came from garden centers to the crops in which the plant will flourish and form the fruits.

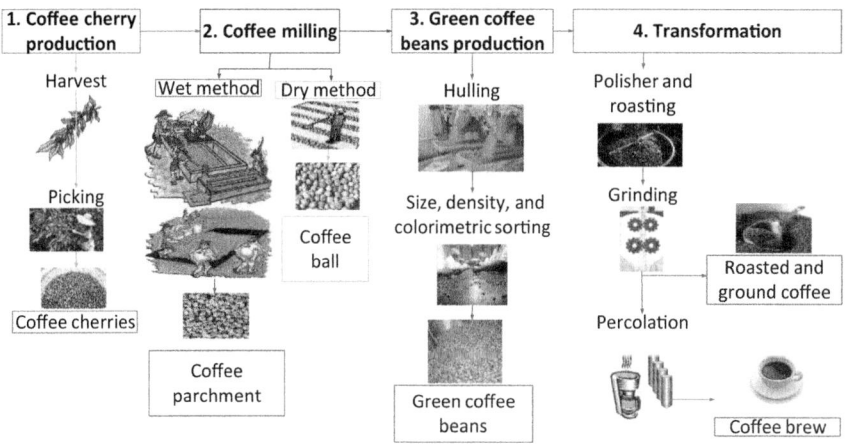

FIGURE 3.1 Structure of Coffee Plant and Coffee Cherry

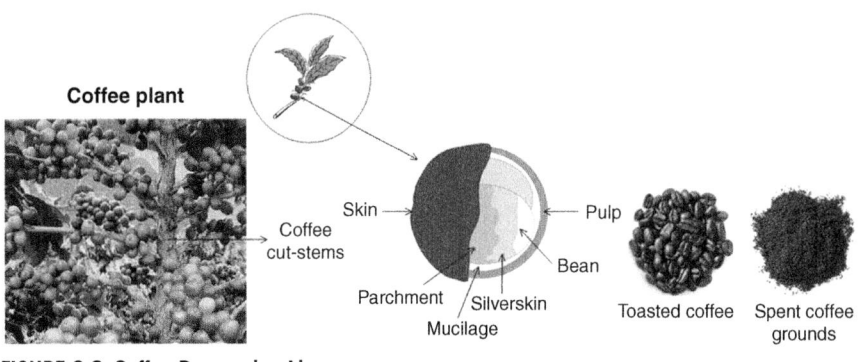

FIGURE 3.2 Coffee Processing Line

This period takes 3–5 years. Coffee fruits are similar in appearance to a small cherry and the change of color to the point of ripeness (i.e., crimson red) reflects the start of harvest time. In harvesting, certain collection techniques are applied so to ensure an excellent quality of the fruits. After the harvest time, renovations of crops are carried out by pruning coffee trees to allow them to recover until the next season of flowering and keep a productive coffee crop. This last stage is where the coffee cut-stems are obtained (Arcila et al., 2007; Arya and Rao, 2007; Murthy and Madhava Naidu, 2012; Riaño Luna, 2010; Vincent, 1987).

After collecting the coffee cherries starts the postharvest, where the second process step is carried out in a period no longer than 24 h after collection to avoid the decomposition. This stage is known as coffee milling where the layers of coffee cherries are removed. It starts with a cleaning to remove the dirt and then the over-ripe or unripe of the coffee cherries. Then, through a dry or wet method, the epicarp (skin), external (pulp), and internal (mucilage) mesocarps are separated from the coffee endocarp. In the dry method, a sun-drying of the cherries is performed for about 20–30 days, reducing its moisture content until values between 10% and 12%. It is commonly used for *Robusta* species. Dried cherries obtained with this method are called coffee balls and the green coffee beans that come from these are distinguished for giving to the brew a sweet aroma, because the coffee grains are impregnated with the sugars and other compounds present in the mucilage. On the other hand, mechanical dryers can be used in regions where the relative humidity is high or to reduce the drying time in regions where solar energy issued. However, its use involves a raise in processing costs affecting the process profitability.

The wet method is applied mainly for processing *Arabica* species, beginning with the mechanical pulping of coffee cherries to remove the skin and pulp of these in single equipment. The pulped beans remain with a slug contexture similar to honey that is the mucilage, which is removed when fermentation is carried out. The fermentation aims to decompose the peptidic substances of mucilage and the time of fermentation is directly affected by the climatic conditions, coffee maturity, amount of mucilage, and volume of processed coffee. Then, it is necessary to perform a wash to remove traces of honey adhered to the coffee grains. Finally, the grains are dried in the sun or mechanically until reaching moisture content from 10% to 12%. The results of this stage are the coffee parchment, which are the green coffee beans surrounded by their endocarp (parchment) and epidermis (silverskin). Due to the high consumption of water in the wet method, the use of a green module has been presented as an alternative. It can be done mechanically or with a dry fermentation to remove the mucilage and then the same process involved in the wet method (i.e., washing and drying) requiring less than one liter of water per kilogram of dry coffee parchment (Arya and Rao, 2007; Mussatto et al., 2011a; Pabón Usaquén et al., 2009; Vincent, 1987).

In the third process stage, the green coffee beans are obtained. First, a cleaning is performed to remove the impurities, followed by the hulling, where the hull (i.e., parchment) of the coffee parchment or husk of the coffee ball is mechanically removed (Bekalo and Reinhardt, 2010). The husk has then the components of the coffee pericarp that are skin, pulp, mucilage, and parchment, obtaining just one waste in contrast with the different wastes generated when the wet method is used. Continuing with the process of the green coffee beans, it is necessary to carry out the sorting by size, density, and colorimetric features to obtain the export and domestic consumption beans.

Eventually, the green coffee is brought to the transformation stage (fourth process step) where the polishing, roasting, and grinding are used for obtaining roasted and ground coffee. This is used in the elaboration of coffee brew or soluble coffee both in domestic or industrial level, respectively (Simoes et al., 2013). During the preparation of the coffee beverage, lixiviation of roasted coffee occurs because hot water extracts soluble coffee components leaving a solid fraction known as spent coffee grounds (SCG) (Clarke, 1987; Cruz et al., 2014; De Melo et al., 2014).

3.3 COFFEE PROCESSING BY-PRODUCTS

Currently, there are policies focused in the decrease of liquids, solids, and pollutants derived from agroindustrial activities. In addition, the industries are searching to have processes environmentally friendly, with an integral use of raw materials and renewal of technologies. All countries are implementing methodologies for recycling the residues generated in agroindustrial activities and crops. For example, the coffee is the second largest product marketed in the world, after the oil. In this sense, this industry generates large amounts of residues among which are coffee silverskin (CS), SCG, coffee pulp, coffee husk (CH), and coffee cut-stems (Ballesteros et al., 2014a; Mussatto et al., 2011a; Nabais et al., 2008).

CS and SCG are the main coffee industry residues (Mussatto et al., 2011a). Besides, this biomass has high content of organic material and demands high amount of oxygen to be degraded. The content of polyphenols, tannins, and caffeine in the residues from coffee can give toxic characteristics. Then, these materials can be considered as a pollution hazard when they are discharged in high amounts into the environment (Mussatto et al., 2011a).

The coffee plant is divided into three parts: stem, leaf, and fruit, as indicated in Fig. 3.1. The coffee fruit is a cherry that is divided in three layers: epicarp or skin (external layer); mesocarp, mucilage, or pulp, which forms a sweet and aromatic pulp protected by a film of cellulose called parchment, hull, or endocarp. Finally, a layer named silverskin or spermoderm covers the bean (Sánchez and Anzola, 2013). As aforementioned, there are two techniques for coffee processing, dry and wet. In the dry method, when 1 ton of clean coffee is obtained, 1 ton of skin, pulp, parchment, and silverskin is produced. While in the wet process to obtain 1 ton of clean coffee, 2 tons of pulp, 0.16 tons of parchment, 0.9 tons of silverskin, and 22.7 tons of effluent are produced (Sánchez and Anzola, 2013).

A brief description of the main residues obtained in the coffee industry is presented in the next sections.

3.3.1 **COFFEE SILVERSKIN**

Coffee silverskin (CS) is a fine film of the external layer of green coffee bean that is obtained as by-product or residue in the roasting process (Mussatto et al., 2011b). CS represents around 4.2% w/w of the coffee bean (Ballesteros et al., 2014a).

CS in some countries is used as fertilizer or fuel (Saenger et al., 2001; Sánchez and Anzola, 2013). Some studies have found healthy properties in this material related mainly to the antioxidant capacity and high content of dietary fiber (Sánchez and Anzola, 2013). CS can be used as substrate for cultivation of microorganism to release phenolic compounds or to produce enzymes (Ballesteros et al., 2014a; Machado et al., 2012; Mussatto et al., 2013). Recently, another study analyzed the solid-state cultivation of seven fungal strains from genus *Aspergillus*, *Mucor*, *Penicillium*, and *Neurospora* using CS as substrate to release phenolic compounds (Mussatto et al., 2011a). CS can be considered as a new potential functional ingredient (Borrelli et al., 2004; Mussatto et al., 2011a).

3.3.2 **SPENT COFFEE GROUNDS**

SCG is a by-product or residue obtained in the treatment of coffee powder with steam or hot water for the preparation of instant coffee. Around 6 million tons of SCG are generated per year considering that almost 50% of the worldwide coffee production is destined to soluble coffee preparation (Ballesteros et al., 2014a; Mussatto et al., 2011b). SCG presents characteristics as high humidity, fine particle size, acidity, and organic load. Around 650 kg of SCG can be obtained from 1 ton of green coffee, and for 1 kg of soluble coffee produced 2 kg of wet SCG can be obtained (Mussatto et al., 2011a; Pfluger, 1975).

SCG has a high calorific power (5000 kcal/kg) and for this reason it can be used as fuel in boilers (Ballesteros et al., 2014a; Mussatto et al., 2011a). In addition, it is a potential source to obtain oil, biodiesel, fuel pellets, and other fuels by fermentation as ethanol and hydrogen. This residue has the possibility to be used as animal feed (ruminants, pigs, chickens, and rabbits) although the lignin content (Table 3.2) is considered as a limiting factor (Givens and Barber, 1986; Mussatto et al., 2011a). In wastewater treatments, SCG can be used as a promising adsorbent for the removal of cationic dyes due to its low cost and availability (Franca et al., 2005; Mussatto et al., 2011a).

3.3.3 **COFFEE PULP**

Coffee pulp (CP) is a residue that represents approximately 29% dry-weight of cherry and it is obtained during coffee wet processing. Two tons of CP are obtained per 1 ton of coffee (Murthy and Madhava Naidu, 2012). CP can be used as animal feed (livestock, pigs, rabbits, fishes) replacing around 20% of traditional animal feed but the high caffeine content can be a negative factor. Additionally, this material can be used as substrate for the culture of fungi (Murthy and Madhava Naidu, 2012). CP can be source of refreshing and alcoholic beverages (Rathinavelu and Graziosi, 2005).

Table 3.2 Chemical Composition of Coffee Processing By-Products in Percent (*Dry Material)

By-Product	CS			SCG				CP	
References	Ballesteros et al. (2014a)	Mussatto et al. (2011a)	Pourfarzad et al. (2013)	Ballesteros et al. (2014a)	Mussatto et al. (2011a)	Uribarri et al. (2014)	Murthy and Madhava Naidu (2012)	Peñaloza et al. (1985)	Ulloa Rojas et al. (2002)
Moisture	ND	ND	7.1 ± 0.21	ND	ND	10.49 ± 0.09	ND	ND	ND
Carbohydrates	ND	ND	65.1 ± 1.16	ND	ND	ND	ND	ND	ND
Cellulose	23.77 ± 0.09	17.8	NC	12.40 ± 0.79	8.6	28.05	63.0 ± 2.5	18.65	286g/kg*
Hemicellulose	16.68 ± 1.30	13.1	NC	39.10 ± 1.94	36.7	18.83	2.3 ± 1.0	0.98	ND
Lignin	28.58 ± 0.46	ND	NC	23.90 ± 1.70	ND	16.21	17.5 ± 2.2	12.20	ND
Insoluble	20.97 ± 0.43	ND	NC	17.59 ± 1.56	ND	ND	ND	ND	ND
Soluble	7.61 ± 0.16	ND	NC	6.31 ± 0.37	ND	ND	ND	ND	ND
Fat	3.78 ± 0.40	ND	2.2± 0.54	2.29 ± 0.30	ND	13.41 ± 0.25	2.0 ± 2.6	ND	29g/kg*
Extractives	ND	ND	NC	ND	ND	ND	ND	2.19	ND
Caffeine	ND	ND	NC	ND	ND	ND	1.5 ± 1.0	0.68	18g/kg*
Tannins	ND	ND	NC	ND	ND	ND	3.0 ± 5.0	2.33	7g/kg*
Total polyphenols	ND	ND	ND	ND	ND	ND	1.5 ± 1.5	ND	20g/kg*
Chlorogenic Acid	ND	ND	ND	ND	ND	ND	2.4 ± 1.0	ND	ND
Ash	5.36 ± 0.20	4.7	7 ±0.16	1.30 ± 0.10	1.6	1.81 ± 0.05	ND	4.99	89g/kg*
Protein	18.69 ± 0.10	16.2	18.6 ± 0.25	17.44 ± 0.10	13.6	11.72 ± 0.28	11.5 ± 2.0	ND	80g/kg*
Nitrogen	2.99 ± 0.10	ND	ND	2.79 ± 0.10	ND	1.88 ± 0.05	ND	1.74	ND
C/N ratio	14.41 ± 0.10	ND	ND	16.91 ± 0.10	ND	ND	ND	ND	ND
Total dietary fiber	54.11 ± 0.10	ND	62.4 ±0.5	60.46 ± 2.19	ND	ND	60.5 ± 2.9	54.15	ND

(Continued)

Table 3.2 Chemical Composition of Coffee Processing By-Products in Percent (*Dry Material) (cont.)

By-Product	CS			SCG				CP	
References	Ballesteros et al. (2014a)	Mussatto et al. (2011a)	Pourfarzad et al. (2013)	Ballesteros et al. (2014a)	Mussatto et al. (2011a)	Uribarrí et al. (2014)	Murthy and Madhava Naidu (2012)	Peñaloza et al. (1985)	Ulloa Rojas et al. (2002)
Insoluble	45.98 ± 0.18	ND	53.7 ± 0.4	50.78 ± 1.58	ND	ND	ND	ND	ND
Soluble	8.16 ± 0.9	ND	8.8 ± 0.6	9.68 ± 2.70	ND	ND	ND	ND	ND
Ca	9.4 mg/kg*	ND	ND	1,2 mg/kg*	ND	0.174%	ND	ND	ND
Mg	3.1 mg/kg*	ND	ND	1,9 mg/kg*	ND	0.138%	ND	ND	ND
P	21.1 mg/kg*	ND	ND	1,8 mg/kg*	ND	0.132%	ND	ND	ND
Na	1.2 mg/kg*	ND	ND	33.7 mg/kg*	ND	0.036%	ND	ND	ND
K	21.1 mg/kg*	ND	ND	11.7 mg/kg*	ND	0.339%	ND	1.39	ND
Zn	22.3 mg/kg*	ND	ND	8.40 mg/kg*	ND	96.09 ppm	ND	ND	ND
Mn	50.0 mg/kg*	ND	ND	28.8 mg/kg*	ND	43.26 ppm	ND	ND	ND
Cu	63.3 mg/kg*	ND	ND	18.66 mg/kg*	ND	8.87 ppm	ND	ND	ND
Fe	843.3 mg/kg*	ND	ND	52.0 mg/kg*	ND	75.42 ppm	ND	ND	ND

By-Product	CH				CCS
References	Gouvea et al. (2009)	Murthy and Madhava Naidu (2012)	Gouvea et al. (2009)	Triana et al. (2011)	Aristizábal et al. (2015)
Moisture	ND	ND	15.0	4.00	ND
Carbohydrates	58–85	ND	72.3	ND	ND
Cellulose	43	43.0 ± 8.0	16.0	31.06	40.39 ± 2.2
Hemicellulose	7	7.0 ± 3.0	11.0	13.28	34.01 ± 1.2
Lignin	9	9.0 ± 1.6	9.0	44.73	10.13 ± 1.3
Insoluble	ND	ND	ND	ND	ND

Soluble	ND	ND	ND	ND	ND
Fat	0.5–5	0.5 ± 5.0	0.3	ND	ND
Extractives	ND	ND	ND	1.62	14.18 ± 0.85
Caffeine	≈1	1.0 ± 0.5	ND	ND	ND
Tannins	≈5	5.0 ± 2.0	ND	ND	ND
Total polyphenols	ND	0.8 ± 5.0	ND	ND	ND
Chlorogenic Acid	ND	2.5 ± 0.6	ND	ND	ND
Ash	ND	ND	5.4	0.88	1.27 ± 0.03
Protein	8–11	8.0 ± 5.0	7.0	4.43	ND
Nitrogen	ND		ND	ND	ND
C/N ratio	ND		ND	ND	ND
Total dietary fiber	ND	24 ± 5.9	ND	ND	ND
Insoluble	ND	ND	ND	ND	ND
Soluble	ND	ND	ND	ND	ND
Ca	ND	ND	ND	ND	ND
Mg	ND	ND	ND	ND	ND
P	ND	ND	ND	ND	ND
Na	ND	ND	ND	ND	ND
K	ND	ND	ND	ND	ND
Zn	ND	ND	ND	ND	ND
Mn	ND	ND	ND	ND	ND
Cu	ND	ND	ND	ND	ND
Fe	ND	ND	ND	ND	ND

ND, *Not Determined.*

3.3.4 COFFEE HUSK

When coffee cherry is processed by dry method, the CH is obtained. This residue represents near 12% of cherry on a dry-weight basis. From 1 ton of coffee fruits 0.18 tons of CH are obtained (Murthy and Madhava Naidu, 2012). CP is mainly composed of lignocellulosic materials and it has no value as fertilizer (Rathinavelu and Graziosi, 2005). Generally, CP is used as fuel of ovens localized in farms and fuel in gasifiers to produce electricity or biogas (Rathinavelu and Graziosi, 2005). This solid residue can be used as supplement for animal feed, and substrate for fermentation processes to obtain bioproducts as organic acids, enzymes, among others (Gouvea et al., 2009).

3.3.5 COFFEE CUT-STEMS

Coffee cut-stems come from the aerial part of coffee tree. It is the product of cutting the tree when the crop renovations are carried out, extending coffee productivity. These residues are produced seasonally and usually the cuts are made every five years or more according to crop area, planting density, time of year, phytosanitary conditions, production system (i.e., in the shade or free sun exposure). About 0.6 kg of coffee cut-stems are obtained per 1 kg of coffee cherry processed. It is burned in the farms for food preparation, auto degradation, or fuel for coffee parchment drying. In the last one activity it is necessary only 4.0 kg of coffee cut-stems for processing 11.34 kg of coffee parchment dry due to its high calorific value (Arcila et al., 2007; Rodríguez and Zambrano, 2010).

3.4 CHARACTERIZATION OF COFFEE PROCESSING BY-PRODUCTS

Table 3.2 indicates the chemical composition of coffee-processing by-products. Lignocellulosic materials as holocellulose (i.e., cellulose + hemicellulose) and lignin are the main components of these residues. The holocellulose content can be determined with the chlorination method described by the ASTM Standard D1104 (Han and Rowell, 1997). The α-cellulose content is a second part of the earlier procedure, in pursuit of the ultimately pure form of fiber (Han and Rowell, 1997). The content of acid-insoluble lignin in wood and pulp is determined according to TAPPI T222 (Han and Rowell, 1997). Also, these materials contain ashes and extractives that are determined according to protocols of the National Renewable Energy Laboratories (NREL/TP-510-42619) (Sluiter et al., 2008) and to National Renewable Energy Laboratories (NREL/TP-510-42622) (Sluuter et al., 2008), respectively. In the characterization of CCS the methods aforementioned are used (Aristizábal et al., 2015; Triana et al., 2011).

In CS and SCG, the mineral content in ashes is estimated by inductively coupled plasma atomic emission spectrometry (ICP-AES), the fat content can be determined using petroleum ether as solvent in a Soxhlet extraction, nitrogen content is estimated by combustion and the protein content is determined using the $N_2 \times 6.25$ conversion factor (Ballesteros et al., 2014a). Finally, total dietary fiber is determined by enzymatic gravimetric method according to the official AOAC standard procedure No.985.29 (Ballesteros et al., 2014a). In addition, official international methods of analysis of the AOAC are used for the determination of moisture (44-16 A), protein (46-13), ash (08-07), fat (30-10), total dietary fiber and soluble and insoluble fractions (32-07) (Gouvea et al., 2009; Pourfarzad et al., 2013).

3.5 POSSIBILITIES OF INTEGRAL VALORIZATION OF COFFEE PROCESSING BY-PRODUCTS

The wastes generated during coffee processing represent 90.5% of the total weight of the fruit. As seen in Table 3.2, these have a composition rich in carbohydrates, proteins, oils, and antioxidant compounds industrially relevant that are not exploited (Mussatto et al., 2011a). From this point of view, these can be considered as by-products of the agroindustrial coffee chain to increase the value chain in the productive sectors and reduce the actual environmental impact. It may be possible to use these residues to produce feed concentrates, fertilizers, compost, biogas, single-cell protein, particleboard, and energy (Bekalo and Reinhardt, 2010; Esquivel and Jiménez, 2012; Murthy and Madhava Naidu, 2012; Shemekite et al., 2014; Ulloa Rojas et al., 2003).

As an alternative energy source, spent coffee ground is mainly used in industrial boilers to generate steam and returns to soluble coffee process. This is due to its high heating value of 20.92–26.9 MJ/kg of spent coffee ground that is higher than other by-products as 18.26 MJ/kg of coffee cut-stems (Caetano et al., 2012; García et al., 2016; Mussatto et al., 2011a; Prata and Oliveira, 2007; Rodríguez and Zambrano, 2010). In the other mentioned productive areas, the coffee pulp and husk represent the 29% wt. and 12 % wt. dry basis of coffee cherry, respectively (Murthy and Madhava Naidu, 2012). However, the presence of fiber and other compounds like tannins, caffeine, and polyphenols have limited their use at the industrial level (Murthy and Madhava Naidu, 2012; Ulloa Rojas et al., 2002, 2003). It is necessary to carry out a detoxification to obtain bioactive molecules. In addition, this is a material with a high content of organic matter that can be exploited in biotechnological processes to get value-added products, as shown in Fig. 3.3. Thus, it is possible to achieve sustainable development during the processing and valorization of coffee by-products (Brand et al., 2000; Murthy and Madhava Naidu, 2012; Prata and Oliveira, 2007). This allows glimpsing the application of coffee by-products under the biorefinery concept, where four of the five product families proposed by Moncada et al. (2014) can be identified.

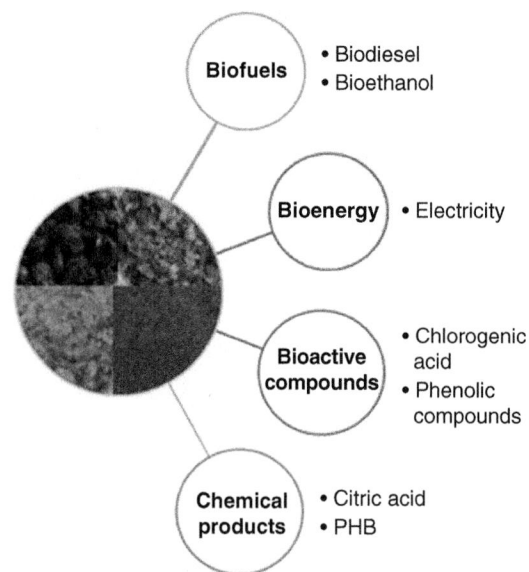

FIGURE 3.3 Possible Uses of Coffee Processing By-Products

3.6 PRODUCTS OBTAINED FROM COFFEE PROCESSING BY-PRODUCTS

3.6.1 POLYPHENOLS

Phenolic compounds present in green coffee bean have also been identified in coffee by-products. Among these are chlorogenic acid, ferulic acid, catechin, rutin, and coumaric acid (Esquivel and Jiménez, 2012; Jeszka-Skowron et al., 2016; López-Barrera et al., 2016). The importance of these bioactive compounds is attributed to their antioxidant capacity and for the prevention of apparition and/or development of diseases, also in aging caused by free radicals that oxidize lipids (Campos-Vega et al., 2015; Cotes et al., 2006; López-Barrera et al., 2016; Panusa et al., 2013; Zuorro and Lavecchia, 2012). Given its health benefits, different extraction methods have been used in the literature to recover phenolic compounds from coffee by-products. Some common methods reported are shown in Table 3.3, with the antioxidant capacity and the content of total phenolic compounds and flavonoids in the obtained extracts.

The concentration of phenolic compounds expressed as Gallic acid equivalents (GAE) in the coffee by-products varies according to the extraction conditions due to the compounds solubility (Pavlović et al., 2013). The use of water in subcritical conditions is more favorable because high concentrations of phenolic compounds are obtained and it does not require the use of solvents and reduces the operating times. This can lead to its viability, but the use of the coffee by-products as raw materials in this process implies more specific analysis for applications at

Table 3.3 Polyphenols and Antioxidant Capacity of Extracts From Coffee Processing By-Products

Methods and Conditions of Polyphenols Extraction	Coffee By-Products				References
	SCG	CS	CP	CH	
Soxhlet					
27°C 10 mL of isopropanol (60%) per gram sample	10.2 mg GAE/g SCG	13.2 mg GAE/g SCG	14.8 mg GAE/g SCG	12.2 mg GAE/g SCG	Murthy and Naidu (2012)
100°C 3 h 16.7 mL of water per gram sample	• 13.58 mg GAE/g SCG. • 58.00 µmol Trolox/g dry matter of SCG	—	—	—	Bravo et al. (2013)
Solid–liquid extraction					
120°C, 20 min, 20 mL of water per g sample	• 32.92 mg GAE/g SCG • 8.29 mg quercetin equivalents/g SCG	• 19.17 mg GAE/g CS • 2.73 mg quercetin equivalents/gram CS	—	—	Conde and Mussatto (2016)
180°C, 10 min, 50 mL of water per gram sample		• 22 mg GAE/g CS	—	—	Narita and Inouye (2012)
80°C, 10 min, 16.7 mL of water per gram sample	• 17.44 mg GAE/g SCG • 66.38 µmol Trolox/g dry matter of SCG.	—	—	—	Bravo et al. (2013)
60–65°C, 90 min, 25 mLof methanol (50%) per gram sample	• 18.2 mg GAE/g SCG. • 0.98 mg Chlorogenic acid/g SCG.	—	—	—	Mussatto et al. (2011b)
60°C, 30 min, 50 mL ethanol (60%) per gram sample	• 28.26 mg GAE/g SCG. • 5.63 mg quercetin equivalents/g SCG.	—	—	—	Panusa et al. (2013)

(Continued)

Table 3.3 Polyphenols and Antioxidant Capacity of Extracts From Coffee Processing By-Products (*cont.*)

Methods and Conditions of Polyphenols Extraction	SCG	Coffee By-Products			References
		CS	CP	CH	
60–65°C, 30 min, 35 mL of ethanol (60%) per gram sample	—	• 12.81 mg GAE/g CS • 17.95 µmol Trolox equivalents/gram dry matter CS	—	—	Ballesteros et al. (2014b)
50°C, 120 min, 40 mL of ethanol (70%) per gram sample	• 17.09 mg GAE/g dry matter SCG	—	—	—	Zuorro and Lavecchia (2012)
80°C, 1 h, 50 mL of water per gram sample	—	7 mg GAE/g CS	—	—	Narita and Inouye (2012)
80°C, 1 h, 50 mL of 0.1 M HCL per gram sample	—	7 mg GAE/g CS	—	—	
80°C, 1 h, 50 mL of 0.1 M NaOH per gram sample	—	8 mg GAE/g CS	—	—	
Biological					
Penicillium purpurogenum GH2 at 30°C during 6 days.	• 7.02 mg GAE/g SCG.	• 3.47 mg GAE/g SCG	—	—	Machado et al. (2012)

From Conde, T., Mussatto, S.I., 2016. Isolation of polyphenols from spent coffee grounds and silverskin by mild hydrothermal pretreatment. Prep. Biochem. Biotechnol. 6068(October).

the industrial level. On the other hand, taking into account the process time and the yields in the phenolic compounds production from fungal strains, it would be better to use this process to detoxify coffee processing by-products and send them as substrates for biotechnological process (Machado et al., 2012; Mussatto et al., 2011a).

3.6.2 BIOFUELS

Bioethanol and biodiesel are among liquid biofuels. Ethanol is produced through the fermentation of hydrolysates of coffee by-products. The coffee parts mainly used are the CH and pulp. Gouvea et al. (2009) use CH and 3 g of *Saccharomyces cerevisiae* per liter of substrate to obtain a concentration and productivity of ethanol of 13.6 and 1.23 g/L h, respectively. Shenoy et al. (2011) carried out the acid hydrolysis of cellulose from dry CP and then fermented it with 5 g of *S. cerevisiae* per liter of substrate obtaining a yield of ethanol/substrate of 8% wt. Other pretreatments were carried out to obtain fermentable sugars. A mechanical pretreatment of CP is performed by Menezes et al. (2013) in a manual press to obtain the liquid fraction. This was fermented to obtain an ethanol concentration of 18.3 g/L and a productivity of 2.99 g/L h (Menezes et al., 2013). A higher ethanol concentration was found when performing the fermentation with a hydrolyzate of pretreated CP with 4% NaOH during 25 min at 121°C (Menezes et al., 2014).

Biodiesel can be obtained from the transesterification of the oil extracted from the SCG (Haile et al., 2013; Haile, 2014; Ilickovic et al., 2012; Rodríguez and Zambrano, 2010). This is because the fatty acid profile of the SCG oil has a high content of palmitic acid that does not allow its use in food industry (Mancini et al., 2015). Al-Hamamre et al. (2012) evaluated the production of biodiesel with an alkali-catalyzed transesterification at different times, temperatures, and ratio of free fatty acid:methanol-KOH, the maximum conversion to fatty acid methyl ester (79.60%) was at 1 h, 85°C and 1:9, respectively. Different solvents and equipment have been used for oil extraction, but extraction with hexane is becoming the most profitable option because the oil extraction yields are between 12 and 19 g of oil per 100 g of SCG (Al-Hamamre et al., 2012; Andrade et al., 2012; Couto et al., 2009; Cruz et al., 2014; De Melo et al., 2014; Haile et al., 2013; Ilickovic et al., 2012; Uribarrí et al., 2014). However, supercritical CO_2 could be used for cosmetic purposes because it does not require toxic solvents (Araus et al., 2009).

Another biofuel of great interest is the biogas obtained from a fermentation process known as anaerobic digestion. Depending on its calorific value, it can be used in different fields, such as chemical synthesis, generating heat or electricity (Rezaiyan and Cheremisinoff, 2005). The coffee by-products that have been studied for this purpose are the CH and pulp. Since they have high protein and carbohydrates content, this allows a good anaerobic fermentation with respect to other materials that require to be supplemented for a proper C/N ratio (Calzada et al., 1984; Qiuxia et al., 2013).

3.6.3 **BIOENERGY**

As mentioned earlier, the combustion of SCG as an alternative energy source is currently performed. As a result of this, different authors have performed the determination of high heating value after extracting the value-added products to continue into an integral valorization of these (Go et al., 2016). Obruca et al. (2014) and Al-Hamamre et al. (2012) found a decrease of 9% (19.61 a 17.86 MJ/kg and 20.79 a 17.86 MJ/kg) in the SCG remainder of oil extraction with hexane while Vardon et al. (2013) found an even bigger decrease of 14.10% (23.4 a 20.1 MJ/kg) (Al-Hamamre et al., 2012; Obruca et al., 2014; Vardon et al., 2013). Another example is Haile et al. (2013), which extracts the oil and hydrolyzes the residue, leaving a remaining solid with a high calorific value of 20.8 MJ/kg. Another proposal has been the pellets formed with crude glycerin, a by-product of biodiesel production. This has a heat of combustion of 21.6 MJ/kg, which is sufficient to cover the energy requirement in small-scale industries (Haile, 2014).

3.6.4 **BIOPRODUCTS**

The bioproducts include citric acid, lactic acid, and polyhydroxyalkanoates, among others. Coffee husk or pulp and SCG have been used as substrate in biochemical pathways (Campos-Vega et al., 2015). In solid-state fermentation, 150 mg of citric acid and 492.5 mg of gallic acid per kilogram of dry CH are obtained (Machado et al., 2002; Shankaranand and Lonsane, 1994). For polyhydroxyalkanoates production, the SCG oil was used obtaining a high performance with respect of yields of soybean oil and waste rapeseed oil (i.e., 0.82 g of poly(3-hydroxybutyrate) per gram of SCG oil) (Campos-Vega et al., 2015; Obruca et al., 2014).

Lactic acid production from coffee pulp at pilot scale was performed by Pleissner et al. (2016). A previous pretreatment was necessary by hydrolyzing the material to use the hemicellulose and cellulose fractions as carbon sources. The fermentation of hydrolyzate was carried out with *Bacillus coagulants* at 52°C, pH 6 achieving a yield of 0.78 g of lactic acid per g of sugars.

3.6.5 **BIOFERTILIZERS**

As a soil amendment, a few studies have realized for exhausted SCG (Murthy and Naidu, 2012; Vardon et al., 2013). Vardon et al. (2013) performed a slow pyrolysis to obtain biochar with a better yield of fertilizer than sorghum−sudangrass.

3.7 **BIOREFINERIES OBTAIN BIOPRODUCTS AND BIOFUELS FROM SCG AND CCS: CASE STUDY**

The bioprocess term is defined as the structure that follows a method or operation of preparationer transformation of biomass into bioproducts. A biorefinery is an ensemble of coupled processes that also uses biomass as raw material. Therefore, bioprocess and biorefinery terms are not synonyms. A biorefinery is analogous to an oil refinery, where many products can be obtained and a bioprocess is a piece that completes the biorefinery scheme.

A biorefinery can integrate technologies and transformation routes to produce biofuels, biochemicals, and bioenergy. The biorefinery design includes economic, environmental, and social assessments and analysis. Then, biorefineries have been identified as a promising alternative to generate an industry based on biomass. When the biomass and crude oil are compared, characteristics as renewability, storage, substitutability, abundance, and carbon neutral (zero emissions) are highlighted, generating remarkable differences between these raw materials. In this way, when a biorefinery and refinery are compared, differences as the origin of raw material, technologies, and generations of building-blocks (products) to obtain others products are emphasized. Finally, biorefineries are a dynamic system due to its wide variety of feedstock with better features than refineries (Chisti, 2008).

The aforementioned concept is applied on coffee-processing by-products in a case study. In this sense, a biorefinery from SCG to obtain oil, ethanol, and xylitol and other biorefineries from CCS to produce ethanol, furfural, and hydroxymethyl-furfural (HMF) are analyzed from a technical and economic point of view. In the biorefinery from CCS, an analysis of the production scale is also made.

3.7.1 SMALL-SCALE VERSUS HIGH-SCALE BIOREFINERIES

Table 3.4 indicates a brief description of biorefineries proposed from some coffee-processing by-products as SCG and CCS. The biorefinery from SCG considers the obtaining of oil as the first product, followed by ethanol and xylitol. The ethanol and xylitol are obtained from the lignocellulosic material by hydrolysis and fermentation. Fig. 3.4 shows the flowsheet of the biorefinery.

Table 3.4 Description of Biorefineries Proposed From Coffee Processing By-Products

| Raw Materials | Products | | Distributions | Scale of Processing (ton/h) |
	Top	By-Products		
Biorefinery 1				
SCG	Oil, ethanol, xylitol	Glucose, xylose	Oil production 100% of SCG. Ethanol production 100% of glucose and 20% of xylose from cellulose and hemicellulose respectively. Xylitol production. 80% of xylose from hemicellulose.	1
Biorefinery 2				
CCS	Ethanol, furfural, HMF	Glucose, xylose	Ethanol production 50% of glucose from cellulose. HMF production 50% of glucose from cellulose. Furfural production. 100% of xylose from hemicellulose.	5, 25, 50, 100

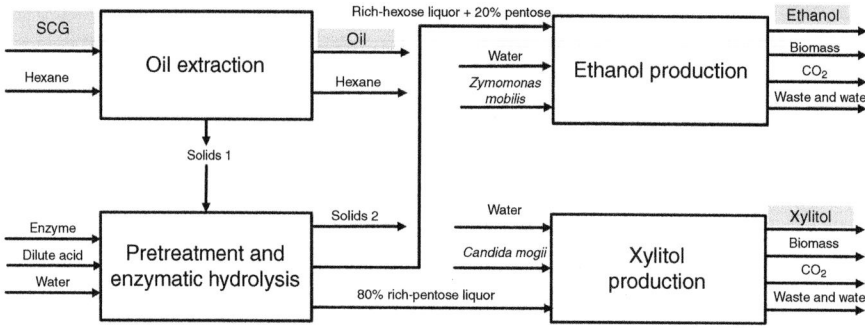

FIGURE 3.4 Flowsheet of the Biorefinery Proposed from SCG

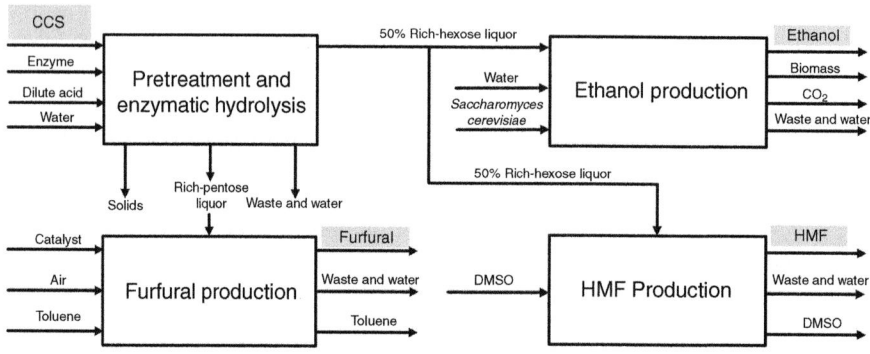

FIGURE 3.5 Flowsheet of the Biorefinery Proposed from CCS

In the second biorefinery, the extraction of sugars (xylose and glucose) and the production of ethanol, furfural, and HMF from CCS are considered. In this case, the economic assessment takes into account the scales of processing of the raw material to analyze the influence in the economic results. Fig. 3.5 shows the flowsheet of biorefinery from CCS.

3.7.2 PROCESS DESCRIPTION

The process begins from a basic characterization of the raw materials to be processed.

Table 3.5 shows the chemical characterization of the residues used as raw material in the formulated biorefineries. These values were determined using the methods mentioned in the early fewlines of Section 3.4. Then, this chemical characterization is fed to the simulator as the initial composition of SCG and CCS in the biorefineries to obtain oil, xylitol, ethanol, furfural, and HMF.

The description of each process considered in the biorefineries proposed is presented as follows.

Table 3.5 Chemical Compositions of SCB and CCS

Component (%)	SCG	CCS
Cellulose	15.95 ± 0.51	40.39 ± 2.2
Hemicellulose	21.24 ± 0.82	34.01 ± 1.2
Lignin	34.53 ± 0.80	10.13 ± 1.3
Ash	0.93 ± 0.06	1.27 ± 0.03
Extractives	27.34 ± 0.42	14.18 ± 0.85

From Aristizábal, M.V., Gómez, P.Á., Cardona, C.A., 2015. Biorefineries based on coffee cut-stems and sugarcane bagasse: furan-based compounds and alkanes as interesting products. Biores. Technol. 196, 480–489.

3.7.2.1 Oil extraction

Two steps compose the extraction process: extraction and solvent recovery. In the extraction step, hexane at 65°C makes contact with the material to obtain an organic phase and a solid residue. The organic phase (oil + hexane) is sent to a distillation column to recover the solvent. The lignocellulosic material (exhausted SCG) undergoes the sugars extraction stage by acid and enzymatic hydrolysis and the oil is used in cosmetic industry or in the biodiesel production.

3.7.2.2 Sugars extraction

Lignocellulosic biomass is submitted to a process divided in three stages: (1) size reduction, (2) dilute-acid pretreatment, and (3) enzymatic hydrolysis. The first stage of the process involves a size reduction of the material. After milling and sieving, in the second stage the hemicellulose fraction is hydrolyzed with sulfuric acid (2% by weight) based on the kinetic expressions reported by Jin et al. (2011) at 100°C. From the dilute-acid hydrolysis, a nonconverted solid fraction and a rich-pentose liquor are obtained. Then, the liquid fraction is separated by filtration from the solid fraction. Finally, in the third stage, the solid fraction rich in cellulose and lignin is submitted to an enzymatic hydrolysis step based on the kinetic expressions reported by Morales-Rodriguez et al. (2011) at 50°C to obtain a hexose-rich liquor and a solid residue rich in lignin and unconverted cellulose.

From the dilute-acid pretreatment, furfural, and HMF are obtained as by-products of the decomposition reactions of sugars. Then, a detoxification technology is applied (Mussatto and Roberto, 2004). This procedure is carried out to avoid poisoning and inhibition by the acids, furfural, and HMF in the fermentation stage where *Zymomonas mobilis* is used as microorganism.

3.7.2.3 Ethanol

The fermentation step is carried out with *Z. mobilis* when hexoses and pentoses are considered as feedstock. On the other hand, as an alternative *S. cerevisiae* is used as fermentation microorganism for ethanol production using hexoses as feedstock (scenario 3). Initially, the sugar-rich liquor is sent to a sterilization process at 121°C in which the biological activity is neutralized. Later the fermentation process is carried out based on the kinetic expressions reported by Rivera et al. (2006) and Leksawasdi et al. (2001), using *S. cerevisiae* at 37°C and *Z. mobilis* at 30°C as microorganisms, respectively. Afterward, cell biomass is separated from the culture broth by a simple gravitational sedimentation technology. After the fermentation stage, the culture broth containing approximately 5%–10% w/w of ethanol is taken to the separation step, which consists of two distillation columns. In the first column, ethanol is concentrated nearly to 45%–50% by weight. In the second column, the liquor is concentrated until the azeotropic point (96% w) to be led to the dehydration step with molecular sieves to obtain an ethanol concentration of 99.6% by weight (Pitt et al., 1983).

3.7.2.4 Furfural
Furfural is obtained from the pentose-rich liquor via xylose dehydration, catalyzed by aluminum and hafnium pillared clays with 86.2% of conversion (Cortés et al., 2013) and using air as stripping agent for removing the product while it is produced (Agirrezabal-Telleria et al., 2013). First, the liquor is sent to a reactor at 170°C and 10 bar. Air is fed into the reactor at a ratio of 30:1 air to feed. The resulting stream is depressurized to recover the liquid fraction. Then, the mixture is sent to a liquid–liquid extraction process with toluene as solvent with 1:1 v/v ratio to recover the furfural from the water. Finally, the furfural-solvent stream is submitted to a distillation process where the furfural is obtained as bottom product (Agirrezabal-Telleria et al., 2013).

3.7.2.5 HMF
HMF is obtained from 50% of the hexose-rich liquor via glucose dehydration, through a noncatalyzed system in water as described by Jing and Lü (2008). Liquor is sent to a reactor at 220°C and 10 MPa. The resulting stream is depressurized. Then, the mixture is sent to a liquid–liquid process with dimethyl sulfoxide (DMSO) as solvent with 1:0.6 molar ratio to recover the HMF from water. Finally, the HMF-solvent stream is submitted to a distillation process where the HMF is obtained as bottom product (Xiong et al., 2014).

3.7.2.6 Xylitol
Xylitol is synthesized from xylose by the yeast *Candida mogii*. Initially the liquor is sent to a sterilization process at 121°C in which the biologic activity is neutralized. Once the culture media is sterilized the fermentation with *C. mogii* is performed according to Tochampa et al. (2005), at 30°C under aerobic conditions (dissolved oxygen concentration of 20%). After fermentation, the resulting stream is filtered to separate the biomass and the temperature is increased at 40°C and a flash operation is used to concentrate the obtained xylitol. The next step for isolating the metabolite from the fermentation broth consists on an evaporation to eliminate the excess of water for facilitating the concentration by crystallization, adding ethanol in order to decrease drastically the xylitol solubility and supersaturate the solution to carry out the crystallization at 5°C (Parajó et al, 1998).

3.7.3 TECHNOECONOMIC ASSESSMENT

The flowsheet synthesis of biorefineries is carried out using process simulation tools to perform the technical and economic assessment. The objective of this procedure is to generate the mass and energy balances from which the requirements for raw materials, consumables, utilities, and energy are calculated. The main simulation tool used for this purpose was the commercial package Aspen Plus v8.2 (Aspen Technology, Inc., USA). Specialized package for programming mathematical calculations, especially for kinetic analysis, such as Matlab, are also used. Nonrandom two-liquid (NRTL) thermodynamic model is applied to calculate the activity coefficients of the liquid phase and the Hayden-O'Connell equation of state was used for the description of the vapor phase.

The capital and operating costs are calculated using the Aspen Process Economic Analyzer v8.2 software (Aspen Technologies, Inc., USA). This analysis was estimated in US dollars for a 10-year period at an annual interest rate of 17% (typical for the Colombian economy), considering the straight-line depreciation method and a 25% income tax. Prices and economic data used in this analysis correspond to Colombian conditions, such as the costs of the raw materials, income tax, and labor salaries. This information is incorporated in order to calculate the production costs per kilogram of product. Table 3.6 summarizes the economic data used in the processes for each raw material.

Table 3.6 Price/Cost of Feedstock and Products Used in the Technoeconomic Assessment

Items	Unit	Price	References
SCG	USD/kg	0.061	*
CCS	USD/ton	18	Quintero et al. (2013)
Oil	USD/kg	0.17	*
Ethanol	USD/kg	1.04	Fedebiocombustibles (2014)
Xylitol	USD/kg	3.6	Alibaba (2016)
Furfural	USD/kg	1.7	Alibaba (2016)
HMF	USD/kg	2.0	Alibaba (2016)
Toluene	USD/kg	0.85	ICIS Chemical Pricing (2010)
DMSO	USD/kg	1.0	Alibaba (2016)
Hexane	USD/kg	1.0	Alibaba (2016)
Electricity	USD kWh	0.14	*
Calcium hydroxide	USD/kg	0.28	Alibaba (2016)
Sulfuric acid	USD/kg	0.094	ICIS Chemical Pricing (2010)
Fuel	USD/MW	24.58	Nme (2013)
Mid P. Steam (30 bar)	USD/ton	8.18	Moncada Botero (2012)
Low P. Steam (3 bar)	USD/ton	1.57	Moncada Botero (2012)
Water	USD/m³	0.74	*
Operator labor	USD/h	2.56	*
Supervisor labor	USD/h	5.12	*

Typical prices in Colombia.

Table 3.7 Flows of Main Products Obtained in the Biorefineries Proposed From SCG y CCS

Biorefinery 1			
Scale of Processing (ton/h)	Oil (kg/h)	Ethanol (kg/h)	Xylitol (kg/h)
1	251.8	113.8	23.1
Biorefinery 2			
Scale of processing (ton/h)	Ethanol (kg/h)	Furfural (kg/h)	HMF (kg/h)
5	329.2	833.1	391.3
25	1640.2	4198.3	1557.0
50	3290.3	7931.2	3916.2
100	6564.7	16650	7824.7

3.7.4 RESULTS AND ANALYSIS

The results obtained in the technical assessment of biorefineries are indicated in Table 3.7. These values are expressed in the mass flow of each product. In the biorefinery that considers the production of oil, ethanol, and xylitol from SCG obtained the following yields: 0.25, 0.11, and 0.023 g of product per gram of raw material, respectively. According to literature, there is no register of a biorefinery from SCG. In this way, this study case is presented as an alternative of use for integral exploitation of this type of residues to generate added-value products. In the biorefinery from CCS, some scales of processing of raw material are evaluated. From a technical point of view, the biorefineries present an average yield of 0.3 kg of product per kilogram of raw material. This calculation is done based on the sum of flows of obtained products and the flow of processed raw material.

The results obtained in the economic assessment are indicated in the Table 3.8. In the case of biorefinery from SCG, the production costs are high in comparison with sale price of each product. In this sense, the individual and total economic margins are negative and the biorefinery is unfeasible. This result can be due to the scale processing. If a major scale is considered, it is possible to find the balance between the amount of processed raw material and the economic assessment to obtain positive economic margins and feasibility.

For the biorefinery from CCS, the production costs decrease as the processing scale increases (Table 3.8). When considering the processing scales of 5 and 25 ton/h, the production costs are higher than the sale price of the product. But, when the processing scales of 50 and 100 ton/h are analyzed, the production costs are lower than the sales price of products. Then, the biorefinery finds the balance in the processing amount to achieve positive economic margins as indicates the Fig. 3.6.

Table 3.8 Production Cost Obtained for Biorefineries Proposed From SCG and CCS

Biorefinery 1			
Scale of Processing (ton/h)	Oil (USD/kg)	Ethanol (USD/kg)	Xylitol (USD/kg)
1	0.87	2.11	10.29
Biorefinery 2			
Scale of processing (ton/h)	Ethanol (USD/kg)	Furfural (USD/kg)	HMF (USD/kg)
5	1.07	1.99	2.12
25	0.97	2.06	2.01
50	0.80	1.34	1.91
100	0.72	0.88	1.66

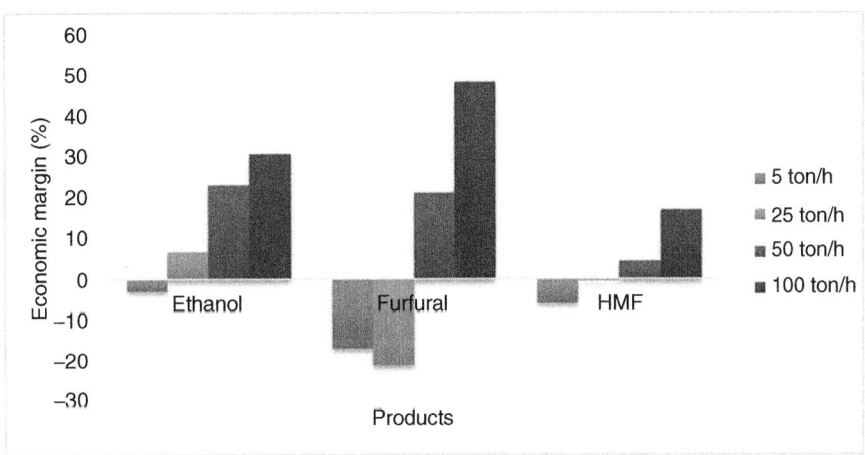

FIGURE 3.6 Economic Margin for the Products Obtained in the Biorefineries Proposed from CCS

The economic margin is the relation between sale price and production cost. Fig. 3.6 shows the individual economic margin for the products obtained in the biorefineries proposed from CCS.

Fig. 3.7 indicates the total economic margin for four processing scales. For the scales of 5, 25, 50, and 100 ton/h total economic margins of −9.3%, −6.3%, 14.6%, and 31.2% are obtained, respectively. According to the Fig. 3.7, the minimum processing amount of CCS is 28.5 ton/h to generate positive values in the economic assessment.

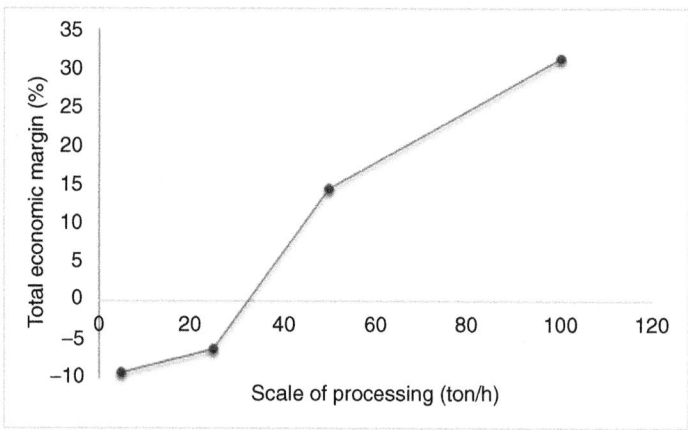

FIGURE 3.7 Total Economic Margin for Biorefineries from CCS

3.8 CONCLUSIONS

Most of the so-called residues from coffee processing at different levels shall be considered as interesting raw materials based on the total production and composition. However, as it was demonstrated in the SCG and CCS a technoeconomic analysis reveals the minimal scales the process (in this case biorefineries) needs to be considered feasible at preliminary design stages. If considered that stand-alone processes usually can be less efficient than biorefineries, the scale would have more influence on the project. The fact that in countries like Colombia most of the coffee crops are located in mountains with difficult access and the size of the field usually (based on the ownership) is one to two hectares, makes it necessary in the future to investigate the logistics and supply-chain strategies to achieve the needed scales for biorefining these residues.

ACKNOWLEDGMENT

The authors express their acknowledgments to Departamento Administrativo de Ciencia, Tecnología e Innovación (Colciencias) call 727–2015. The authors express their acknowledgments to the project entitled "Development of modular small-scale integrated biorefineries to produce an optimal range of bioproducts from a variety of rural agricultural and agroindustrial residues/wastes with a minimum consumption of fossil energy—SMBIO" from ERANET LAC 2015.

REFERENCES

Agirrezabal-Telleria, I., Gandarias, I., Arias, P.L., 2013. Production of furfural from pentosan-rich biomass: analysis of process parameters during simultaneous furfural stripping. Biores. Technol. 143, 258–264.

Al-Hamamre, Z., Foerster, S., Hartmann, F., Kröger, M., Kaltschmitt, M., 2012. Oil extracted from spent coffee grounds as a renewable source for fatty acid methyl ester manufacturing. Fuel 96, 70–76.

Alibaba, 2016. International Prices. Available from: https://spanish.alibaba.com.

Andrade, K.S., Goncalvez, R.T., Maraschin, M., Ribeiro-Do-Valle, R.M., Martínez, J., Ferreira, S.R.S., 2012. Supercritical fluid extraction from spent coffee grounds and coffee husks: antioxidant activity and effect of operational variables on extract composition. Talanta 88, 544–552.

Araus, K., Uquiche, E., del Valle, J.M., 2009. Matrix effects in supercritical CO_2 extraction of essential oils from plant material. J. Food Eng. 92 (4), 438–447.

Arcila, J., Farfán, F., Moreno, A., Salazar, L., Hincapié, E., 2007. Sistemas de Producción de Café en Colombia. 1. Editorial Blanecolor Ltda., Colombia.

Aristizábal, M.V., Gómez, P.Á., Cardona, C.A., 2015. Biorefineries based on coffee cut-stems and sugarcane bagasse: furan-based compounds and alkanes as interesting products. Biores. Technol. 196, 480–489.

Arya, M., Rao, L.J.M., 2007. An impression of coffee carbohydrates. Crit. Rev. Food Sci. Nutr. 47 (1), 51–67.

Ballesteros, L.F., Teixeira, J.A., Mussatto, S.I., 2014a. Chemical, functional, and structural properties of spent coffee grounds and coffee silverskin. Food Bioproc. Technol. 7 (12), 3493–3503.

Ballesteros, L.F., Teixeira, J.A., Mussatto, S.I., 2014b. Selection of the solvent and extraction conditions for maximum recovery of antioxidant phenolic compounds from coffee silverskin. Food Bioproc. Technol. 7 (5), 1322–1332.

Bekalo, S.A., Reinhardt, H.W., 2010. Fibers of coffee husk and hulls for the production of particle board. Mater. Struct. 43, 1049–1060.

Borrelli, R.C., Esposito, F., Napolitano, A., Ritieni, A., Fogliano, V., 2004. Characterization of a new potential functional ingredient: coffee silverskin. J. Agric. Food Chem. 52 (5), 1338–1343.

Brand, D., Pandey, A., Roussos, S., Soccol, C.R., 2000. Biological detoxification of coffee husk by filamentous fungi using a solid state fermentation system. Enzyme Microb. Technol. 27 (1–2), 127–133.

Bravo, J., Monente, C., Juániz, I., De Peña, M.P., Cid, C., 2013. Influence of extraction process on antioxidant capacity of spent coffee. Food Res. Int. 50 (2), 610–616.

Caetano, N.S., Silva, V.F.M., Mata, T.M., 2012. Valorization of coffee grounds for biodiesel production. Chem. Eng. Trans. 26, 267–272.

Calzada, J.F., de Porres, E., Yurrita, A., de Arriola, M.C., de Micheo, F., Rolz, C., Menchú, J.F., Cabello, A., 1984. Biogas production from coffee pulp juice: one- and two-phase systems. Agric. Wastes 9 (3), 217–230.

Campos-Vega, R., Loarca-Piña, G., Vergara-Castañeda, H., Oomah, B.D., 2015. Spent coffee grounds: a review on current research and future prospects. Trends Food Sci. Technol. 45 (1), 24–36.

Chisti, Y., 2008. Biodiesel from microalgae beats bioethanol. Trends Biotechnol. 26 (3), 126–131.

Clarke, R.J., 1987. Extraction. Clarke, R.J., Macrae, R. (Eds.), Coffee, vol. 2, Elsevier Science Publishing Ltd, New York, pp. 109–144.

Conde, T., Mussatto, S.I., 2016. Isolation of polyphenols from spent coffee grounds and silverskin by mild hydrothermal pretreatment. Prep. Biochem. Biotechnol. 46, 406–409.

Cortés, W., Piñeros-Castro, Y., Campos, A.M., 2013. Conversion of D-xylose into furfural with aluminum and hafnium pillared clays as catalyst. Dyna 80, 105–112.

Cotes, A.M., Cristancho, E., García, X.M., 2006. Antioxidantes, Oportunidades para la producción agropecuaria y agroindustria. Minist. Agric. y Desarro. Rural. Corpoica, 1–2.

Couto, R.M., Fernandes, J., da Silva, M.D.R.G., Simões, P.C., 2009. Supercritical fluid extraction of lipids from spent coffee grounds. J. Supercrit. Fluids 51 (2), 159–166.

Cruz, M.V., Paiva, A., Lisboa, P., Freitas, F., Alves, V.D., Simões, P., Barreiros, S., Reis, M.A.M., 2014. Production of polyhydroxyalkanoates from spent coffee grounds oil obtained by supercritical fluid extraction technology. Biores. Technol. 157, 360–363.

De Melo, M.M.R., a Barbosa, H.M., Passos, C.P., Silva, C.M., 2014. Supercritical fluid extraction of spent coffee grounds: measurement of extraction curves, oil characterization and economic analysis. J. Supercrit. Fluids 86, 150–159.

Esquivel, P., Jiménez, V.M., 2012. Functional properties of coffee and coffee by-products. Food Res. Int. 46 (2), 488–495.

Fedebiocombustibles, Federación Nacional de Biocombustibles de Colombia 2014. Biofuel prices in Colombia 2016-2017. Available from: http://www.fedebiocombustibles.com/v3/estadistica-precios-titulo-Alcohol_Carburante_(Etanol).htm

Federacion Nacional de Cafeteros, 2010. Available from: http://www.cafedecolombia.com/particulares/es/sobre_el_cafe/el_cafe/el_arbol_y_el_entorno/

Franca, A.S., Mendonça, J.C.F., Oliveira, S.D., 2005. Composition of green and roasted coffees of different cup qualities. LWT Food Sci. Technol. 38 (7), 709–715.

García, C.A., Gomez, A., Betancourt, R., Cardona, C.A., 2016. Environmental comparison of thermochemical and biochemical ways for producing energy from agricultural solid residues: coffee cut-stems case. Congress: Cyprus 2016 proceedings. Cyprus, pp. 1–24.

Givens, D.I., Barber, W.P., 1986. In vivo evaluation of spent coffee grounds as a ruminant feed. Agric. Wastes 18 (1), 69–72.

Go, A.W., Conag, A.T., Cuizon, D.E.S., 2016. Recovery of sugars and lipids from spent coffee grounds: a new approach. Waste Biomass Valor. 7, 1047–1053.

Gouvea, B.M., Torres, C., Franca, A.S., Oliveira, L.S., Oliveira, E.S., 2009. Feasibility of ethanol production from coffee husks. Biotechnol. Lett. 31 (9), 1315–1319.

Haile, M., 2014. Integrated volarization of spent coffee grounds to biofuels. Biofuel Res. J. 1 (2), 65–69.

Haile, M., Asfaw, A., Asfaw, N., 2013. Investigation of waste coffee ground as a potential raw material for biodiesel production. Int. J. Renew. Energ. Res. 3 (4), 854–860.

Han, J.S., Rowell, R.M., 1997. Chemical composition of fibers. In: Rowell, M.R., Young, R.A., Rowell, J.K. (Eds.), Paper and Composites From Agro-Based Resources. CRC Press, New York, pp. 83–134.

ICIS Chemical Pricing, 2010. Price reports for chemicals. Available from: http://www.icis.com/about/price-reports/.

Ilickovic, Z., Redzic, E., Andrejas, F., Avdic, G., Stuhli, V., 2012. Analysis of the possibility for obtaining oil from the spend coffee grounds as potential raw material for biodiesel production. Chem. Ind. 66 (4), 581–586.

Jeszka-Skowron, M., Sentkowska, A., Pyrzyńska, K., de Peña, M.P., 2016. Chlorogenic acids, caffeine content and antioxidant properties of green coffee extracts: influence of green coffee bean preparation. Eur. Food Res. Technol. 242, 1403–1409.

Jin, Q., Zhang, H., Yan, L., Qu, L., Huang, H., 2011. Kinetic characterization for hemicellulose hydrolysis of corn stover in a dilute acid cycle spray flow-through reactor at moderate conditions. Biomass Bioenerg. 35, 4158–4164.

Jing, Q., Lü, X., 2008. Kinetics of non-catalyzed decomposition of glucose in high-temperature liquid water. Chinese J. Chem. Eng. 16, 890–894.

Leksawasdi, N., Joachimsthal, E.L., Rogers, P.L., 2001. Mathematical modelling of ethanol production from glucose/xylose mixtures by recombinant *Zymomonas mobilis*. Biotechnol. Lett. 23, 1087–1093.

López-Barrera, D.M., Vázquez-Sánchez, K., Loarca-Piña, M.G.F., Campos-Vega, R., 2016. Spent coffee grounds, an innovative source of colonic fermentable compounds, inhibit inflammatory mediators in vitro. Food Chem. 212, 282–290.

Machado, C.M.M., Soccol, C.R., de Oliveira, B.H., Pandey, A., 2002. Gibberellic acid production by solid-state fermentation in coffee husk. Appl. Biochem. Biotechnol 102–103 (1–6), 179–191.

Machado, E.M.S., Rodriguez-Jasso, R.M., Teixeira, J.A., Mussatto, S.I., 2012. Growth of fungal strains on coffee industry residues with removal of polyphenolic compounds. Biochem. Eng. J. 60, 87–90.

Mancini, A., Imperlini, E., Nigro, E., Montagnese, C., Daniele, A., Orrù, S., Buono, P., 2015. Biological and nutritional properties of palm oil and palmitic acid: effects on health. Molecules 20 (9), 17339–17361.

Menezes, E.G.T., Do Carmo, J.R., Menezes, A.G.T., Alves, J.G.L.F., Pimenta, C.J., Queiroz, F., 2013. Use of different extracts of coffee pulp for the production of bioethanol. Appl. Biochem. Biotechnol. 169 (2), 673–687.

Menezes, E.G.T., do Carmo, J.R., Alves, J.G.L.F., Menezes, A.G.T., Guimaraes, I.C., Queiroz, F., Pimenta, C.J., 2014. Optimization of alkaline pretreatment of coffee pulp for production of bioethanol. Biotechnol. Prog. 30 (2), 451–462.

Moncada Botero, J., 2012. Design and evaluation of sustainable biorefineries from feedstock in tropical regions, Master's Thesis. Universidad Nacional de Colombia, Departamento de Ingeniería Química, Colombia.

Moncada, J., Tamayo, J.A., Cardona, C.A., 2014. Integrating first, second, and third generation biorefineries: incorporating microalgae into the sugarcane biorefinery. Chem. Eng. Sci. 118, 126–140.

Morales-Rodriguez, R., Gernaey, K.V., Meyer, A.S., Sin, G., 2011. A Mathematical model for simultaneous saccharification and co-fermentation (SSCF) of C6 and C5 sugars. Chinese J. Chem. Eng. 19, 185–191.

Murthy, P.S., Madhava Naidu, M., 2012. Sustainable management of coffee industry by-products and value addition: a review. Resour. Conserv. Recycl. 66, 45–58.

Murthy, P.S., Naidu, M.M., 2012. Recovery of phenolic antioxidants and functional compounds from coffee industry by-products. Food Bioproc. Technol. 5 (3), 897–903.

Mussatto, S.I., Roberto, I.C., 2004. Alternatives for detoxification of diluted-acid lignocellulosic hydrolyzates for use in fermentative processes: a review. Biores. Technol. 93, 1–10.

Mussatto, S.I., Machado, E.M.S., Martins, S., Teixeira, J.A., 2011a. Production, composition, and application of coffee and its industrial residues. Food Bioproc. Technol. 4 (5), 661–672.

Mussatto, S.I., Ballesteros, L.F., Martins, S., Teixeira, J.A., 2011b. Extraction of antioxidant phenolic compounds from spent coffee grounds. Sep. Purif. Technol. 83 (1), 173–179.

Mussatto, S.I., Ballesteros, L.F., Martins, S., Maltos, D.A.F., Aguilar, C.N., Teixeira, J.A., 2013. Maximization of fructooligosaccharides and β-fructofuranosidase production by aspergillus japonicus under solid-state fermentation conditions. Food Bioproc. Technol. 6 (8), 2128–2134.

Nabais, J.M.V., Nunes, P., Carrott, P.J.M., Ribeiro Carrott, M.M.L., García, A.M., Díaz-Díez, M.A., 2008. Production of activated carbons from coffee endocarp by CO_2 and steam activation. Fuel Process. Technol. 89 (3), 262–268.

Narita, Y., Inouye, K., 2012. High antioxidant activity of coffee silverskin extracts obtained by the treatment of coffee silverskin with subcritical water. Food Chem. 135 (3), 943–949.

Nme, 2013. LyD considers risky the propose of an energetic development based on shale gas.

Obruca, S., Petrik, S., Benesova, P., Svoboda, Z., Eremka, L., Marova, I., 2014. Utilization of oil extracted from spent coffee grounds for sustainable production of polyhydroxyalkano-ates. Appl. Microbiol. Biotechnol. 98 (13), 5883–5890.

Pabón Usaquén, J.P., Sanz Uribe, J.R., Oliveros Tascón, C.E., 2009. Manejo del café desmuci-laginado mecánicamente. Av. Técnic. Cenicafé 388, 1.

Panusa, A., Zuorro, A., Lavecchia, R., Marrosu, G., Petrucci, R., 2013. Recovery of natural antioxidants from spent coffee grounds. J. Agr. Food Chem. 61, 4162–4168.

Parajó, J.C., Domínguez, H., Domínguez, J.M., 1998. Biotechnological production of xyli-tol. Part 3: Operation in culture media made from lignocellulose hydrolysates. Bioresour. Technol. 66, 25–40.

Pavlović, M.D., Buntić, A.V., Šiler-Marinković, S.S., Dimitrijević-Branković, S.I., 2013. Eth-anol influenced fast microwave-assisted extraction for natural antioxidants obtaining from spent filter coffee. Sep. Purif. Technol. 118, 503–510.

Peñaloza, W., Molina, M.R., Brenes, R.G., Penaloza, W., Bressani, R., 1985. Solid-state fer-mentation: an alternative to improve the nutritive value of coffee pulp. Appl. Environ. Microbiol. 49 (2), 388–393.

Pfluger, R.A., 1975. Solid wastes: origin, collection, processing, and disposal. John Wiley & Sons, New York.

Pitt, W.W., Haag, G.L., Lee, D.D., 1983. Recovery of ethanol from fermentation broths using selective sorption-desorption. Biotechnol. Bioeng 25, 123–131.

Pleissner, D., Neu, A.-K., Mehlmann, K., Schneider, R., Puerta-Quintero, G.I., Venus, J., 2016. Fermentative lactic acid production from coffee pulp hydrolysate using *Bacillus coagulans* at laboratory and pilot scales. Biores. Technol. 218, 167–173.

Pourfarzad, A., Mahdavian-Mehr, H., Sedaghat, N., 2013. Coffee silverskin as a source of dietary fiber in bread-making: optimization of chemical treatment using response surface methodology. LWT Food Sci. Technol. 50 (2), 599–606.

Prata, E.R.B.A., Oliveira, L.S., 2007. Fresh coffee husks as potential sources of anthocyanins. LWT Food Sci. Technol. 40 (9), 1555–1560.

Qiuxia, W., Rui, X., Jianchang, L., Huanyun, D., Yage, Y., Jiahong, H., 2013. One study on biogas production potential character of coffee husks. Int. Conf. Mater. Renew. Energy Environ. 1, 188–192.

Quintero, J.A., Moncada, J., Cardona, C.A., 2013. Techno-economic analysis of bioethanol production from lignocellulosic residues in Colombia: a process simulation approach. Biores. Technol. 139, 300–307.

Rathinavelu, R., Graziosi, G., 2005. Posibles usos alternativos de los residuos y subproductos del café. Cent. Eur. J. Chem. 942, 4.

Rezaiyan, J., Cheremisinoff, N.P., 2005. Gasification Technologies: A Primer for Engineers and Scientists. CRC Press, Taylor & Francis.

Riaño Luna, C.E., 2010. Tecnología del Café. Universidad Nacional Abierta y a Distancia, Bogota, Colombia.

Rivera, E.C., Costa, A.C., Atala, D.I.P., Maugeri, F., Maciel, M.R.W., Filho, R.M., 2006. Evaluation of optimization techniques for parameter estimation: application to ethanol fermentation considering the effect of temperature. Proc. Biochem. 41, 1682–1687.

Rodríguez, N., Zambrano, D., 2010. Los subproductos del café: fuente de energía renovable. Av. Técnic. Cenicafé 3, 8.

Saenger, M., Hartge, E.U., Werther, J., Ogada, T., Siagi, Z., 2001. Combustion of coffee husks. Renew. Energy 23 (1), 103–121.

Sánchez, D.A., Anzola, C., 2013. Caracterización química de la película plateada del café (Coffee arábica) en variedades Colombia y Caturra. Rev. Colomb. Química 41 (2), 211–225.

Shankaranand, V.S., Lonsane, B.K., 1994. Coffee husk: an inexpensive substrate for production of citric-acid by aspergillus-niger in a solid-state fermentation system. World J. Microbiol. Biotechnol. 10 (2), 165–168.

Shemekite, F., Gomez-Brand, M., Franke-Whittle, I.H., Praehauser, B., Insam, H., Assefa, F., 2014. Coffee husk composting: an investigation of the process using molecular and non-molecular tools. Waste Manag. 34 (3), 642–652.

Shenoy, D., Pai, A., Vikas, R.K., Neeraja, H.S., Deeksha, J.S., Nayak, C., Rao, C.V., 2011. A study on bioethanol production from cashew apple pulp and coffee pulp waste. Biomass Bioenerg. 35 (10), 4107–4111.

Simoes, J., Nunes, F.M., Domingues, M.R., Coimbra, M.A., 2013. Extractability and structure of spent coffee ground polysaccharides by roasting pre-treatments. Carbohydr. Polym. 97 (1), 81–89.

Sluiter, A., Ruiz, R., Scarlata, C., Sluiter, J., Templeton, D., 2008. Determination of Extractives in Biomass Laboratory Analytical Procedure (LAP). Technical report NREL/TP-510-42619.

Sluiter, A., Hames, B., Ruiz, R., Scarlata, C., Sluiter, J., Templeton, D., 2008. Determination of Ash in Biomass Laboratory Analytical Procedure (LAP). Technical report NREL/TP-510-42622.

Tochampa, W., Sirisansaneeyakul, S., Vanichsriratana, W., Srinophakun, P., Bakker, H.H.C., Chisti, Y., 2005. A model of xylitol production by the yeast Candida mogii. Bioprocess. Biosyst. Eng. 28 (3), 175–183.

Triana, C.F., Quintero, J.A., Agudelo, R.A., Cardona, C.A., Higuita, J.C., 2011. Analysis of coffee cut-stems (CCS) as raw material for fuel ethanol production. Energy 36 (7), 4182–4190.

Ulloa Rojas, J.B., Verreth, J.A.J., Van Weerd, J.H., Huisman, E.A., 2002. Effect of different chemical treatments on nutritional and antinutritional properties of coffee pulp. Anim. Feed Sci. Technol. 99 (1–4), 195–204.

Ulloa Rojas, J.B., Verreth, J.A.J., Amato, S., Huisman, E.A., 2003. Biological treatments affect the chemical composition of coffee pulp. Biores. Technol. 89 (3), 267–274.

Uribarrí, A., Zabala, A., Sáchez, J., Arenas, E., Chandler, C., Rincón, M., González, E., Aiello Mazzarri, C., 2014. Evaluación del potencial de la borra de café como materia prima para la producción de biodiesel. Multiciencias 14 (2), 129–139.

Vardon, D.R., Moser, B.R., Zheng, W., Witkin, K., Evangelista, R.L., Strathmann, T.J., Rajagopalan, K., Sharma, B.K., 2013. Complete utilization of spent coffee grounds to produce biodiesel, bio-oil, and biochar. ACS Sustain. Chem. Eng. 1 (10), 1286–1294.

Vincent, J.C., 1987. Green coffee processing. In: R. J. Clarke, R.J., and Macrae, R. (Eds.), Coffee. New York, pp. 1–33.

Xiong, R., León, M., Nikolakis, V., Sandler, S.I., Vlachos, D.G., 2014. Adsorption of HMF from water/DMSO solutions onto hydrophobic zeolites: experiment and simulation. ChemSusChem 7, 236–244.

Zuorro, A., Lavecchia, R., 2012. Spent coffee grounds as a valuable source of phenolic compounds and bioenergy. J. Clean. Prod. 34, 49–56.

FURTHER READING

Chheda, J.N., Dumesic, J.A., 2007. An overview of dehydration, aldol-condensation and hydrogenation processes for production of liquid alkanes from biomass-derived carbohydrates. Catal. Today 123, 59–70.

Villarreal, M.L., Prata, A.M., Felipe, M.G., Silva, J.A.E., 2006. Detoxification procedures of eucalyptus hemicellulose hydrolysate for xylitol production by *Candida guilliermondii*. Enzyme Microb. Technol. 40, 17–24.

Xiong, R., León, M., Nikolakis, V., Sandler, S.I., Vlachos, D.G., 2014. Adsorption of HMF from water/DMSO solutions onto hydrophobic zeolites: experiment and simulation. ChemSusChem 7, 236–244.

Extraction and formulation of bioactive compounds

4

Ana Belščak-Cvitanović, Draženka Komes
University of Zagreb, Zagreb, Croatia

ABSTRACT

The chapter presents the state-of-the-art recovery of bioactive compounds from coffee processing by-products using conventional extraction techniques and approaches. Comprehensive literature overviews of the employed lab- or pilot-scale recovery procedures and yields of the most potent bioactive compounds (polyphenols, chlorogenic acids, caffeine, anthocyanins, lipids, diterpens, sterols) from coffee-processing by-products are provided. Commonly used extraction procedures, in particular solid–liquid solvent extraction, Soxhlet extraction, supercritical fluid extraction, and microwave-assisted extraction are reviewed, and the advantages and limitations of each technique, as well as the optimal recovery parameters (pretreatments, extraction solvent, temperature, pressure, duration) for each group of bioactives are emphasized. The approaches used for the formulation of recovered bioactives are summarized, focusing on encapsulation as a means of delivery of those bioactive compounds to food, thus ensuring complex properties (such as delayed release, stability, thermal protection, suitable sensorial profile) to the final application. Encapsulation techniques including microgel formation by ionic gelation of biopolymers, spray drying, coacervation, yeast encapsulation, or microemulsion nanoencapsulation used for delivery of coffee processing by-products bioactives are addressed. An overview of patented solutions for the recovery and potential applications resulting from the use of the recovered bioactives and their commercial potential is also provided.
Keywords: antioxidants; bioactive compounds; coffee by-products; encapsulation; extraction

4.1 INTRODUCTION

Due to the countless beneficial health effects of naturally derived phytochemicals and other macroconstituent bioactive compounds, the number of scientific studies evaluating the potential of agroindustrial residues as sources of functional ingredients and their recovery has been markedly intensified (Laufenberg et al., 2003). Coffee by-products, including spent coffee grounds (SCG), coffee husks, pulp, and coffee silverskin contain several classes of health-related chemicals, such as phenolic compounds, melanoidins, diterpenes, methylxanthines, and vitamin precursors. In the case of SCG, these compounds are only partially extracted during the brewing process, so this substrateand also all other coffee by-products represent very valuable sources of bioactive compounds that can have a wide range of applications in the food, cosmetic, and pharmaceutical industries. However, although scientific studies have clearly demonstrated the potential use of coffee processing by-products for extracting bioactive compounds, these processes are not yet being conducted on an industrial scale.

Just like drugs of synthetic origin, bioactive phytocompounds range from simple to complex structures, because they are present as conjugates or mixtures in extracts, which require labor-intensive and time-consuming purification procedures (Lam, 2007).

In order to obtain a targeted active constituent, extraction is the most important recovery stage starting with the solid–liquid extraction of dried and powdered material, most often by employing conventional techniques that are based on the extracting power of different solvents and application of heat and/or mixing. In order to obtain bioactive compounds from plants, the most common conventional technologies include solvent extraction, steam distillation, and acid and alkali extraction (Prado et al., 2005). All of those extraction technologies are conducted most often by means of several techniques and procedures, such as the Soxhlet extraction, maceration, and hydrodistillation. Extraction efficiency of any conventional method mostly depends on the choice of solvents (Cowan, 1999), influenced by the polarity of the targeted compound. Molecular affinity between solvent and solute, mass transfer, use of cosolvent, environmental safety, human toxicity, and financial feasibility should also be considered in selection of solvent for bioactive compounds extraction (Azmir et al., 2013). Solvent extraction is advantageous compared to other methods due to low processing cost and ease of operation. However, this method uses toxic solvents, usually calls for large amounts of solvent and extended time to be carried out, requires an evaporation/concentration step and carries the possibility of thermal degradation of recovered bioactive compounds, due to high temperatures of the solvents during long extraction times. At the same time conventional extraction methods, such as Soxhlet, are still considered as reference methods to compare the success of newly developed methodology (Azmir et al., 2013). All these techniques have some common objectives:

1. to extract targeted bioactive compounds from complex plant sample,
2. to increase selectivity of analytical methods,
3. to increase sensitivity of bioassay by increasing the concentration of targeted compounds,
4. to convert the bioactive compounds into a more suitable form for detection and separation, and
5. to provide a strong and reproducible method that is independent of variations in the sample matrix (Azmir et al., 2013; Smith, 2003).

Because the final purpose of recovered bioactive compounds is their formulation and further implementation, the issues arising during the food or pharmaceutical applications are their limited stability, uptake, and bioavailability. For that reason, the food and pharmaceutical industries are searching for solutions to achieve a stable and preserved formulated form of recovered bioactives, while at the same time ensuring minimal impact on the organoleptic and qualitative properties of the final products. Encapsulation is a powerful technique, which allows it to overcome many of the aforementioned issues, because it enables the protection of a wide range of materials of biological interest by their embedding into a polymeric matrix (Thies, 2005). So far, numerous encapsulation techniques have been developed and employed for encapsulation of different compounds, among which, some are already established and employed in industrial proportions (e.g., spray drying) while some have been increasingly evaluated for their performance and applicability. From a technological point of view, an efficient system for the encapsulation and delivery of bioactive compounds to be suitable for incorporation into functional foods and nutraceutical formulations, must comply with the following characteristics (McClements et al., 2007):

1. it must be formulated with food-grade, possibly natural ingredients using solvent-free production methods (Acosta, 2009),
2. it should be able to incorporate the bioactive compounds into food matrices with high physicochemical stability and minimal impact on the organoleptic properties of the product (Donsi et al., 2011a),
3. should be able to protect the encapsulated compounds from interaction with other food ingredients and from degradation due to temperature, light or pH (McClements et al., 2007),
4. should maximize the uptake of encapsulated compounds upon consumption, their transport to the sites of action (Acosta, 2009) and ensure controlled release in response to a specific environmental stimulus (McClements et al., 2007), and
5. production of delivery systems should be easily scalable to industrial production (Donsi et al., 2010).

Additionally, the interest in encapsulated bioactive compounds for food purposes also relies on the possibilities to modify physical properties of food materials, for example, rheological properties and to overcome solubility incompatibilities between ingredients, for example, bioactive compounds and the food matrices (Dube et al., 2010; Helgason et al., 2009).

In this chapter, recent developments and applications of coffee by-products bioactive compounds extraction and separation procedures, as well as formulation aimed at food and pharmaceutical application purposes, are reviewed. Emphasis is placed on the brief description of the unique capabilities and advantages and disadvantages of the extraction and formulation approaches used. A brief overview of patented recovery and application solutions of bioactive ingredients recovered from or present in coffee by-products is provided with respect to a number of literature sources.

4.2 RECOVERY OF BIOACTIVE CONSTITUENTS OF COFFEE BY-PRODUCTS

4.2.1 POLYPHENOLIC COMPOUNDS

Phenolic compounds are the major contributors to the strong antioxidant activity of coffee brews and corresponding coffee processing by-products. These bioactive compounds have attracted much interest in recent years due to their strong antioxidant and metal-chelating properties, which are believed to provide in vivo protection against free radical damage and reduce the risk of degenerative diseases associated with oxidative stress (Dai and Mumper, 2010). However, the recovery of these phenolic compounds from the coffee industry by-products and their antioxidant activity has been more intensively investigated only in the past few years.

The recovery of polyphenolic antioxidants from coffee processing by-products, according to the undertaken scientific studies, is initiated by the preparation of materials involving drying (especially in case of SCG) and size reduction of the waste materials by milling and grinding. Brazinha et al. (2015) recommended mandatory drying of the raw material (at around 40°C), in order to increase the stability and storage time of the spent coffee. Also, several studies reported the application of defatting with organic solvents by Soxhlet extraction (Acevedo et al., 2013; Bravo et al., 2012; Panusa et al., 2013), prior to polar solvent extraction of hydrophilic bioactives. The defatting step was shown to increase the antioxidant capacity of spent coffee extracts probably due to the prevention of fat rancidity and radical formation during long extract storage (Bravo et al., 2013).

Because the studies performed so far were mainly conducted for analytical purposes (with the final aim of characterization of polyphenolic yield), pretreatment steps for macromolecules separation have been rarely performed. Only one study (Murthy and Naidu, 2012) evaluated the efficiency of steam- and enzymatic treatment prior to extraction of coffee processing by-products to evaluate the recovery efficiency of chlorogenic acids (CGA) (this is further discussed in the following section). For the purpose of extracting and obtaining highly valuable polyphenolic antioxidants from coffee processing by-products, the main employed techniques are based on solid–liquid solvent extraction using a range of solvents and maceration or heat/mixing procedures. Solvent extraction is very convenient, as the solvent provides a physical carrier to transfer molecules between different phases (i.e., solid, liquid, and vapor). Polyphenolic compounds are easily solubilized in polar protic solvents like hydroalcoholic mixtures, resulting to different fractions obtained by varying alcohol concentration based on polarity (Prado et al., 2005). A literature overview of different solvent-extraction procedures from coffee processing by-products (coffee pulp, husk, silver skin, and SCG) is displayed in Table 4.1, providing the employed pretreatment, extraction solvent, main extraction parameters and obtained contents of total polyphenolic compounds. For comparison of the extraction efficiency of polyphenolic antioxidants the parameter of total polyphenol content (TPC), most often determined by the Folin–Ciocalteu reagent is used as the preliminary indicator of the

Table 4.1 Overview of the Employed Conventional Extraction Techniques for Recovery of Polyphenolic Compounds from Coffee By-Products

Coffee By-Product	Pretreatment	Extraction Solvent	Extraction Parameters	Polyphenolic Yield	References
Spent coffee grounds (SCG)					
Arabica, Robusta, commercial mixture	Sterilization (autoclave), 134°C, 40 min	Water 100%	Mixing (600 rpm) at room temperature, 60 min	14.97, 22.56 mg GAE/g, 11.04	Sousa et al. (2015)
Filter, espresso, plunger, mocha	Oven-drying to constant weight, 2 h, 102°C; defatting: Soxhlet extraction, petroleum ether, 3 h, 60°C	Water 100%	Filter coffeemaker extraction: 6 min, 90°C	>95%, ~70%, ~40%, ~5% of original coffee brews TPC	Bravo et al. (2012)
40% Arabica + 60% robusta, 70% Arabica + 30% robusta, regular arabica coffee capsules, decaffeinated arabica coffee capsules	Drying (forced-air food dehydrator), 40°C, 12–15 h	Water 100%	60°C, 30 min constant stirring	19.62, 17.43 mg GAE/g, 7.43, 6.33	Panusa et al. (2013)
Espresso SCG	Oven-drying, 45°C	Ethanol/water (50:50, v/v)	Orbital shaker at 300 rpm, 5 h	13.0 ± 5.6 mg GAE/g	Páscoa et al. (2013)
Not specified	Drying, 35°C, grinding in a coffee grinder, sieving <2 mm; Soxhlet defatting, hexane	Ethanol/water (50:50, v/v)	Room temperature, mortar and pestle, filtering	273.34 mg GAE/g dm, *255.61 mg GAE/g dm	Acevedo et al. (2013)
Blends of arabica and robusta varieties	Sterilization, 121°C, 20 min, powdering, extraction by Soxhlet	Isopropanol/water, 60/40 (v/v)	Extraction in a glass column, 27°C, evaporation in a rotary evaporator	10.2 mg GAE/g	Murthy and Naidu (2012)

(Continued)

Table 4.1 Overview of the Employed Conventional Extraction Techniques for Recovery of Polyphenolic Compounds from Coffee By-Products (*cont.*)

Coffee By-Product	Pretreatment	Extraction Solvent	Extraction Parameters	Polyphenolic Yield	References
40% Arabica + 60% Robusta, 70% Robusta, Arabica + 30% Robusta, regular Arabica coffee capsules, decaffeinated Arabica coffee capsules	Drying (forced-air food dehydrator), 40°C, 12–15 h	Ethanol/water (50:50, v/v)	Three-stage extraction, 60°C, 30 min constant stirring	35.52, 31.92 mg GAE/g, 17.45, 17.07	Panusa et al. (2013)
40% Arabica + 60% robusta, 70% arabica + 30% robusta, regular arabica coffee capsules, decaffeinated arabica coffee capsules	Drying (forced-air food dehydrator), 40°C, 12–15 h	Ethanol/water (60:40, v/v)	Three stage extraction, 60°C, 30 min constant stirring	28.26, 23.90 mg GAE/g, 12.58, 11.83	Panusa et al. (2013)
Arabica filter coffee	Freeze drying	Ethanol/water (60:40, v/v)	Shaking on an orbital shaker for 30 min (100 oscillation) at room temperature	19.2 mg/100 g dry matter	Jimenez-Zamora et al. (2015)
Espresso coffee capsules	Partial drying (air-oven)	100% ethanol, 80% ethanol, 60% ethanol, 40% ethanol, 20% ethanol, 0% ethanol (water)	Batch extraction, temperature 20°C–60°C Extraction duration 30–180 min, liquid to solid ratio 10–50 mL/g	7.96–15.40 mg GAE/g	Zuorro and Lavecchia (2011)
Arabica, Robusta	Extraction with distilled water in a column, 92°C, 6 h; Drying (hot air oven), 60°C, 3 h, defatted by Soxhlet, hexane	Methanol	Extraction duration: 8 h	63.2 mg/g GAE, 48.1 mg/g GAE	Ramalakshmi et al. (2009)

Not specified	Drying, 60°C until 5% of moisture	0%–100% methanol	Batch extraction by magnetic stirring, 60–65°C, extraction duration 30–90 min, liquid to solid ratio 10–30 mL/g	6.0–18.2 mg GAE/g	Mussatto et al. (2011)
Guatemala arabica filter coffee SCG	Drying (oven), 102°C, 2 h; Defatting (Soxhlet extraction), petroleum ether, 3 h, 60°C	Distilled water	Solvent extraction at 80°C for 10 min, filter coffeemaker extraction for 6 min at 90°C, Soxhlet extraction for 1 h, Soxhlet extraction for 3 h	17.44, 13.94 mg GAE/g DM, 10.20, 13.58	Bravo et al. (2013)
Guatemala arabica filter coffee SCG	Drying (oven), 102°C, 2 h; Defatting (Soxhlet extraction), petroleum ether, 3 h, 60°C	pH 4.5, pH 7.0, pH 9.5	Filter coffeemaker extraction: 6 min, 90°C	8.17, 13.94 mg GAE/g DM, 19.01	Bravo et al. (2013)
Guatemala arabica filter coffee SCG	Drying (oven), 102°C, 2 h; Defatting (Soxhlet extraction), petroleum ether, 3 h, 60°C	Ethanol, 0%–100%; methanol, 0%–100%	Filter coffeemaker extraction: 6 min, 90°C	2.65–17.48 mg GAE/g DM, 7.37–16.03 mg GAE/g DM	Bravo et al. (2013)
Cherry husk					
Blends of Arabica and Robusta varieties	Sterilization, 121°C, 20 min, powdering, extraction by Soxhlet	Isopropanol/water, 60/40 (v/v)	Extraction in a glass column, 27°C, evaporation in a rotary evaporator	12.2 mg GAE/g	Murthy and Naidu (2012)
Coffee pulp					
Arabica	Drying, 60°C until the moisture content ~13%	Ethanol, methanol, distilled water	Maceration for 3 days at room temperature	13.21, 9.09, 20.02	Jaisan et al. (2015)
Blends of Arabica and Robusta varieties	Sterilization, 121°C, 20 min, powdering, extraction by Soxhlet	Isopropanol/water, 60/40 (v/v)	Extraction in a glass column, 27°C, evaporation in a rotary evaporator	14.8 mg GAE/g	Murthy and Naidu (2012)

(Continued)

Table 4.1 Overview of the Employed Conventional Extraction Techniques for Recovery of Polyphenolic Compounds from Coffee By-Products (*cont.*)

Coffee By-Product	Pretreatment	Extraction Solvent	Extraction Parameters	Polyphenolic Yield	References
Coffee silverskin					
Commercial coffee blend: ~40% of Arabica and ~60% of robusta coffee beans	Roasting and grinding	100%, 75%, 50%, 25% ethanol, 100% water	Different temperatures (25, 30, 40, 50, and 60°C); 30, 60, 90, and 180 min; extractions performed on a heating plate with constant stirring (600 rpm)	4–6.5, 10.5–14.5, 9.5–16.5 mg GAE/g, 9–14, 7–12	Costa et al. (2014)
Not specified	*Not specified*	Ethanol, 20%–90%	60°C, solid/liquid ratio (1/10 to 1/40 g CS/mL solvent); extraction time 30–90 min	5.26–13.53 mg/g	Ballesteros et al. (2011)
Blends of Arabica and Robusta varieties	Sterilization, 121°C, 20 min, powdering, extraction by Soxhlet	Isopropanol/water, 60/40 (v/v)	Extraction in a glass column, 27°C; evaporation in a rotary evaporator	13.2 mg GAE/g	Murthy and Naidu (2012)
Arabica coffee	Collecting by aspiration	Ethanol/water (60:40, v/v)	Shaking on an orbital shaker for 30 min (100 oscillation) at room temperature	17.3 mg/100 g dry matter	Jimenez-Zamora et al. (2015)
Whole coffee fruit					
Arabica, Robusta	Grinding, sieving (0.5 mm)	Ethanol/water (50:50, v/v)	Vortexinng for 10 min in a tube, repetition 5-fold	33–50 16–84 mg/g	Mullen et al. (2013)

substrate's bioactive quality. As seen in Table 4.1, the employed solvents for extraction of polyphenolic compounds consist of pure distilled water, ethanol, methanol, or their hydroalcoholic mixtures in different ratios, as well as a mixture of isopropanol and water (Murthy and Naidu, 2012).

The best extraction efficiency of total polyphenols is normally achieved by using an equimolar (50:50, v/v) mixture of ethanol and water as the solvent. Generally using aqueous ethanol mixtures as the extraction solvents enable to provide the highest polyphenolic yield, up to threefold higher than using ispopropanol/water mixture. Increasing ethanol concentration in the extraction solvent augments the polyphenolic yield, however, only up to a critical concentration of ethanol (of 60%) in the hydroalcoholic mixture. Further increases of ethanol content in the extraction solvent results in a lower polyphenolic yield, while pure ethanol as the extraction solvent provides the lowest polyphenols content, markedly lower than using pure distilled water. According to the results of Murthy and Naidu (2012) methanol is the least effective in extraction of polyphenolic compounds when compared to water and ethanol, although other studies using methanol and its aqueous mixtures as solvents (Bravo et al., 2013; Mussatto et al., 2011), determined no significant differences in the polyphenolic yield when compared to ethanol and hydroethanolic mixtures as extraction solvents. Only the study conducted by Ramalakshmi et al. (2009) revealed much higher polyphenolic yield (63.2 mg/g GAE) obtained by extracting SCG with methanol, however this may be attributed to an extremely long extraction duration (8 h) and no provided details on the extraction procedure. Bravo et al. (2013) also determined the effect of using acidic (pH = 4.5) or alkaline (pH = 9.5) conditions for extraction of SCG and revealed that alkaline conditions (pH = 9.5) enable the highest polyphenolic yield, even higher than using hydroethanolic or hydromethanolic mixtures or plain (neutral) distilled water. Such alkaline conditions can be achieved using sodium bicarbonate and sodium carbonate, however, their further purification and removal from the extract is then required in future recovery stages.

Generally, alcohols (polar protic mediums) as the solvents are characterized by the presence of a hydroxyl group and an aliphatic part within their molecule, thus allowing the solubilization of natural products with intermediate polarity and also reducing polyphenoloxidase activity (Abad-Garcia et al., 2007). Methanol contains a smaller and more flexible aliphatic fragment compared to other alcohols and thus it surrounds polyphenols with substituted carbons inside their aromatic ring more easily (e.g., ferulic, vanillic, syringic, and synaptic acids that contain three or four substituted carbons). However, the major disadvantage of methanol is that it is toxic and should be completely removed from the extract. Bigger phenolics (e.g., oleuropein and rosmarinic acid) as well as acids with two antidiametric substituted carbons (e.g., p-hydroxybenzoic and p-hydroxyphenyl acetic) or phenolics similar to the ones deriving from coffee processing by-products are preferably solubilized in ethanol. This medium could better surround the above compounds since it can "cover" the gaps between the hydrogen bonds (Galanakis, 2013). This is the reason for a better extraction efficiency of polyphenolics from coffee by-products using ethanol rather than methanol. Moreover, the best extraction efficiency of polyphenolics using hydroethanolic mixtures as solvents is that water swells the plant material and allows the solvent to increase extractability by penetrating solid matrices more easily. Ethanol may be able to penetrate the capillary porous structure of tissues and dissociate molecular clusters or complexes with a subsequent precipitation of macromolecules in the alcohol insoluble residue and solubilization of micromolecules in the ethanolic extract. The only disadvantage of hydroethanolic mixtures is that their efficacy is extended to other extractable compounds (e.g., sugars, organic acids) and thus additional purification steps are required (Galanakis, 2015). In summary, by comparing the extraction efficiencies of different solvents in the SE processes of coffee by-products, hydroethanolic mixtures are preferred because pure ethanol also has the advantage of being potable, cheap (taxes are removed in case of industrial applications), recyclable, and corresponding extracts could be applied directly in the beverage industry.

Because the effectiveness of extraction procedures relies not only on the employed solvents, but also likely on other extraction parameters, such as the particle size, solid to liquid ratio, employed temperature, mixing/agitation speed and extraction time, these parameters must also be taken into account. Based on the data overview displayed in Table 4.1, most of the conducted solvent-extraction procedures were based on stirring or percolation during a certain period or in a defined number of repetitions. Only the study of Bravo et al. (2013) compared the polyphenolic yield obtained by simple solvent extraction, filter coffeemaker, and Soxhlet extraction and found the simple solvent extraction by stirring to be the most effective, while Soxhlet extraction provided very low yields. Extraction optimization studies were conducted on SCG by Zuorro and Lavecchia (2011) and Mussatto et al. (2011), as well as on coffee silverskin by Ballesteros et al. (2011) and Costa et al. (2014). All research groups studied the extraction of polyphenolics as a function of either extraction temperature, extraction time or solid-to-liquid ratio and found the optimal extraction conditions to be the application of a lower temperature (40° and 60°C). Most other studies evaluating the polyphenolic compounds of coffee processing by-products were conducted either at ambient or defined temperature (27°C), while only one study conducted the extractions at 60°C (Panusa et al., 2013). By comparing those results it can be deduced that the application of a higher temperature (60°C) in comparison to the ambient one, results with an augmented polyphenolic yield. Namely, an increase in the extraction temperature can promote higher analyte solubility by increasing both solubility and mass transfer rate. In addition, the viscosity and the surface tension of the solvents are decreased at higher temperature, which helps the solvents to reach the sample matrices, improving the extraction rate. With regard to extraction duration, Ballesteros et al. (2011) found no significant effect of extraction time (ranging from 30 to 90 min), while Costa et al. (2014) suggested 60 min of extraction as the optimal time. The results of other studies (Table 4.1) conducted in variable extraction durations, ranging from 30 min to several hours or even days (maceration) also revealed no marked differences in the polyphenolic yield with increase of the extraction duration.

Another important factor that influences the solvent extraction yield of polyphenols from coffee processing by-products is the continuous or sequential solvent extractions, which increases the yield, but also the consumed time and cost of the process. According to the results of Panusa et al. (2013) a three-stage extraction performed at the same temperature markedly increases the polyphenolic yield. Thus, performing sequential or multiple (repeated) extractions may also be a good approach for maximizing the polyphenolic yield from coffee processing by-products.

4.2.1.1 Chlorogenic acids

The main polyphenolic compounds of coffee beans, as well as all coffee processing by-products are CGA, formed by esterification of one molecule of quinic acid and one to three molecules of *trans*-hydroxycinnamic acids, mainly caffeic, ferulic, and *p*-coumaric (Clifford et al., 2003). Although 71 different species of CGA have been identified the vast majority of compounds belong to three classes: caffeoylquinic acids (CQA; 3CQA, 4CQA, and 5CQA), dicaffeoylquinic acids (diCQA; 3,4diCQA, 3,5diCQA, and 4,5diCQA) and feruloylquinic acids (FQA). CGAs are of great interest to scientists and food and pharmaceutical engineers as they exhibit several biological and functional properties: antimicrobial, antiviral, antimycotoxigenic, anticarcinogenic, antioxidant, and chelating (Suárez-Quiroz et al., 2014). In addition, the main CGA present in coffee are highly bioavailable, being easily absorbed and/or metabolized throughout the gastrointestinal tract (Farah et al., 2008). CGAs could be used as multifunctional natural antioxidants in food, feed, pharmaceutical, cosmetics, or nutraceutical industries.

The first report on polyphenols, and thus also CGAs recovery from coffee processing by-products dates from 1988, from a study conducted by Ramirez-Martinez (1988). In that study the authors extracted arabica coffee pulp with 80% methanol by a successive (threefold) extraction, evaporation of methanol, and then liquid–liquid extraction of the aqueous suspension with ethyl acetate or a mixture of methanol and ethyl acetate. In that way, the authors identified CGA (5-caffeoylquinic acid) to constitute 42.2% of the total identified phenolic compounds in fresh coffee pulp. The authors also identified 3,4-dicaffeoylquinic acid, (5.7%), 3,5-dicaffeoylquinic acid (19.3%), 4,5-dicaffeoylquinic acid (4.4%), and ferulic acid (1.0%). Later, Ramirez-Coronel et al. (2004) also developed a successive solvent extraction procedure of different polyphenolic compounds from arabica coffee pulp. For that purpose freeze-dried pulp was first defatted by hexane, and then extracted 3 times with methanol acidified with acetic acid, where after the insoluble residue was extracted again 3 times with acetic acid acidified aqueous acetone. All obtained solvent fractions were evaporated and freeze-dried, revealing that methanol fraction (extract) solubilized a higher content of hydroxycinnamic acids (66% of the total freeze-dried extract), while acetone solubilized a higher proanthocyanidins content (63.5% of the total freeze-dried extract).

By taking into account all coffee processing by-products, the recovery of CGAs was so far conducted in the same way as in case of total polyphenolics recovery, mainly by solvent extraction with the majority of studies conducted on SCG. Only the study of Murthy and Naidu (2012) compared the CGAs yields of all coffee processing by-products, recovered by extraction with a mixture of isopropanol and water (60:40) in a glass column at 27°C (Table 4.1). By using that extraction approach, coffee silverskin revealed to be the most abundant source of CGAs (15.82%), followed by coffee cherry husk (12.59%), SCG (11.45%), and coffee pulp as the least abundant by-product containing CGAs. However, in the stated study the authors also employed a pretreatment step of coffee by-products, using steam-treatment and enzyme treatment to check whether the CGAs recovery can be increased in that way. The coffee by-products were pretreated with steam at 121°C for 30 min or treated with viscozyme, air-dried, and used for extraction as previously mentioned. The results revealed that enzyme-treated samples gave the highest yields in comparison to control or steam-treated ones, providing even up to 23.82% of CGAs in case of coffee silverskin.

In comparison to the yield of Murthy and Naidu (2012), Mullen et al. (2013) recovered a much lower CGAs yield from Arabica and Robusta coffee husks deriving from distinct geographical origins. Those authors extracted CGAs by a successive 5-fold extraction by vortexing with 50% aqueous ethanol as the solvent and yielded a range of 0.2–2.6 mg/g (much lower than 25 mg/g obtained by Murthy and Naidu, 2012). This indicates a much higher extraction efficiency of CGAs from coffee husks by using the isopropanol/water mixture. The extraction efficiency of other solvents like ethanol and methanol was also not as effective for obtaining a high extraction yield. By using methanol, Ramalakshmi et al. (2009) obtained much higher values of total CGAs from Arabica (56.2 mg/g) and Robusta (48.7 mg/g) SCGs, however, details on the extraction procedure were not provided other than a longer extraction time than usual (8 h). Mussatto et al. (2011) conducted an optimization study of extraction of SCG by using varying concentrations of methanol and obtained a recovery of CGA of only 0.37–1.39 mg/g. Panusa et al. (2013) obtained a slightly higher content of total CGAs from SCG using 60% ethanol as the solvent than just plain distilled water, by conducting the extraction at 60°C and during 30 min. In that way, all SCG samples had the total CGAs contents ranging from 1.65 to 6.09 mg/g. Jimenez-Zamora et al. (2015) also used the same extraction solvent; 60% ethanol and extraction time (30 min) at room temperature and obtained only 0.0118 mg of total CGAs per g of dry matter. The observed discrepancy could be due to the different type of coffee, Arabica or Robusta used in the studies, but also due to other technological factors, such as roasting degree, extraction methodology.

Bravo et al. (2012) used pure distilled water for SCG extracts preparation from all four regular coffeemakers, by using the filter coffee machine as the preparation technique at 92°C during 6 min and determined a 2-fold higher total CGAs content in the extracts (in comparison to Panusa et al., 2013). According to the results of Bravo et al. (2012) the SCG extracts with the exception of those from mocha coffeemaker, had relevant amounts of total caffeoylquinic acids ranging from 11.05 (espresso) to 13.24 (filter) mg/g of Arabica spent coffee, and from 6.22 (filter) to 7.49 (espresso) mg/g of Robusta spent coffee. These amounts corresponded to 56.6%–86.6% of those caffeoylquinic acids obtained for the coffee brews. Cruz et al. (2012) also used distilled water for espresso derived SCG extract preparation, by twofold boiling of SCG for 5 min and obtained a range of 2.12 to 7.65 mg of total CGAs/g of sample.

This may indicate the fact that the extraction temperature is the determinant parameter of extraction of CGAs from coffee processing by-products, with higher extraction temperature being more beneficial and preferred for maximizing the extraction yield of total CGAs. Also, the use of distilled water, instead of methanol or organic solvent mixtures combined with extraction systems with turbulences during few minutes facilitates the contact of grounds and water, a polar solvent, and favors the CGAs extraction (Bravo et al., 2012; Ludwig et al., 2012).

4.2.2 CAFFEINE

Caffeine (1,3,7-trimethylxanthine) is an alkaloid of the xanthine group widely known due to its occurrence in extensively consumed beverages, drinks, and food and as an ingredient in pharmaceuticals and beverages (Ogita et al., 2004). Numerous studies have reported the effects of caffeine consumption in humans, such as the well-known stimulant effect of low doses of caffeine on the nervous system, which enhances concentration capacity and counteracts tiredness. High amounts of caffeine in the coffee by-products extracts should be considered in the evaluation of possible fields of application of these extracts, which could include, among others, the energy drink industry and the cosmetic sector. On the other hand, caffeine-free SCG, as are those from the manufacturing of decaffeinated soluble-coffee products, could be used to produce phenolic-rich extracts for a variety of food and nutraceutical applications (Panusa et al., 2013).

The efforts of scientific community to extract and quantify caffeine in coffee by-products has so far been focused mainly on the application of polar solvents, mostly distilled water and hydroethanolic mixtures (Table 4.2). The majority of studies performed with the purpose of obtaining and characterizing polar, hydrophilic bioactive compounds in coffee processing by-products, primarily polyphenolic antioxidants, also displayed the contents of caffeine in thus obtained extracts. By using the same solid–liquid solvent extraction approach aided by water or ethanol and mixing/stirring during a certain time period at a defined temperature, or just by repeating the extraction stage for multiple times, a substantial amount of caffeine can be recovered in polyphenolic-containing extracts. The yield of caffeine obtained in that way ranges up to 12.4 mg/g of dry sample, which constitutes roughly half of usually recovered total polyphenolic yield of the extracts. According to the results of Bravo et al. (2012) the caffeine contents in spent coffee extracts obtained in that way were two- to threefold lower than those obtained for their respective coffee brews. Taking into account that a second successive aqueous extraction of SCG extracted less than 3% of the caffeine concentration found in the first extraction, the total content of caffeine present in roasted coffee has been practically extracted with the preparation of the coffee brew and the first aqueous extraction of spent coffee (Bravo et al., 2012). The exceptions in the extraction studies performed with the aim of caffeine isolation are the subsequent steps of liquid–liquid extraction with a certain organic solvent, such as chloroform:isopropanol (3:1) in a study by Sousa et al. (2015) or chloroform in a study of Ramalakshmi et al. (2009), which recovers the caffeine and enables much higher caffeine yields than just polar solvent solid–liquid extraction.

Table 4.2 Overview of the Employed Conventional Extraction Techniques for Recovery of Caffeine From Coffee By-Products

Coffee By-Product	Pretreatment	Extraction Solvent	Extraction Parameters	Caffeine Yield	References
Spent coffee grounds (SCG)					
Espresso SCG	Oven-drying, 45°C	Ethanol/water (50:50, v/v)	Orbital shaker at 300 rpm for 5 h	0.74–12.4 mg/g, mean: 3.49 mg/g	Magalhães et al. (2016)
Arabica, robusta, commercial mixture	Sterilization (autoclave), 134°C, 40 min	Ebullition water (30 min)	Fourfold liquid/ liquid extraction with chloroform:isopropanol (3:1), drying with CaCl₂, filtering, evaporation to dryness	3.0, 11.4, 8.2 mg/g	Sousa et al. (2015)
Filter, espresso, plunger, mocha	Oven-drying to constant weight, 2 h, 102°C; defatting: Soxhlet extraction, petroleum ether, 3 h, 60°C	Water 100%	Filter coffeemaker extraction: 6 min at 90°C	5.20–7.53 mg/g, 3.59–8.09 mg/g, 3.76–5.73 mg/g n.d.	Bravo et al. (2012)
Espresso SCG	Oven-drying, 80°C until 5% of moisture	Water 100%	Twofold extraction with boiling water for 5 min under constant stirring	2.81–7.88 mg/g DW	Cruz et al. (2012)
40% Arabica + 60% robusta, 70% Arabica + 30% Robusta, regular Arabica coffee capsules, decaffeinated Arabica coffee capsules	Drying (forced-air food dehydrator), 40°C, 12–15 h	Water 100%	60°C, 30 min constant stirring	11.23, 6.00g, 0.97 mg/g n.d.	Panusa et al. (2013)

Where $CaCl_2$.

40% Arabica + 60% Robusta, 70% Robusta, Arabica + 30% Robusta, Regular Arabica coffee capsules, decaffeinated Arabica coffee capsules	Drying (forced-air food dehydrator), 40°C, 12–15 h	Ethanol/water (60:40, v/v)	Three stage extraction, 60°C, 30 min constant stirring	11.50, 5.99, 0.96 mg/gn.d.	Panusa et al. (2013)
Arabica, Robusta	Extraction with distilled water in a column, 92°C, 6 h; drying (hot air oven), 60°C, 3 h, defatted by Soxhlet, hexane	Distilled water for 45 min with magnesium oxide, filtration	Liquid/liquid extraction with chloroform, evaporation	25, 10 mg/g	Ramalakshmi et al. (2009)
Cherry husk					
Arabica, Robusta	Grinding, sieving (0.5 mm)	Ethanol/water (50:50, v/v)	Vortexing for 10 min in a tube, repetition fivefold	n.d.–2.2, n.d.–1.0 mg/g	Mullen et al. (2013)
Whole coffee fruit					
Arabica, Robusta	Grinding, sieving (0.5 mm)	Ethanol/water (50:50, v/v)	Vortexing for 10 min in a tube, repetition fivefold	1.3–5.2, 4.9–8.2 mg/g	Mullen et al. (2013)

Usually caffeine is extracted by using organic solvents, such as acetone, methanol, ethanol, and acetonitrile or their aqueous suspensions as extraction solvents, not just from coffee but also tea and other plant substrates (Wang and Helliwell, 2000). However, even though effective decaffeination can be achieved using organic solvents, the use of organic solvents is not appropriate because residual organic solvents have potential adverse effects on human health. Therefore, nontoxic and effective decaffeination alternatives, such as supercritical fluid extraction using carbon dioxide as a solvent, have been extensively explored in recent years (Gokulakrishnan et al., 2005; Liang et al., 2007). The use of that extraction technique for the recovery of hydrophilic bioactive compounds, including caffeine, is described in the following paragraph.

4.2.2.1 Application of other conventional extraction techniques for the extraction of polyphenols and caffeine from coffee processing by-products

As revealed by the stated overviews of polyphenols and caffeine extraction from coffee processing by-products, those compounds are commonly recovered by solid–liquid extraction methods using organic solvents, which are often harmful to human health and the environment. Additionally, under severe processing and extraction conditions they can lead to thermal degradation of the targeted compounds (particularly when steam distillation is involved). Despite the high extraction yield of these extraction processes, their selectivity is often low and subsequent purification of the targeted compounds may reach very high economic values (Reverchon et al., 2000). Supercritical CO_2 extraction is among the environmentally friendly technologies for the processing of food and pharmaceutical products. Alkaloids (Santana et al., 2006) and phenolics (Okuno et al., 2002) have been extracted from diverse plant materials using supercritical CO_2. However, this technique strongly depends on the solubility of low volatile substances in supercritical fluids, usually CO_2, a nonpolar solvent, with low affinity for polar substances. So, the solubility of substances in supercritical CO_2 decreases with the increase in the number of polar functional groups (e.g., hydroxyl, carboxyl, amino, and nitro). Small additions of polar cosolvents are usually employed to increase the solubility of polar and high molecular weight substances, despite a possible decrease in selectivity (Brunner, 1994). Two major effects are associated with the addition of a cosolvent:

1. Contribution to the enhancement of physical interactions between solute and solvent molecules, which, depending on the nature of the solute, can lead to chemical interactions, such as hydrogen bonding, and a consequent increase of the overall solubility.
2. Higher critical temperature of the mixed solvent when compared to pure solvent. In the vicinity of the critical point, the isothermal compressibility assumes high values, which leads to the clustering of solvent molecules around the solute molecule and thereby enhancing the solubility (Brunner, 1994).

Supercriticalfluid extraction (SFE) extraction aimed for production of hydrophilic bioactive compounds from coffee processing by-products was so far conducted only on coffee husks (Andrade et al., 2012; Tello et al., 2011) and SCGs (Andrade et al., 2012). In a study by Andrade et al. (2012) optimization of SFE extraction was conducted as a function of temperature (40°–60°C) and pressure (100–300 bar), at a constant flow rate of 11 g/min and during 4.30 h extraction for the coffee husks, and 2.30 h extraction for SCG. As the pretreatments the SCG were dried at 45°C for 5 h up to a final content of 14% of moisture, while coffee husks contained 13.04% of moisture and were ground for further extractions. For comparison of the oil yield and extraction efficiency of SFE, the authors conducted Soxhlet extraction using four different polar solvents (hexane, dichloromethane, ethyl acetate, and ethanol), according to the A.O.A.C. 920.39C method, for 6 h at the boiling temperature of the solvent used (AOAC, 1990). The obtained results revealed that the best extraction yields were obtained by the Soxhlet extraction using ethanol as solvent, for both materials studied (coffee husk: 2.7%–4.8% and SCG: 10.8%–15.0%). SFE extraction provided much lower extraction yields, ranging from 0.55% to 1.97% for coffee husk and 0.43% to 10.5% for SCG. The obtained results and the higher efficiency of Soxhlet extraction may be explained by the operating temperature of the recycle solvent and the interactions between solvent and plant matrix, which contribute to increasing the solubility of compounds of different types thereby raising the extraction yield (Benelli et al., 2010; Michielin et al., 2011). In case of SFE, at constant temperature the yield increases with the pressure enhancement, which is justified by the increase in the supercritical solvent density with pressure. However, at constant pressure, increasing the temperature decreases the yield due to the fact that an increase in temperature reduces the solubility of the solute caused by reduction in the density of the supercritical solvent (Mezzomo et al., 2009). The addition of ethanol as the cosolvent in the extraction process with supercritical CO_2 increased the values of extraction yields, for both coffee husk and SCG, in all concentrations and conditions tested.

The extracts of coffee husk exhibited a TP content of 65–151 mg CAE/g of extract obtained by the Soxhlet extraction and 16.1–36 mg CAE/g of extract obtained by the SFE extraction, respectively. However, despite the higher TPC of Soxhlet extracts, surprisingly, the SFE extracts contained markedly higher contents of CGA (943 µg/g extract) and caffeine (684 µg/mg extract) than in comparison to Soxhlet extraction. In case of SCG, the TPC of extracts obtained by Soxhlet (119–183 mg CAE/g of extract) and SFE (30.9–57 mg CAE/g of extract) extraction were higher than for the coffee husk, however, the amounts of CGA and caffeine were much lower ranging to 41.3 µg/g of extract and 41.3 µg/mg of extract. The results presented in the stated study revealed that supercritical CO_2 extraction of coffee husks is an appropriate technique for obtaining bioactive extracts containing high CGAs and caffeine contents.

In a study by Tello et al. (2011) the feasibility of caffeine extraction from robusta coffee husks using supercritical carbon dioxide was examined, as a means of using this residue to obtain high-value caffeine that may be subsequently used in the food, pharmaceutical, veterinary, or cosmetic industries. To evaluate the influence of the raw material condition on the extraction rate, grinding, and humidification to the desired percentage of humidity with ultrapure water as the pretretments were evaluated. In industrial supercritical CO_2 decaffeination processes, previous prewetting of the coffee beans is needed in order to make the process feasible (Ramalakshmi and Raghavan, 1999). Similarly, Tello et al. (2011) found that it was necessary to wet the coffee husks prior to extraction, because hardly any caffeine was removed when working with coffee husks as received. However, while prior wetting of the coffee husks was needed, milling was not required to extract the caffeine.

Caffeine solubility showed a direct relationship with the extraction conditions. By increasing the pressure (from 100 to 300 bar), the extract yield (<65%) and the caffeine yield (<84%) increased. The effect of temperature depended upon the working pressure, with retrograde behavior at pressures lower than about 200 bar. This pressure, where the solubility isotherms cross, is known as the crossover pressure (Kopcak and Mohamed, 2005). Below it, the caffeine solubility decreases as temperature increases. Above this value, an increase in temperature increases the caffeine solubility (de Azevedo et al., 2008). The use of higher flow rates and/or operational times also resulted in higher extraction rates. An increment in operational pressure, temperature, and solvent to raw material mass ratio resulted in higher extraction rates, achieving the removal of more than 80% of the initial caffeine when working under the most extreme conditions. Maximum extraction yield of caffeine of 84% was obtained when working at 99.85°C and 300 bar, using 197 kg CO_2/kg husks. Caffeine was not extracted pure, but in a mixture of fats and small quantities of pigments, which could be easily separated by simple water washing, generating two immiscible phases: a fatty one with the vast majority of the undesirable compounds, and an aqueous one in which the caffeine was selectively soluble. After a subsequent evaporation, high purity extracts of >94% of caffeine were obtained. By comparing the global production data, the initial caffeine content and global extraction yield data of other natural sources; this process could be very advantageous for its technological application.

Among other conventional extraction techniques, microwave assisted extraction (MAE) is a relatively new extraction technology, but recent literature reports indicate its increasing use for obtaining of polyphenols from some different types of natural materials and agroindustrial wastes. MAE is a process that uses microwave energy along with solvent to extract target compounds from various matrices. The volumetric heating, localized temperature, and pressure can cause selective migration of target compounds from the material to the surroundings at a more rapid rate and with similar or better recoveries compared to other solvent extractions. Moreover, the main advantages of MAE are that it can considerably reduce both extraction time and solvent consumption compared to other conventional methods (Rao and Ramalakshmi, 2015). So far only the study of Pavlović et al. (2013) evaluated the use of MAE for the extraction of natural antioxidants from SCG. For that purpose predried filter coffee SCG are extracted in batch experiments by a household microwave oven operating at 80 W, for a variable extraction time (40–360 s) and with different hydroethanolic mixtures (20%–80% ethanol). By varying the extraction parameters, the extraction yields from 11.93 to 53.80 mg/g of SCG were achieved with a TPC of 192.32 to 398.95 mg GAE/g of extract. The highest TPC of 398.95 mg GAE/g of extract dry matter was obtained by using 20% aqueous ethanol solution under just 40 s of microwave radiation (80 W), which implies that method was very effective and thereby saved time and chemicals. This study represented rapid, effective and cheap method for obtaining natural antioxidants from spent filter coffee residues.

4.2.3 **PIGMENTS**

4.2.3.1 Anthocyanins

The research and development efforts on value addition and efficient utilization of nutritionally rich coffee by-products and residues also revealed the potential of coffee pulp as a source of anthocyanins, plant pigments widely distributed in colored fruits and flowers. Anthocyanins are flavonoid compounds responsible for the red/blue coloration of many fruits and flowers (Stintzing and Casle, 2004). In the wet process, coffee pulp is removed prior to drying, while still fresh, and its color rapidly gets degraded by the action of enzymes (peroxidases and polyphenoloxidases) liberated by the damaged cells of the outer skin and pulp during the dehulling process or by other oxidizing agents, such as oxygen. Thus, fresh coffee husk, comprising of outer skin and pulp, can be exploited as potential source of anthocyanins for application as natural food colorant. Coffee anthocyanins have shown multiple biological effects resulting in effective α-glucosidase and α-amylase inhibitory activities. It was concluded that coffee skin/pulp are potential sources of colorants and bioactive ingredients to be used in formulated foods (Murthy et al., 2012).

Characterization of anthocyanins from coffee pulp was investigated by Emille and Oliveira (2007). Cyanidin 3-rutinoside was characterized as the dominant anthocyanin in fresh coffee pulp and its quantification recommended fresh coffee pulp to be a good candidate as a pigment source. In a study of Murthy et al. (2012) coffee skin/pulp was evaluated as a potential source of those compounds, with a thorough characterization of their polyphenolic profile and biological activity. For that purpose the anthocyanin pigments were extracted conventionally from coffee pulp by mixing with 0.01 HCl solutions in methanol for 18 h at cold temperature (4°C). The anthocyanin yield obtained in that way amounted to 24 mg of monomeric anthocyanins/100 g of fresh pulp. In a study involving coffee pulp derived from wet-processed fruits, Esquivel et al. (2010) identified cyanidin-3-rutinoside, cyanidin-3-glucoside and its aglycone as the major anthocyanins present before and after tissue browning. Jaisan et al. (2015) recovered a range of 1.08 to 7.02 mg cyanidin-3-glucoside/100 g of crude extract by macerating in different solvents for 3 days. According to their results, plain distilled water and methanol recovered the lowest total monomeric anthocyanins content, while ethanol was the most effective, providing a yield of anthocyanins of up to 7.02 mg. This is consistent with previous studies, which reported that ethanol has the ability to extract higher anthocyanin content than methanol (Vannini et al., 2009; Wang et al., 2011).

4.2.3.2 Melanoidins

During roasting of coffee, phenolic compounds may react with free radicals from the Maillard reaction and be incorporated in the brown colored Maillard reaction products (MRPs), named melanoidins and formed during the thermal treatment. Melanoidins are one of the major components of coffee brews, accounting for up to 25% of dry matter. They are responsible for the strong antioxidant properties and metal-chelating ability showed by coffee brews. Some authors have confirmed that this effect is due to their ability to break the radical chain by donation of hydrogen, effectivity as metal-chelating agents, capacity to reduce hydroperoxide to nonradical products, or to scavenge hydroxyl radicals (Delgado-Andrade and Morales, 2005). However, in the case of SCG low amounts of those compounds are found since studies revealed a lower browning index in aqueous extracts from soluble SCGs in comparison to that from the initial coffee brew, where they reach up to an 80%.

According to the results, Bravo et al. (2012) spent coffee extracts obtained from filter coffeemaker showed the highest browning index, similar to those of the filter coffee brews. In contrast, browned compounds in spent coffee extracts from espresso and plunger coffeemakers were three- to fivefold lower than in their respective coffee brews. Similar to the other bioactive compounds (CGAs and caffeine), aqueous soluble browned compounds were mainly extracted in coffee brews when mocha coffeemaker was used, releasing only traces in spent coffee. Therefore, the extraction of melanoidins from SCG would not present a feasible process.

4.2.4 LIPID COMPOUNDS

Although coffee pulp, husks and silverskin resulting from coffee processing are mainly composed of macronutrient compounds, such as carbohydrates (up to 35%), proteins (up to 5.2%), fibers (up to 30.8%), and minerals (up to 10.7%) (Esquivel and Jiménez, 2012), the SCG still contains a significant portion of coffee oil, varying from 11% to 20% (Al-Hamamre et al., 2012), which is retained in the spent grounds during the brewing process. The lipid composition of SCG may vary analogous to those of green coffee oil depending on the source, although generally up to 80%–90% of the oil will be glycerides, including free fatty acids, with the rest of the lipids containing terpenes, sterols, and tocopherols (Jenkins et al., 2014). SCG oils consist predominantly of linoleic, palmitic, stearic, and oleic acids, however, despite the favorable compositional profile and a naturally very high content of antioxidants it was so far not used for food purposes. The relatively low levels of saponified matter make the oil remain viscous and not congeal easily (Oliveira et al., 2007), making it very well suited for biodiesel production, however recent studies display the interest in its phenolic and diterpenic constituents and its promising potential in cosmetic industry (Campos-Vega et al., 2015).

According to the overview of extraction procedures of lipid compounds and oil from SCG conducted so far (Table 4.3), it can be seen that three main extraction approaches have been employed: simple solvent extraction, Soxhlet extraction, and supercritical fluid extraction. Previous to the extraction step, drying of SCG is

Table 4.3 Overview of the Employed Conventional Extraction Techniques for Spent Coffee Grounds Oil Recovery

Samples	Pretreatment	Extraction Solvent	Extraction Parameters	Oil/Lipid Yield	References
Solid–liquid solvent extraction					
Not specified	Drying, 35°C, grinding in a coffee grinder, sieving <2 mm	*n*-Hexane	Mixing at 250 rpm, 5 h, 35°C; evaporation and drying: 50°C	11.2 g/100 g SCG	Acevedo et al. (2013)
French press coffeemaker SCG	Drying (oven), 67°C, 24 h	Heptane	Solid–liquid solvent extraction, 180 min, by stirring, room temperature: 2-fold	c. 7–13 g/100 g SCG	Jenkins et al. (2014)
Soxhlet extraction					
SCG from a pressurized bean to cup coffee machine	Drying (oven), 102°C, 5 h	Iso-propanol ethanol, acetone, toluene chloroform, *n*-hexane, *n*-pentane	Soxhlet extraction, 15–70 min	10.52–11.43, 9.18–11.90, 12.30–12.92, 11.25–14.32 g/100 g SCG, 8.60–11.15, 11.20–15.28, 11.57–15.18	Al-Hamamre et al. (2012)
Residual coffee capsules and espresso SCG	Air drying, several days; oven-drying, 105°C	Ethanol, *n*-octane, *n*-heptane, *n*-hexane Isopropanol, Hex:IsoP 50:50, Hex: soP 60:40, Hex:IsoP 7C:30, Hex:IsoP 80:20	Soxhlet extraction, 2.5–9.5 h (until three consecutive measurements of the solvent refraction index constant)	16.0, 26.5, 18.0, 16.5, 21.0 g/100 g SCG, 22.0, 17.0, 21.0, 19.5	Caetano et al. (2012)
Not specified	Drying, 60°C, 2 days, sieving	Petroleum benzene *n*-hexane	Soxhlet extraction: *conditions not specified*	14.97–17.55 g/100 g SCG and 12.29–14.88	Akgün et al. (2014)
Not specified	Drying, 35°C, grinding, sieving < 2 mm	*n*-hexane	Extraction duration 5 h, evaporation and drying at 50°C	26.4 g/100 g SCG	Acevedo et al. (2013)

(Continued)

Table 4.3 Overview of the Employed Conventional Extraction Techniques for Spent Coffee Grounds Oil Recovery (*cont.*)

Samples	Pretreatment	Extraction Solvent	Extraction Parametres	Oil/Lipid Yield	References
Espresso SCG	Oven-drying, 79.85°C, 5 days	*n*-hexane	Soxhlet–68.85°C until the feedstock was exhausted of oil	14.0 g/100 g SCG	Calixto et al. (2011)
Automatic coffee machine	Drying to a constant weight (80°C, 24 h),	*n*-hexane	Soxhlet extraction until the solvent reflux was clear	15.1 g/100 g SCG	Obruca et al. (2014)
Indian coffee varieties SCG	Drying	Petroleum ether	Soxhlet extraction: *conditions not specified*	8.9–16.7 g/100 g SCG	Ravindranath et al. (1972)
Espresso SCG	Oven-drying, 80°C, to about 5% moisture	Petroleum ether	Soxhlet extraction for 6 h	9.5–14.5 g/100 g SCG	Cruz et al. (2012)
Supercritical fluid extraction					
Not specified	Drying, 35°C, grinding, sieving < 2 mm	Sc–CO$_2$	40°C/98 bar–80°C/379 bar, superficial velocity of 1 mm/s, total extraction time 60 min, expansion valve temperature 100°C	1.48–24.53 g/100 g SCG	Acevedo et al. (2013)
Not specified	Drying, 60°C, 2 days, sieving	Sc–CO$_2$	Temperature 33.18 to 66.81°C, pressure 11.59 to 28.40 MPa, time 19.09 to 220.90 min	10.62–11.41 g/100 g SCG	Akgün et al. (2014)
Espresso ground SCG	Oven-drying, 79.85°C, 5 days	Sc–CO$_2$, Sc–CO$_2$ + ethanol, Soxhlet$_{hexane}$	39.85–59.85°C, 150–300 bar, constant flow rate of 10 g/min, total extraction time <3 h, expansion valve temperature 100°C	4.2–15.4 19.4 g/100 g SCG 18.3	Couto et al. (2009)

Sample	Pretreatment	Extraction	Conditions	Yield	Reference
Espresso SCG	Oven-drying, 105°C, 8 h	Soxhlet$_{hexane}$, Sc–CO$_2$	Soxhlet: 4 h extraction time, constant flow rate of 12 g/min, 40–55°C, 190 bar	15.0 g/100 g SCG 0.6–0.8 of Soxhlet extraction yield	de Melo et al. (2014)
Espresso SCG	Oven-drying, 105°C, 8 h	Soxhlet$_{hexane}$, Sc–CO$_2$ + ethanol	Soxhlet: 4 h extraction time, 80°C, constant flow rate of 12 g/min, 40–70°C, 140–190 bar	15.03 g/100 g SCG 1.99–11.97 g/100 g SCG	Barbosa et al. (2014)
Coffee grounds extracted with hot water for extracting flavor substances	Drying (oven), 39.85°C, pulverization until 0.7 mm	Sc–CO$_2$; modifier: water, ethanol, hexane	Temperature 40 to 60°C, pressure 200 to 300 bar, Modifier volume 0–18 mL/100 g	13.8–16.39 g/100 g SCG	Ahangari and Sargolzaei (2013)

Ultrasonic and microwave extraction

Sample	Pretreatment	Extraction	Conditions	Yield	Reference
Coffee grounds extracted with hot water for extracting flavor substances	Drying (oven), 39.85°C, pulverization until 0.7 mm	Petroleum benzene	Ultrasound (45 min), micro–200 W (10 min), micro–800 W (10 min), Soxhlet (6 h)	13.9, 12.9 g/100 g SCG, 13.7, 16.6	Ahangari and Sargolzaei (2013)
Coffee grounds extracted with hot water for extracting flavor substances	Drying (oven), 39.85°C, pulverization until 0.7 mm	n-hexane	Ultrasound (45 min), micro–200 W (10 min), micro–800 W (10 min), Soxhlet (6 h)	14.0, 13.0 g/100 g SCGm, 13.8, 16.7	Ahangari and Sargolzaei (2013)

mandatory, because the presence of bounded moisture, due to inefficient drying, in the SCG sample can reduce the solvent extraction efficiency resulting in low oil recoveries (Al-Hamamre et al., 2012). In case of oil extraction from SCG by simple extraction, stirring at a defined time and temperature is conducted using an appropriate solvent, such as n-hexane or heptane. However, low oil yields from SCG are obtained by that approach (7–13 g/100 g SCG) as evidenced by Acevedo et al. (2013) and Jenkins et al. (2014) (with no significant differences observed between the oil extraction efficiency of the solvents). Soxhlet extraction is the most often employed and conducted technique for oil extraction, supported also as a standard technique for that purpose by the AOAC (AOAC, 1990), and often employed as a reference technique while developing and using SFE (Sc-CO$_2$) for oil extraction from SCG, as evidenced by the studies of Acevedo et al. (2013) and Akgün et al. (2014). Several studies evaluated the oil extraction efficiency from SCG by Soxhlet extraction using a range of nonpolar and polar solvents (Akgün et al., 2014; Al-Hamamre et al., 2012; Caetano et al., 2012). Solvents with high boiling points, such as toluene and isopropanol consume more energy than solvents with low boiling points (e.g., hexane and pentane) during the extraction process. On the other hand, using solvents with low boiling points result in substantial solvent losses during extraction. Therefore, such solvents may be uneconomic when used for extraction (Al-Hamamre et al., 2012). The most widely used solvent for that purpose is n-hexane, although the conducted studies revealed that n-octane is by far the most efficient solvent providing a very high oil yield of up to 26.5 g of oil/100 g of SCG (Caetano et al., 2012). Polar solvents, such as ethanol and isopropanol also result with a good total oil yield of up to 16 g/100 g SCG. Commercial ethanol (99%) has been used to recover lipids from industrial SCGs containing 25.6% oil (dry weight petroleum ether extraction). Maximum oil yield (82%) was obtained at 1:7 SCG: alcohol ratio, 75°C and was not affected by extraction time (1 or 2 h) and pretreatment (milling or extrusion) (Freitas et al., 2000). However, when polar solvents are used a black gummy material is observed in the extraction flask beside the extracted oil. Such material could be proteins, carbohydrates, and other compounds produced due to complex formation between fatty acids and carbohydrate breakdown components, which are likely to cause difficulties in oil extraction (Badrie and Mellowes, 1992).

The total oil recovery can be improved by employing a mixture of polar and nonpolar solvents, such as n-hexane and isopropanol. Binary solvent mixtures in different ratios, especially in equivalent ones (50:50, v/v) can enhance the oil recovery up to even 22% (Caetano et al., 2012). However, the yield of SCG oil extracted using Soxhlet, apart from being a function of the type of solvent is also markedly affected by the duration of extraction, because increasing the extraction time provides higher oil yield as evidenced by Al-Hamamre et al. (2012) and Caetano et al. (2012). Also, generally the highest oil yield recovered from SCG (26.4%) was obtained in a study of Acevedo et al. (2013) by Soxhlet extraction using n-hexane during 5 h, however the distinction in the study was the sieving stage of starting SCG material indicating that size reduction of SCG should be a prerequisite for oil recovery.

Environmentally friendly technologies, such as SFE are increasingly being used for SCG oil extraction. In comparison to Soxhlet extraction, SFE extraction usually provides lower extraction yields (Akgün et al., 2014; Barbosa et al., 2014; Couto et al., 2009; de Melo et al., 2014) of up to c. 80%–90% of Soxhlet-recovered oil yield. In optimization experiments of oil extraction from SCG by SFE, Akgün et al. (2014) determined that pressure is the most important operative parameter that significantly influences the yield of lipids obtained from SCG, which was also confirmed by Ahangari and Sargolzaei (2013), who stated that through increasing pressure because of an increase in the fluid density and also the solvent power of CO_2, the oil extraction yield increases (Gupta and Shim 2007; Freitas et al., 2008). According to Akgün et al. (2014) it is not feasible to extract the lipids at higher temperature by SFE, but the extraction should be operated under the pressure of 25 MPa. As stated by de Azevedo et al. (2008a), the solubility behavior of lipids is independent of the temperature at pressures higher than 30 MPa. Yet, because the effect of temperature on the solvent power is not good, raising the temperature leads to decrease in the extraction yield (Ahangari and Sargolzaei, 2013). This is attributed to the decrease of the CO_2 density, which dominates over the increase of the solute vapor pressure at this certain pressure (Arai et al., 2002; Gupta and Shim, 2007). Similarly, de Azevedo et al. (2008b) also observed that an increase in extraction temperature results in a decreased amount of caffeine and lipids when pure CO_2 is used as solvent. However, the oil yield obtained by SFE can be increased by using cosolvents or modifiers, such as ethanol (Ahangari and Sargolzaei, 2013; Couto et al., 2009), because Sc-CO_2 modified with aliphatic alcohols decreases both the extraction time and the consumption of CO_2 (Ahangari and Sargolzaei, 2013; Akgün et al., 2014; de Azevedo et al., 2008b) by solubility improvement of the lipids in the solvent phase. Using modifiers, such as water, ethanol, and n-hexane causes an increase in the extraction yield in the order hexane > ethanol > water (Ahangari and Sargolzaei, 2013). It can be considered that such a result is because of the decrease in the solvent polarity, which brings about the extraction of less polar components of coffee oil (Bernardo-Gil and Lopes, 2004). In addition, a positive effect on the extraction yield through increasing the volume of the modifier is established (Ahangari and Sargolzaei, 2013; de Azevedo et al., 2008).

Based on the reviewed literature, the conducted optimization studies of SFE extraction of oil from SCG recommends the use of higher pressure and temperature conditions to achieve the highest lipid content during SFE extraction—50–55°C, 190–300 bar (Barbosa et al., 2014; Couto et al., 2009; de Melo et al., 2014) and the addition of a polar cosolvent, such as ethanol (5 wt.%) (Ahangari and Sargolzaei, 2013; Barbosa et al., 2014). Only one study (Akgün et al., 2014) recommended the use of lower temperature for SFE extraction from SCG (33.18°C/284 bar/extraction time of 220.90 min). Andrade and Ferreira (2013) estimated the manufacturing cost of US\$ 48.60 kg^{-1} for spent coffee oil obtained by supercritical technology (200 bar, 50°C, 90 min) and may reach US\$ 460 kg^{-1} depending on the processing conditions. According to the optimization of SFE extraction by de Melo et al. (2014) a preliminary economic evaluation revealed that 300 bar/50°C/30 kg CO_2 per kg of

SCG and per 1 h are the most advantageous operating conditions, independently of the unit arrangement considered.

The application of alternative extraction methods for oil recovery from SCG was conducted only in one study (Ahangari and Sargolzaei, 2013), where the authors compared the oil yield of ultrasound and MAE using petroleum benzene and *n*-hexane as the solvents. Applying petroleum benzene and *n*-hexane as the solvents caused no significant differences among the extraction yields, but different extraction methods showed significant differences. Neither ultrasound nor microwave extraction were not efficient to extract the maximum oil yield as recovered by Soxhlet, however, increasing the microwave power (200–800 W) provides an oil yield similar to the one of ultrasound assisted extraction. The benefits of the stated extraction techniques is certainly the shortening of extraction time (10 min in case of microwave extraction) and low temperatures, which aroused only as the consequence of physical changes induced during the extraction process.

4.2.4.1 Diterpenes

The unsaponifiable part of SCG oil contains the diterpenes kahweol and cafestol, which are exclusively found in coffee (Silva et al., 2012) and are known for their beneficial physiological effects (UVB skin protection, anticarcinogenic, antiinflammatory, and antioxidant activities) (Silva et al., 2014). So far, several methods have been devised to extract/prepare the diterpenes, cafestol, and kahweol from coffee oil because of their potential use and applications in pharmacological and cosmetic preparations.

In a study by Acevedo et al. (2013) to isolate diterpenes kahweol and cafestol from SCG, two different extraction systems were used: (1) direct saponification on the SCG (Dias et al., 2010) and (2) saponification in oil extracted by Soxhlet procedure, simple method, or supercritical extraction (Araújo and Sandi, 2006). Diterpenes were generally extracted by the described principles by saponification using 50% KOH solution and ethanol 95% as solvent, 21 followed by extraction with diethyl ether and cleanup with water. The method that reached the highest level of these diterpenes (2.14 mg/g SCG for kahweol and 4.67 mg/g SCG for cafestol) was direct saponification of SCG. As kahweol and cafestol are highly unstable molecules that easily form oxides, this particular extraction procedure promotes the highest diterpene recovery yields.

Saponification of oil extracted by solid–liquid extraction, Soxhlet, or supercritical extraction (Acevedo et al., 2013) provided lower diterpene yields. By means of a simple extraction method, diterpene isolation was the least efficient resulting with kahweol and cafestol values of 0.34 mg/g SCG and 0.44 mg/g SCG, respectively. This simple extraction method makes it possible to work at low temperatures, positively affecting the extracted oil and diterpene stability. However, the requirement of several cleanup steps, including filtration may generate diterpene losses during the process. In case of Soxhlet extraction, the concentrations of kahweol and cafestol were 1.64 mg/g SCG and 2.49 mg/g SCG, respectively. This process may be considered an efficient and robust method to extract lipids and diterpenes. However, this method requires long extraction times, uses generally large volumes of toxic solvents, high temperature and reflux, favoring peroxidation reactions and hydrolysis, whereby thermolabile compounds may be degraded and oxidized. In the case of supercritical CO_2 extraction diterpene yield from SCG depends on the processing conditions thus concentration of cafestol and kahweol are lower at 40°C/98 bar (0.207 mg/g SCG and 0.114 mg/g SCG, respectively) than those at 80°C/379 bar (0.828 and 0.425 mg/g SCG for cafestol and kahweol, respectively) (Acevedo et al., 2013).

Barbosa et al. (2014) evaluated the extraction efficiency of total diterpenes after SFE extraction of SCG. The content of diterpenes was determined after saponification with KOH/ethanol, using diethyl ether for the solvent extraction and employing two additional washing steps with water (Rafael et al., 2010). The contents of kahweol obtained by that way ranged from 14.63 to 47.56 mg/g oil, for cafestol from 9.21 to 32.06 mg/g oil, which was even lower than the contents recovered by the reference, Soxhlet extraction of SCG oil (29.49 mg/g for kahweol and 16.67 mg/g for cafestol). Overall, experimental data clearly denote a negative impact of adding ethanol as a cosolvent during SFE extraction on diterpenes concentration, which must be due to the higher affinity of the modified supercritical solvent to other compounds of nonditerpenic nature. The operating conditions that provide the maximization of diterpenes concentration in SCG extracts (102.90 mg/g oil) arise as the combination of a lower pressure with no cosolvent addition, specifically 140 bar, 40°C and 0 wt.% EtOH.

4.2.4.2 Sterols and tocopherols

SCG oil also contains minor lipid components, such as sterols well known for their serum cholesterol lowering effect by reducing intestinal absorption of cholesterol. Sterols constitute about 5.4% of the total lipids in Arabica coffee (Spiller, 1998). Sterol content of SCG depends on the origin and source of roasted coffee with sitosterol, stigmasterol, and campesterol as the most abundant sterols. So far, only one study describing the extraction of oil from SCG by SFE was conducted (Akgün et al., 2014). It was observed that pressures higher than 20 MPa, near-critical temperatures, and prolonged extraction times are required for the extraction of sterols and tocopherols (yielding 19.4 g/100 g), in a system without cosolvents.

4.3 FORMULATION AND DESIGN OF COFFEE PROCESSING BY-PRODUCTS BIOACTIVES

The administration of recovered bioactive compounds from coffee processing by-products requires the use of suitable formulations able to preserve the structural integrity of the molecules and increase their water solubility, bioavailability and bioactivity (Parisi et al., 2014). In order to overcome drawbacks associated with the low stability of bioactive compounds, caused by environmental factors including physical, chemical, and biological conditions (heat, light, and pH) employed in food and pharmaceutical applications, several systems have been developed over the past few years. In most experimental studies conducted so far, coffee by-products bioactive compounds, such as polyphenolic compounds, especially CGA, caffeine, and diterpens have been used in their free form, either as recovered liquid extracts (Bresciani et al., 2014; Costa et al., 2014; Cruz et al., 2012; Jimenez-Zamora et al., 2015; Magalhães et al., 2016; Mussatto et al., 2011; Panusa et al., 2013; Pavlović et al., 2013; Zuorro and Lavecchia, 2011) or as solids obtained by concentrating and drying.

After the extraction steps, for separation purposes only centrifugation and filtering procedures are mainly applied. The filtration step is usually conducted by means of common laboratory filter papers (Whatman) or sintered glass filter (Ramirez-Coronel et al., 2004). In the case of solvent extraction of hydrophilic bioactive compounds (i.e., polyphenols, CGAs, anthocyanins, caffeine) using organic solvents or their mixtures with water, after separation of the extracted solid residue, the organic solvents are removed by evaporation to enable concentration of the extract for analytical characterization (Jaisan et al., 2015; Murthy and Naidu, 2012; Palomino García et al., 2015) or the solvent is completely removed to dryness to obtain a crude extract (Andrade et al., 2012; Jaisan et al., 2015; Ramalakshmi et al., 2009; Sousa et al., 2015). In a study by Murthy and Naidu (2012), after removal of pigments and caffeine to obtain a purified CGAs isolate from coffee processing by-products, the isolate was distilled by evaporation and air-dried to obtain a crude preparation. Freeze-drying of recovered bioactive compounds was conducted mainly on obtained aqueous or hydroethanolic extracts (Bravo et al., 2012, 2013; Mullen et al., 2013; Palomino García et al., 2015; Ramirez-Coronel et al., 2004).

For purification purposes, the recent study of Brazinha et al. (2015) reported a sustainable process of producing a natural extract from SCG using membrane technology, with no organic solvents or adsorbents involved. The use of membranes has been proposed previously for the concentration of water coffee extracts (roasted coffee) in the production of instant coffee (Pan et al., 2013; Vincze and Vatai 2004). The advantage of using membrane processing is the production of fractionated extracts:

1. enriched in small molecules (particularly caffeine), which are potentially more bioavailable;
2. without large molecules with detrimental activity (toxic compounds or compounds that may react with the target compounds);
3. assuring the removal of contaminants that may be present in crude extracts, for example, heavy metals when using nanofiltration membranes (Brazinha and Crespo 2010).

Therefore, the fractionated extracts purified in caffeine, with health benefits, are expected to have enhanced overall bioactivities properties than the corresponding crude extracts. By first optimizing the extraction procedure of SCG, the authors established the optimal extraction parameters [25°C (saving energy to heat the process), a liquid to solid ratio of 4 and 30 min of extraction time] for lab-scale extraction in an oil bath with the aim of achieving extracts with a maximum polyphenolic yield. The use of citric acid as the extraction solvent was selected due to the increased caffeine solubility when adding citric acid (Moffat, 1986). The results of the SCG extraction revealed that the presence of citric acid (3 g/L) in water as the extraction solvent also significantly improves the yields and concentrations of phenolic compounds along with caffeine. After obtaining the extracts, they were decanted and prefiltered and the nanofiltration/ultrafiltration experiments were conducted using polyethersulfone (NP010) and polyamide (GE and GH) membranes operating at 5 bar and at 40°C. Upon the membrane processing to either fractionate (with nano-, ultrafiltration membranes) or concentrate (with reverse osmosis membranes) the extracts, they were spray-dried. The dry fractionated extracts showed higher concentrations of caffeine (1.7%) than the dry nonfractionated extract (slightly better when membrane GE was used), the highest percentage of phenolic compounds (2.2%), as well as the best bioactivities at lab scale. The performed economic analysis for the production of the extract revealed the final selling price of the dry extract to be 70 €/kg. Also, the economic evaluation carried out in this work shows that the process is economically viable with a reasonable payback period of 3.5 years.

Coffee oil and lipid compounds recovered from SCG are in case of Soxhlet and simple solid–liquid extraction, concentrated by evaporation or vacuum distillation (Ahangari and Sargolzaei, 2013; Akgün et al., 2014; Al-Hamamre et al., 2012; Barbosa et al., 2014; Caetano et al., 2012; de Melo et al., 2014) and then further oven-dried (Akgün et al., 2014) or dried in a desiccator (Acevedo et al., 2013). In the case of using SFE for the purpose of lipid compounds recovery, eventual drying in a desiccator with silica gel to remove coextracted water is conducted (Acevedo et al., 2013), while in the case of using a cosolvent, such as ethanol, the cosolvent residues must be removed from the SFE-recovered oil by evaporation (Barbosa et al., 2014; Couto et al., 2009).

4.3.1 ENCAPSULATION AS A FORMULATION APPROACH OF COFFEE BY-PRODUCTS BIOACTIVES

In recent years, several studies have focused on novel formulation approaches to stabilize and protect polyphenolic bioactives, caffeine, or lypophilic bioactives from degradation, increase their solubility in water in order to improve its bioavailability, to achieve a sustained release, and ultimately to target their delivery to specific locations via multiparticulate forms and colloidal carriers (Marques, 2010; Munin and Edwards-Lévy, 2011). Among the existing stabilization methods, encapsulation appears to be a promising approach to protect sensitive compounds from the external environment and degradation.

4.3.1.1 Encapsulation of hydrophilic bioactive compounds

The encapsulation of recovered coffee processing by-products bioactive compounds was up to now scarcely performed and evaluated. Only one study reported the encapsulation of polyphenolic compounds from SCG extract, by means of formulating spherical beads composed of oxidized tapioca starch as an alginate alternative. The approach used in the study was a double extraction of SCG by ethanol and water during 30 min of stirring, followed by maceration and concentration of the extract by evaporation. The carrier suspension composed of oxidized tapioca starch, alginate, and sodium caseinate (in variable proportions) was prepared in the previously obtained coffee extract, and beads produced based on ionic gelation of alginate using calcium chloride as the ionic crosslinker. Beads produced in that way were in the macrosized range (1630–2200 μm) and after drying at room temperature had mean water content of up to 13.6%. By formulating the beads in this way, the authors were able to entrap 0.83%–1.28% of total polyphenols on dry weight basis of the extract, corresponding to 33.18% loading capacity of total polyphenolics (Palupi and Praptiningsih, 2016). Similarly, Lozano-Vazquez et al. (2015) prepared calcium crosslinked macrosized beads composed of alginate and tapioca starch entrapping CGA and revealed a very high encapsulation efficiency of CGA (<98%) in such delivery systems. The authors revealed that CGA enabled to retain its antioxidant capacity during storage at ambient conditions [exposure to daily light at room temperature (~18°C) and relative humidity (66%)] for 1 month (28 days), because the stored beads exhibited 80%–88% of the initial antioxidant capacity.

The advantage of using ionotropic gelation as the encapsulation technique is the ability to produce spherical beads ranging from nano- to macrosized delivery systems, and composed of different biopolymers as wall/carrier material, which often posses functional properties (e.g., proteins) and may act synergistically in combination with the active ingredient. By using ionic gelation of natural biopolymers, numerous plant extracts recovered from botanically distinct origins were encapsulated, revealing that a variety of plant polyphenolic compounds can be entrapped in that way, thus characterizing this encapsulation approach as a nonselective entrapment method. With respect to coffee by-products bioactives, several studies revealed that both polyphenolic compounds and caffeine could be entrapped simultaneously from plant extracts using the ionotropic gelation of certain biopolymers or their combination. In a study of Belščak-Cvitanović et al. (2015a) simultaneous encapsulation of both, green tea extract polyphenolics and caffeine was achieved by using ionic gelation of alginate-protein matrices as the carrier materials. In this study, potential interactions of alginate-proteins and protein-polyphenols were utilized for the formulation of microparticles by electrostatic assisted extrusion of alginate-protein (whey proteins, bovine serum albumine, calcium caseinate, soy proteins, hemp proteins) formulations. Microsized particles were obtained ranging up to 1100 μm in size and enabling even up to 80% loading capacity of total polyphenols in comparison to initial green tea TPC, and even up to 12.59 mg of caffeine per g of beads. Apart from the macro- and microsized beads produced by ionic gelation from natural biopolymers, CGA loaded chitosan nanoparticles with sustained release property, retained antioxidant activity, and enhanced bioavailability were recently produced in a study by Nallamuthu et al. (2015). The advantage of nanosized delivery systems is that due to the greater surface area of nanoparticles per mass unit, they are expected to be more biologically active than larger-sized particles of the same chemical composition. Namely, upon ingestion, the nanoparticles entrapping active compounds adhere to the mucosa of gastrointestinal tract (GIT), which is a prerequisite before transit into the body, and then transported via circulation to different organs. Such systems could prolong the therapeutic effect of nutraceuticals at their specific target sites (McClements, 2013). In a study by Nallamuthu et al. (2015), CGA-loaded chitosan nanoparticles were produced using a simple stirring procedure by ionic crosslinking of chitosan with tripolyphosphate (TPP) and recovered by centrifugation and freeze-drying. The suspended nanoparticles obtained in that way ranged from 150 to 650 nm in size, exhibited 59% of CGA encapsulation efficiency and showed a controlled release profile and a preserved antioxidant activity under in vitro conditions. They also showed a considerable heat stability and excellent storage stability for one month at ambient condition in aqueous environment, demonstrating its usage in various types of thermally processed foods. Nanoparticles are well known to transport bioactive compounds across the mucosal barrier and therefore the synthesized chitosan-loaded CGA nanoparticles with increased bioavailability can be a suitable carrier for better delivery of CGA in food and pharmaceutical applications (Nallamuthu et al., 2015).

Beside ionic gelation, other encapsulation technologies have been employed to encapsulate bioactives deriving from coffee by-products. CGA was encapsulated by inclusion complexation using β-cyclodextrin (Zhao et al., 2010) or by yeast-cell-based microencapsulation (Shi et al., 2007). The yeast encapsulated CGA was found to be highly stable under wet and thermal stresses, with the release profiles suggesting that the yeast cells could prevent CGA from change, without significantly slowing down the release. Another obvious benefit of this technique is that no additives apart from water, yeast, and core materials are used during processing, thereby ensuring its safety in the food industries (Blanquet et al., 2005). Also, the study of Feng et al. (2016) recently reported about a formulation system consisting of cholesterol and phosphatidyl choline, which was used to prepare effective chlorogenic acid-loaded liposomes (CAL) with an improved oral bioavailability and increased antioxidant activity. The developed liposomal formulation produced regular, spherical, and multilamellar-shaped distribution nanoparticles with a mean particle diameter of 132.15 nm.

Simultaneous entrapment and encapsulation of both polyphenolic compounds and caffeine from green tea extract was also achieved by spray drying using various carrier materials as encapsulants in a study by Belščak-Cvitanović et al. (2015b). Spray drying is usually the main process used during soluble (instant) coffee production and consists of atomization of preconcentrated coffee extract (40%–50% solid content) in a cocurrent flow-drying chamber. In the scope of bioactive compounds encapsulation and formulation of nutraceutical compounds for functional food applications, spray drying has been the preferred encapsulation technique because it is an economical, flexible, continuous operation, and produces particles of good quality (Desai and Park, 2005). For encapsulation purposes, modified starch, maltodextrin, gum, or other substances are usually hydrated to be used as the wall materials. However, the high temperatures used for spray drying, most often above 160°C, may cause degradation of heat-sensitive compounds like polyphenolic compounds (Sun-Waterhouse et al., 2013). The study of Belščak-Cvitanović et al. (2015b) revealed that these drawbacks can be overcome by using a wide range of natural biopolymers as the wall materials and by conducting the spray-drying process at lower temperatures (130°C) as to enable preservation and prevent thermal degradation of bioactive compounds. In the stated study, microsized encapsulates ranging from 3 to 10 μm, with diverse color, physical, and sensory properties, were produced by varying the used encapsulant (carrier material) powders. Screening of carriers revealed that inulin and whey proteins provide the highest product yields (67.04% and 65.18%, respectively) and, accompanied with pectin, also the highest total polyphenols (67.5%–82.2%) loading capacity, while low caffeine contents (<5 mg/g) indicated the potential of obtaining low-caffeine functional ingredients. Employing alginate, carrageenan, and gums (acacia gum and xanthan) enabled the best color yield preservation and highest chlorophylls content. Reconstituted green tea microencapsulates comprising modified starch, inulin, or carageenan exhibited the lowest bitterness and astringency and the highest green tea flavor intensity as the most favorable sensory properties (Belščak-Cvitanović et al., 2015b).

Similarly as in the previously stated study, other research groups used spray drying in the largest extent for encapsulation of anthocyanin pigments. The ethanol extracts of black carrots, containing high levels of anthocyanins, have been spray-dried using maltodextrins as a carrier and coating agents (Ersus and Yurdagel, 2007). High air inlet temperatures (>160–180°C) caused greater anthocyanin losses, while the maltodextrin of 20–21 DE gave the highest anthocyanin content powder at the end of drying process (Ersus and Yurdagel, 2007). The same approach using maltodextrin or inulin as wall materials was also used to encapsulate bayberry anthocyanins (Fang and Bhandari, 2011), black currant anthocyanins (Bakowska-Barczak and Kolodziejczyk, 2011), cactus pear fruit (*Opuntia indica*) (Saénz et al., 2009), and *Opuntia stricta* fruit polyphenolic pigments (Obon et al., 2009). All of the stated studies, enabled to produce stable powders with good flowability properties, which could serve as a ready-to-use food additive for incorporation into functional foods, due to both the presence of anthocyanin pigments (colorant) and polyphenolic antioxidants deriving from the fruit extracts. This indicates that spray drying could be a potential encapsulation technique for the formulation of coffee pulp anthocyanins, which could achieve the simultaneous entrapment of other polyphenolics deriving from the coffee pulp extract, such as CGA and caffeine.

4.3.1.2 Encapsulation of lypophilic bioactive compounds

Apart from the hydrophilic bioactive compounds, spray drying is the major technique used for encapsulation of coffee lypophilic compounds, most often roasted coffee oil for cosmetic purposes or aimed for aroma compounds retention. Namely, there has been some effort to use coffee oil to improve the sensory properties of instant coffee preparations (Bhumiratana et al., 2011; Weiss et al., 2006). Since up to now, there are no studies on the encapsulation of lypophilic compounds recovered from coffee processing by-products available, spray drying of green or roasted coffee oil is the closest formulation technique, which could be used for the formulation of SCG oil and related bioactive compounds, such as diterpenes and sterols. Therefore, SCG coffee oil encapsulation could be a promising alternative to stabilize the recovered compounds and promote their controlled release.

The microencapsulation of green or roasted coffee oil was previously carried out by spray-drying atomization (Carvalho et al., 2014; Frascareli et al., 2012; Silva et al., 2014; Yu et al., 2012). In the case of spray drying, previous emulsification of the oil is required by formulating the wall and the core materials, followed by the atomization of this emulsion in a drying chamber with hot air circulation. Different emulsion types and wall materials can be used for that purpose. In a study by Nosari et al. (2015), for spray-drying encapsulation of green coffee oil, an oil-in-water emulsion was prepared containing arabic gum as the wall material, similarly as in the study of Frascareli et al. (2012). Carvalho et al. (2014) encapsulated green coffee oil by spray drying using emulsions stabilized by lecithin and chitosan through electrostatic layer-by-layer deposition technique. Emulsions formulated with coffee oils can be a delivery system by themselves, without the further spray-drying step, because they have been particularly utilized for the entrapment of lipophilic actives that can be further incorporated into aqueous-based foods or beverages (especially those that need to stay transparent, e.g., soft drinks, fortified waters, sauces, and dips) or in foods with high lipid content. Oliveira et al. (2011) formulated a hydrogel-thickened nanoemulsion with green coffee seed oil for topical delivery of vitamin A. In that study, the attainment of hydrogel-thickened nanoemulsions with vitamin A palmitate (retinyl palmitate) and green coffee seed oil by phase inversion was proposed, with nanoemulsion particle size ranging between 77 and 110 nm.

However, in the case of using spray drying for microencapsulation of flavors, most of the flavoring compounds are highly volatile and are easily lost during spray drying (Madene et al., 2006). High temperatures (90–190°C) employed during atomization cause decrease of the volatile content of final microcapsules. As an alternative, Tilahun and Chun (2016) used polyethyleneglycol (PEG) and gas-saturated solution (PGSS) process for microparticles formation encapsulating SFE obtained coffee oil. In that way microsized particles (average 78 μm) were obtained enabling up to 79.78% of oil encapsulation efficiency. Flavor profiles of particles obtained using optimal conditions showed very good preservation of flavors, indicating that microencapsulation using PGSS process could be employed for production of freely flowing powdered particles that can be used in food processing industries.

Coacervation/crosslinking technique (Gaonkar et al., 1998) was also proposed for roasted coffee oil microencapsulation but the crosslinking agent used was glutaraldehyde, which is considered toxic. As an alternative, Veiga et al. (2015) used complex coacervation for encapsulation of roasted coffee oil in enzymatically crosslinked gelatin and gum arabic microcapsules. The microcapsules crosslinked by transglutaminase had an average diameter up to 21 μm and allowed 80% to 100% of oil recovery and 33% to 68% encapsulation efficiency.

However, it is suggested that flavor nanoencapsulation provides better results than microsized capsules, because nanocapsules can promote a faster release from the encapsulated material and present higher degradability, due to its high surface area (Leimann et al., 2013; Sanguansri and Augustin, 2006). In a study by Freiberger et al. (2015) nanocapsules were obtained by the miniemulsification-solvent evaporation technique using poly(L-lactic acid) (PLLA) and poly(hydroxybutyrate-co-hydroxyvalerate) (PHBV) as encapsulant polymers. The characterization of particles suggested the formation of stable nanosized particles varying from 100 to 360 nm in size, with the major aroma constituents of coffee oil successfully encapsulated, demonstrating that miniemulsification-solvent evaporation is an appropriate technique to encapsulate roasted coffee oil, being advantageous in that it generally allows an efficient entrapment and the solvent can be readily removed from the final nanoparticles powder.

4.4 PATENTED AND POTENTIAL APPLICATIONS OF RECOVERED BIOACTIVES

As displayed in this chapter, the majority of conventional extraction processes of coffee by-products bioactives have been performed in lab-scale. The industrial recovery of bioactive compounds from coffee processing by-products is still not sufficiently present, however, due to the wide application range of the recovered bioactives, a growing number of patented extraction and formulation related techniques can be found. Some examples of patented applications for bioactive compounds from all coffee processing by-products are shown in Table 4.4.

With respect to different coffee processing by-products, all four substrates have been the subject of patented recovery and application procedures mostly for production of bioactive extracts or functional ingredients. From those ingredients, the production of coffee oil and lipid compounds has the oldest reported patent application (Bertholet et al., 1996; Gottesman, 1985), aimed for cosmetic and food industries. For the purpose of oil recovery, primarily solvent extraction and in recent years as evidenced by the patents, SFE is also used. According to the study of Acevedo et al. (2013) the oil from SCG contains high levels of linoleic and palmitic acids, which have excellent emollient properties and is an essential component for the organization and perpetuation of the skin barrier and protection (UV radiation filtering). The polyunsaturated fatty acids in the coffee residue extracted oils can crosslink and form polymeric networks with hydrophobic barrier properties that are suitable as protective coating layers (compare linseed oil, varnish, coatings). Such drying oils may be interesting for copolymerization with vinyl or for the manufacturing of alkyd resins, polyamide, and polyurethanes production. Good film-forming properties of lipids may find use in novel paint formulations. Among the lipid fraction, sterols found in recovered coffee oil have potential in biomedical, pharmaceutical, hygienic, or toxic effects when right or wrongly applied. Searching in the patent literature on stigmasterol more than 300 patents and on camposterol alone more than 200 patents are found with applications in cosmetics (dermopharmaceutical moisturizing composition), nutraceutic (antiobesitas), immunomodulating, and herbicide preparations.

The remaining patented applications suggest the production of functional extracts containing high contents of polyphenolic antioxidants and caffeine. For their extraction, acid or alkali hydrolyses or simple solid–liquid solvent extractions are performed, most often just by using water. These processes include simple and low-cost extraction techniques, aided by high temperatures (65–100°C), enzyme treatments or by using percolation to aid the extraction of hydrophilic extractables and maximize their yield. Only recent patented applications also suggest the use of emerging

Table 4.4 Overview of Patented Methodologies and Potential Applications of Bioactive Compounds Recovered from Coffee Processing By-Products

Patent Application Number	Applicant/ Company	Title	Extraction/ Treatment Steps	Products/Brand Names	Potential Applications	References
Spent coffee grounds (SCG)						
PCT/ ES2014/070062	Consejo Superior de Investigaciones Cientificas/CIAL (Madrid, Spain)	Healthy bakery products with high level of dietary Antioxidant fiber	Hot water extraction; 100°C/10 min	Antioxidant insoluble dietary fiber	Functional food ingredient	Del Castillo et al. (2014)
US 8,591,605 B2	Board of Regents of the Nevada System of Higher Education, on behalf of the University of Nevada, Reno	methods, systems, and apparatus for obtaining biofuel from coffee and fuels	Countercurrent or Soxhlet extraction with a mixture of polar and nonpolar solvents	Triglyceride and antioxidant extracts	Cosmetics, medicinal products, food products, or combustible materials	Misra et al. (2013)
TW201524573	HAIR O RIGHT INTERNAT CORP	Recycled coffee grounds extracts and their extraction method	Supercritical carbon dioxide extraction	Coffee oil and antioxidant extract	Cosmetics	Wang-Ping and Tsang-Chi (2015)
US 4,544,567	General Foods Corporation	Simultaneous coffee hydrolysis and oil extraction	Acid hydrolisis and solvent extraction in a plug flow reactor	Coffee oil	Cosmetics, food industry	Gottesman (1985)
EP 0 693 547	SOCIETE DES PRODUITS NESTLE S.A.	Antioxidant composition and process for the preparation thereof	Extraction using an organic solvent, such as hexane or methylene chloride	Antioxidant complex composition	Food industry, aroma protection, and stabilization	Bertholet et al. (1996)

(Continued)

Table 4.4 Overview of Patented Methodologies and Potential Applications of Bioactive Compounds Recovered from Coffee Processing By-Products (*cont.*)

Patent Application Number	Applicant/ Company	Title	Extraction/ Treatment Steps	Products/Brand Names	Potential Applications	References
Coffee husks						
WO2016/097450 (PCT/ ES2015/070915)	Consejo Superior de Investigaciones Cientificas/CIAL (Madrid, Spain)	Use of products from the husk of coffee for the prevention and treatment of the diseases that form metabolic syndrome and the risk factors thereof	Hot water extraction, 100°C/10 min; subcritical water extraction, 50–200°C, 103 bar, 20 min	Bioactive coffee husk extract: chlorogenic acid and caffeine	Nutrition and health	Del Castillo Bilbao et al. (2016)
WO2006101807	Kraft Foods Holdings, Inc.	A beverage formulation and method of making such beverage that is derived from extract from coffee cherry husks and coffee cherry pulp	Water extraction at 65–80°C, enzymatic treatment or ultrafiltration	Functional food extract: caffeine	Functional food ingredient	Velissariou et al. (2006)
Coffee silverskin						
WO 2013/004873	Consejo Superior, de Investigaciones, Cientificas/CIAL (Madrid, Spain)	Application of products from coffee silverskin: antiaging cosmetics and functional food	Conventional hot water extraction (100°C), subcritical water extraction	Bioactive silverskin extract: chlorogenic acid and caffeine	Cosmetics, nutrition, and health	Del Castillo et al. (2014)

Coffee cherry pulp

WO 2014083032	Pectcof B.V.	Pectin extract on from coffee pulp	Acid or alkali hydrolisis, enzymatic treatment	Polyphenol functionalized coffee pectin extract	Food or pharmaceutical industry	Otalora and Belalcazar (2014)
WO 2004098320	Vdf Futureceuticals	Methods for coffee cherry products	Water extraction at boiling temperature	Nutrient (bioactive) extract: polyphenol, caffeine, polysaccharide	Functional food ingredient	Miljkovic et al. (2004)
WO2006101807	Kraft Foods Holdings, Inc.	A beverage formulation and method of making it that is derived from the extract of coffee cherry husks and coffee cherry pulp	Water extraction at 65–80°C, enzymatic treatment or ultrafiltration	Functional food extract: caffeine	Functional food ingredient	Velissariou et al. (2006)
EP2772133	Sustainable Agro Solutions, SA	Coffee extract and its agrochemical use against plant pathogens	Percolation with lower alcohols, reextraction by acetone/water	Antifungal activity extract: protocatehuic acid	Agrochemical industry	Gil et al. (2014)

extraction technologies as an alternative in the extraction step of coffee by-products bioactives (Del Castillo et al., 2014; Del Castillo Bilbao et al., 2016). In the majority of patents, the final recovery stages after the extraction involve the product formulation by concentrating (ultrafiltration), freeze-drying or spray-drying and its further implementation into a certain product. One of the main applications of thus recovered bioactives, primarily polyphenols and caffeine is their direct use or addition in production of enriched, functional food products, as evidenced by a patent reported by Velissariou et al. (2006), who formulated an extract-based beverage derived from coffee cherry husk or cherry pulp. Another patented food application of SCG extract was reported by Del Castillo et al. (2014), where the aqueous antioxidant-rich SCG extract was used in the production of bakery products. Although industrial commercialization of recovered bioactive compounds from coffee by-products is still not viable, the wide application range in case of food products justifies the recovery of hydrophilic bioactive compounds from coffee processing by-products, since previous studies revealed the potential of implementing the biologically potent polyphenolic antioxidants, as well as caffeine, in diverse food products.

4.5 CONCLUSIONS

The recovery of bioactive compounds, especially hydrophilic ones, such as polyphenolic compounds and in particular CGAs, caffeine and anthocyanin pigments has, despite a comprehensive lab-scale characterization, not yet being industrially exploited. By reviewing the available scientific studies, in case of using conventional extraction techniques for the recovery of bioactive compounds, the following operational parameters for maximization of their yield can be recommended:

- *Total polyphenols*: prior drying and size reduction of raw material, hydroethanolic solvent (50%–60% of ethanol), moderate extraction temperature (60°C), 30–60 min of duration and preferably sequential (repeated) extractions,
- *Chlorogenic acids*: application of organic solvents or acidified water, higher extraction temperatures, enzyme pretreatment,
- *Caffeine*: liquid–liquid extraction following solid–liquid extraction, application of SFE, membrane processes (ultrafiltration) for concentration and purification.

Soxhlet extraction using ethanol as a solvent can be employed to provide polyphenolic antioxidants and caffeine, and in comparison to other conventional techniques it provides high recovery yields, however SFE and MAE also present promising extraction approaches to recover those compounds from any coffee processing by-product. In case of SFE, higher operating pressure (300 bar) and the addition of a polar cosolvent, such as ethanol, will increase the yield, while in case of MAE, higher microwave radiation strength during shorter exposure time will have an equal effect.

In case of oil and lipid compounds recovery from SCG, the highest oil yields are achieved by using Soxhlet extraction with *n*-hexane or a mixture of polar and nonpolar solvent (such as *n*-hexane and isopropanol), by prolonging the extraction time and grinding the sample prior to extraction. SFE has become increasingly used for recovery of oil from SCG, ensuring 80%–90% of oil yield recovered by Soxhlet extraction. The conducted optimization studies of SFE extraction of oil from SCG recommend the use of higher pressure and temperature conditions to achieve the highest lipid content; 50–55°C, 190–300 bar and the addition of a polar cosolvent, such as ethanol (5 wt.%). For the recovery of diterpenes from SCG, lower temperature and pressure (140 bar, 40°C) and no cosolvents are preferred followed by saponification of the recovered oil, while additionally in case of sterols and tocopherols prolonged extraction is required. In order to achieve the economic feasibility of the bioactive compounds recovery, a promising route would be a combined extraction primarily of the lipid fraction, including the bioactive unsaponifiable fraction, followed by the extraction of hydrophilic fraction by an appropriate solvent-extraction technique (as suggested in Fig. 4.1).

Following the extraction, the recovered bioactive compounds are available as liquid extracts (hydrophilic compounds) or are further concentrated and dried to obtain solid preparations by freeze-drying or preferably encapsulation. Encapsulation techniques, such as hydrogel formation by ionic or covalent crosslinking of natural biopolymers enable simultaneous entrapment of recovered hydrophilic bioactives and provide morphologically diverse encapsulates (nano- to macrosized). Spray drying is the most often employed technique for the encapsulation-formulation of coffee oil and could potentially

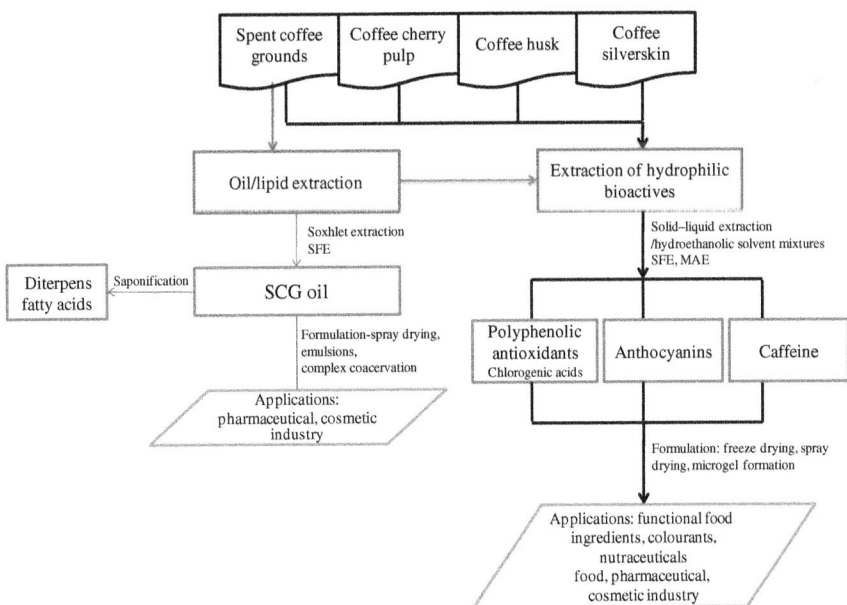

FIGURE 4.1 Schematic Display of Recovery Approaches and Resulting Potential Applications of Coffee Processing By-Products Bioactives

used for encapsulation of coffee cherry pulp anthocyanins when used with appropriate carrier materials. Nanoencapsulation using microemulsification or liposome entrapment can improve the functionality of formulated delivery systems, and facilitate their potential food implementations. Although commercial applications of the recovered bioactive compounds from coffee processing by-products are still not available, patents and solutions for their extraction and implementation in various foods, cosmetics, medicinals, and agricultural applications reveal their promising future potential, for technical use as antioxidants, or as functional food additives.

REFERENCES

Abad-Garcia, B., Barrueta, L.A., Lopez-Marquez, D.M., Crespo-Ferrer, I., Gallo, B., Vicente, F., 2007. Optimization and validation of a methodology based on solvent extraction and liquid chromatography for the simultaneous determination of several polyphenolic families in fruit juices. J. Chromatogr. A 1154, 87–96.

Acevedo, F., Rubilar, M., Scheuermann, E., Cancino, B., Uquiche, E., Garcés, M., Inostroza, K., Shene, C., 2013. Spent coffee grounds as a renewable source of bioactive compounds. J. Biobased Mater. Bioenerg. 7, 420–428.

Acosta, E., 2009. Bioavailability of nanoparticles in nutrient and nutraceutical delivery. Curr. Opin. Colloid. Interf. Sci. 14, 3–15.

Ahangari, B., Sargolzaei, J., 2013. Extraction of lipids from spent coffee grounds using organic solvents and supercritical carbon dioxide. J. Food Proc. Preserv. 37, 1014–1021.

Akgün, N.A., Bulut, H., Kikic, I., Solinas, D., 2014. Extraction behavior of lipids obtained from spent coffee grounds using supercritical carbon dioxide. Chem. Eng. Technol. 37, 1975–1981.

Al-Hamamre, Z., Foerster, S., Hartmann, F., Kröger, M., Kaltschmitt, M., 2012. Oil extracted from spent coffee grounds as a renewable source for fatty acid methyl ester manufacturing. Fuel 96, 70–76.

Andrade, K.S., Ferreira, R.S., 2013. Cost manufacturing of oil extraction of spent coffee grounds obtained by supercritical technology. 16th World Congress of Food Science and Technology. Available from: http://iufost.org.br/sites/iufost.org.br/files/anaid/03671.pdf

Andrade, K.S., Goncalvez, R.T., Maraschin, M., Ribeiro-do-Valle, R.M., Martinez, J., Ferreira, S.R.S., 2012. Supercritical fluid extraction from spent coffee grounds and coffee husks: antioxidant activity and effect of operational variables on extract composition. Talanta 88, 544–552.

Association of Official Agricultural Chemists (AOAC), 1990. Official Methods of Analysis, vol. 1141, eleventh ed., AOAC, Washington, DC.

Arai, Y., Sako, T., Takebayashi, Y., 2002. Supercritical Fluids: Molecular Interactions, Physical Properties and New Applications. Springer, New York.

Araújo, J., Sandi, D., 2006. Extraction of coffee diterpenes and coffee oil using supercritical carbon dioxide. Food Chem. 101, 1087–1094.

Azmir, J., Zaidul, I.S.M., Rahman, M.M., Sharif, K.M., Mohamed, A., Sahena, F., Jahurul, M.H.A., Ghafoor, K., Norulaini, N.A.N., Omar, A.K.M., 2013. Techniques for extraction of bioactive compounds from plant materials: a review. J. Food Eng. 117, 426–436.

Badrie, N., Mellowes, W.A., 1992. Cassava starch or amylose effects on characteristics of cassava (Manihot esculenta Crantz) extrudate. J. Food Sci. 57, 103–107.

Bakowska-Barczak, A.M., Kolodziejczyk, P.P., 2011. Black currant polyphenols: their storage stability and microencapsulation. Ind. Crop Prod. 34, 1301–1309.

Ballesteros, L.F., Conde, T., Teixeira, J.A., Mussatto, S.I., 2011. Optimization of the extraction conditions for antioxidant phenolic compounds recovery from coffee silverskin. In: Book of Abstracts of the 11th International Chemical and Biological Engineering Conference. Lisabon, Portugal, pp. 223–224.

Barbosa, H.M.A., Melo, M.M.R., Coimbra, M.A., Passos, C.P., Silva, C.M., 2014. Optimization of the supercritical fluid coextraction of oil and diterpenes from spentcoffee grounds using experimental design and response surface methodology. J. Supercrit. Fluids 85, 165–172.

Belščak-Cvitanović, A., Đorđević, V., Karlović, S., Pavlović, V., Komes, D., Ježek, D., Bugarski, B., Nedović, V., 2015a. Protein-reinforced and chitosan-pectin coated alginate microparticles for delivery of flavan-3-ol antioxidants and caffeine from green tea extract. Food Hydrocoll. 51, 361–374.

Belščak-Cvitanović, A., Lević, S., Kalušević, A., Špoljarić, I., Đorđević, V., Komes, D., Mršić, G., Nedović, V., 2015b. Efficiency assessment of natural biopolymers as encapsulants of green tea (*Camellia sinensis* L.) bioactive compounds by spray drying. Food Bioproc. Technol. 8, 2444–2460.

Benelli, P., Riehl, C.A.S., Smânia, Jr., A., Smânia, E.F.A., Ferreira, S.R.S., 2010. Bioactive extracts of orange (*Citrus sinensis* L. Osbeck) pomace obtained by SFE and low pressure techniques: mathematical modeling and extract composition. J. Supercrit. Fluids 55, 132–141.

Bernardo-Gil, M.G., Lopes, L.M.C., 2004. Supercritical fluid extraction of *Cucurbita ficifolia* seed oil. Eur. Food Res. Technol. 219, 593–597.

Bertholet, R., Colarow, L., Kusy, A., Rivier, V., 1996. Antioxidant composition and process for the preparation thereof. EP 0693547B1.

Bhumiratana, N., Adhikari, K., Chambers, E., 2011. Evolution of sensory aroma attributes from coffee beans to brewed coffee. LWT Food Sci. Technol. 44, 2185–2192.

Blanquet, S., Garrait, G., Beyssac, E., Perrier, C., Denis, S., Hébrard, G., Alric, M., 2005. Effects of cryoprotectants on the viability and activity of freeze dried recombinant yeasts as novel oral drug delivery systems assessed by an artificial digestive system. Eur. J. Pharm. Biopharm 61, 32–39.

Bravo, J., Juániz, I., Monente, C., Caemmerer, B., Kroh, L.W., M. De Peña, P., Cid, C., 2012. Evaluation of spent coffee obtained from the most common coffeemakers as a source of hydrophilic bioactive compounds. J. Agr. Food Chem. 60, 12565–12573.

Bravo, J., Monente, C., Juániz, I., De Peña, M.P., Cid, C., 2013. Influence of extraction process on antioxidant capacity of spent coffee. Food Res. Int. 50, 610–616.

Brazinha, C., Crespo, J.G., 2010. Membrane processing: natural antioxidants from winemaking by-products. Filtr. Sep. 47, 32–35.

Brazinha, C., Cadima, M., Crespo, J.G., 2015. Valorisation of spent coffee through membrane processing. J. Food Eng. 149, 123–130.

Bresciani, L., Calani, L., Bruni, R., Brighenti, F., Del Rio, D., 2014. Phenolic composition, caffeine content and antioxidant capacity of coffee silverskin. Food Res. Int. 61, 196–201.

Brunner, G., 1994. Gas Extraction: An Introduction to Fundamentals of Supercritical Fluids and the Applications to Separation Processes. Steinkopff Darmstadt Springer, New York.

Caetano, N.S., Silva, V.F., Mata, T.M., 2012. Valorization of coffee grounds for biodiesel production. Chem. Eng. Trans. 26, 267–272.

Calixto, F., Fernandes, J., Couto, R., Hernández, E.J., Najdanovic-Visak, V., Simões, P.C., 2011. Synthesis of fatty acid methyl esters via direct transesterification with methanol/carbon dioxide mixtures from spent coffee grounds feedstock. Green Chem. 13, 1196–1202.

Campos-Vega, R., Loarca-Piña, G., Vergara-Castañeda, H., Oomah, B.D., 2015. Spent coffee grounds: a review on current research and future prospects. Trends Food Sci. Technol. 45, 24–36.

Carvalho, A.G.S., Silva, V.M., Hubinger, M.D., 2014. Microencapsulation by spray drying of emulsified green coffee oil with two-layered membranes. Food Res. Int. 61, 236–245.

Clifford, M.N., Johnston, K.L., Knight, S., Kuhnert, N., 2003. Hierarchical scheme for LC-MSn identification of chlorogenic acids. J. Agr. Food Chem. 51, 2900–2911.

Costa, A.S.G., Alves, R.C., Vinha, A.F., Barreira, S.V.P., Nunes, M.A., Cunha, L.M., Oliveira, M.B.P.P., 2014. Optimization of antioxidants extraction from coffee silverskin, a roasting by-product, having in view a sustainable process. Ind. Crops Prod. 53, 350–357.

Couto, R.M., Fernandes, J., da Silva, M.D.R., Simões, P.C., 2009. Supercritical fluid extraction of lipids from spent coffee grounds. J. Supercrit. Fluids 51, 159–166.

Cowan, M.M., 1999. Plant products as antimicrobial agents. Clin. Microbiol. Rev. 12, 564–582.

Cruz, R., Cardoso, M.M., Fernandes, L., Oliveira, M., Mendes, E., Baptista, P., Morais, S., Casal, S., 2012. Espresso coffee residues: a valuable source of unextracted compounds. J. Agr. Food Chem. 60, 7777–7784.

Dai, J., Mumper, R.J., 2010. Plant phenolics: extraction, analysis and their antioxidant and anticancer properties. Molecules 15, 7313–7352.

de Azevedo, A.B.A., Kieckbusch, T.G., Tashima, A.K., Mohamed, R.S., 2008. Supercritical CO_2 recovery of caffeine from green coffee oil: new experimental solubility data and modeling. Quim. Nova 31, 1319–1323.

de Azevedo, A.B.A., Kieckbush, T.G., Tashima, A.K., Mohamed, R.S., Mazzafera, P., Melo, S.A.B., 2008a. Extraction of green coffee oil using supercritical carbon dioxide. J. Supercrit. Fluids 44, 186–192.

de Azevedo, A.B.A., Mazzafera, P., Mohamed, R.S., Vieira de Melo, S.A.B., Kieckbusch, T.G., 2008b. Extraction of caffeine, chlorogenic acids and lipids from green coffee beans using supercritical carbon dioxide and co-solvents. Braz. J. Chem. Eng. 25, 543–552.

de Melo, M.M.R., Barbosa, H.M.A., Passos, C.P., Silva, C.M., 2014. Supercritical fluid extraction of spent coffee grounds: measurement ofextraction curves, oil characterization and economic analysis. J. Supercrit. Fluids 86, 150–159.

Del Castillo Bilbao, M.D., Fernández, G.B., Ullate Artiz, M., Mesa García, M.D., 2016. Use of products of the husk of coffee for the prevention and treatment of the diseases that form the metabolic syndrome and the risk factors thereof. WO 2016097450 A1.

Del Castillo, M.D., Martinez-Saez, N., Ullate, M., 2014. Healthy bakery products with high level of dietary antioxidant fibre. PCT/ES2014/07 0062.

Delgado-Andrade, C., Morales, F.J., 2005. Unraveling the contribution of melanoidins to the antioxidant activity of coffee brews. J. Agr. Food Chem. 53, 1403–1407.

Desai, K.G.H., Park, H.J., 2005. Recent developments in microencapsulation of food ingredients. Dry Technol. 23, 1361–1394.

Dias, R.C.E., Campanha, F.G., Vieira, L.G.E., Ferreira, L.P., Pot, D., Marraccini, P., Benassi, M.T., 2010. Evaluation of kahweol and cafestol in coffee tissues and roasted coffee by a new high-performance liquid chromatography methodology. J. Agr. Food Chem. 58, 88–93.

Donsi, F., Sessa, M., Ferrari, G., 2010. Nanoencapsulation of essential oils to enhance their antimicrobial activity in foods. J. Biotechnol. 150, S67.

Donsi, F., Annunziata, M., Sessa, M., Ferrari, G., 2011a. Nanoencapsulation of essential oils to enhance their antimicrobial activity in foods. LWT Food Sci. Technol. 44, 1908–1914.

Dube, A., Ng, K., Nicolazzo, J.A., Larson, I., 2010. Effective use of reducing agents and nanoparticle encapsulation in stabilizing catechins in alkaline solution. Food Chem. 122, 662–667.

Emille, R.B.A.P., Oliveira, S.L., 2007. Fresh coffee husks as potential sources of anthocyanins. LWT Food Sci. Technol. 40, 1555–1560.

Ersus, S., Yurdagel, U., 2007. Microencapsulation of anthocyanin pigments of black carrot (*Daucus carota* L.) by spray drier. J. Food Eng. 80, 805–812.

Esquivel, P., Jiménez, V.M., 2012. Functional properties of coffee and coffee by-products. Food Res. Int. 46, 488–495.

Esquivel, P., Kramer, M., Carle, R., Jiménez, V.M., 2010. Anthocyanin profiles and caffeine contents of wet-processed coffee (*Coffea arabica*) husks by HPLCDAD-MS/MS. 28th International Horticultural Congress, Book of Abstracts. Elsevier, Amsterdam, pp. 129–130.

Fang, Z., Bhandari, B., 2011. Effect of spray drying and storage on the stability of bayberry polyphenols. Food Chem. 129, 1139–1147.

Farah, A., Monteiro, M., Donangelo, C.M., Lafay, S., 2008. Chlorogenic acids from green coffee extract are highly bioavailable in humans. J. Nutr. 138, 2309–2315.

Feng, Y., Sun, C., Yuan, Y., Zhu, Y., Wan, J., Firempong, C.K., Omari-Siaw, E., Xu, Y., Pu, Z., Yu, J., Xu, X., 2016. Enhanced oral bioavailability and in vivo antioxidant activity of chlorogenic acid via liposomal formulation. Int. J. Pharm. 501, 342–349.

Frascareli, E.C., Silva, V.M., Tonon, R.V., Hubinger, M.D., 2012. Effect of process conditions on the microencapsulation of coffee oil by spray drying. Food Bioprod. Process. 90, 413–424.

Freiberger, E.B., Kaufmann, K.C., Bona, E., de Araújo, P.H.H., Sayer, C., Leimann, F.V., Gonçalves, O.H., 2015. Encapsulation of roasted coffee oil in biocompatible nanoparticles. LWT Food Sci. Technol. 64, 381–389.

Freitas, S.P., Monteiro, P.L., Lago, R.C. A., 2000. Extração do óleo da borra de café solúvel-com etanol comercial. Simpósio de Pesquisa dos Cafés do Brasil, Brasil, pp. 740–743.

Freitas, L.D.S., De Oliveira, J.V., Dariva, C., Jacques, R.A., Caramaeo, E.B., 2008. Extraction of grape seed oil using compressed carbon dioxide and propane: extraction yields and characterization of free glycerol compounds. J. Agr. Food Chem. 56, 2558–2564.

Galanakis, C.M., 2013. Emerging technologies for the production of nutraceuticals from agricultural by-products: a viewpoint of opportunities and challenges. Food Bioprod. Process 91, 575–579.

Galanakis, C.M., 2015. The universal recovery strategy. In: Galanakis, C.M. (Ed.), Food Waste Recovery: Processing Technologies and Industrial Techniques. Elsevier, Academic Press, Oxford, UK, pp. 59–81.

Gaonkar, A.G., Nicholson, V.J., Tufts, H.M., 1998. Preparation of microcapsule flavour delivery system—useful for incorporation into food product, especially low fat product, to improve flavour perception. EP815743-A2.

Gil, R.J.F., Echeverri, L.L.F., Justribo, A.F.X., 2014. Coffee extract and its agrochemical use against plant pathogens. EP 2772133 A1.

Gokulakrishnan, S., Chandraraj, K., Gummadi, S.N., 2005. Microbial and enzymatic methods for the removal of caffeine. Enzyme Microb. Technol. 37, 225–232.

Gottesman, M., 1985. Simultaneous Coffee Hydrolysis and Oil Extraction. EP 0216971A1.

Gupta, R.B., Shim, J., 2007. Solubility in Supercritical Carbon Dioxide. CRC Press, Boca Raton, USA.

Helgason, T., Awad, T.S., Kristbergsson, K., Decker, E.A., McClements, D.J., Weiss, J., 2009. Impact of surfactant properties on oxidative stability of β-carotene encapsulated within solid lipid nanoparticles. J. Agr. Food Chem. 57, 8033–8040.

Jaisan, C., Chase, S., Punbusayakul, N., 2015. Antioxidant and antimicrobial activities of various solvents extracts of arabica coffee pulp. J. Process. Energ. Agr. 19, 224–227.

Jenkins, R.W., Stageman, N., Fortune, C., Chuck, C.J., 2014. Effect of the type of bean, processing and geographical location on the biodiesel produced from waste coffee grounds. Energ. Fuels 28, 1166–1174.

Jimenez-Zamora, A., Pastoriza, S., Rufian-Henares, J.A., 2015. Revalorization of coffee by-products: prebiotic, antimicrobial and antioxidant properties. LWT Food Sci. Technol. 61, 12–18.

Kopcak, U., Mohamed, R.S., 2005. Caffeine solubility in supercritical carbon dioxide/cosolvent mixtures. J. Supercrit. Fluids 34, 209–214.

Lam, K., 2007. New aspects of natural products in drug discovery. Trends Microbiol. 15, 279–289.

Laufenberg, G., Kunz, B., Nystroem, M., 2003. Transformation of vegetable waste into value added products. Biores. Technol. 87, 167–198.

Leimann, F.V., Cardozo, L., Sayer, C., Araújo, P.H.H., 2013. Poly(3-hydroxybutyrate-co-3-hydroxyvalerate) nanoparticles prepared by a miniemulsion/solvent evaporation technique: effect of PHBV molar mass and concentration. Brazilian J. Chem. Eng. 30, 369–377.

Liang, H.L., Liang, Y.R., Dong, J.J., Lu, J.L., Zu, H.R., Wang, H., 2007. Decaffeination of fresh green tea leaf (*Camellia sinensis*) by hot water treatment. Food Chem. 101, 1451–1456.

Lozano-Vazquez, G., Lobato-Calleros, C., Escalona-Buendia, H., Chavez, G., Alvarez-Ramirez, J., Vernon-Carter, E.J., 2015. Effect of the weight ratio of alginate-modified tapioca starch on the physicochemical properties and release kinetics of chlorogenic acid containing beads. Food Hydrocoll. 48, 301–311.

Ludwig, I.A., Sanchez, L., Caemmerer, B., Kroh, L.W., De Peña, M.P., Cid, C., 2012. Extraction of coffee antioxidants: Impact of brewing time and method. Food Res. Int. 48, 57–64.

Madene, A., Jacquot, M., Scher, J., Desobry, S., 2006. Flavour encapsulation and controlled release: a review. Int. J. Food Sci. Technol. 41, 1–21.

Magalhães, L.M., Machado, S., Segundo, M.A., Lopes, J.A., Páscoa, R.N.M.J., 2016. Rapid assessment of bioactive phenolics and methylxanthines in spent coffee grounds by FT-NIR spectroscopy. Talanta 147, 460–467.

Marques, H.M.C., 2010. A review on cyclodextrin encapsulation of essential oils and volatiles. Flavour Frag. J. 25, 313–326.

McClements, D.J., 2013. Edible lipid nanoparticles: digestion, absorption, and potential toxicity. Prog. Lipid Res. 52, 409–423.

McClements, D.J., Decker, E.A., Weiss, J., 2007. Emulsion-based delivery systems for lipophilioc bioactive components. J. Food Sci. 72, 109–124.

Mezzomo, N., Martínez, J., Ferreira, S.R.S., 2009. Supercritical fluid extraction of peach (*Prunus persica*) almond oil: kinetics, mathematical modeling and scale-up. J. Supercrit. Fluids 51, 10–16.

Michielin, E.M.Z., Wiese, L.P.L., Fereira, E.A., Pedrosa, R.C., Ferreira, S.R.S., 2011. Radical-scavenging activity of extracts from *Cordia verbenacea* DC obtained by different methods. J. Supercrit. Fluids 56, 89–96.

Miljkovic, D., Duell, B., Miljkovic, V., 2004. Methods for coffee cherry products. WO 2004098320 A1.

Misra, M., Mohapatra, S.K., Kondamudi, N.V., 2013. Methods, systems, and apparatus for obtaining biofuel from coffee and fuels produced therefrom. US 8,591,605 B2.

Moffat, A.C., 1986. Clarke's isolation and identification of drugs. In: Moffatt, A.C. et al., (Ed.), Clarke's Isolation and Identification of Drugs. second ed. The Pharmaceutical Press, London, pp. 421–423.

Mullen, W., Nemzer, B., Stalmach, A., Ali, S., Combet, E., 2013. Polyphenolic and hydroxy-cinnamate contents of whole coffee fruits from China, India, and Mexico. J. Agr. Food Chem. 61, 5298–5309.

Munin, A., Edwards-Lévy, F., 2011. Encapsulation of natural polyphenolic compounds: a review. Pharmaceutics 3, 793–829.

Murthy, P.S., Naidu, M.M., 2012. Recovery of phenolic antioxidants and functional compounds from coffee industry by-products. Food Bioproc. Tech. 5, 897–903.

Murthy, P.S., Manjunatha, M.R., Sulochannama, G., Naidu, M.M., 2012. Extraction, characterization and bioactivity of coffee anthocyanins. Eur. J. Biol. Sci. 4, 13–19.

Mussatto, S.I., Ballesteros, L.F., Martins, S., Teixeira, J.A., 2011. Extraction of antioxidant phenolic compounds from spent coffee grounds. Sep. Purif. Technol. 83, 173–179.

Nallamuthu, I., Devi, A., Khanum, F., 2015. Chlorogenic acid loaded chitosan nanoparticles with sustained release property, retained antioxidant activity and enhanced bioavailability. Asian J. Pharm. Sci. 10, 203–211.

Nosari, A.B.F.L., Lima, J.F., Serra, A.O., Freitas, L.A.P., 2015. Improved green coffee oil antioxidant activity for cosmetical purpose by spray drying microencapsulation. Rev. Bras. Farmacogn. 25, 307–311.

Obon, J.M., Castellar, M.R., Alacid, M., Fernandez-lopez, J.A., 2009. Production of a red purple food colorant from *Opuntia stricta* fruits by spray drying and its application in food model systems. J. Food Eng. 90, 471–479.

Obruca, S., Petrik, S., Benesova, P., Svoboda, Z., Eremka, L., Marova, I., 2014. Utilization of oil extracted from spent coffee grounds for sustainable production of polyhydroxyalkanoates. Appl. Microbiol. Biotechnol. 98, 5883–5890.

Ogita, S., Uefuji, H., Morimoto, M., Sano, H., 2004. Application of RNAi to confirm theobromine as the major intermediate for caffeine biosynthesis in coffee plants with potential for construction of decaffeinated varieties. Plant Mol. Biol. 54, 931–941.

Okuno, S., Yoshinaga, M., Nakatani, M., Ishiguro, K., Yoshimoto, M., Morishita, T., Uehara, T., Kawano, M., 2002. Extraction of antioxidants in sweetpotato waste powder with supercritical carbon dioxide. Food Sci. Technol. Res. 8, 154.

Oliveira, L.S, Franca, A.S., Camargos, R.R.S., Ferraz, V.P., 2007. Coffee oil as a potential feedstock for biodiesel production. Biores. Technol. 99, 3244–3250.

Oliveira, J.S., Aguira, T.A., Mezadri, H., Santos, D.H.O., 2011. Hydrogel-thickened nanoemulsion with green coffee seed oil for topical delivery of vitamin A. Lat. Am. J. Pharm. 30, 1999–2003.

Otalora, A., Belalcazar, F., 2014. Pectin extraction from coffee pulp. WO 2014083032 A1.

Palomino García, L.R., Biasetto, C.R., Araujo, A.R., Del Bianchi, V.L., 2015. Enhanced extraction of phenolic compounds from coffee industry's residues through solid state fermentation by *Penicillium purpurogenum*. Food Sci. Technol. 35 (4), 704–711.

Palupi, N.W., Praptiningsih, Y., 2016. Oxidized tapioca starch as an alginate substitute for encapsulation of antioxidant from coffee residue. Agr. Agr. Sci. Procedia 9, 304–308.

Pan, B., Yan, P., Zhu, L., Li, X., 2013. Concentration of coffee extract using nanofiltration membranes. Desalination 317, 127–131.

Panusa, A., Zuorro, A., Lavecchia, R., Marrosu, G., Petrucci, R., 2013. Recovery of natural antioxidants from spent coffee grounds. J. Agr. Food Chem. 61, 4162–4168.

Parisi, O.I., Casaburi, I., Sinicropi, M.S., Avena, P., Caruso, A., Givigliano, F., Pezzi, V., Puoci, F., 2014. Most relevant polyphenols present in the Mediterranean diet and their incidence in cancer diseases. In: Watson, R.R., Preedy, V.R., Zibadi, S. (Eds.), Polyphenols in Human Health and Disease. Elsevier, UK, pp. 1341–1352.

Páscoa, R.N.M.J., Magalhaes, L.M., Lopes, J.A., 2013. FT-NIR spectroscopy as a tool for valorization of spent coffee grounds: application to assessment of antioxidant properties. Food Res. Int. 51, 579–586.

Pavlović, M.D., Buntić, A.V., Šiler-Marinković, S.S., Dimitrijević-Branković, S.I., 2013. Ethanol influenced fast microwave-assisted extraction for natural antioxidants obtaining from spent filter coffee. Sep. Purif. Technol. 118, 503–510.

Prado, J.M., Vardanega, R., Debien, I.C.N., de Almeida Meireles, M.A., Gerschenson, L.N., Sowbhagya, H.B., Chemat, S., 2005. Conventional extraction. In: Galanakis, C.M. (Ed.), Food Waste Recovery: Processing Technologies and Industrial Techniques. Elsevier, Academic Press, Oxford, UK, pp. 127–148.

Rafael, C.E.D., Campanha, F.G., Vieira, L.G.E., Ferreira, L.P., David, P.O.T., Marraccini, P., De Benassi, M.T., 2010. Evaluation of kahweol and cafestol in coffee tissues and roasted coffee by a new high-performance liquid chromatography methodology. J. Agr. Food Chem. 58, 88–93.

Ramalakshmi, K., Raghavan, B., 1999. Caffeine in coffee: its removal. Why and how? Crit. Rev. Food Sci. Nutr. 39, 441–456.

Ramalakshmi, K., Rao, L.J.M., Takano-Ishikawa, Y., Goto, M., 2009. Bioactivities of low-grade green coffee and spent coffee in different in vitro model systems. Food Chem. 115, 79–85.

Ramirez-Coronel, M.A., Marnet, N., Kolli, V.K., Roussos, S., Guyot, S., Augur, C., 2004. Characterization and estimation of proanthocyanidins and other phenolics in coffee pulp (*Coffea arabica*) by thiolysis-high performance liquid chromatography. J. Agr. Food Chem. 52, 1344–1349.

Ramirez-Martinez, J.R., 1988. Phenolic compounds in coffee pulp: quantitative determination by HPLC. J. Sci. Food Agr. 43, 135–144.

Rao, L.J.M., Ramalakshmi, K., 2015. Use of microwaves to extract chlorogenic acids from green coffee beans. In: Preedy, V. (Ed.), Processing and impact on active components in food. Academic Press Elsevier, Oxford, UK, pp. 583–590.

Ravindranath, R., Khan, R., Obi Reddy, T., Thirumala Rao, S.D., Reddy, B.R., 1972. Composition and characteristics of Indian coffee bean, spent grounds and oil. J. Sci. Food Agric. 23, 307–310.

Reverchon, E., Kaziunas, A., Marrone, C., 2000. Supercritical CO_2 extraction of hiprose seed oil: experiments and mathematical modeling. Chem. Eng. Sci. 55, 2195.

Saénz, C., Tapia, S., Chávez, J., Robert, P., 2009. Microencapsulation by spray drying of bioactive compounds from cactus pear (*Opuntia ficus-indica*). Food Chem. 114, 616–622.

Sanguansri, P., Augustin, M.A., 2006. Nanoscale materials development: a food industry perspective. Trends Food Sci. Technol. 17, 547–556.

Santana, L.L.B., Cardos, L.A., Druzian, J.I., Souza, V.F., Costa, T.A.C., Nobrega, D.A., Hohlemwerger, S.V.A., Velozo, E.S., 2006. Selectivity in the extraction of 2-quinolone alkaloids with supercritical CO_2. Braz. J. Chem. Eng. 23, 525.

Shi, G., Rao, L., Yu, H., Xiang, H., Pen, G., Long, S., Yang, C., 2007. Yeast-cell-based microencapsulation of chlorogenic acid as a water-soluble antioxidant. J. Food Eng. 80, 1060–1067.

Silva, J.A., Borges, N., Santos, A., Alves, A., 2012. Method validation for cafestol and kahweol quantification in coffee brews by HPLC-DAD. Food Anal. Method. 5, 1404–1410.

Silva, V.M., Vieira, G.S., Hubinger, M.D., 2014. Influence of different combinations of wall materials and homogenisation pressure on the microencapsulation of green coffee oil by spray drying. Food Res. Int. 61, 132–143.

Smith, R.M., 2003. Before the injection—modern methods of sample preparation for separation techniques. J. Chromatogr. A 1000, 3–27.

Sousa, C., Gabriel, C., Cerqueira, F., Manso, M.C., Vinha, A.F., 2015. Coffee industrial waste as a natural source of bioactive compounds with antibacterial and antifungal activities. In: Méndez-Vilas, A. (Ed.), The Battle Against Microbial Pathogens: Basic Science. Technological Advances and Educational Programs. Formatex Research Center, Badajoz, Spain, pp. 131–136.

Spiller, G.A., 1998. Basic metabolism and physiological effects of the methylxanthines. In: Spiller, G.A. (Ed.), Caffeine. CRC Press, Boca Raton, pp. 225–231.

Stintzing, F.C., Casle, R., 2004. Functional properties of anthocyanins and betalins in plants, food, and in human nutrition. Trends Food Sci. Technol. 15, 1938.

Suárez-Quiroz, M.L., Campos, A.A., Alfaro, G.V., González-Ríos, O., Villeneuve, P., Figueroa-Espinoza, M.C., 2014. Isolation of green coffee chlorogenic acids using activated carbon. J. Food Comp. Anal. 33, 55–58.

Sun-Waterhouse, D., Wadhwa, S.S., Waterhouse, G.I.N., 2013. Spray-drying microencapsulation of polyphenol bioactives: a comparative study using different natural fibre polymers as encapsulants. Food Bioproc. Tech. 6, 2376–2388.

Tello, J., Viguera, M., Calvo, L., 2011. Extraction of caffeine from Robusta coffee (*Coffea canephora* var. Robusta) husks using supercritical carbon dioxide. J. Supercrit. Fluids 59, 53–60.

Thies, C., 2005. A survey of microencapsulation processes. In: Benita, S. (Ed.), Microencapsulation. Marcel Dekker Inc, New York, pp. 1–20.

Tilahun, A., Chun, B.S., 2016. Optimization of coffee oil flavor encapsulation using response surface methodology. LWT Food Sci. Technol. 70, 126–134.

Vannini, L.S., Hirata, T.A., Kwiatkowski, A., Clemente, E., 2009. Extraction and stability of anthocyanins from the Benitaka grape cultivar (*Vitis vinifera* L.). Braz. J. Food Technol. 12, 213–219.

Veiga, C.C., Simoni, R.C., Gonçalves, O.H., Shirai, M.A., Leimann, F.V., 2015. Effect of experimental parameters on the encapsulation of roasted coffee oil by complex coacervation. IX Simpósio de Pesquisa dos Cafés do Brasil, Brasil.

Velissariou, M., Laudano, R.J., Edwards, P.M., Stimpson, S.M., Jeffries, R.L., 2006. Beverage derived from the extract of coffee cherry husks and coffee cherry pulp. US 7,833,560 B2.

Vincze, I., Vatai, G., 2004. Application of nanofiltration for coffee extract concentration. Desalination 162, 287–294.

Wang, H., Helliwell, K., 2000. Epimerisation of catechins in green tea infusions. Food Chem. 70, 337–344.

Wang, Z., Pan, H.M., Atungulu, G.G., 2011. Extract of Phenolics From Pomegranate Peels. Open Food Sci. J. 5, 17–25.

Wang-Ping, K., Tsang-Chi, T., 2015. Recycled coffee grounds extracts and their extraction method. TW201524573.

Weiss, J., Takhistov, P., McClements, D.J., 2006. Functional materials in food nanotechnology. J. Food Sci. 71, 107–116.

Yu, T., Macnaughtan, B., Boyer, M., Linforth, R., Dinsdale, K., Fisk, I.D., 2012. Aroma delivery from spray dried coffee containing pressurised internalised gas. Food Res. Int. 49, 702–709.

Zhao, M., Wang, H., Yang, B., Tao, H., 2010. Identification of cyclodextrin inclusion complex of chlorogenic acid and antimicrobial activity. Food Chem. 120, 1138–1142.

Zuorro, A., Lavecchia, R., 2011. Polyphenols and energy recovery from spent coffee grounds. Chem. Eng. Trans. 25, 285–290.

FURTHER READING

Del Castillo Bilbao, M.D., Ezequiel Ibáñez, M.E., Amigo Benavent, M., Calleja Herrero, M., Del Moral Plaza, M., Ullate Artiz, M., 2014. Application of products of coffee silverskin in anti-ageing cosmetics and functional food. EP 2730171 A1.

Zuorro, A., Lavecchia, R., 2012. Spent coffee grounds as a valuable source of phenolic compounds and bioenergy. J. Cleaner Prod. 34, 49–56.

Emerging technologies for the recovery of valuable compounds from coffee processing by-products

Marcelo M.R. de Melo, Armando J.D. Silvestre, Inês Portugal, Carlos M. Silva

University of Aveiro, Aveiro, Portugal

ABSTRACT

Spent coffee grounds (SCG) and coffee silverskins (CSS) are two high-volume industrial by-products rich in valuable components, such as alkaloids, diterpenic, and phenolic compounds, triglycerides (oils), and polysaccharides. Their recovery yield and purity are highly dependent on the selected extraction method and operating conditions. This chapter discusses recent advances on the following separation technologies: solid–liquid extraction (SLE) using conventional solvents and ionic liquids, supercritical fluid extraction (SFE), subcritical water extraction, ultrasound assisted extraction, and microwave assisted extraction. SLE and SFE are the most mature approaches for the extraction of coffee by-products, although promising results have been reported for the other technologies. Solvents like ionic liquids (ILs) and deep eutectic solvents are expected to play an important role in the near future, along with other approaches, such as pervaporation, nanofiltration, sorption (adsorption and ion exchange), hydrothermal liquefaction, pulsed electric fields extraction, and products synthesis under the scope of nanotechnology. Overall, these methods produce oils/extracts rich in phenolic compounds exhibiting valuable antioxidant activity levels, with encouraging extraction yields. The optimization of both operating conditions and unit's arrangement is crucial to ensure their technological and economic viability. Currently, a considerable number of scientific studies and patents demonstrate or claim their use for various applications, including dermatological formulations, pest control formulations, growth enhancer substrates for bacterial cultures, biodiesel production, among others.

Keywords: coffee; spent coffee grounds; coffee silverskins; extraction; supercritical fluids; subcritical water; ionic liquids; hybrid methods; microwaves; ultrasounds

Handbook of Coffee Processing By-Products. http://dx.doi.org/10.1016/B978-0-12-811290-8.00005-0

5.1 INTRODUCTION

Spent coffee grounds (SCG) and coffee silverskins (CSS) are two industrial by-products generated during the processing of coffee for beverages, in amounts that surpass 6 million tons per year worldwide (Tokimoto et al., 2005). While CSS may be considered almost exclusively an industrial residue (generated during the roasting of green coffee), SCG has a rather ubiquitous existence being produced abundantly in industrial facilities (e.g., in the production of powdered soluble coffee) and also wherever coffee beverages are served (in private residences, coffee shops, restaurants, etc.).

The opportunity to recover bioactive compounds from coffee by-products is challenging, not only for diversification and enrichment of the value chain but also to reduce waste management costs associated to the disposal of industrial large volume residues. Although chemically suitable for soil amendment (Cruz et al., 2015) or even for pest control (Hollingsworth et al., 2003), the presence of caffeine above certain levels in coffee by-products may be counterproductive for the development of crops (Batish et al., 2008). Furthermore, inappropriately stored coffee residues can increase the release of mycotoxins associated with fungal development (such as ochratoxin A, classified by the International Agency for Research on Cancer as a possible human carcinogen (International Agency for Research on Cancer (IARC), 1993) causing significant safety risks for their direct use in agricultural applications (Toschi et al., 2014). Consequently, the extraction of valuable components from SCG and CSS may decrease waste management costs and provide additional revenues from the novel products that match natural and sustainable market expectations.

Within the scientific literature, the most relevant target components from coffee by-products have been alkaloids, diterpenic, and phenolic compounds, triglycerides (oils), and polysaccharides, as summarized in Table 5.1. Although caffeine is the most popular compound linked to coffee, one may cite other examples, such as the diterpenes cafestol and kahweol, the phenolic compounds coumaric and caffeic acids, or even the polysaccharides galactomannans and arabinogalactans. These compounds have been individually linked to a vast range of bioactivity features, such as antidepressant, immunostimulant, anticancer, serum cholesterol booster, or glucose metabolism improvement as described in Table 5.1. Overall, these features justify the increasing interest on the recovery of natural extracts from coffee by-products.

Considerable efforts have been made in recent years toward the development of adequate separation technologies for the recovery of valuable compounds from SCG and CSS (Fig. 5.1). The most investigated ones can be systematized into: (1) classic solid–liquid extraction (SLE) with organic or aqueous solvents (though ILs and deep eutectic solvents may be promising alternatives in the near future); (2) high-pressure extractions, such as supercritical fluid extraction (SFE), and subcritical water extraction (SWE); (3) hybrid approaches where SLE is assisted by microwaves (MAE) or ultrasounds (UAE). These technologies lead to extracts

Table 5.1 Most Investigated Target Compounds Extracted From Coffee By-Products, and Associated Biological Properties

Extractive Class	Compounds	Structure	Abundance (g/kg$_{SCG}$)	Biological Properties	References
Alkaloid	Caffeine		0.4–41.6	• Increase alertness • Increase anxiety • Antiinflammatory • Immunosuppressant • Antidepressant	Choi et al. (2015); López-Barrera et al. (2016); Bravo et al. (2012); Smith (2002); Daly (2007)
Diterpenes	Kahweol		4.4	• Increase serum cholesterol	Barbosa et al. (2014); Urgert et al. (1995)
Diterpenes	Cafestol		2.5	• Increase serum cholesterol • Anticancer	Barbosa et al. (2014); Urgert et al. (1995); Kotowski et al. (2015)

(Continued)

Table 5.1 Most Investigated Target Compounds Extracted From Coffee By-Products, and Associated Biological Properties (cont.)

Extractive Class	Compounds	Structure	Abundance (g/kg$_{SCG}$)	Biological Properties	References
Diterpenes	16-O-methylcafestol		1.0	• Increase serum cholesterol	Barbosa et al. (2014); Urgert et al. (1995)
Phenolics	Coumaric acid		0.01	• Antioxidant	López-Barrera et al. (2016); Rice-Evans et al. (1996)
Phenolics	Caffeic acid		0.03–0.07	• Antioxidant • Anticancer • Neuroprotective	López-Barrera et al. (2016); Rice-Evans et al. (1996); Kang et al. (2009); Kalonia et al. (2009)

| Polysaccharides | Galactomannans | | n.a. | • Immunostimulant | Simões et al. (2010) |
| Polysaccharides | Arabinogalactans | | n.a. | • Immunostimulant | Simões et al. (2010) |

n.a., Not available.

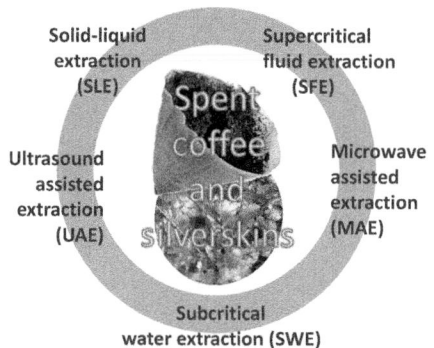

FIGURE 5.1 Extraction Technologies for the Recovery of Valuable Compounds From Spent Coffee Grounds (SCG) and Coffee Silverskins (CSS)

of variable complexity depending on the chosen extraction method and operating conditions, thus providing a great margin for optimization. The tuning of extraction conditions enables better performance toward the type of extracts that are sought, namely high yield bulk extracts, bioactive extracts, target compounds enriched extracts.

Several examples of oils/extracts obtained by the referred extraction techniques, predominantly from SCG, can be found in the scientific literature (Choi et al., 2015; Huang et al., 2016; Marto et al., 2016; Khelil et al., 2016; Obruca et al., 2014; Ramalakshmi et al., 2009) and patents (Bilbao et al., 2013; Gaonkar et al., 2007; Mi et al., 2015; Misra and Kondamudi, 2008; Mok, 2013). Four key applications identified are: dermatological formulations (Bilbao et al., 2013; Choi et al., 2015; Huang et al., 2016; Marto et al., 2016; Ramalakshmi et al., 2009), enhancement of substrates for bacterial cultures (Khelil et al., 2016; Obruca et al., 2014), biodiesel production (Misra and Kondamudi, 2008), and pest control formulations (Hollingsworth et al., 2003; Mok, 2013). Expectedly, many other applications will emerge as SCG and CSS oils/extracts become further investigated. For instance, purification of coffee oils/extracts to isolate a single fraction of compounds or a specific molecule is an open field providing new opportunities for scientific studies and patents. As a matter of fact, only one patent (Mi et al., 2015) has been filed claiming a method for the preparation of purified tagatose (a functional sweetener) from coffee by-products.

A final word should be dedicated to the exhausted coffee matrices left behind after the extraction processes, apparently with no added value. In fact, they can be incorporated in edible products as a source of insoluble dietary fiber (Bilbao et al., 2014) or used to produce heating pellets (bioenergy) because the calorific value of the coffee matrixes is practically unchanged by the extraction process (Zuorro and Lavecchia, 2012).

5.2 SEPARATION TECHNOLOGIES

5.2.1 SOLID–LIQUID EXTRACTION

5.2.1.1 Solid–liquid extraction with conventional solvents

SLE with conventional solvents is a technically mature unit operation used as a reference method for evaluating newer technologies or as a chemical characterization tool. Accordingly, much research relating to coffee by-products extractives has been accomplished through SLE as discussed in the following paragraphs.

When assessing different separation methods it is important to know a priori the maximum extraction yield of a given family of compounds from a specific vegetal source or by-product. This information is often based on SLE data (e.g., from Soxhlet extraction). Furthermore, this boundary indicates how promising a by-product can be as a source of natural bulk oils/extracts or of a particular compound. For example, Regazzoni et al. (2016) published a comparative study on the potential of SCG and CSS from different sources (nonextracted roasted coffee and green coffee) in terms of total extraction yield (Fig. 5.2A), total phenolic content (Fig. 5.2B), and IC_{50}, an antioxidant activity indicator (Fig. 5.2C). In this study, SLE was performed at room temperature under mechanical stirring, using an ethanol/water mixture (70:30, v/v) as solvent, a solvent/biomass ratio of 10 mL/g, and batches of 18 h. The results indicate lower extraction yields from SCG and CSS (5 and 13 wt.%, respectively) when compared to roasted coffee and green coffee (ca. 20 wt.% yield for both). The higher yield of CSS in relation to SCG is in accordance with the results reported by Murthy and Naidu (2012). On the other hand, SCG extracts are considerably richer in phenolics exhibiting a total phenolic content of 12.3 wt.%, which is similar to the levels in roasted coffee and 1.14 times higher than for CSS (Fig. 5.2B). SCG extracts are also those exhibiting highest antioxidant activity as revealed by the lower IC_{50} value, equal to 10.6 µg/mL (IC_{50} is the concentration required to inhibit 50% of the free radicals). Noteworthy, SCG and CSS extracts (IC_{50} 11.3 µg/mL) are both stronger antioxidants than roasted and green coffee extracts (IC_{50} 12.0 and 14.2 µg/mL, respectively, Fig. 5.2B). Overall, the results presented in Fig. 5.2 highlight the potential of CSG and CSS as a source of extracts rich in phenolic compounds with significant antioxidant properties.

A crucial aspect to consider when using SLE (e.g., Soxhlet extraction) to assess the potential of coffee by-products is the choice of solvent. Apart from different physicochemical properties like density, viscosity, boiling point, or polarity, other aspects linked to their impact on the health of living organisms and on environmental sustainability should also be pondered, particularly if the ensuing extracts are targeted for human or livestock applications. Furthermore, the claim on "green and natural" extracts that is important for marketing reasons may be blocked by a wrong solvent selection. In this respect, agents like toluene or chloroform are inadequate although typically used in nonfood extraction processes in the chemical industry. To emphasize the importance of this item, Fig. 5.3 provides results (yield and free fatty acids (FFA) content) for CSG extraction with pentane, hexane, toluene, chloroform

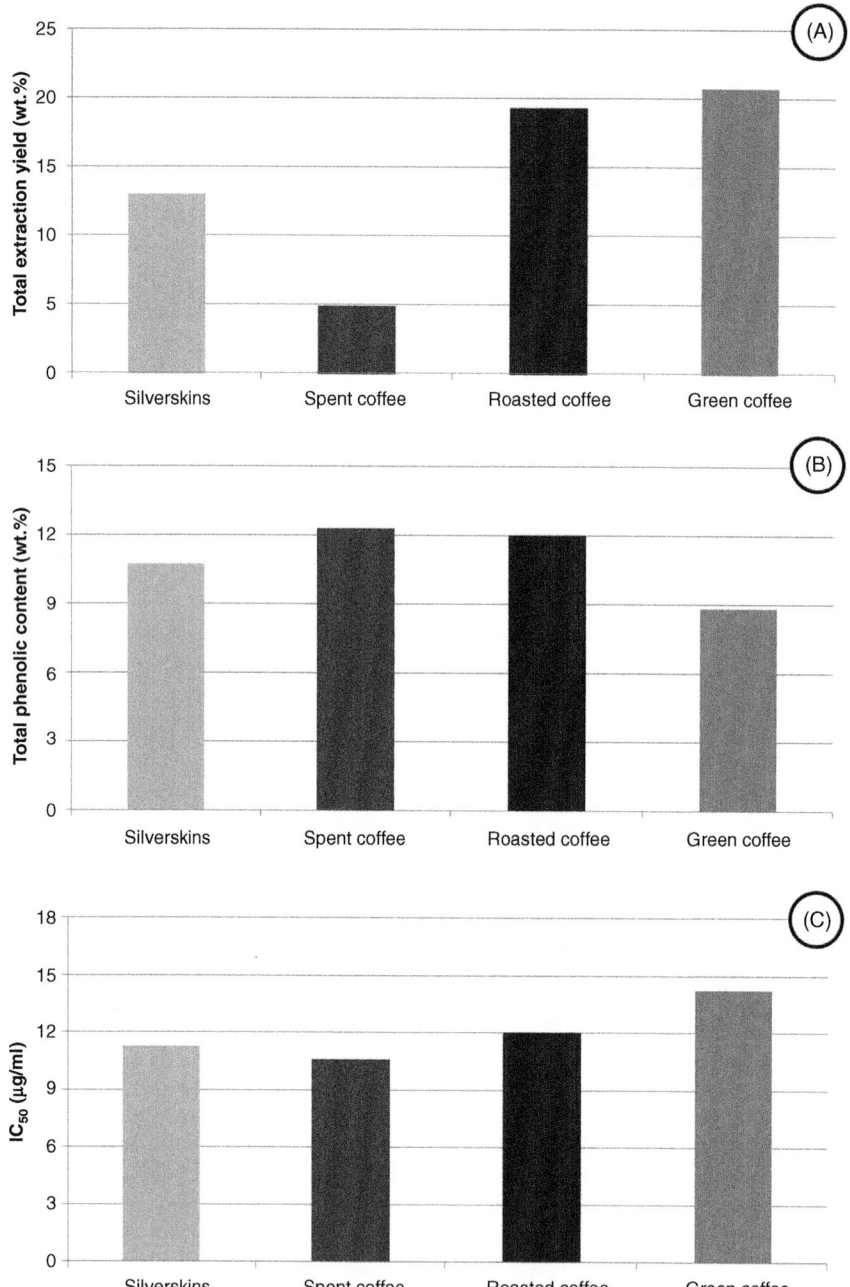

FIGURE 5.2 Comparative Study of Soxhlet Extracts (30 min, ethanol:water 70:30 v/v mixture) Obtained From Coffee by-Products (Silverskins and Spent Coffee Grounds) in Contrast With Roasted and Green Coffee

(A) Total extraction yield, (B) total phenolic content, and (C) antioxidant activity (IC_{50}).

Graphs built using data from Regazzoni, L., Saligari, F., Marinello, C., Rossoni, G., Aldini, G., Carini, M., Orioli, M. 2016. Coffee silver skin as a source of polyphenols: High resolution mass spectrometric profiling of components and antioxidant activity. J. Func. Foods 20, 472–485.

acetone, isopropanol, and ethanol (Al-Hamamre et al., 2012). Total extraction yields (Fig. 5.3A) depend on the chosen solvent, going from 8.6 wt.% when using chloroform to 15.3 wt.% for hexane, with the polar solvents (acetone, isopropanol, and ethanol) affording lower yields in comparison to pentane, hexane, and toluene. Since coffee by-product extracts are mainly composed of triglycerides, their affinity to weakly polar fluids is higher thus resulting in enhanced extraction yields. In addition, the solvent impacts the FFA content of the resulting oils/extracts (Fig. 5.3B), with ethanol giving the lowest value (3.3 wt.%) in contrast to isopropanol, which provides the highest (6.4 wt.%). Considering that FFA content is a common quality parameter of edible oils (Erickson, 1990), with lower values being preferable because they indicate lower degrees of decomposition (Desai, 2000; Erickson, 1990), in the

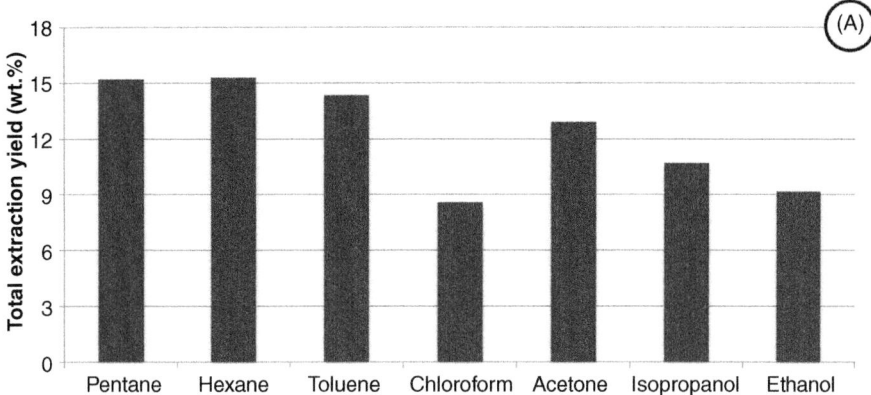

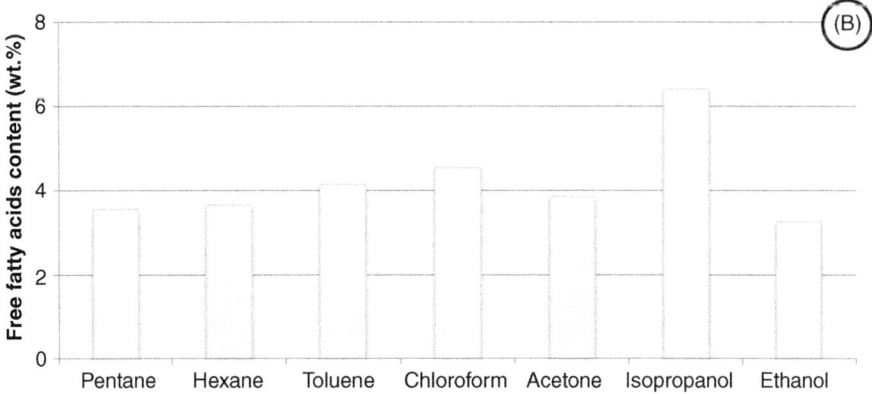

FIGURE 5.3 Extraction of SCG with Different Solvents

(A) Total extraction yield, and (B) free fatty acids (FFA) content.

Graphs built using data from Al-Hamamre, Z., Foerster, S., Hartmann, F., Kröger, M., Kaltschmitt, M., 2012.
Oil extracted from spent coffee grounds as a renewable source for fatty acid methyl ester manufacturing.
Fuel, 96, 70–76.

present example the CSG extract/oil obtained with ethanol is expected to exhibit the highest quality. However, the much lower extraction yield may be too costly and thus prohibitive for an industrial implementation of the process. On the other hand, the oils resulting from CSG extraction with hexane and pentane have similar FFA contents (ca. 3.7 vs. 3.3 wt.% for ethanol) and better yields (above 15 wt.%), hence a more favorable compromise between quality and productivity. Furthermore, FFA content can influence the viscosity of the resulting oils with implications in their use as biofuels, taking into account that vegetal oils with higher viscosities cause mediocre fuel atomization levels in the engines and decrease their performance (Esteban et al., 2012). Al-Hamamre et al. (2012) reported that kinematic viscosity of CSG oil jumped from ca. 60 mm^2/s to almost 250 mm^2/s when FFA content increased from 3.3 to 6.4 wt.%.

Globally, these results evidence the decisive role of solvent selection on productivity, composition, and properties of the SLE oils/extracts. Moreover, cautiousness is required when solvent mixtures are used because their composition can alter significantly the attainable yields and the oil/extract properties. For instance, Mussatto et al. (2011) investigated the impact of extraction time, solvent/biomass ratio, and methanol concentration in methanol/water systems, and reported that intermediate concentrations of methanol (from 35% to 75%) maximize the total content of phenolic compounds in SCG extracts.

Pretreatment of coffee by-products prior to SLE stage, using fungi or enzymes to improve extractability, has been examined by some researchers in recent years (Machado et al., 2012; Simões et al., 2013). For example, Machado et al. (2012) studied the ability of several fungal strains to improve the release of phenolic compounds from CSS and SCG. Simões et al. (2013) used an enzymatic pretreatment (with a mannanase) to investigate the extractability of galactomannans from SCG submitted to a roasting pretreatment and then sequentially extracted with hot water (90°C, for 1 h), and alkali solutions (20°C, 2 h, NaOH 4 M).

A complete description of the multiple aspects influencing the yields and properties of oils and extracts from coffee by-products is beyond the scope of this chapter. The more curious readers may consult the bibliography, especially Conde and Mussatto (2016) who published a compilation of SLE methods and operating conditions for the extraction of SCG and CSS residues with an emphasis on the recovery of polyphenols. A correct selection of the SLE method (Soxhlet or batch) and operation parameters like biomass pretreatment (none, by microorganisms or hydrothermally), solvent (water, methanol, and ethanol), solvent/biomass ratio (6–50 mL/g), temperature (60–120°C), and extraction time (10–180 min) can provide recoveries in the range 6.5–32.9 mg$_{GAE}$/g of SCG, where subscript GAE stands for gallic acid equivalents. Similar results were reported for CSS extracts (Conde and Mussatto, 2016).

A final word concerns the analysis of oils/extracts obtained by solvent extraction from coffee by-products. Jeszka-Skowron et al. (2015) systematized the analytical methods used for characterization and quantification of bioactive compounds in coffee extracts, including a full description of chromatographic techniques (e.g., the most suitable columns and eluent systems for HPLC and LC-MS are provided), as well as the type of compounds that can be detected in each case. Magalhães et al. (2016) proposed an alternative method to analyze bioactive phenolics, caffeine, and related compounds existing in SCG, based on Fourier-transform near infrared (FT-NIR) spectroscopy, which is faster and simpler than the chromatographic methods. FT-NIR spectroscopy is also a suitable tool for the prediction of the antioxidant capacity of SCG extracts in a fast, reliable and accurate way (Páscoa et al., 2013).

5.2.1.2 Extraction with ionic liquids

ILs are an emerging class of solvents comprising noncorrosive molten salts that remain fluid below 100°C. One of the key advantages of ILs is the vast margin for tuning their polarity and selectivity by changing the anion/cation combinations as desired, which makes them "tailor-made" solvents (Dupont, 2005; Keskin et al., 2007). In comparison to conventional organic solvents, ILs offer extremely low (nonmeasurable) vapor pressures, higher polarity, density, the ability to dissolve a wide range of materials (salts, fats, proteins, amino acids, surfactants, sugars, and polysaccharides), high thermal stability (in many cases up to 300°C), and immiscibility with many organic solvents (Dupont, 2005; Han and Row, 2010). Such attracting features have raised interest in many research topics, such as solvent replacement, gas purification, homogenous, and heterogeneous catalysis, biological reactions/processing, and metal ions removal (Keskin et al., 2007).

The industrial attractiveness of ILs is very dependent on their costs and recyclability. Accordingly, processes, such as distillation, adsorption, nanofiltration, ion exchange, and liquid–liquid extraction have been suggested for ILs recovery from aqueous media (Pereira et al., 2016). In any case, significant research contributions are still required toward clarification of the environmental impact and potential health risks of ILs, which ultimately will give rise to the design of entirely safe and benign ILs (Heckenbach et al., 2016).

The use of ILs for the extraction of natural matrices can also be cited (Han and Row, 2010; Passos et al., 2014b; Pereira et al., 2016), including the recovery of caffeine from guarana seeds (Claudio et al., 2013) or from green tea extracts (Tian et al., 2009). For noncoffee biomass, the application of ILs on the solid–liquid and liquid–liquid separation of essential oils and bioactive compounds can be referred (Han and Row, 2010). Even though ILs have not been explored for the valorization of coffee by-products, their use can be expected to rise in the next years. The same can be foreseen for deep eutectic solvents, which are also considered novel green fluids (Duan et al., 2016; Kumar et al., 2016).

5.2.2 SUPERCRITICAL FLUID EXTRACTION

SFE exploits thermodynamic and kinetic features of fluids above their critical points. In the last decades, SFE has been applied to more than 300 vegetal species predominantly for the production of natural extracts including oils (De Melo et al., 2014b). This technology is historically tied to the coffee industry (namely for coffee decaffeination), which implies that its extension to the valorization of coffee by-products can be easily anticipated with expectable synergies on the integration of processes.

Carbon dioxide is the preferred solvent, both for research and industrial applications due to the milder P–T conditions required to attain the critical point ($T_c = 31.1°C$ and $P_c = 73.8$ bar). Supercritical CO_2 possesses liquid-like densities, gas-like viscosities, diffusivities up to two orders of magnitude greater than those of liquids, and its solvent power can be tuned by small changes of pressure and/or temperature (De Melo et al., 2014b). Besides, explosion-proof operating licenses are not required (contrarily to operation with supercritical alkanes like propane and ethane, increasingly studied in the last years), and it offers complementary advantages, such as being chemically inert, noncorrosive (in contrast to supercritical water), and nontoxic (Mukhopadhyay, 2000). Finally, CO_2 has a practically null surface tension, which is advantageous to wet and easily penetrate most solid materials, including biomass (Sangeeta and Lagraff, 2004).

Concerning the industrial layout of SFE units (Fig. 5.4) it can be described as follows. A storage tank (1) is used to feed, maintain, and return pressurized CO_2 (in

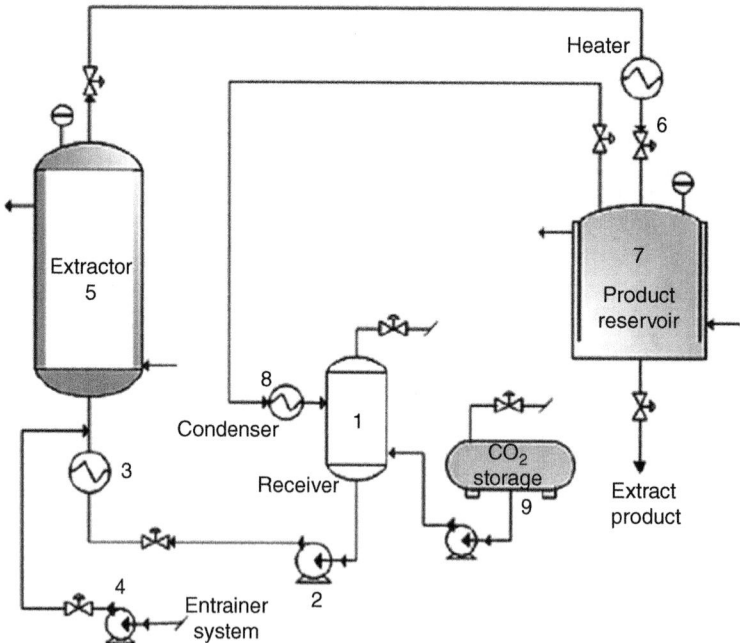

FIGURE 5.4 Scheme of a Typical SFE Layout Comprising the Use of an Entrainer (Cosolvent)

(1) Storage tank/receiver, (2) pump, (3) heat exchanger, (4) cosolvent pump, (5) extractor, (6) backpressure regulator, (7) product reservoir, (8) condenser, (9) storage tank for make-up liquid CO_2.

the liquid state) to and from the process. At the onset of the operation, liquid CO_2 is pumped (2), heated (3), mixed with a cosolvent in the defined proportions (if required) (4), and fed to the extractor (5) already filled with the coffee by-products. Afterwards, supercritical solvent percolates the extractor bed and exits the vessel with the dissolved solutes, being then depressurized by a backpressure regulator (6) down to a few dozens of bar. As the solvent power of the supercritical mixture becomes negligible at this point, the extracts precipitate in a collection vessel (7) with the cosolvent. The gaseous CO_2 can be recycled to the storage tank (1), after condensation (8). Make-up liquid CO_2 is stored in a side tank (9) and periodically added to the system to compensate the losses linked to full decompression of the extractor for replacement of the raw biomass. Depending on the specificities of the process, the final expansion may be performed in a cascade mode allowing the fractional recovery of the extracts in multiple collection vessels submitted to a stepwise pressure profile (Čretnik et al., 2005; Simándi et al., 2000).

In recent years, several studies on SFE of SCG have been published (Akgün et al., 2014; Andrade et al., 2012; Barbosa et al., 2014; Couto et al., 2009; Cruz et al., 2014; De Melo et al., 2014a; Tello et al., 2011), being split into two broad categories: those aiming oil production and those focusing on the selective removal of key molecules, such as caffeine or diterpenic compounds. Couto et al. (2009) studied the influence of solvent pressure and temperature (the key operating conditions) on the extraction rate and SCG oil composition. At 300 bar and 55°C (the highest pressure and temperature considered in their study) a maximum extraction yield of 15 wt.% was attained, which corresponds to ca. 85% of the maximum oil recovery from SCG (assessed by Soxhlet extraction) (Couto et al., 2009). Andrade et al. (2012) compared the extraction of SCG by SFE and SLE, and revealed that under the most favorable $P–T$ conditions (300 bar and 40°C) SFE provides a total extraction yield comparable to the one obtained by Soxhlet with dichloromethane, that is, 10.5 wt.%. This yield was increased to 14 wt.% after ethanol addition (15 wt.%) as CO_2 modifier for the conditions of 100 bar and 60°C. However, the ensuing SFE extracts had lower total phenolic content and weaker antioxidant activity than those obtained by Soxhlet extraction with dichloromethane (typically, this solvent is only able to remove the simpler phenolics).

Caffeine, being a key coffee component, its extraction from coffee husks has been investigated (Tello et al., 2011). Accordingly, different pretreatments (humidification and milling) and SFE operating conditions (pressure, temperature, time, and flow rate) were considered. The results revealed that coffee husks with moisture levels ranging from 32 to 48 wt.% are desirable for a privileged removal of caffeine, and that milling the matrix into a powder is irrelevant, at least for the $P–T$ conditions considered for SFE (60–300 bar, and 40–100°C), which imply, for example, that solvent density varied from 100 to 910 kg/m^3. Overall, it is possible to remove up to 84% of the initial caffeine in coffee husks using a 5 h long SFE, at 300 bar and 100°C, with a total percolation of 197 $kg_{CO_2} / kg_{coffee\ husks}$ (Tello et al., 2011).

With identical objectives but different target compounds, Barbosa et al. (2014) optimized the selective uptake of diterpenic compounds from SCG using design of experiments and response surface methodologies. It was shown that the conditions of pressure, temperature, and content of cosolvent (ethanol) that maximize the total extraction yield are distinct from those that favor the uptake of the target compounds kahweol, cafestol, and 16-O-methylcafestol. In fact, the highest extraction yield (11.97 wt.%) was attained at 190 bar, 55°C, and 5 wt.% ethanol, providing an extract with 96.3 $mg_{diterpenes}/kg_{extract}$ (Fig. 5.5B) whereas with pure CO_2, at 140 bar and 55°C, the extract was richer in diterpenic compounds providing 121 $mg_{diterpenes}/kg_{extract}$ (Fig. 5.5A). These results emphasize how the optimal SFE operating conditions can vary with the objective pursued. Furthermore, comparing the SFE and Soxhlet results (horizontal lines, in Fig. 5.5) reveals that SFE can be much more selective than SLE with dichloromethane. De Melo et al. (2014a) carried out a technoeconomic study, involving cost of manufacturing and net income calculations, pondering different operating conditions and SFE unit arrangements for oil production from SCG. The preferable conditions comprise an arrangement of three beds of 1 m^3, with extraction cycles of 2.0 h and 30 $kg_{CO_2}/kg_{SCG}/h$, and operation at 300 bar and 50°C. For these conditions, oil production can reach 454 ton/year, with an expected annual cost of manufacturing of 2.4 M€ and an estimated process net income of 56.6 M€/year (De Melo et al., 2014a). A sensitivity analysis varying the unit capacity, extraction time and precipitation pressure (in the extract vessel) showed that process economics remains viable.

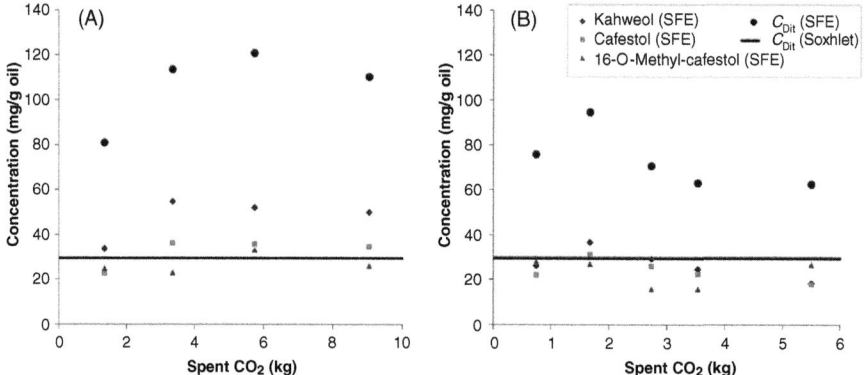

FIGURE 5.5 Concentration of the Most Important Diterpenes (Kahweol, Cafestol, and 16-O-Methyl-Cafestol) and Total Diterpenes (C_{dit}) in the Cumulative Supercritical Extracts of Spent Coffee Grounds Along Time, Here Expressed in Terms of Mass of Spent CO_2

Experimental conditions: (A) 140 bar, 55°C, pure CO_2; (B) 190 bar, 55°C, 5 wt. % ethanol. The Soxhlet results are the horizontal lines graphed.

From Barbosa, H.M.A., De Melo, M.M.R., Coimbra, M.A., Passos, C.P., Silva, C.M., 2014.
Optimization of the supercritical fluid coextraction of oil and diterpenes from spent coffee grounds using
experimental design and response surface methodology. J. Supercrit. Fluids 85, 165–172.

On the whole, the aforementioned essays foster the industrial recovery of valuable compounds from coffee by-products using SFE.

5.2.3 SUBCRITICAL WATER EXTRACTION

SWE comprises the use of pressurized water, heated at temperatures above its boiling point, as extraction solvent. The interest in subcritical water is linked to the possibility of boosting the separation process by modifying the dielectric constant and polarity of the solvent, without losses in the biological properties of the extracts (Chemat and Vian, 2014; Zakaria and Kamal, 2016). Moreover, it is claimed that the mass transfer rate in SWE is improved (Zakaria and Kamal, 2016). Research on SWE for the removal of bioactive compounds from plants and algae has been reported, with maximum pressures and temperatures of 400 bar and 360°C, respectively (Zakaria and Kamal, 2016).

A typical extraction procedure consists on a sequence of steps performed in a SWE apparatus (see scheme in Fig. 5.6), namely: (i) charging of the biomass of interest into the extraction vessel (4) placed in a heating furnace (5); (ii) transferring deionized water from the reservoir (1) through the preheater (3) into the extractor (4) using the high pressure pump (2) until the desired pressure is reached (control is ensured by the back-pressure regulator (7)); (iii) after increasing the temperature in the extraction vessel to the desired set point a static extraction is carried out during a precise extraction time; (iv) the solubilized extract is removed from the extractor, cooled (6) and recovered with the solvent in the extract collecting vessel (9). Alternatively, the mode of operation can be semicontinuous, in which case a back-pressure regulator is employed as previously explained for SFE (Apibalsri et al., 2012).

Narita and Inouye (2012) studied the production of SWE extracts from CSS at four different temperatures (180, 210, 240, and 270°C) and pressures (10, 19, 32, and 53 bar). Process performance was evaluated in terms of total extraction yields, total phenolics content, and antioxidant activity. The results show that yields of 16–29 wt.% can be achieved solely by tuning P–T conditions, and further improved to 37–44 wt.% by the addition of NaOH 0.1 M. Nevertheless, the antioxidant activity of CSS extracts decreased when an acid (HCl 0.1 M) or a base (NaOH 0.1 M) was added to the SWE liquid medium (Narita and Inouye, 2012). Globally, the study demonstrates that SWE is effective for the extraction of reducing sugars, proteins, and phenolic components from CSS.

Xu et al. (2015) used response surface methodology to optimize SWE of phenolic components from SCG, changing the temperature (160, 170,180°C), extraction time (35, 45, 55 min), and solvent/biomass ratio (70.9, 49.5, 38.9 mL/g), while keeping a constant pressure of 50 bar. The results revealed the importance of using higher temperatures and lower liquid/solid ratios to maximize the extraction of phenolics (up to 89.3 mg_{GAE}/g_{SCG}), as well as a negative synergy when the product of the solvent/biomass ratio and extraction time is increased. In contrast, shorter extraction times and higher liquid/solid ratios favor the antioxidant activity of the extracts (assessed

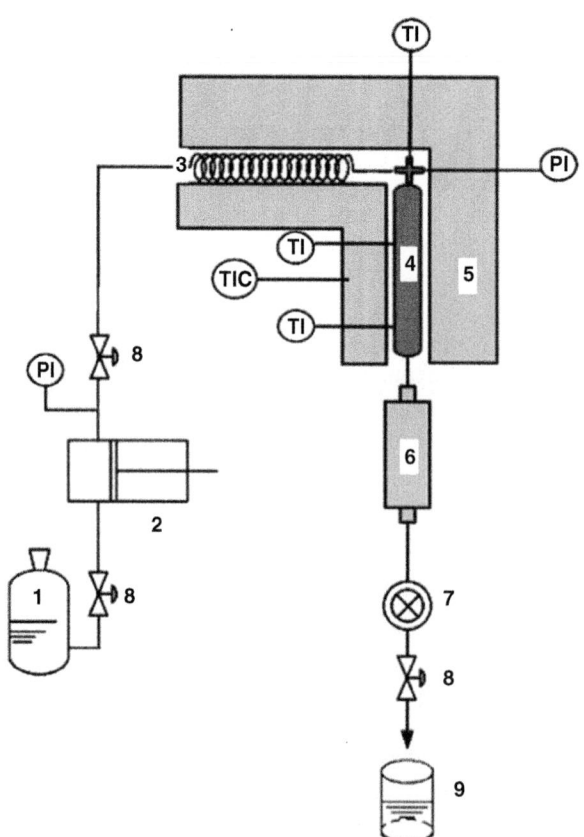

FIGURE 5.6 Schematic Diagram of a Subcritical Water Extraction (SWE) Apparatus

(1) Deionized water reservoir, (2) high pressure pump, (3) preheater, (4) high pressure extraction vessel, (5) heat furnace, (6) heat exchanger, (7) back-pressure regulator, (8) on/off valve, and (9) extract collecting vessel.

Retrieved with permission from Kim, W.J., Kim, J., Veriansyah, B., Kim, J.D., Lee, Y.W., Oh, S.G., Tjandrawinata, R.R., 2009. Extraction of bioactive components from Centella asiatica *using subcritical water. J. Supercrit. Fluids 48, 211–216.*

by the DPPH method). These results evidence that optimal SWE conditions for SCG extraction depend strongly on the key response to be optimized, as previously mentioned for SFE of coffee by-products.

Given the limited number of SWE studies reported in the literature, comparison between the attained results and those of SFE and classic SLE is not straightforward. Yet, given the "green" connotation of water, SWE might gain relevance for the valorization of coffee by-products in the near future.

5.2.4 MICROWAVE ASSISTED EXTRACTION

MAE adapts conventional solvent extraction with microwave (MW) heating. Its effect is strongly dependent on the dielectric susceptibility of both the solvent and the biomass matrix (Wang and Weller, 2006). When polar solvents (e.g., water and ethanol) are used, heating of the liquid media is more efficient, which may shorten the required extraction time (Chan et al., 2011; Wang and Weller, 2006). In addition, the heating efficiency ensured by MW can be combined with pressurized systems enabling SLE to be carried out at temperatures higher than the normal boiling point. Such approaches are very much like SWE (where the solvent is water), although faster heating may be expected in MAE. Moreover, the thermal effect of MW contributes to higher diffusivity and lower viscosity of the solvent, which is by itself advantageous for the transport mechanisms behind extraction processes.

When the solvent is MW transparent, as is the case of hexane, the impact of MW radiation is exclusively felt on the biomass matrix. In such circumstances, changes in the vegetal structures are expected to increase the accessibility of the solvent to the target solutes. Accordingly, a mechanistic explanation on the underlying effect of MW in vegetal matrices has been recently proposed by Chan et al. (2016), following other MAE modeling studies published by the same authors (Chan et al., 2013, 2015a,b). The mechanism is schematized in Fig. 5.7 and can be summarized as follows: the moisture located in plant cells (more specifically inside vacuoles) absorbs part of the incident energy and undergoes vaporization promoted by the heating effect of MW. This change of physical state inside the cells induces stretching and expansion movements, and eventually leads to cell wall rupture caused by excessive internal pressure. As a result, bioactive compounds are released from the cells and become more accessible to the extraction solvent. This implies a specific heating time for each MAE process, in order to ensure the rupture of the plant cells.

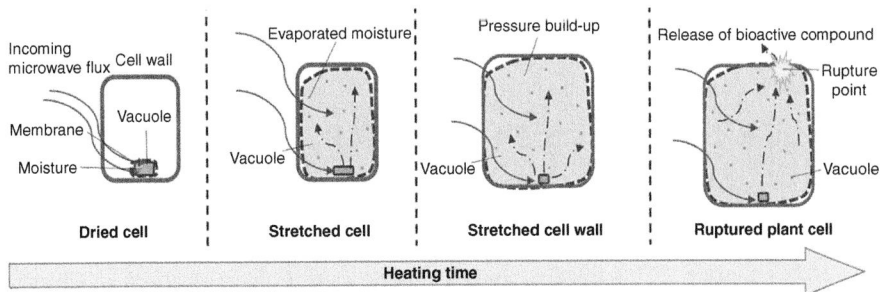

FIGURE 5.7 Schematic Illustration of the MAE Cell Rupture Model

Retrieved with permission from Chan, C.-H., Yeoh, H.K., Yusoff, R., Ngoh, G.C., 2016.
A first-principles model for plant cell rupture in microwave-assisted extraction
of bioactive compounds. J. Food Eng. 188, 98–107.

To the best of our knowledge, MAE has been applied to the extraction of SCG (Passos and Coimbra, 2013; Passos et al., 2014a; Pavlović et al., 2013; Ranic et al., 2014) and not for CSS. Ranic et al. (2014) used design of experiments to study the joint effect of MW power (240, 400, and 500 W), time (11, 40, 110, 180, and 209 s), and solvent/biomass ratio (4.76, 6, 9, 12, and 13.24 mL/g) on SCG extractions carried at atmospheric pressure, using aqueous ethanol (20 wt.%) as solvent. The results showed that MW power, extraction time, and solvent/biomass ratio were equally important in what concerns total extraction yield, but a negative synergy exists between extraction time and MW power. Consequently, maximum yield (3.1 wt.%) was attained with a power of 550 W, for 180 s and solvent/biomass ratio of 12 mL/g (the temperature reached by the system was not reported). On the other hand, the best conditions to obtain enriched phenolic extracts (79.8 wt.% phenolic content) were 240 W, 40 s, and 6 mL/g. The authors expect their work may provide useful guidelines for an upcoming industrial process for the extraction of valuable compounds from coffee by-products.

Passos et al. (Passos and Coimbra, 2013; Passos et al., 2014a) investigated MAE for the removal of polysaccharides (mannans) from SCG, using superheated water as solvent and solvent/biomass ratios in the range 5–30 mL/g (Fig. 5.8A). Instead of imposing a given MW power (their apparatus had a maximum output of 1 kW, 2.45 GHz), the latter was adjusted in order to attain a temperature of 200°C in 3 min, and then maintain it for 2 min. The highest extraction yield obtained was 42% for a solvent/biomass ratio of 10 mL/g. The total yield distribution and the specific yields of monosaccharides, oligosaccharides, and polysaccharides (Fig. 5.8) (Passos and Coimbra, 2013), prompted the authors to develop a two-step sequential MAE process to fractionate the different mannans (Passos et al., 2014a).

Finally, Du et al. (2007) combined MAE and ILs to extract *trans*-resveratrol from a Chinese traditional medicine herb taking advantage of the efficient absorption of MW energy by ILs. The authors used a 1-butyl-3-methylimidazolium bromide solution (IL) as solvent and optimized several operating conditions of the process, namely sample size, solvent/solid ratio, temperature, and extraction time. This study evidences the benefit of MAE for the extraction of coffee by-products.

5.2.5 ULTRASOUND ASSISTED EXTRACTION

UAE is carried out in a vessel submitted to ultrasound irradiation at frequencies above 20 kHz (Kadam et al., 2015). The ultrasound waves are typically produced by transducers in contact with the liquid medium, which cause bubble cavitation in the vegetal matrices. This phenomenon imposes the disruption of plant cell walls and, consequently, promotes an easier access of the solvent to the intracellular compounds, and increases the effective diffusitivities of solutes in general (Michail et al., 2016; Teh and Birch, 2014).

Michail et al. (2016) investigated UAE for the recovery of valuable polyphenols from SCG, using aqueous glycerol as solvent and fixed ultrasound conditions (a power of 200 W, a frequency of 37 kHz, and an acoustic energy density of 50 W/L).

(A)

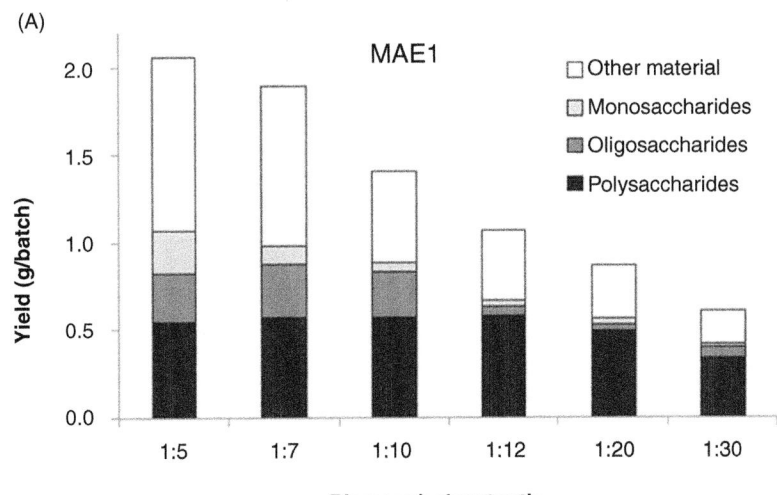

(B)

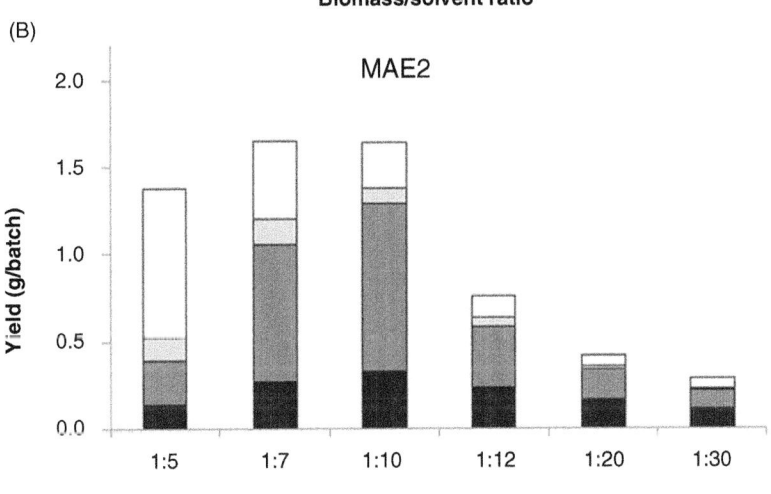

FIGURE 5.8 Total and Polysaccharides (Mannans) Extraction Yields Obtained by Sequential MAE as Function of the Solvent/Biomass Ratio

(A) In the first extraction (MAE1), the biomass is SCG; and, (B) in the second extraction (MAE2), the biomass is the unextracted insoluble material from MAE1.

Retrieved with permission from Passos, C.P., Coimbra, M.A., 2013.
Microwave superheated water extraction of polysaccharides
from spent coffee grounds. Carbohydr. Polym. 94, 626–633.

They applied response surface methodology to determine the optimal glycerol concentration (0%–8%, w/v) and extraction time (5–175 min) in order to maximize the yield in total phenolics, under constant solvent/biomass ratio (50 mL/g) and temperature (45°C). The results demonstrated that extraction time (preferably 175 min) and glycerol concentration (preferably 3.6%, w/v) were independently important for the extraction of phenolic compounds reaching yields around 9 mg_{CAE}/g_{SCG}, where subscript CAE stands for caffeic acid equivalents.

5.2.6 PULSED ELECTRIC FIELD EXTRACTION

This technology (PEFE) can be seen as a water SLE process assisted by the intermittent application of short electric pulses (voltage <50 kV and current <1 kA). In practice, the vegetal matrix and the solvent are placed in the extraction vessel between two electrodes, one connected to a high voltage generator and the other a ground electrode. By subjecting the vegetal biomass to pulsed electric fields, small pores are locally formed in cell membranes, reversibly or irreversibly depending on the treatment intensity with the latter being responsible for the improvement of the extraction process (Puértolas and Barba, 2016). The application of this technique to coffee by-products has not been reported in the literature. However, its use can be foreseen in the near future, for PEFE has been widely investigated as a method to enhance the recovery of polyphenols, proteins, and carbohydrates from vegetal by-products, such as orange, mango, grape, or papaya peels (López et al., 2008; Luengo et al., 2013; Parniakov et al., 2014, 2016). A more powerful version of this technique—high voltage pulse discharge—has been applied to food wastes as an enhancer of anaerobic digestion, being capable of improving methane production by 153%, almost 2 times higher than the results attained by acid, alkali, or ultrasonic pretreatments (Puértolas and Barba, 2016). Overall, PEFE and related technologies can be seen as a prospective pathway to enhance the recovery of valuable compounds from coffee by-products.

5.2.7 OTHER TECHNOLOGICAL OPPORTUNITIES FOR COFFEE BY-PRODUCTS VALORIZATION

In addition to the emerging separation methods discussed earlier, other approaches may be mentioned in order to expand and boost the range of application possibilities for coffee by-products valorization. A list of additional technologies is provided in Table 5.2, together with supporting examples from the literature on the application to coffee related raw materials.

Pervaporation owes its name to a combination of two phenomena, permeation, and evaporation. This is an advantageous membrane separation process that may offer energy savings in relation to distillation (due to operation at mild temperatures), high selectivities toward the target compounds, and compatibility with thermo labile raw materials (Lipnizki et al., 2002; Pereira et al., 2006). Moreover, since solvents are not needed, pervaporation can be considered especially suitable for the food

Table 5.2 Additional Technologies and Opportunities for Coffee By-Products Processing

Technology	Raw Material	Target Product	References
Pervaporation	Coffee processing wastewater	Volatile organic compounds	Oliveira et al. (2014)
	Soluble coffee solution	Aroma compounds	Weschenfelder et al. (2015)
Nanofiltration	SCG	Caffeine	Brazinha et al. (2015)
		Phenolic compounds	
	—	Coffee extract	Pan et al. (2013)
	Roasted coffee	Coffee extract	Vincze and Vatai (2004)
Sorption	SCG	Cadmium, copper, and lead	Davila-Guzman et al. (2016); Cerino-Córdova et al. (2013); Tokimoto et al. (2005)
Hydrothermal liquefaction	SCG	Bio-oil	Yang et al. (2016)
Nanotechnology applications	SCG	Carbon nanosheets	Yun et al. (2015)
	SCG	Silver nanoparticles	Baiocco et al. (2016)

SCG, *Spent coffee grounds.*

industry. Pervaporation relies on the existence of a permselective membrane separating the liquid feed mixture from which one or more target compounds are to be preferentially removed, and a vapor phase (permeate side) containing the molecules that permeate the membrane, which are thereafter condensed and recovered (Nunes and Peinemann, 2006). The feed pressure is usually ambient or elevated enough to ensure liquid state, while the permeate side is maintained at or below the dew point of the permeate stream, making it vapor (it is usually under vacuum).

In the context of coffee by-products, pervaporation can be used to remove valuable or undesired volatile compounds. For instance, it has been applied to the removal of noxious volatile organic compounds (e.g., pyrrolles, furans) from coffee processing wastewaters (Oliveira et al., 2014), and specific aroma compounds (e.g., 2,3-butanedione and 2,5-dimethyl pirazine) from soluble coffee with the purpose of increasing the enjoyability and attractiveness of coffee beverages (Weschenfelder et al., 2015).

Nanofiltration is a pressure-driven process in which a membrane with small pore size (typically <2 nm) separates two liquid phases, enabling the separation of inorganic salts and small organic molecules (Mcnaught and Wilkinson, 1997). It can be considered as an alternative or complementary to reverse osmosis, which has been used to concentrate coffee extracts (Bélafi-Bakó et al., 2012). Nanofiltration and reverse osmosis membranes are the same, though the network structure of the former is more open. Although compatible with the food processing industry requirements, the underlying principles of nanofiltration demand that the material to be processed is available in the liquid state. Consequently, for the specific case of coffee by-products a preliminary extraction process is necessary before this technology can be applied. Brazinha et al. (2015) showed that commercial nanofiltration membranes can be used to increase the concentration of caffeine up to 3.4 times in permeate extracts obtained from an aqueous bulk coffee extract, but with a significant loss of total yield (ca. 73% in relation to nonfractioned extract).

Sorption, an abundant low-cost materials with metal-binding capacity, such as agricultural wastes or industrial by-products, can be used for heavy metals recovery via ion exchange and adsorption (Kurniawan et al., 2006). Accordingly, coffee by-products are natural candidates for this quest. One may cite Davila-Guzman et al. (2016) who used SCG for the adsorption of cadmium, copper, and lead, and reported solid capacities of 0.12, 0.21, and 0.32 mmol/g, respectively. These results were better than those obtained with unprocessed coffee grounds (ca. 0.10, 0.03, and 0.24 mmol/g, respectively) and, for some particular metal ions, even better than those reported for activated carbon and commercial ion exchange resins (Davila-Guzman et al., 2016). This achievement was made possible by a preliminary treatment of SCG with citric acid, which increased the content of carboxylic groups and thus the metal binding capacity (Cerino-Córdova et al., 2013).

Hydrothermal liquefaction (HTL) is a compressed water technique (alternative to pyrolysis and gasification) used to convert vegetal biomass into liquid fuels and chemicals at temperatures between ca. 150 and 420°C and pressures up to ca. 240 bar. Its main advantage is the possibility to process wet biomass directly (Tran, 2016). Yang et al. (2016) used HTL for the production of bio-oil from SCG, reporting crude bio-oil yields up to 47.3 wt.% (at 275°C, 20 bar, 10 min of retention time, and 20:1 water/feedstock mass ratio). This result is higher than the yields reported for other biomass materials, such as, for example, pulp/paper sludge, swine manure, corn stalks, tea waste, Jack pine wood sawdust, and microalgae (Yang et al., 2016).

Nanotechnology applications, for the valorization of coffee by-products, might arise from the chemical nature of the extracts components, and also from the lignocellulosic carbon structures these by-products contain. With relation to the former approach, the phenolic compounds obtained from SCG and CSS may be employed as reducing agents for the production of high purity nanoparticles of noble metals, such as silver (Baiocco et al., 2016). The motivation behind this approach is the scientific search for simpler, cheaper, and "greener" synthesis routes of nanostructured materials (Kulkarni and Muddapur, 2014). Baiocco et al. (2016) demonstrated the feasibility of producing silver nanoparticles using ethanol:water extracts of SCG at 40°C, with an extraction time of 160 min and a solvent/biomass ratio of 10 mL/g. Nevertheless, the extraction conditions influence the characteristics of the resulting nanoparticles, possibly because of changes in the phenolics profile of the obtained extracts (Baiocco et al., 2016). Concerning the second approach, Yun et al. (2015) tested SCG as raw material for the synthesis of hierarchically porous carbon nanosheets, using in situ carbonization and KOH activation to produce nanomaterials with encouraging performances as supercapacitors.

5.3 CONCLUSIONS

The recovery of valuable compounds from high-volume coffee by-products through emerging separation technologies has become a topic of interest in recent years, with space for further improvements and consolidation in the research field. The scientific focus has been mostly on solid–fluid separation technologies that allow either the production of a bulk oil/extract, or the recovery of nonpure caffeine, phenolic and diterpenic compounds, galactomannans, and so on. Moreover, research has been mainly devoted to SCG with CSS being a weakly exploited matrix with high valorization potential.

Classic SLE is the reference approach to assess novel separation methods, and also for the characterization of natural extracts. Within the emerging separation technologies, SFE is the most investigated, followed by SWE, which is considered a promising approach. UAE and MAE are theoretically sound methods but further research is required before they can be considered reliable alternatives for adding value to coffee by-products. Technoeconomic analysis and optimization models are necessary for scale-up studies demonstrating the suitability of these processes for industrial implementation in the coffee industry.

The results compiled and analyzed in this chapter confirm the potential of these emerging technologies to produce extracts rich in phenolic compounds and valuable antioxidant properties, with extraction yields far from discouraging. These outcomes justify the existence of a considerable number of scientific studies and patents demonstrating/claiming the usefulness of the ensuing coffee by-products oils/extracts. In the upcoming years one may expect the rise of further technical attempts to recover valuable compounds from SCG and CSS by the said technologies, and the emergence of novel solvents (such as ILs and deep eutectic solvents), eventually combined with or coupled to other technologies, such as pervaporation, nanofiltration, sorption, ion exchange, and pulsed electric field extraction. Furthermore, efforts to isolate key compounds of coffee by-products may appear, as there seems to be a vast window of opportunities both for scientific publications and patents. Here one may cite the application of very selective solvents (e.g., ILs, deep eutectic solvents) and the large potential offered by continuous chromatographic techniques like temperature swing adsorption (TSA) and simulated moving bed (SMB). Additional applications for the valorization of coffee by-products are the hydrothermal liquefaction of biomass and the synthesis of products under the scope of nanotechnology.

ACKNOWLEDGMENTS

This work was developed under the scope of the project CICECO-Aveiro Institute of Materials (Ref. FCT UID/CTM/50011/2013), financed by national funds through the FCT/MEC and when applicable co-financed by FEDER under the PT2020 Partnership Agreement.

REFERENCES

Akgün, N., Bulut, H., Cortesi, A., De Zordi, N., Kikic, I., Moneghini, M., Procida, G., Solinas, D., 2014. Supercritical fluid extraction of fatty acids from spent coffee grounds. 14th European Meeting on Supercritical Fluids, Marseille, France.

Al-Hamamre, Z., Foerster, S., Hartmann, F., Kröger, M., Kaltschmitt, M., 2012. Oil extracted from spent coffee grounds as a renewable source for fatty acid methyl ester manufacturing. Fuel 96, 70–76.

Andrade, K.S., Gonçalvez, R.T., Maraschin, M., Ribeiro-Do-Valle, R.M., Martínez, J., Ferreira, S.R.S., 2012. Supercritical fluid extraction from spent coffee grounds and coffee husks: antioxidant activity and effect of operational variables on extract composition. Talanta 88, 544–552.

Apibalsri, A., Tantayanont, S., Ngamprasertsith, S., 2012. Semi-continuous extraction of Agarwood oil with subcritical water: Optimization of plug flow by response surface methodology. In: Liu, X.H., Jiang, Z., Han, J.T., (Eds.), Advanced Materials Research (Materials Processing Technology, ICAMMP2001), pp. 1721–1724.

Baiocco, D., Lavecchia, R., Natali, S., Zuorro, A., 2016. Production of metal nanoparticles by agro-industrial wastes: a green opportunity for nanotechnology. Chem. Eng. Trans. 47, 67–72.

Barbosa, H.M.A., De Melo, M.M.R., Coimbra, M.A., Passos, C.P., Silva, C.M., 2014. Optimization of the supercritical fluid coextraction of oil and diterpenes from spent coffee grounds using experimental design and response surface methodology. J. Supercrit. Fluids 85, 165–172.

Batish, D.R., Singh, H.P., Kaur, M., Kohli, R.K., Yadav, S.S., 2008. Caffeine affects adventitious rooting and causes biochemical changes in the hypocotyl cuttings of mung bean (*Phaseolus aureus* Roxb.). Acta Physiol. Plant. 30, 401–405.

Bélafi-Bakó, K., Gubicza, L., Mulder, J., 2012. Integration of Membrane Processes into Bioconversions. Springer, New York, US.

Bilbao, M.D.C., Ezequiel, M.E.I., Benavent, M.A., Calleja, M.H., Del Moral, M.P., Artiz, M.U., 2013. Application of products of coffee silverskins in anti-ageing cosmetics and function food, WO 2013/004873.

Bilbao, M.D.C., Saez, N.M., Artiz, M.U., 2014. Food formulation comprising spent coffee grounds, WO 2014/128320 A1.

Bravo, J., Juániz, I., Monente, C., Caemmerer, B., Kroh, L.W., De Peña, M.P., Cid, C., 2012. Evaluation of spent coffee obtained from the most common coffeemakers as a source of hydrophilic bioactive compounds. J. Agr. Food Chem. 60, 12565–12573.

Brazinha, C., Cadima, M., Crespo, J.G., 2015. Valorisation of spent coffee through membrane processing. J. Food Eng. 149, 123–130.

Cerino-Córdova, F.J., Díaz-Flores, P.E., García-Reyes, R.B., Soto-Regalado, E., Gómez-González, R., Garza-González, M.T., Bustamante-Alcántara, E., 2013. Biosorption of Cu(II) and Pb(II) from aqueous solutions by chemically modified spent coffee grains. Int. J. Environ. Sci. Technol. 10, 611–622.

Chan, C.-H., Yusoff, R., Ngoh, G.-C., Kung, F.W.-L., 2011. Microwave-assisted extractions of active ingredients from plants. J. Chromat. A 1218, 6213–6225.

Chan, C.H., Yusoff, R., Ngoh, G.C., 2013. Modeling and prediction of extraction profile for microwave-assisted extraction based on absorbed microwave energy. Food Chem. 140, 147–153.

Chan, C.H., Lim, J.J., Yusoff, R., Ngoh, G.C., 2015a. A generalized energy-based kinetic model for microwave-assisted extraction of bioactive compounds from plants. Sep. Purif. Technol. 143, 152–160.

Chan, C.H., Yusoff, R., Ngoh, G.C., 2015b. Assessment of scale-up parameters of microwave-assisted extraction via the extraction of flavonoids from cocoa leaves. Chem. Eng. Technol. 38, 489–496.

Chan, C.-H., Yeoh, H.K., Yusoff, R., Ngoh, G.C., 2016. A first-principles model for plant cell rupture in microwave-assisted extraction of bioactive compounds. J. Food Eng. 188, 98–107.

Chemat, F., Vian, M.A., 2014. Alternative Solvents for Natural Products Extraction. Springer, Berlin, Heidelberg.

Choi, H.-S., Park, E.D., Park, Y., Suh, H.J., 2015. Spent coffee ground extract suppresses ultraviolet B-induced photoaging in hairless mice. J. Photochem. Photobiol. B 153, 164–172.

Claudio, A.F.M., Ferreira, A.M., Freire, M.G., Coutinho, J.A.P., 2013. Enhanced extraction of caffeine from guarana seeds using aqueous solutions of ionic liquids. Green Chem. 15, 2002–2010.

Conde, T., Mussatto, S.I., 2016. Isolation of polyphenols from spent coffee grounds and silverskin by mild hydrothermal pretreatment. Prep. Biochem. Biotechnol. 46, 406–409.

Couto, R.M., Fernandes, J., Da Silva, M.D.R.G., Simões, P.C., 2009. Supercritical fluid extraction of lipids from spent coffee grounds. J. Supercrit. Fluids 51, 159–166.

Čretnik, L., Škerget, M., Knez, Ž., 2005. Separation of parthenolide from feverfew: performance of conventional and high-pressure extraction techniques. Sep. Purif. Technol. 41, 13–20.

Cruz, M.V., Paiva, A., Lisboa, P., Freitas, F., Alves, V.D., Simões, P., Barreiros, S., Reis, M.A.M., 2014. Production of polyhydroxyalkanoates from spent coffee grounds oil obtained by supercritical fluid extraction technology. Biores. Technol. 157, 360–363.

Cruz, R., Mendes, E., Torrinha, Á., Morais, S., Pereira, J.A., Baptista, P., Casal, S., 2015. Revalorization of spent coffee residues by a direct agronomic approach. Food Res. Int. 73, 190–196.

Daly, J.W., 2007. Caffeine analogs: biomedical impact. Cell. Mol. Life Sci. 64, 2153–2169.

Davila-Guzman, N.E., Cerino-Córdova, F.J., Loredo-Cancino, M., Rangel-Mendez, J.R., Gómez-González, R., Soto-Regalado, E., 2016. Studies of adsorption of heavy metals onto spent coffee ground: equilibrium, regeneration, and dynamic performance in a fixed-bed column. Int. J. Chem. Eng. 2016, 11.

De Melo, M.M.R., Barbosa, H.M.A., Passos, C.P., Silva, C.M., 2014a. Supercritical fluid extraction of spent coffee grounds: measurement of extraction curves, oil characterization and economic analysis. J. Supercrit. Fluids 86, 150–159.

De Melo, M.M.R., Silvestre, A.J.D., Silva, C.M., 2014b. Supercritical fluid extraction of vegetable matrices: applications, trends and future perspectives of a convincing green technology. J. Supercrit. Fluids 92, 115–176.

Desai, B.B., 2000. Handbook of Nutrition and Diet. Marcel Dekker Inc, New York.

Du, F.-Y., Xiao, X.-H., Li, G.-K., 2007. Application of ionic liquids in the microwave-assisted extraction of trans-resveratrol from Rhizma Polygoni Cuspidati. J. Chroma. A 1140, 56–62.

Duan, L., Dou, L.L., Guo, L., Li, P., Liu, E.H., 2016. Comprehensive evaluation of deep eutectic solvents in extraction of bioactive natural products. ACS Sust. Chem. Eng. 4, 2405–2411.

Dupont, J., 2005. Ionic Liquids: Structure, Properties and Major Applications in Extraction/ Reaction Technology. In: Afonso, C.A.M., Crespo, J.G. (Eds.), Green Separation Processes. Wiley VCH, Weinheim.

Erickson, D.R., 1990. Edible fats and oils processing: Basic principles and modern practices. World Conference Proceedings, American Oil Chemists' Society, Champaign, Illinois, p. 267.

Esteban, B., Riba, J.-R., Baquero, G., Rius, A., Puig, R., 2012. Temperature dependence of density and viscosity of vegetable oils. Biomass Bioenerg. 42, 164–171.

Gaonkar, A.G., Zeller, B.L., Bradbury, A., Wragg, A., 2007. Coffee-derived surfactants. US 2007/0259084 A1.

Han, D., Row, K.H., 2010. Recent applications of ionic liquids in separation technology. Molecules 15, 2405.

Heckenbach, M.E., Romero, F.N., Green, M.D., Halden, R.U., 2016. Meta-analysis of ionic liquid literature and toxicology. Chemosphere 150, 266–274.

Hollingsworth, R.G., Armstrong, J.W., Campbell, E., 2003. Caffeine as a novel toxicant for slugs and snails. Annal. Appl. Biol. 142, 91–97.

Huang, H.-C., Wei, C.-M., Siao, J.-H., Tsai, T.-C., Ko, W.-P., Chang, K.-J., Hii, C.-H., Chang, T.-M., 2016. Supercritical fluid extract of spent coffee grounds attenuates melanogenesis through downregulation of the PKA, PI3K/Akt, and MAPK signaling pathways. Evid. Based Compl. Altern. Med. 2016, 11.

International Agency for Research on Cancer (IARC), 1993. IARC Monographs on the Evaluation of Carcinogenic Risks to Humans, Vol. 56, IARC Publications, France, p. 489.

Jeszka-Skowron, M., Zgoła-Grześkowiak, A., Grześkowiak, T., 2015. Analytical methods applied for the characterization and the determination of bioactive compounds in coffee. Eur. Food Res. Technol. 240, 19–31.

Kadam, S.U., Tiwari, B.K., Álvarez, C., O'donnell, C.P., 2015. Ultrasound applications for the extraction, identification and delivery of food proteins and bioactive peptides. Trends Food Sci. Technol. 46, 60–67.

Kalonia, H., Kumar, P., Kumar, A., Nehru, B., 2009. Effect of caffeic acid and rofecoxib and their combination against intrastriatal quinolinic acid induced oxidative damage, mitochondrial and histological alterations in rats. Inflammopharmacology 17, 211–219.

Kang, N.J., Lee, K.W., Shin, B.J., Jung, S.K., Hwang, M.K., Bode, A., Heo, Y.S., Lee, H.J., Dong, Z., 2009. Caffeic acid, a phenolic phytochemical in coffee, directly inhibits fyn kinase activity and UVB-induced COX-2 expression. Carcinogenesis 30 (2), 321–330.

Keskin, S., Kayrak-Talay, D., Akman, U., Hortaçsu, Ö., 2007. A review of ionic liquids towards supercritical fluid applications. J. Supercrit. Fluids 43, 150–180.

Khelil, O., Choubane, S., Cheba, B.A., 2016. Polyphenols content of spent coffee grounds subjected to physico-chemical pretreatments influences lignocellulolytic enzymes production by *Bacillus* sp. R2. Biores. Technol. 211, 769–773.

Kotowski, U., Heiduschka, G., Seemann, R., Eckl-Dorna, J., Schmid, R., Kranebitter, V., Stanisz, I., Brunner, M., Lill, C., Thurnher, D., 2015. Effect of the coffee ingredient cafestol on head and neck squamous cell carcinoma cell lines. Strahl. Onkol. 191, 511–517.

Kulkarni, N., Muddapur, U., 2014. Biosynthesis of metal nanoparticles: a review. J. Nanotechnol. 2014, 8.

Kumar, A.K., Parikh, B.S., Pravakar, M., 2016. Natural deep eutectic solvent mediated pretreatment of rice straw: bioanalytical characterization of lignin extract and enzymatic hydrolysis of pretreated biomass residue. Environ. Sci. Poll. Res. 23, 9265–9275.

Kurniawan, T.A., Chan, G.Y.S., Lo, W.H., Babel, S., 2006. Comparisons of low-cost adsorbents for treating wastewaters laden with heavy metals. Sci. Total Environ. 366, 409–426.

Lipnizki, F., Olsson, J., Tragardh, G., 2002. Scale-up of pervaporation for the recovery of natural aroma compounds in the food industry. Part 1: simulation and performance. J. Food Eng. 54, 183–195.

López, N., Puértolas, E., Condón, S., Álvarez, I., Raso, J., 2008. Effects of pulsed electric fields on the extraction of phenolic compounds during the fermentation of must of Tempranillo grapes. Innov. Food Sci. Emerg. Technol. 9, 477–482.

López-Barrera, D.M., Vázquez-Sánchez, K., Loarca-Piña, M.G.F., Campos-Vega, R., 2016. Spent coffee grounds, an innovative source of colonic fermentable compounds, inhibit inflammatory mediators in vitro. Food Chem. 212, 282–290.

Luengo, E., Álvarez, I., Raso, J., 2013. Improving the pressing extraction of polyphenols of orange peel by pulsed electric fields. Innov. Food Sci. Emerg. Technol. 17, 79–84.

Machado, E.M.S., Rodriguez-Jasso, R.M., Teixeira, J.A., Mussatto, S.I., 2012. Growth of fungal strains on coffee industry residues with removal of polyphenolic compounds. Biochem. Eng. J. 60, 87–90.

Magalhães, L.M., Machado, S., Segundo, M.A., Lopes, J.A., Páscoa, R.N.M.J., 2016. Rapid assessment of bioactive phenolics and methylxanthines in spent coffee grounds by FT-NIR spectroscopy. Talanta 147, 460–467.

Marto, J., Gouveia, L.F., Chiari, B.G., Paiva, A., Isaac, V., Pinto, P., Simões, P., Almeida, A.J., Ribeiro, H.M., 2016. The green generation of sunscreens: using coffee industrial subproducts. Ind. Crops Prod. 80, 93–100.

Mcnaught, A.D., Wilkinson, A., 1997. IUPAC. Compendium of Chemical Terminology. Blackwell Scientific Publications, Oxford.

Mi, L.Y., Bo, K.S., Hae, K.M., Jae, Y.S., Won, P.S., 2015. Method for preparing tattoos from residue after extracting coffee. WO2015080501 (A1).

Michail, A., Sigala, P., Grigorakis, S., Makris, D.P., 2016. Kinetics of ultrasound-assisted polyphenol extraction from spent filter coffee using aqueous glycerol. Chem. Eng. Commun. 203, 407–413.

Misra, M., Kondamudi, N.V., 2008. Methods, systems, and apparatus for obtaining biofuel from coffee and fuels produced therefrom. US 8,591,605 B2.

Mok, K.S., 2013. Pest control composition and method for preparing same. WO2013085345 (A2).

Mukhopadhyay, M., 2000. Natural Extracts Using Supercritical Carbon Dioxide. CRC Press, USA.

Murthy, P.S., Naidu, M.M., 2012. Recovery of phenolic antioxidants and functional compounds from coffee industry by-products. Food Bioproc. Technol. 5, 897–903.

Mussatto, S.I., Ballesteros, L.F., Martins, S., Teixeira, J.A., 2011. Extraction of antioxidant phenolic compounds from spent coffee grounds. Sep. Purif. Technol. 83, 173–179.

Narita, Y., Inouye, K., 2012. High antioxidant activity of coffee silverskin extracts obtained by the treatment of coffee silverskin with subcritical water. Food Chem. 135, 943–949.

Nunes, S.P., Peinemann, K.V., 2006. Membrane Technology: In the Chemical Industry. Wiley, Weinheim.

Obruca, S., Petrik, S., Benesova, P., Svoboda, Z., Eremka, L., Marova, I., 2014. Utilization of oil extracted from spent coffee grounds for sustainable production of polyhydroxyalkanoates. Appl. Microbiol. Biotechnol. 98, 5883–5890.

Oliveira, A., Cabral, L.M.C., Bizzo, H., Arruda, N.P., Freitas, S.P., 2014. Identification and recovery of volatiles organic compounds (VOCs) in the coffee-producing wastewater. J. Water Res. Prot. 06 (04), 6.

Pan, B., Yan, P., Zhu, L., Li, X., 2013. Concentration of coffee extract using nanofiltration membranes. Desalination 317, 127–131.

Parniakov, O., Barba, F.J., Grimi, N., Lebovka, N., Vorobiev, E., 2014. Impact of pulsed electric fields and high voltage electrical discharges on extraction of high-added value compounds from papaya peels. Food Res. Int. 65, 337–343, Part C.

Parniakov, O., Barba, F.J., Grimi, N., Lebovka, N., Vorobiev, E., 2016. Extraction assisted by pulsed electric energy as a potential tool for green and sustainable recovery of nutritionally valuable compounds from mango peels. Food Chem. 192, 842–848.

Páscoa, R.N.M.J., Magalhães, L.M., Lopes, J.A., 2013. FT-NIR spectroscopy as a tool for valorization of spent coffee grounds: application to assessment of antioxidant properties. Food Res. Int. 51, 579–586.

Passos, C.P., Coimbra, M.A., 2013. Microwave superheated water extraction of polysaccharides from spent coffee grounds. Carbohy. Polym. 94, 626–633.

Passos, C.P., Moreira, A.S.P., Domingues, M.R.M., Evtuguin, D.V., Coimbra, M.A., 2014a. Sequential microwave superheated water extraction of mannans from spent coffee grounds. Carbohydr. Polym. 103, 333–338.

Passos, H., Freire, M.G., Coutinho, J.A.P., 2014b. Ionic liquid solutions as extractive solvents for value-added compounds from biomass. Green Chem. 16, 4786–4815.

Pavlović, M.D., Buntić, A.V., Šiler-Marinković, S.S., Dimitrijević-Branković, S.I., 2013. Ethanol influenced fast microwave-assisted extraction for natural antioxidants obtaining from spent filter coffee. Sep. Purif. Technol. 118, 503–510.

Pereira, C.C., Ribeiro, JR., C.P., Nobrega, R., Borges, C.P., 2006. Pervaporative recovery of volatile aroma compounds from fruit juices. J. Memb. Sci. 274, 1–23.

Pereira, M.M., Coutinho, J.A.P., Freire, M.G., 2016. Ionic liquids as efficient tools for the purification of biomolecules and bioproducts from natural sources. In: Bogel-Lukasik, R. (Ed.), Ionic Liquids in the Biorefinery Concept—Challenges and Perspectives. The Royal Society of Chemistry, UK.

Puértolas, E., Barba, F.J., 2016. Electrotechnologies applied to valorization of by-products from food industry: main findings, energy and economic cost of their industrialization. Food Bioprod. Process. 100, 172–184, Part A.

Ramalakshmi, K., Rao, L.J.M., Takano-Ishikawa, Y., Goto, M., 2009. Bioactivities of low-grade green coffee and spent coffee in different in vitro model systems. Food Chem. 115, 79–85.

Ranic, M., Nikolic, M., Pavlovic, M., Buntic, A., Siler-Marinkovic, S., Dimitrijevic-Brankovic, S., 2014. Optimization of microwave-assisted extraction of natural antioxidants from spent espresso coffee grounds by response surface methodology. J. Clean. Prod. 80, 69–79.

Regazzoni, L., Saligari, F., Marinello, C., Rossoni, G., Aldini, G., Carini, M., Orioli, M., 2016. Coffee silver skin as a source of polyphenols: high resolution mass spectrometric profiling of components and antioxidant activity. J. Funct. Foods 20, 472–485.

Rice-Evans, C.A., Miller, N.J., Paganga, G., 1996. Structure-antioxidant activity relationships of flavonoids and phenolic acids. Free Rad. Biol. Med. 20, 933–956.

Sangeeta, D., Lagraff, J.R., 2004. Inorganic Materials Chemistry Desk Reference, second ed. CRC Press, Boca Raton, USA.

Simándi, B., Sass-Kiss, Á., Czukor, B., Deák, A., Prechl, A., Csordás, A., Sawinsky, J., 2000. Pilot-scale extraction and fractional separation of onion oleoresin using supercritical carbon dioxide. J. Food Eng. 46, 183–188.

Simões, J., Nunes, F.M., Domingues, M.D.R.M., Coimbra, M.A., 2010. Structural features of partially acetylated coffee galactomannans presenting immunostimulatory activity. Carbohy. Polym. 79, 397–402.

Simões, J., Nunes, F.M., Domingues, M.R., Coimbra, M.A., 2013. Extractability and structure of spent coffee ground polysaccharides by roasting pre-treatments. Carbohy. Polym. 97, 81–89.

Smith, A., 2002. Effects of caffeine on human behavior. Food Chem. Toxicol. 40, 1243–1255.

Teh, S.-S., Birch, E.J., 2014. Effect of ultrasonic treatment on the polyphenol content and antioxidant capacity of extract from defatted hemp, flax and canola seed cakes. Ultrason. Sonochem. 21, 346–353.

Tello, J., Viguera, M., Calvo, L., 2011. Extraction of caffeine from Robusta coffee (*Coffea canephora* var. Robusta) husks using supercritical carbon dioxide. J. Supercrit. Fluids 59, 53–60.

Tian, M., Yan, H., Row, K.H., 2009. Solid-phase extraction of caffeine and theophylline from green tea by a new ionic liquid-modified functional polymer sorbent. Analy. Lett. 43, 110–118.

Tokimoto, T., Kawasaki, N., Nakamura, T., Akutagawa, J., Tanada, S., 2005. Removal of lead ions in drinking water by coffee grounds as vegetable biomass. J. Colloid Interf. Sci. 281, 56–61.

Toschi, T.G., Cardenia, V., Bonaga, G., Mandrioli, M., Rodriguez-Estrada, M.T., 2014. Coffee silverskin: characterization, possible uses, and safety aspects. J. Agr. Food Chem. 62, 10836–10844.

Tran, K.-Q., 2016. Fast hydrothermal liquefaction for production of chemicals and biofuels from wet biomass: the need to develop a plug-flow reactor. Biores. Technol. 213, 327–332.

Urgert, R., Van Der Weg, G., Kosmeijer-Schuil, T.G., Van De Bovenkamp, P., Hovenier, R., Katan, M.B., 1995. Levels of the cholesterol-elevating diterpenes cafestol and kahweol in various coffee brews. J. Agr. Food Chem. 43, 2167–2172.

Vincze, I., Vatai, G., 2004. Application of nanofiltration for coffee extract concentration. Desalination 162, 287–294.

Wang, L., Weller, C.L., 2006. Recent advances in extraction of nutraceuticals from plants. Trends Food Sci. Technol. 17, 300–312.

Weschenfelder, T.A., Lantin, P., Viegas, M.C., De Castilhos, F., Scheer, A.D.P., 2015. Concentration of aroma compounds from an industrial solution of soluble coffee by pervaporation process. J. Food Eng. 159, 57–65.

Xu, H., Wang, W., Liu, X., Yuan, F., Gao, Y., 2015. Antioxidative phenolics obtained from spent coffee grounds (*Coffea arabica* L.) by subcritical water extraction. Ind. Crops Prod. 76, 946–954.

Yang, L.X., Nazari, L., Yuan, Z.S., Corscadden, K., Xu, C.B., He, Q., 2016. Hydrothermal liquefaction of spent coffee grounds in water medium for bio-oil production. Biomass Bioenerg. 86, 191–198.

Yun, Y.S., Park, M.H., Hong, S.J., Lee, M.E., Park, Y.W., Jin, H.J., 2015. Hierarchically porous carbon nanosheets from waste coffee grounds for supercapacitors. ACS Appl. Mater. Interf. 7, 3684–3690.

Zakaria, S.M., Kamal, S.M.M., 2016. Subcritical water extraction of bioactive compounds from plants and algae: applications in pharmaceutical and food ingredients. Food Eng. Rev. 8, 23–34.

Zuorro, A., Lavecchia, R., 2012. Spent coffee grounds as a valuable source of phenolic compounds and bioenergy. J. Clean. Prod. 34, 49–56.

Applications of recovered compounds in food products

6

Maria D. del Castillo*, Amaia Iriondo-DeHond*, Nuria Martinez-Saez*,
Beatriz Fernandez-Gomez*, Maite Iriondo-DeHond**, Jin-Rong Zhou†

*Institute of Food Science Research (UAM–CSIC), Madrid, Spain; **Madrid Institute
for Research and Rural Development, Agriculture, and Food, Alcalá de Henares, Spain;
†Beth Israel Deaconess Medical Center, Boston, MA, United States

ABSTRACT

Approximately 90% of the edible parts of the cherry are discarded during its conversion into coffee brew. This chapter summarizes applications of coffee by-products as novel ingredients possessing biological, nutritional, and technological functions. Coffee cherries and their derivatives have potential as superfoods because they are composed of several phytochemicals with synergic biological effects and nutrients. They also have components with technological interest, which may be used as natural colorants, aromas, and texturizers, among others. Novel food products based on coffee wastes are being welcomed by Western consumers due to their well-known health-promoting and sensorial properties. Nowadays, several tasty foods and beverages from roasted coffee beans and coffee cherry by-products are available for commercialization. The number of commercial products based on coffee by-products has increased in the last few years, with the aim to achieve sustainable production and consumption of the coffee cherry.

Keywords: additive; bioactive compounds; coadjuvant; coffee by-products; food ingredients; functional foods; functional beverages; nutrients; phytochemicals.

6.1 INTRODUCTION

Coffee brew has taken a long journey to arrive in your cup. There are generally 10 steps of coffee processing that end in the coffee cup: planting, harvesting, and processing the cherries, drying, milling, and exporting the beans, tasting the coffee, roasting the green coffee beans, grinding the roasted coffee beans, and brewing coffee powder. Coffee is a cherry made up of different anatomic parts with different

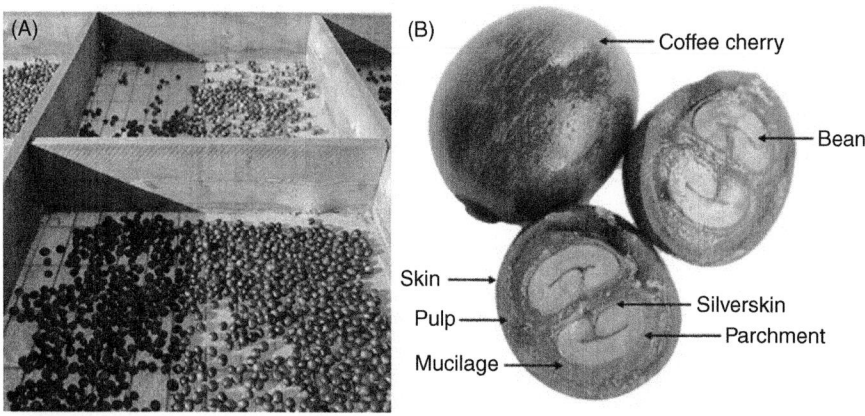

FIGURE 6.1

Coffee cherries (A), coffee cherry Aanatomy (B).

appearances and compositions of nutrients and bioactive phytochemicals possessing diverse preventive and therapeutic effects for chronic diseases (Fig. 6.1).

Coffee is mainly commercialized as a beverage obtained from ground, roasted beans. During this processing, over 90% of the coffee cherry is discarded as an agricultural waste or by-product, which contains health-promoting phytochemicals and nutrients. According to the European Commission, a food waste is defined as "food (including inedible parts) lost from the food supply chain, not including food diverted to material uses such as bio-based products, animal feed, or sent for redistribution" (e.g., food donation). The members of the European Union are required to establish frameworks to collect and report levels of food waste across all sectors in a comparable way. The latest data are requested to develop national food-waste prevention plans, aimed toward reaching the objective to reduce food waste by at least 30% between January 1, 2017 and December 31, 2025. As a consequence, the European coffee sector should find new ways for improving its sustainability. The future of coffee depends on sustainability since it is one of the most frequently consumed drinks in the planet. Total production in 2014–15 was estimated at 141.9 million 60 kg bags (International Coffee Organization, 2015). A wide range of initiatives impact every part of the supply coffee chain: improving the conditions at origin, recycling packaging materials, reducing emissions, developing ecofriendly facilities, and designing new coffee products. Interest in that field has greatly stimulated collaboration between the coffee industry and academics. An updated review of the application and marketing of products from coffee wastes has been recently published (Galanakis et al., 2015). The definition, chemical composition, and potential applications of coffee by-products are provided in Chapter 1 of the present book as well as in Chapter 12 of the book entitled *Coffee: Chemistry, Quality and Health Implications* (del Castillo et al., 2017). The potential of coffee silverskin (CS) (unique roasting coffee waste) as a functional ingredient in treatments for chronic diseases has also been reported (del Castillo et al., 2016; Fernandez-Gomez et al., 2016).

New applications and innovative products from coffee wastes are frequently proposed in scientific publications and symposia, patent applications, and marketing materials. The number of patents and scientific papers on the value of coffee wastes has increased in the last decade. The present chapter intends to provide an updated overview of all those food products developed from coffee wastes or by-products with the final goal of improving human health and/or achieving sustainability in the coffee sector. Table 6.1 provides information about areas of interest with respect to the use of coffee by-products in food applications, either as novel food ingredients or as functional foods. Alternative sustainable solutions for upgrading coffee processing by-products, such as husks, silverskin, and spent coffee grounds (SCGs) are described too.

6.2 COFFEE BY-PRODUCTS AS HIGH VALUE-ADDED SUSTAINABLE FOOD INGREDIENTS

6.2.1 NUTRIENTS

Humans require a wide range of essential nutrients for normal growth and development. Essential nutrients, including vitamins, minerals, amino acids, fatty acids, water, and fiber, must be obtained through foods and beverages because they cannot, for the most part, be endogenously synthesized. Nutrients are divided into two groups: macronutrients (dietary components that provide energy, or proteins, fats, and carbohydrates) and micronutrients (comprising vitamins and minerals that are required in small quantities to ensure normal metabolism, growth, and physical wellbeing) (US Department of Agriculture, 2015). Dietary fiber is also a very important nutrient for human health. As a consequence, according to the US Food and Drug Administration, dietary fiber is a mandatory label nutrient, and as a nondigestible carbohydrate, it is considered part of total carbohydrates for US labeling (www.fda. gov/Food/GuidanceRegulation/GuidanceDocumentsRegulatoryInformation/LabelingNutrition/ucm064894.htm). Vegetable food wastes provide very important natural sources of dietary fiber.

Coffee pulp (the mesocarp) represents 29% of the dry weight of the whole coffee cherry and it is mechanically removed by pressing the coffee fruit in a depulper. Coffee pulp is mainly composed of carbohydrates (44%–50%), proteins (10%–12%), and fibers (18%–21%), and it also contains appreciable amounts of polyphenols (1.48%) and caffeine (1.3%) (del Castillo et al., 2017). This coffee by-product has been characterized as a source of soluble dietary fiber for use in food and pharmaceutical applications (www.pectcof.com). Due to its properties as an emulsifier and stabilizer, this extracted coffee dietary fiber is a promising new ingredient for the food and beverage industry.

Table 6.1 Coffee By-Products' Applications in Foods due to Their Chemical Composition

By-Product	Nutrients	Bioactive Compounds	Others	Application in Foods
Skin		X	X	Ingredient
Pulp	X	X	X	Additive, ingredient
Husks		X	X	Additive, ingredient
Mucilage		X		Coadjuvant

Table 6.1 Coffee By-Products' Applications in Foods due to Their Chemical Composition (*cont.*)

By-Product	Nutrients	Bioactive Compounds	Others	Application in Foods
Parchment	X	X		—
Silverskin	X	X		Ingredient
Spent coffee grounds	X	X	X	Coadjuvant, ingredient
Wastewater[a]	—	—	—	—

[a]Coffee wastewater, *also known as coffee effluent, is another by-product of coffee processing.*

Esquivel and Jiménez (2012) defined coffee husks as the product obtained from the dry processing of coffee berries, composed of the outer skin, pulp, and parchment of the coffee berry. However, other authors also called husks as the by-product formed by skin and pulp produced during wet coffee processing (Prata and Oliveira, 2007). Coffee husks enclose coffee beans and comprise nearly 45% of the berry. They are high in carbohydrates (35%–85%), soluble fiber (30.8%), minerals (3%–11%), and proteins (5%–11%). They are also rich in insoluble dietary fiber and can be a source of phytochemicals such as tannins (5%–9%) and cyanidins (20%) for the food and pharmaceutical industries (del Castillo et al., 2017). Therefore, coffee husks have great potential as a food ingredient and as a natural source of nutrients and bioactive compounds.

The coffee mucilage portion (pectin layer) is located between the pulp and the parchment (Fig. 6.1), and represents 5% of the dry weight of the berry. Mucilage is composed of water (84.2%), protein (8.9%), sugar (4.1%), pectic substances (0.91%), and ash (0.7%) (del Castillo et al., 2017). According to the International Coffee Organization, coffee mucilage could be used in foods as an unrefined source of pectin, antioxidants, and flavonoids. All these compounds have raised special interest in the food industry (Rathinavelu and Graziosi, 2005).

The parchment is a strong fibrous endocarp that covers both hemispheres of the coffee seed and separates them from each other. It represents 5.8% of the dry weight of the berry. Coffee parchment is formed by (α-)cellulose (40%–49%), hemicellulose (25%–32%), lignin (33%–35%), and ash (0.5%–1%) (del Castillo et al., 2017). No applications of coffee parchment as a food ingredient have been found. However, considering the chemical composition of this by-product, new applications as a natural source of dietary fiber in the food industry should be considered.

CS represents 4.2% (w/w) of the coffee cherry. It is a thin tegument of the outer layer of the two beans forming the green coffee seed and is a by-product of the roasting process. CS has a high dietary fiber content (68%–80%) (Pourfarzad et al., 2013); polysaccharides are also abundant components (60%–70%) and total sugar content varies greatly in this by-product (1.6%–12%). CS contains protein, fat, and ash, at 16.2%–19.0%, 1.56%–3.28%, and 5%–7%, respectively (del Castillo et al., 2017). Borrelli et al. (2004) reported that the composition of CS, obtained from several Italian roasting plants, showed a high amount of soluble dietary fiber (about 14% of the total fiber) and very high antioxidant activity.

SCGs make up the residual material obtained during the treatment of coffee powder with hot water to prepare coffee infusion or steam for instant coffee preparation. Polysaccharides are the main components of SCGs derived from instant coffee production (75%). SCGs also contain dietary fiber (43%–54%), protein (11.2%–17.4%), fat (2.3%–24.3%), ash (0.5%–1.6%), minerals, caffeine, chlorogenic acids, and so on (del Castillo et al., 2017; Martinez-Saez et al., 2017).

The use of SCGs as antioxidant dietary fiber in healthy bakery products has been proposed (del Castillo et al., 2014b; Martinez-Saez et al., 2017). This ingredient could be directly applied in the manufacture of pastry and confectionery foods such as bread, cookies, and breakfast cereals, among others, making it a simple, low-cost alternative. SCGs from the manufacture of instant coffee are a natural source of antioxidant insoluble dietary fiber, proteins, essential amino acids, and low-glycemic sugars (Martinez-Saez et al., 2017).

6.2.2 BIOACTIVE COMPOUNDS

The high content of phenolic compounds and inhibitors of α-glucosidase and α-amylase and powerful antioxidant capacity of coffee fruit skin suggest its potential as a food ingredient with health-promoting properties (Murthy et al., 2012).

Coffee pulp has been proposed as a potential source of anthocyanins. They are plant pigments widely distributed in colored fruits and flowers. Interest in anthocyanins has emerged because of their potential health benefits as antioxidant, anticarcinogenic, antiinflammatory, and hypoglycemic agents and enhancers of insulin sensitivity (Lila, 2004). Murthy et al. (2012) reported cyanadin-3-rutinoside as the major anthocyanin in coffee pulp.

Mullen et al. (2013) studied the composition of the husks of coffee fruits. In this study, the authors analyzed the profile of polyphenolic and hydroxycinnamate compounds contained in six coffee husks grown in three different countries (China, India, and Mexico). The variation in both the quantitative and qualitative profiles of the husk samples was large. Robusta husks from India had the highest total polyphenolic content at 553.1 μg/g, mainly due to the high level of flavan-3-ol procyanidin A type trimers. The arabica sample from Mexico had the highest level of flavonols at 260.6 μg/g, but only 114.6 μg/g of flavan-3-ols. All husks samples contained low levels of chlorogenic acid and caffeine (Mullen et al., 2013).

Coffee husks have also been used in the production of other food ingredients, such as fructooligosaccharides (FOSs), which are fructose oligomers with low caloric values, noncarcinogenicity properties, and decreasing levels of phospholipids, triglycerides, and cholesterol, helping gut absorption of calcium and magnesium and stimulating the bifidobacterial growth in the human colon (Mussatto et al., 2009). The increasing demand for FOSs, mainly as an ingredient in food applications due to their functional properties, has led to the development of different production processes. FOSs are produced on an industrial scale from sucrose by microbial enzymes that have transfructosylating activity (β-fructofuranosidase, or FFase). Mussatto et al. (2009) analyzed different lignocellulosic materials, including coffee husks, brewers' spent grain, wheat straw, corncobs, cork oak, and loofah sponges, for evaluating FOS and FFase production from submerged fermentation of sucrose (200 g/L) by immobilized *Aspergillus japonicus*. Although corncobs returned the highest production of FOSs and FFase, both corncobs and coffee husks returned the highest results in terms of microorganism immobilization (1.49 and 1.46 g/g carrier, respectively) and the transfructosylating activities that are responsible for the conversion of sucrose to FOSs.

CS is considered a good source of bioactive compounds, particularly chlorogenic acids. The compounds 5-caffeoylquinic acid and 3-caffeoylquinic acid are the most common chlorogenic acids in CS, with levels of 1.99 mg/g and 1.48 mg/g, respectively (Bresciani et al., 2014). As a consequence, CS has been proposed as a natural source of several ingredients besides antioxidants, such as prebiotic carbohydrates (Borrelli et al., 2004). Coffee silverskin extract (CSE), the aqueous extract obtained from CS described by del Castillo et al. (2013) in patent number WO 2013/004873

FIGURE 6.2 Coffee Silverskin Extract (CSE)

(Fig. 6.2), has also been proposed as a bioactive ingredient. CSE is obtained through environmentally friendly technologies. The extraction of phytochemicals and nutrients from the food matrix can take place with water at a temperature of about 100°C for at least 10 min (low technology). The extraction process can be performed under subcritical water conditions at a temperature of 50°C and a pressure of about 1500 psi, without the prior grinding step, spraying, or equivalent method (high technology). For long-term storage, the extracts obtained by the process described in the present patent application can be frozen and lyophilized or spray-dried. CSE enriched with chlorogenic acid and caffeine was proposed as a natural antioxidant and/ or natural preservative for food. In addition, it can be used as a nutritional supplement to prevent pathological processes (del Castillo et al., 2014a; Fernandez-Gomez et al., 2016; Martinez-Saez et al., 2014) and as an antiaging cosmetic ingredient (Iriondo-DeHond et al., 2016).

CS has also been used in the production of FOSs by solid-state fermentation (SSF), due to its great potential for immobilizing cells, as well as its serving as a nutrient source for microorganisms. Mussatto et al. (2013) optimized the conditions for moisture content, temperature, and inoculum rate (*A. japonicus*) to maximize the production of FOSs and the FFase enzyme using CS in an SSF system. Temperatures ranges of 26–30°C and an inoculum rate of 2×10^7 spores/gram dry material produced the highest content of FOSs (208.8 g/L) and FFase (64.12 units U/mL).

In addition, Mussatto et al. (2015) analyzed different fermentation techniques for FOS production in terms of their economic aspects and environmental impact. The fermentation processes evaluated were submerged free-cell fermentation (FCF) of a sucrose solution using *A. japonicus*, submerged immobilized-cell fermentation (ICF) using corncobs, and SSF using CS as a support material and nutrient source. The analysis identified SSF using CS as the most interesting process as it provided the highest FOS annual productivity (232.6 t) and purity (98.6%), and was economically and environmentally more favorable than the other processes evaluated as it reached the highest annual profit and lowest payback time, and generated the lowest waste-water volume and carbon footprint, respectively. Moreover, SSF using CS produced greater amounts of the shorter chains of FOSs, which have more prebiotic activity and stronger sweetness (Mussatto et al., 2015)

SCGs can be directly employed as a sustainable natural source of dietary fiber (del Castillo et al., 2014a). However, it may be also extracted from the raw material employing different processes, such as ohmic technology (Campos-Vega et al., 2016; Vazquez Sanchez et al., 2015), alkaline hydrogen peroxide treatment (Vilela et al., 2016), and autohydrolysis (Ballesteros et al., 2017). The antioxidant character of the coffee fiber remains after the application of any of these technological processes to the raw material. SCG dietary fiber can be fermented by colon microbiota, producing short-chain fatty acids (SCFAs) with the ability to prevent inflammation (López-Barrera et al., 2016). Metabolites produced by colonic fermentation of SCGs exhibited strong antiinflammatory potential by suppressing nitric oxide production and inhibiting inflammatory mediators such as IL-10, CCL-17, CXCL9, IL-1β, and IL-5 cytokines. These effects were associated with SCFAs being released from SCG dietary fiber during colonic fermentation. As a consequence, SCGs have been proposed as a protective agent against the onset and/or progression of chronic inflammatory diseases, such as inflammatory bowel disease and rheumatoid arthritis. Coffee dietary fiber has recently been reported to stimulate the release of the satiety hormone, serotonin, and glucagon-like peptide-1 (Martinez-Saez et al., 2016). In addition, the antioxidant polysaccharides extracted from SCGs by autohydrolysis can be employed in encapsulating other bioactive food ingredients, such as prebiotics (Ballesteros et al., 2017). Antioxidant polysaccharides can be recovered from raw SCGs by thermal treatment at 160°C for 10 min, using a liquid/solid ratio of 15 mL water/g SCGs. A material containing 29.29% (w/w) of polysaccharides was obtained, from which galactose was the most representative sugar, followed by mannose, glucose, and arabinose. The polysaccharides presented thermostability in a large range of temperatures. They are therefore of great interest for industrial applications, mainly in the food industry (for example, for the encapsulation of additives or as prebiotics), due to their high antioxidant potential and other functional properties (Ballesteros et al., 2017). Moreover, SCGs have been proposed as a raw material for producing a bioactive extract enriched by caffeine (Brazinha et al., 2015).

Taking into account the chemical composition of coffee by-products, each anatomic part could be considered a natural source of bioactive compounds due to the presence of phytochemicals with potential beneficial health effects (del Castillo et al., 2017; Esquivel and Jiménez, 2012).

6.2.3 ADDITIVES AND COADJUVANTS

According to the US Food and Drug Administration, a food additive is "any substance the intended use of which results or may reasonably be expected to result, directly or indirectly, in its becoming a component or otherwise affecting the characteristics of any food." This definition includes any substance used in the production, processing, treatment, packaging, transportation, or storage of food (http://www.fda.gov/Food/IngredientsPackagingLabeling/FoodAdditivesIngredients/ucm094211.htm - foodadd). On the other hand, the Spanish Agency for Consumer Affairs, Food Safety, and Nutrition defines food coadjuvants as substances intentionally added to food products in order to improve some technological process (http://www.aecosan.msssi.gob.es/AECOSAN/web/seguridad_alimentaria/subdetalle/coadyuvantes_tecnologicos.htm).

The natural food-colorant industry is growing 10%–15% annually, and awareness is increasing in developed countries about the harmful effects and consequences of using synthetic colorants (Hartati et al., 2012). Coffee pulp has been proposed as a potential source of anthocyanins for natural food colorants. In the wet process, coffee pulp is removed before drying and its color is rapidly degraded by the action of enzymes or oxygen. As a result, large amounts of natural colorants are wasted. Cyanidin 3-rutinoside is characterized as the dominant anthocyanin, responsible for the red color observed in the outside of the coffee berry (Murthy et al., 2012). Prata and Oliveira (2007) carried out the extraction of monomeric and polymeric anthocyanins in five varieties of coffee cherries. There were no significant differences in anthocyanin content among the studied varieties, and the average content of monomeric anthocyanins in coffee husks was 19.2 mg of pigment per 100 g of fresh husks. Due to the extensive amounts of coffee produced annually in Brazil, the world's largest coffee producer, coffee husks possess great potential as an economical source of anthocyanins (Prata and Oliveira, 2007).

The constituents of coffee pulp, such as fermented sugars, protein, and cellulose, indicate their high potential for biotechnological processes. Research proves the use of coffee pulp as an excellent carbon source for enzymatic production of β-glucosidase by *Bacillus subtilis* CCMA 0087 (Bhoite and Murthy, 2015). The β-glucosidase enzymes are generally used in the ethanol, juice, pharmaceutical, beverage, and cosmetic industries (Krisch et al., 2010). Bearing in mind that these enzymes are extensively used in industrial processes and that they are obtained at low concentrations, it is important to search for new natural substrates for production and for alternative strains, as well as to establish better production conditions for β-glucosidases. Dias et al. (2015) showed that a new isolate of bacteria could produce β-glucosidase from coffee pulp under ideal conditions. The highest β-glucosidase production (22.59 UI/mL) was reached in 24 h of culturing, at a coffee-pulp concentration of 36.8 g/L, a temperature of 36.6°C, and a pH of 3.64. This study also achieved promising results in the generation of value-added products.

Recently, coffee pulp was fermented with yeast strains to produce flavor compounds. Thirty-five compounds corresponding to the following six groups of volatile compounds were detected: higher alcohols, acetates, ethyl esters, aldehydes, terpenes, and volatile acids. These compounds can contribute to floral and fruity flavors and have commercial interest for their use as food ingredients (Bonilla-Hermosa et al., 2014). Soares et al. (2000) studied the production of fruity flavor from *Ceratocystis fimbriata* grown on steam-treated coffee husks supplemented with glucose, leucine, soybean oil, and a salt solution. Strong pineapple and banana aroma compounds were formed during fermentation, when different concentrations of glucose (20%–46%) were used (Soares et al., 2000). In this sense, natural flavor from coffee by-products could also be used in the food industry.

Coffee husks have been employed in the production of food additives such as citric acid. An SSF system was carried out on coffee husks by employing *Aspergillus niger*. Data indicated that 1.5 g citric acid was produced per 10 g dry coffee husk, with a conversion of 80% of the sugar consumed. It was estimated that a commercial SSF plant could process 5 tons of coffee husk and produce 0.75 tons of citric acid/day, thereby making coffee husks an attractive substrate for the production of citric acid, an additive widely used in juices, candies, and sauces (Shankaranand and Lonsane, 1994).

SCGs have been employed as enzyme immobilization solid carriers instead of expensive commercialized solid carriers for a natural and economical catalytic system. β-glucosidase was covalently immobilized onto SCGs for the conversion of isoflavone glycosides into their aglycones in black soymilk. The total aglycone content in black soymilk was incremented by 67% with an enzymatic treatment duration of 60 min (Chen et al., 2013). The efficiency of an SCG-immobilized enzyme system was evaluated based on the changes in isoflavone content (genistin, daidzin, genistein, and daidzein) in black soymilk before and after deglycosylation. Aglycone forms of isoflavones exhibited higher biological activity (Zheng et al., 2003; Zubik and Meydani, 2003) and are metabolized faster than their glycosides (Izumi et al., 2005; Kawakami et al., 2005). For this reason, the intake of isoflavone aglycone–rich soyfoods might be more effective for purposes of health enhancement. The use of SCGs as coadjuvants can be considered an economic way to prepare aglycone-enriched black soymilk.

SCGs presented an antimicrobial effect on *Staphylococcus aureus* and *Escherichia coli* (Jiménez-Zamora et al., 2015). This antimicrobial activity might be related to the presence of coffee melanoidins in the SCG structure. In fact, the antimicrobial property of coffee melanoidins extracted from SCGs was 2–5 times higher than that assayed alone. Therefore, they could also be used as preservatives in foods at large concentrations. In this vein, other authors obtained spent coffee extract powder with a high antioxidant capacity after defatting and freeze-drying the extract (Bravo et al., 2013). This powder could be used in the food industry as an ingredient or additive with potential preservation and functional properties.

6.2.4 SAFETY CONCERNS ABOUT COFFEE BY-PRODUCTS AS FOOD INGREDIENTS

Limitations on the use of coffee by-products, such as that of pulp in animal feed, as well as in particular populations, such as children and pregnant women, are connected to its high content of tannins and caffeine, respectively. However, coffee husks and pulp are rich in carbohydrates, soluble fibers, minerals, proteins, and bioactive compounds, so they can be considered a good source of phytochemicals for animal feed and for the human food industry. The caffeine content present in coffee husks can be diminished through its degradation by microorganisms (Mazzafera, 2002). Aparicio García and del Castillo (2016) generated a coffee-husk extract following the procedure indicated in patent number WO 2013/004873, for use as a food ingredient. The caffeine content of this extract was below the European Food Safety Authority (EFSA) safety level for daily caffeine consumption of 400 mg for the general population and 200 mg for lactating women. For children and adolescents, the available information is insufficient to derive a safe level of caffeine intake (European Food Safety Authority (EFSA), 2015). Therefore, no limitations on the use of this extract as a food ingredient for human nutrition need be considered. Coffee by-products present different levels of caffeine. However, in all cases their content is lower than that found in green and roasted coffee beans. Although further studies should be conducted in this field, so far published results suggest caffeine content need not be considered as a safety concern in the application of coffee by-products as food ingredients.

All foods can contain chemical and/or biological contaminants. Numerous commonly consumed foods, such as french fries, chips, bread, cookies, and coffee contain acrylamide, a chemical-processing contaminant. Acrylamide is formed during the roasting of coffee beans through the Maillard reaction, during which the aroma and the color of coffee beans are also produced. This reaction takes place when precursors, for example, reducing sugars, such as glucose and fructose, and asparagine, are present in raw materials, in combination with a high temperature and long cooking time (Pedreschi et al., 2014). Recent studies from Garcia-Serna et al. (2014) have shown that decreasing the quantity of sugar added to cookies containing CS and stevia might be a good strategy for obtaining a safe, low-sugar product. To date, no other studies have focused on the impact of CS addition on acrylamide content. The addition of CS improved the physical properties and the nutritional value of stevia cookies, and no bioaccesible acrylamide was detected in the digests of these new innovative cookies. This indicates that CS could be used as a natural coloring and source of dietary fiber to achieve a healthier, nutritious, and safe cookie.

Ochratoxin A (OTA) is a biological contaminant that occurs in various foodstuffs and beverages, including cereals, wine, coffee, spices, beer, cocoa, dried fruits, and pork meat. OTA is a mycotoxin produced by *Aspergillus ochraceus* and *Penicillium verrucosum* that tends to bioaccumulate along the food chain. This mycotoxin can induce renal toxicity, nephropathy, and immunosuppression, representing a risk to human safety. Therefore, its content in foods should be determined (Toschi et al., 2014). OTA is already present in coffee before storage, as contamination can occur due to several climatic conditions, coffee fruits falling onto the soil, transportation, and so on. Therefore, the critical steps leading to the accumulation of this mycotoxin are the harvesting and the postharvest handling of coffee cherries (Napolitano et al., 2007). However, Ferraz et al. (2010) demonstrated that OTA can be destroyed during roasting. Coffee is considered a secondary source of OTA in the human diet. Even when the coffee beverage is prepared from highly contaminated green beans, the coffee-transforming process is able to reduce the amount of OTA that presents a risk to human health (Napolitano et al., 2007). Research carried out by Toschi et al. (2014) suggests that CS could be safe as a source of bioactive compounds (such as fiber and polyphenols) that could be used as ingredients in the pharmaceutical/cosmetic industries or in the development of functional foods. However, as in other food ingredients, it is very important to establish rigorous quality controls along the coffee-bean processing chain to reduce OTA and ensure the healthiness and the sensory quality of the coffee and its by-products (Napolitano et al., 2007).

Heavy metals are widely dispersed in the environment and can be found in varying concentrations in human food. Food contamination is a serious problem, as heavy metals exert a harmful influence on many tissues. Metals disturb ionic balance and mineral regulation, induce oxidative damage to cell structures, produce injury to DNA, and induce cancer transformations (Waalkes, 2003). Nędzarek et al. (2013) studied Mn, Co, Ni, Cr, and Ag levels in coffee. Such levels were shown to be too low to influence human health. However, some coffees had high levels of Pb, which might be harmful if accumulated in the body. This indicates that such products need to be controlled for metal contamination. As with any ingredient, the proper management of by-products is a key step to ensure the safety of these products. Critical control-point procedures for supervising the collection of coffee by-products, cooling/freezing the material, drying or thermal stabilization, and/or addition of chemical preservatives can provide solutions in the case of coffee by-products.

6.3 APPLICATIONS IN SUSTAINABLE AND HEALTH-PROMOTING BEVERAGES

6.3.1 COFFEE HUSKS AND COFFEE PULP

In coffee producing countries, coffee husks (coffee skin and pulp) constitute a source of severe contamination and serious environmental problems. For this reason, efforts are being made to develop alternatives for their utilization as raw materials for the production of foods and beverages, among other applications. The use of fresh or processed coffee pulp has been the subject of numerous studies, which in general lead to the conclusion that coffee by-products and wastes can be used in a variety of applications (Rathinavelu and Graziosi, 2005).

At least 1100 years ago, traders brought coffee across the Red Sea into Arabia (now Yemen), where Muslims started making wine from the pulp of the fermented coffee berries. This beverage was known as qishr and was used during religious

FIGURE 6.3

Coffee skin known as cascara (A); cascara tea beverage (B).

ceremonies. Today, qishr is known as an infusion, made of spiced coffee husks, ginger, and cinnamon. In Yemen, it is usually used as a drink instead of coffee, because it is cheaper (Hestler, 2000).

A decade ago, Aida Batlle, a coffee grower in El Salvador, prepared an infusion made from coffee skins. She named this beverage "cascara tea," a reference to the word for coffee skin in Spanish. Aida Batlle first used the skin of the coffee cherry in "naturals," which are dried and then milled. This cascara product is in the form of fine flakes, similar to the tea in a tea bag. Then she began to make cascara using the skin of coffee berries that were washed to remove the pulp before drying, a process that is used in the production of most coffee in Central America and Colombia. In this way, the cascara dries like a raisin and when brewed reveals the shape of the coffee cherry (Fig. 6.3) (http://www.npr.org/sections/thesalt/2015/12/01/456796760/cascara-tea-a-tasty-infusion-made-from-coffee-waste). According to Batlle, cascara has a fruity taste, while others describe it as smelling like herbal tea, rose hips, hibiscus, mango, or cherry. The coffee husk is brewed alone or with spices, such as cinnamon. It is suggested that this tea contains caffeine, based on the perceived energized experience after consumption. However, no scientific studies have yet determined the presence of caffeine in coffee husks (http://theroasterspack.com/blogs/news/14918821-cascara-the-coffee-cherry-tea-with-a-how-to-brew-guide).

Two commercial beverages based on coffee husks are attracting the interest of Western consumers. Bai Brands is a beverage company founded in 2009 in Princeton, New Jersey. Bai uses what it calls superfruit extract as an ingredient in its beverages. The term "superfruit" refers to the outside part of the coffee cherry, known as the coffee husk, which contain antioxidants and caffeine (http://www.drinkbai.com/). Bai beverages can be found in different supermarkets in the United States.

KonaRed Corporation is a Hawaiian coffee company that also produces beverages from coffee husks. Hawaiian Coffeeberry juice is a beverage containing the outside parts of the coffee cherry. This beverage contains the majority of the nutrients from the coffee plant and the dominant polyphenols found in the coffee fruit, which are chlorogenic acid, quinic acid, and ferulic acid (https://www.konared.com/).

6.3.2 COFFEE SILVERSKIN

CS has been used together with roasted coffee powder, cocoa powder, and golden coffee to obtain innovative coffee blends (Ribeiro et al., 2014). The new blends are rich in bioactive compounds, such as chlorogenic acids, trigonelline, theobromine, and caffeine. Moreover, the espresso obtained from the functional coffee blend shows a higher content of these bioactive compounds compared to regular blends (coffee blends in a sealed package with a one-way degassing valve and commercial coffee blends in capsules). It also presented a high antioxidant capacity and favorable sensory characteristics.

Novel antioxidant beverages based on raw CS and CS extract (CSE) from Arabica and Robusta coffees have been developed to study their inhibitory effect on in vivo fat accumulation using *Caenorhabditis elegans* as an in vivo model (Martinez-Saez et al., 2014) (Fig. 6.4).

Chlorogenic acids and coffee melanoidins may play a role in reducing and controlling body weight, and may therefore be of interest in treating and reducing the risk of obesity (Cho et al., 2010; Murase et al., 2011). A significant dose-dependent effect on reducing body-fat accumulation was found for pure chlorogenic acid (3.54 mg/L) and caffeine (4.85 mg/L), achieving a 30% and a 29% reduction of lipid deposits, respectively. As was expected, the brews of Arabica and Robusta CSE (100 µg/mL),

(A)   (B)

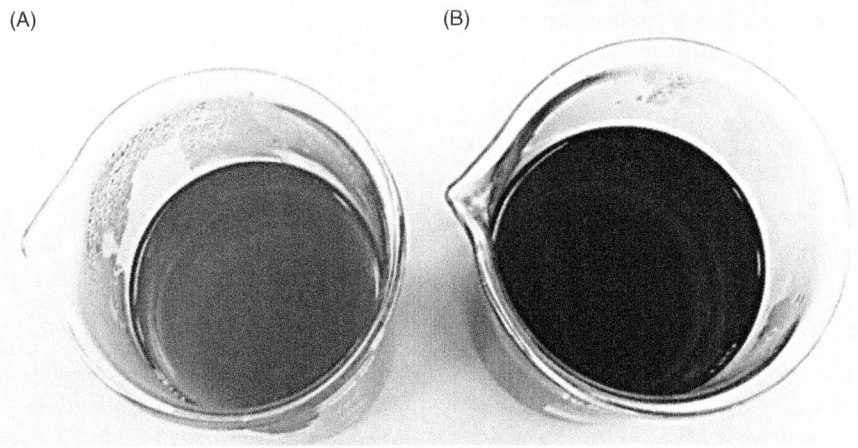

FIGURE 6.4

Visual appearance of beverages prepared with arabic (A) and robusta (B) coffee silverskin extract at 2.5 mg/mL.

which contained physiologically active doses of these compounds, reduced body fat by 21% and 24%, respectively (Martinez-Saez et al., 2014). Furthermore, both Robusta CSE beverages and a commercial dietary supplement that is made from Robusta decaffeinated green coffee extract showed a similar effect on body-fat reduction. Therefore, a novel functional beverage has been made from roasted CSE as a natural alternative to dietary supplements for the prevention of overweight and obesity (Fig. 6.4).

6.3.3 SPENT COFFEE GROUNDS

There is a rising interest in searching for added-value uses for SCGs. Sampaio et al. (2013) successfully used SCGs for the production of a distilled beverage with a coffee aroma. The process was based on the aqueous extraction of aromatic compounds from SCGs, supplemented with sugar and the production of ethanol. Seventeen volatile compounds were identified in the distillate (including alcohols, esters, aldehydes, and acids), all of them in concentrations able to promote pleasant characteristics in the product. A spirit made from SCGs was considered to have the features of a pleasant beverage, with the smell and taste of coffee. Its organoleptic properties were acceptable and different from those of commercial spirits (Sampaio et al., 2013).

6.4 APPLICATIONS IN SUSTAINABLE AND HEALTHIER FOODS
6.4.1 COFFEE HUSKS AND COFFEE PULP

Several studies have evaluated the use of coffee husks as a dietary supplement for cattle, swine, fish, sheep, chicken, and horses (Oliveira and Franca, 2014). The use of coffee husks as a supplement in cattle diets is considered feasible, with the limits for feed substitution in the range of 30%–40%. In the case of pigs, the use of coffee husks was also considered to be technically and economically feasible at inclusion levels in the range of 5%–10% (Franca and Oliveira, 2009). Mullen et al. (2013) suggested the use of the whole coffee husk in energy drinks, since not all polyphenols may be extractable.

The pulp has also been proposed for silage to reduce meat production costs. A study by de Carvalho Couto Filho et al. (2007) prepared residue silage made up of a mixture of mango husks and pulp, with 30% coffee husks. The addition of coffee husks improved the fermentative standard for silages of good quality (de Carvalho Couto Filho et al., 2007).

A new interesting ingredient, coffee flour, has been developed from coffee pulp. It presents high fiber and ash content (18% and 8%, respectively) and a low fat level (1.6%) (Ramirez Velez and Jaramillo Lopez, 2015). This recently developed coffee flour has been proposed for use in different food formulations, such as breads, cookies, muffins, squares, brownies, pastas, sauces, and beverages. Coffee flour possesses 5 times more fiber than whole-grain wheat flour, 84% less fat, and

42% more fiber than coconut flour, and is gluten free. This product does not have a coffee taste, but rather expresses floral, citrus, and roasted-fruit notes (http://www.coffeeflour.com/).

Coffee flour is a revolutionary new ingredient made from dried and ground coffee cherries, the fruit that grows around the coffee bean and is traditionally discarded during the coffee harvest. This flour is nutritious, delicious, gluten free, beneficial to coffee farmers, and good for the environment. Coffee flour contains more iron than spinach, more antioxidants than pomegranates, more protein than kale, more potassium than bananas, and more fiber than whole-grain wheat flour. It also contains less fat and more fiber than coconut flour (http://www.coffeeflour.com/).

The high content of dietary fiber in coffee husks constitutes a problem for the development of a beverage. With higher density, the fibers would probably generate a nonhomogeneous drink with a fibrous layer at the bottom. To include the fiber in a new food product, the development of energy bars through grinding the whole coffee husk and thereby including all the antioxidants and fiber in the product should be considered (Bondesson, 2015). As concentrations of both chlorogenic acid (Farah and Donangelo, 2006) and procyanidins (Esatbeyoglu et al., 2015) are decreased by heating, there is an advantage in making a raw, uncooked, and gluten-free bar.

Coffee mucilage contains high quantities of carbohydrates, reducing and nonreducing sugars, and pectin compounds (del Castillo et al., 2017). The particular chemical composition of mucilage was used to produce honey coffee. The process under patent to obtain honey coffee describes the use of pectolytic enzymes to facilitate the release of the cellular content of mucilage and pectin degradation. This enzymatic treatment increases the stability of mucilage and the content of phenolic compounds. The mucilage concentrate, also called honey coffee (for its high sugar content), was produced by dehydration under vacuum at 65°C, creating a product with minimum nutrient loss, high digestibility, and palatability. The dry matter of honey coffee is mainly composed of reduced amino acids and nonreducing sugars (WO 2013088203 A1). Honey coffee has a shelf life of at least six months at a temperature range of 18–30°C; these conditions will maintain its organoleptic and microbiological quality. The patented treatment begins with mucilage at 17–11 °Bx and reaches 55 °Bx after the removal of water.

Earnest Eats has a line of energized hot cereal that contains antioxidants and natural caffeine from the coffee cherry (http://www.earnesteats.com/product/a-m-trail-mix-energized-hot-cereal/). There is another commercial product containing coffee flour available in supermarkets: Jcoco, a chocolate line crafted by the Seattle Chocolate Company, which was the first to produce a new product containing coffee flour: an infused chocolate bar. Expressing smoky, citrus, and roasted-fruit notes, coffee flour enhances the distinct flavor profile and nutritional value of the new bar. Jcoco Arabica cherry espresso in dark chocolate delivers a smooth taste with a slightly grainy texture from dried Arabica cherries and freshly ground espresso-bean inclusions (http://www.jcocochocolate.com/arabicacherry).

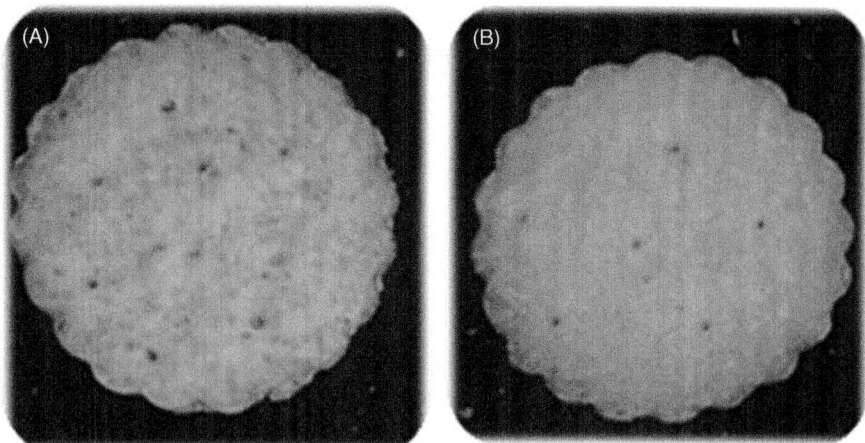

FIGURE 6.5

Visual appearance of cookies prepared with arabica coffee silverskin (CS) (A) and aqueous coffee silverskin extract (B).

6.4.2 COFFEE SILVERSKIN

CS has been employed as dietary fiber for the formulation of innovative bakery products, specifically bread (Pourfarzad et al., 2013). Results showed the feasibility of using alkaline hydrogen peroxide CS as a food ingredient to reduce caloric density and increase the dietary fiber content of bread.

CS has also been used in the formulation of novel cookies (Garcia-Serna et al., 2014) (Fig. 6.5). Cookie formulations were designed using stevia as a sweetener and CS as a natural colorant and source of dietary fiber. CS improved some of the quality attributes of the cookies, such as moisture, texture, thickness, and color. Regarding the processing of chemical contaminants, hydroxymethylfurfural was greatly reduced and no bioaccessible acrylamide was detected in the digests of the new innovative cookies. The nutritional value of the cookies was also improved.

6.4.3 SPENT COFFEE GROUNDS

SCGs can be directly applied in the manufacture of pastry and confectionery foods such as bread, cookies, and breakfast cereals, among others, resulting in a simple, low-cost method (Martinez-Saez et al., 2017). The developed formulation employed SCGs in diverse combinations with other innovative ingredients, such as a hypocaloric sweetener (stevia) and soluble fiber (FOS). SCGs (4% w/w) can be used directly as a food ingredient in solid foods such as cookies (Fig. 6.6) without affecting conventional food preparation and the final quality of the product. Furthermore, these food formulations presented low content in Maillard-reaction products (acrylamide, hydroxymethylfurfural, and advanced glycation end products). Therefore, the novel cookies containing SCGs might be consumed by people with reduced energetic

FIGURE 6.6 Visual Appearance of Cookie Prepared with Robusta SCG

intake and/or other particular requirements. The application of SCGs represents a value-added opportunity for coffee by-product utilization at a very low cost.

SCGs have been also employed in the creation of foods and beverages containing mannooligosaccharides, which possess blood-pressure reducing and/or elevation-suppressing effects (Takao et al., 2009) and also abdominal fat-reducing effects (Asano et al., 2006). These mannooligosaccharides may be produced by hydrolysis of mannans from coffee materials and then incorporated into foods and beverages. This way, an economical and simple food or drink with beneficial effects on the reduction of blood pressure and fat can be incorporated into ordinary eating habits in the general population.

6.5 FUTURE RESEARCH

Considering the chemical composition of the coffee cherry, there are other by-products (mucilage and parchment, see Fig. 6.1) generated during coffee processing that also have potential to be used in the food industry. Novel food legislation in the European Union must be considered when developing a new ingredient from coffee by-products (Turck et al., 2016).

There are no applications for the use of some coffee by-products in the food industry found in the literature. Considering their chemical composition, new applications for mucilage and parchment should be considered, as has been done with other by-products. In addition, coffee wastewater (the by-product generated when coffee beans are processed by the wet method) should also be considered for use in the food industry, besides its use in the production of electricity (Rattan et al., 2015).

6.6 CONCLUSIONS

Coffee cherry wastes or by-products can be transformed into several healthy and tasty foods and beverages, which can significantly reduce waste production and make coffee processing sustainable. Systematic studies are needed for a better understanding of the health-promoting properties of compounds forming different anatomic parts of the cherry and also inedible parts of the plant. The total recycling of the cherry and inedible parts of the coffee plant into health-promoting foods and beverages with particular health claims may contribute to a sustainable nutrition for the population. Results so far obtained support the feasibility of this goal.

ACKNOWLEDGMENTS

The SUSCOFFEE Project (AGL2014-57239-R): Sustainable coffee production and consumption: Valorization of coffee waste into food ingredients funded this work. N. Martinez-Saez thanks the Autonomous University of Madrid, Spain, for her FPI predoctoral fellowship. A. Iriondo-DeHond is a fellow of the FPI-predoctoral program of the Ministry of Economy and Competitiveness (BES-2015-072191). M. Iriondo-DeHond thanks IMIDRA for her predoctoral fellowship. M.D. del Castillo thanks the Ministry of Education, Culture, and Sport for the grant with reference PRX16/00386.

The authors thank Finca Los Castaños, Valle de Agaete, Gran Canaria, Spain, for the coffee cherry photographs (Fig. 6.1); Austin Leih, Trio Craft Coffee, Flower Mound, Texas, for the cascara tea photograph (Fig. 6.3), and Andres Belalcazar (Pectcof B.V., Netherlands) for the pulp picture (Table 6.1).

REFERENCES

Aparicio García, N., del Castillo, M.D., 2016. Evaluación del potencial de subproductos de café como fuente de compuestos bioactivos. Estudio comparativo. Master's thesis. Universidad Autónoma de Madrid, Madrid, Spain.

Asano, I., Fujii, S., Mutoh, K., Takao, I., Ozaki, K., Nakamuro, K., Matsushima, T., 2006. Mannooligosaccaride composition for body fat reduction. WO 2006/036208.

Ballesteros, L.F., Teixeira, J.A., Mussatto, S.I., 2017. Extraction of polysaccharides by autohydrolysis of spent coffee grounds and evaluation of their antioxidant activity. Carbohydr. Polymers 157, 258–266.

Bhoite, R.N., Murthy, P.S., 2015. Biodegradation of coffee pulp tannin by *Penicillium verrucosum* for production of tannase, statistical optimization and its application. Food Bioprod. Process. 94, 727–735.

Bondesson, E., 2015. A nutritional analysis on the by-product coffee husk and its potential utilization in food production. Bachelor's thesis, series 415. Swedish University of Agricultural Sciences, Uppsala, Sweden.

Bonilla-Hermosa, V.A., Duarte, W.F., Schwan, R.F., 2014. Utilization of coffee by-products obtained from semi-washed process for production of value-added compounds. Biores. Technol. 166, 142–150.

Borrelli, R.C., et al., 2004. Characterization of a new potential functional ingredient: coffee silverskin. J. Agric. Food Chem. 52 (5), 1338–1343.

Bravo, J., et al., 2013. Influence of extraction process on antioxidant capacity of spent coffee. Food Res. Int. 50 (2), 610–616.

Brazinha, C., Cadima, M., Crespo, J.G., 2015. Valorisation of spent coffee through membrane processing. J. Food Eng. 149, 123–130.

Bresciani, L., et al., 2014. Phenolic composition, caffeine content and antioxidant capacity of coffee silverskin. Food Res. Int. 61, 196–201.

Campos-Vega, R. et al., 2016. Proceso de obtención de fibra dietaria antioxidante natural de subproductos mediante calentamiento ohmico y compuesto alto en fibra dietaria antioxidante natural de café usado. MX/a/2016008578.

Chen, K.I., et al., 2013. Enrichment of two isoflavone aglycones in black soymilk by using spent coffee grounds as an immobiliser for β-glucosidase. Food Chem. 139 (1–4), 79–85.

Cho, A.-S., et al., 2010. Chlorogenic acid exhibits anti-obesity property and improves lipid metabolism in high-fat diet-induced-obese mice. Food Chem. Toxicol. 48 (3), 937–943.

de Carvalho Couto Filho, C.C., et al., 2007. Quality of mango residue silage with different additives. Ciência e Agrotecnologia 31 (5), 1537–1544.

del Castillo, M.D. et al., 2013. Application of products of coffee silverskin in anti-ageing cosmetics and functional food. WO 2013/004873.

del Castillo, M.D., Fernandez-Gomez, B., Ullate, M., Mesa, M.D., 2014a. Uso de productos de la cascarilla de café para la prevención y tratamiento de las patologías que conforman el síndrome metabólico y de sus factores de riesgo. P201431848.

del Castillo, M.D., Martinez-Saez, N., Ullate, M., 2014b. Healthy bakery products with high level of dietary antioxidant fibre. PCT/ES2014/070062.

del Castillo, M.D., et al., 2016. Coffee silverskin extract for aging and chronic diseases. In: Martirosyan, D.M. (Ed.), Functional Foods in Health and Disease. Colorado Springs, CreateSpace Independent Publishing Platform, Colorado.

del Castillo, M.D., et al., 2017. Coffee by-products. In: Farah, A. (Ed.), Coffee: Chemistry, Quality and Health Implications. RSC Publishing, Cambridge.

Dias, M., et al., 2015. A new alternative use for coffee pulp from semi-dry process to β-glucosidase production by *Bacillus subtilis*. Lett. Appl. Microbiol. 61 (6), 588–595.

Esatbeyoglu, T., Wray, V., Winterhalter, P., 2015. Isolation of dimeric, trimeric, tetrameric and pentameric procyanidins from unroasted cocoa beans (*Theobroma cacao* L.) using countercurrent chromatography. Food Chem. 179, 278–289.

Esquivel, P., Jiménez, V.M., 2012. Functional properties of coffee and coffee by-products. Food Res. Int. 46 (2), 488–495.

European Food Safety Authority (EFSA), 2015. Scientific opinion on the safety of caffeine. EFSA J. 13 (5), 1–21.

Farah, A., Donangelo, C.M., 2006. Phenolic compounds in coffee. Braz. J. Plant Physiol. 18 (1), 23–36.

Fernandez-Gomez, B., Ramos, S., et al., 2016. Coffee silverskin extract improves glucose-stimulated insulin secretion and protects against streptozotocin-induced damage in pancreatic INS-1E beta cells. Food Res. Int. 89, 1015–1022.

Ferraz, M.B.M., et al., 2010. Kinetics of ochratoxin A destruction during coffee roasting. Food Control 21 (6), 872–877.

Franca, A.S., Oliveira, L.S., 2009. Coffee processing solid wastes: current uses and future perspectives. In: Ashworth, G.S., Azevedo, P. (Eds.), Agricultural Wastes. Nova Science Publishers, New York (Chapter 8).

Galanakis, C., et al., 2015. Patented and commercialized applications. In: Galanakis, C. (Ed.), Food Waste Recovery. Academic Press-Elsevier, London, pp. 337–360.

Garcia-Serna, E., et al., 2014. Use of coffee silverskin and stevia to improve the formulation of biscuits. Pol. J. Food Nutr. Sci. 64 (4), 243–251.

Hartati, I., et al., 2012. Potential production of food colorant from coffee pulp. Prosiding SNST ke-3 Tahun 1 (1), 66–71.

Hestler, A., 2000. Yemen. Cavendish Square Publishing, New York.

International Coffee Organization, 2015. Annual Review 2014–2015. Report. International Coffee Organization, London.

Iriondo-DeHond, A., et al., 2016. Coffee silverskin extract protects against accelerated aging caused by oxidative agents. Molecules 21 (6), 1–14.

Izumi, T., et al., 2005. Soy isoflavone aglycones are absorbed faster and in higher amounts than their glucosides in humans. J. Nutr. 130 (7), 1695–1699.

Jiménez-Zamora, A., Pastoriza, S., Rufián-Henares, J.A., 2015. Revalorization of coffee by-products. Prebiotic, antimicrobial and antioxidant properties. LWT Food Sci. Technol. 61 (1), 12–18.

Kawakami, Y., et al., 2005. Comparison of regulative functions between dietary soy isoflavones aglycone and glucoside on lipid metabolism in rats fed cholesterol. J. Nutr. Biochem. 16 (4), 205–212.

Krisch, J., et al., 2010. Characteristics and potential use of β-glucosidases from Zygomycetes. Current Research, Technology and Education Topics in Applied Microbiology and Microbial Biotechnology (2), 891–896.

Lila, M.A., 2004. Anthocyanins and human health: an in vitro investigative approach. J. Biomed. Biotechnol. 2004 (5), 306–313.

López-Barrera, D.M., et al., 2016. Spent coffee grounds, an innovative source of colonic fermentable compounds, inhibit inflammatory mediators in vitro. Food Chem. 212, 282–290.

Martinez-Saez, N., et al., 2014. A novel antioxidant beverage for body weight control based on coffee silverskin. Food Chem. 150, 227–234.

Martinez-Saez, N. et al., 2016. Efecto de la formulación de galletas en la secreción de hormonas de saciedad (Effect of biscuit formulations in the release of satiety hormones). Innotec, pp. 65–72.

Martinez-Saez, N., et al., 2017. Use of spent coffee grounds as food ingredient in bakery products. Food Chem. 216, 114–122.

Mazzafera, P., 2002. Degradation of caffeine by microorganisms and potential use of decaffeinated coffee husk and pulp in animal feeding. Sci. Agric. 59 (4), 815–821.

Mullen, W., et al., 2013. Polyphenolic and hydroxycinnamate contents of whole coffee fruits from China, India, and Mexico. J. Agric. Food Chem. 61 (22), 5298–5309.

Murase, T., et al., 2011. Coffee polyphenols suppress diet-induced body fat accumulation by downregulating SREBP-1c and related molecules in C57BL/6J mice. Am. J. Physiol-Endoc. M. 300 (1), E122–E133.

Murthy, P.S., et al., 2012. Extraction, characterization and bioactivity of coffee anthocyanins. Eur. J. Biol. Sci. 4 (1), 13–19.

Mussatto, S.I., et al., 2009. Fructooligosaccharides and β-fructofuranosidase production by *Aspergillus japonicus* immobilized on lignocellulosic materials. J. Mol. Catal. B-Enzym. 59 (1–3), 76–81.

Mussatto, S.I., et al., 2013. Maximization of fructooligosaccharides and β-fructofuranosidase production by *Aspergillus japonicus* under solid-state fermentation conditions. Food Bioprocess Technol. 6 (8), 2128–2134.

Mussatto, S.I., et al., 2015. Economic analysis and environmental impact assessment of three different fermentation processes for fructooligosaccharides production. Biores. Technol. 198, 673–681.

Napolitano, A., et al., 2007. Natural occurrence of ochratoxin A and antioxidant activities of green and roasted coffees and corresponding by-products. J. Agric. Food Chem. 55 (25), 10499–10504.

Nędzarek, A., et al., 2013. Concentrations of heavy metals (Mn, Co, Ni, Cr, Ag, Pb) in coffee. Acta Biochim. Pol. 60 (4), 623–627.

Oliveira, L.S., Franca, A.S., 2014. An overview of the potential uses for coffee husks. Coffee in Health and Disease Prevention. Academic Press, Elsevier, London.

Pedreschi, F., Mariotti, M.S., Granby, K., 2014. Current issues in dietary acrylamide: formation, mitigation and risk assessment. J. Sci. Food Agric. 94 (1), 9–20.

Pourfarzad, A., Mahdavian-Mehr, H., Sedaghat, N., 2013. Coffee silverskin as a source of dietary fiber in bread-making: optimization of chemical treatment using response surface methodology. LWT Food Sci. Technol. 50 (2), 599–606.

Prata, E.R.B.A., Oliveira, L.S., 2007. Fresh coffee husks as potential sources of anthocyanins. LWT Food Sci. Technol. 40 (9), 1555–1560.

Ramirez Velez, A., Jaramillo Lopez, J.C., 2015. Process for obtaining honey and/or flour of coffee from the pulp or husk and the mucilage of the coffee bean. US 20,150,017,270 A1.

Rathinavelu, R., Graziosi, G., 2005. Potential alternative use of coffee wastes and by-products. Paper. International Coffee Organization, WP-Board No. 942/03, ICS-UNIDO, Italy, pp.1–4.

Rattan, S., et al., 2015. A comprehensive review on utilization of wastewater from coffee processing. Environ. Sci. Poll. R. 22 (9), 6461–6472.

Ribeiro, V.S., et al., 2014. Chemical characterization and antioxidant properties of a new coffee blend with cocoa, coffee silverskin and green coffee minimally processed. Food Res. Int. 61, 39–47.

Sampaio, A., et al., 2013. Production, chemical characterization, and sensory profile of a novel spirit elaborated from spent coffee ground. LWT Food Sci. Technol. 54 (2), 557–563.

Shankaranand, V.S., Lonsane, B.K., 1994. Coffee husk: an inexpensive substrate for production of citric acid by *Aspergillus niger* in a solid-state fermentation system. World J. Microbiol. Biotechnol. 10 (2), 165–168.

Soares, M., et al., 2000. Fruity flavour production by *Ceratocystis fimbriata* grown on coffee husk in solid-state fermentation. Process Biochem. 35 (8), 857–861.

Takao, I., Asano, I., Fujii, S., Kaneko, M., Nielson, J.R., Steffen, D.G., Hatzold, T., 2009. Composition having effect of lowering blood pressure and/or inhibiting increase in blood pressure and food and drink containing the same. EP20060712241.

Toschi, T.G., et al., 2014. Coffee silverskin: characterization, possible uses, and safety aspects. J. Agric. Food Chem. 62 (44), 10836–10844.

Turck, D., et al., 2016. Guidance on the preparation and presentation of an application for authorisation of a novel food in the context of Regulation (EU) 2015/2283. EFSA J. 14 (11), 4594.

US Department of Agriculture, 2015. Scientific Report of the 2015 Dietary Guidelines Advisory Committee. Report. Available from: https://health.gov/dietaryguidelines/2015-scientific-report/pdfs/scientific-report-of-the-2015-dietary-guidelines-advisory-committee.pdf

Vazquez Sanchez, K. et al., 2015. Antioxidant coffee dietary fiber for gastrointestinal health and diabetes. In Martirosyan, D.M. (Ed.), Functional and Medical Foods for Chronic Diseases: Bioactive Compounds and Biomarkers. 20th International Conference of the Functional Food Center, 22–23 September 2016, Boston, MA.

Vilela, W.F., et al., 2016. Effect of peroxide treatment on functional and technological properties of fiber-rich powders based on spent coffee grounds. Int. J. Food Eng. 2 (1), 42–47.

Waalkes, M.P., 2003. Cadmium carcinogenesis. Mutat. Res. Fund. Mol. Mech. Mut. 533 (1–2), 107–120.

Zheng, Y., et al., 2003. Rapid gut transit time and slow fecal isoflavone disappearance phenotype are associated with greater genistein bioavailability in women. J. Nutr. 133 (10), 3110–3116.

Zubik, L., Meydani, M., 2003. Bioavailability of soybean isoflavones from aglycone and glucoside forms in American women. Am. J. Clin. Nutr. 77 (6), 1459–1465.

FURTHER READING

Ballesteros, L.F., Teixeira, J.A., Mussatto, S.I., 2014. Chemical, functional, and structural properties of spent coffee grounds and coffee silverskin. Food Bioprocess Technol. 7 (12), 3493–3503.

Applications of recovered bioactive compounds in cosmetics and other products

7

Francisca Rodrigues, Maria Antónia Nunes, Rita C. Alves, M. Beatriz P.P. Oliveira

REQUIMTE/LAQV, University of Porto, Porto, Portugal

ABSTRACT

The world consumption of coffee generates large amounts of by-products. However, the use of these wastes for animal feed has not been possible due to the presence of antiphysiological and antinutritional factors, such as tannins and caffeine. Nevertheless, coffee silverskin (CS), spent coffee grounds (SCGs), coffee husks, and immature/defective coffee beans, the industrial by-products of coffee processing, have a potential use in skin-care formulations, based on their high content of bioactive compounds. On the other hand, considering that life expectancy has increased in developed countries, raising new concerns about skin and body appearance, the demand for cosmetics with natural ingredients is stronger than ever, representing a challenge for the worldwide economy and society. This chapter revises the potential application of bioactive compounds from coffee by-products to active ingredients for skin-care products. Their potential UV protective action, emollient capacity, antiwrinkle, and antimicrobial activities are critically reviewed and discussed.

Keywords: coffee by-products; cosmetic; skin; epidermis; dermis; aging

7.1 INTRODUCTION

In the last years, new active ingredients have been introduced into a number of cosmetic formulations due to consumer demand for more effective products that improve not only the appearance, but also the health of the skin (Park et al., 2016). Particularly, the use of natural ingredients for cosmetic purposes has been attracting significant attention, due to their potential safeness, multiple potential biological actions, and cost-effectiveness (Marto et al., 2016). In addition to bioactivity, natural products are, in general, considered to be not harmful for humans, not expensive, suitable to be used in a wide range of applications, and, most of the time, obtained from reusable sources (Marto et al., 2016). Nowadays, the complex biological process characterized by a gradual loss of the physiological integrity of the skin—called aging—represents one of the principal concerns of modern society. Skin has historically been used for the topical delivery of compounds. It is much more than a static, impenetrable shield against external insults. Rather, skin is a dynamic, complex, integrated arrangement of cells, tissues and matrix elements that regulates body heat and water loss, while preventing the invasion of toxic substances and microorganisms. However, the mechanisms of action of these bioactive compounds and their effectiveness on the skin have to be verified, since the *stratum corneum* provides a significant barrier to the transport of those ingredients into the body. The *stratum corneum* is the outermost nonviable layer of the skin, interfacing with the outside world as the barrier that prevents unwanted materials from entering, and excessive loss of water from exiting, the body. Structurally, the *stratum corneum* is a thin (~15 µm) heterogeneous layer and it comprises layers of terminally differentiated and keratinized epidermal cells separated by an intercellular lipid domain (Menon et al., 2012). This so-called brick wall results from the arrangement of the corneocytes within the lipid protein matrix, which is "welded" together by corneodesmosomes and is embedded in an intercellular matrix of a complex mixture of lipids (mortar). The lipid-enriched component mainly consists of ceramides, cholesterol, and fatty acids (Schurer and Elias, 1991). Besides keratinocytes (which synthesize the protein keratin), the viable epidermis is composed of melanocytes, Merkel cells, and Langerhans cells (Prow et al., 2011). The epidermis is composed of four separate layers formed by the differing stages of keratin maturation: *stratum basale* (the basal or germinativum cell layer), *stratum spinosum* (the spinous or prickle cell layer), *stratum granulosum* (the granular cell layer), and *stratum corneum* (the horny layer). The *stratum lucidum* represents the transition from the *stratum granulosum* and *stratum corneum* and is not usually seen in thin epidermis.

Deeper skin layers make up the viable epidermis: the dermis (that provides the structural support of the skin) and the subcutis, or hypodermis (the connective tissue layer, which is an important depot of fat). The dermis is composed of two layers: a thin papillary layer and a thicker reticular layer. The papillary layer is composed of loosely arranged collagen fibers while the reticular layer is mainly composed of dense collagen fibers arranged in parallel to the surface of the skin. The dermis is also rich in fibroblasts, elastin, proteoglycans, and immunological cells (such as mast cells and macrophages). The collagen fibers are responsible for dermis strength and toughness, while elastin maintains the elasticity and flexibility of skin. It is also possible to observe appendages, such as sweat glands and hair roots, and dermal vasculature. Finally, the hypodermis is constituted of adipocytes. This skin layer is mainly responsible for insulation, body heat, and acting as a shock absorber.

It is generally accepted that the safety evaluation of cosmetics is based on the safety evaluation of each individual ingredient. According to European Commission Regulation 1223/2009, a cosmetic product available on the market must be safe for human health when used normally or under reasonably foreseeable conditions (European Commission Regulation, 2009). The safeness is guaranteed by an independent safety assessor with proof of qualification. The assessment of the toxicological potential is the first step in the hazard evaluation of an ingredient that consists of a series of distinct toxicity studies, specific to distinct toxicological endpoints. Animal testing of cosmetic products in the European Union (EU) has been prohibited since 2004, and animal testing of cosmetic ingredients since 2009 (the testing ban) (European Commission Directive, 2003). Since March 2009, it has also been prohibited in the EU to market cosmetic products containing ingredients that have been tested on animals (the marketing ban). For the most complex human-health effects (repeated-dose toxicity, including skin sensitization and carcinogenicity, reproductive toxicity, and toxicokinetics), the deadline for the marketing ban was extended to March 2013 (European Commission Directive, 2003). The European Commission's Scientific Committee on Consumer Safety (SCCS) is responsible for recommending guidelines for the cosmetics and raw material industries to develop adequate studies for the safety evaluation of cosmetics. Safety requirements for cosmetic ingredients are listed in "Notes of Guidance for the Testing of Cosmetic Ingredients and Their Safety Evaluation" (European Commission SCCS, 2012). General safety requirements for a regulated cosmetic ingredient include the following parameters: (1) acute toxicity, (2) corrosivity and irritation, (3) skin sensitization, (4) dermal/percutaneous absorption, (5) repeated dose toxicity, (6) reproductive toxicity, (7) mutagenicity/genotoxicity, (8) carcinogenicity, (9) toxicokinetics studies, (10) photo-induced toxicity, and (11) human data (European Commission SCCS, 2012). All the new cosmetic ingredients such as those obtained from food by-products, have to be evaluated prior to going to market. As far as we know, until now few studies have been performed to evaluate the toxicity and safety of coffee by-products as potentially new cosmetic ingredients.

Coffee consumption is growing worldwide, as it is extremely embedded in the cultural habits of many countries, increasing the total import and export of coffee around the world. *Coffea arabica* and *Coffea canephora* var *robusta* are the two plant coffees most cultivated, especially in those countries localized in the equatorial Latin America, South Asia, India, and Africa (Barbulova et al., 2015). In this context, the amount of coffee by-products is extremely high, being mainly made up of immature/defective coffee beans, coffee husks, coffee silverskin (CS), and spent coffee grounds (SCGs) (Fig. 7.1).

These wastes have arisen as potential candidates to replace synthetic chemicals as active ingredients in skin-care and cosmetic formulations, since they are a rich source of antioxidants and polyphenols, as well as caffeine (Ribeiro et al., 2013). For example, phenolic compounds are excellent candidates for the prevention of the harmful effects of ultraviolet (UV) radiation to the skin. More specifically, flavonoids have photoprotection potential due to their UV-absorbing capacity, their ability to act as antioxidants, and as antiinflammatory and immunomodulatory agents. Nevertheless, the efficacy of cosmetic products has been recognized as an important public issue and it is usually evaluated by in vivo assays performed in accordance with various guidelines. Currently cosmetic and cosmeceutical formulations with extracts from the coffee fruit, which contains chlorogenic acid (CGA), condensed proanthocyanidins, and quinic and ferulic acids, have been studied with positive results for skin care (Esquivel and Jiménez, 2012; Rodrigues et al., 2016c).

Coffee grounds

Coffee husks

Coffee silverskin

Spent coffee grounds

Immature/defective
Coffee beans

FIGURE 7.1 Coffee By-Products With Cosmetic Interest

Research on the use of coffee by-products as cosmetic ingredients has greatly accelerated recently, generating different papers on strategies to use food by-products to improve the activities of natural compounds on the skin and the possible mechanisms regarding the process. This chapter aims to provide a critical review on the aspects related to the use of phytochemicals obtained from coffee by-products as cosmetic ingredients. In addition, it discusses the obstacles found in bringing these new active ingredients to market and analyzes future perspectives.

7.2 ACTIVE COMPOUNDS OF COFFEE PROCESSING BY-PRODUCTS OF INTEREST FOR COSMETIC PROPOSES

Coffee by-products have been extensively studied by different research groups to identify and quantify the most representative bioactive compounds. Thus, a brief description of the composition of each coffee by-product is of huge importance to allow the evaluation of these wastes as potential active ingredients for cosmetic purposes. Table 7.1 summarizes the predominant compounds identified in coffee by-products that could have interesting cosmetic activities.

7.2.1 COFFEE SILVERSKIN

CS has been intensively studied by different research groups around the world (Bresciani et al., 2014; Costa et al., 2014; Martinez-Saez et al., 2014; Narita and Inouye, 2012; Pourfarzad et al., 2013; Rodrigues et al., 2015a,b, 2016b; Toschi et al., 2014). This coffee waste is the only by-product obtained during the coffee-bean roasting process, and large amounts are produced by roasters in coffee-consuming countries (Narita and Inouye, 2014). Regarding its macronutritional composition, CS presents a high content of dietary fiber (50%–60%), followed by protein (16%–19%), fat (1.56%–3.28%), and ash (7%) (Borrelli et al., 2004; Napolitano et al., 2007; Pourfarzad et al., 2013). The mineral composition has not been totally clarified so far,

Table 7.1 Cosmetic Activities Potential Cosmetic Applications of Coffee By-Products Based on the Bioactivities of Their Major Chemical Compounds.

Compound	Cosmetic Activities	Coffee By-Product	References
Caffeine	Anticellulite activity, antiaging activity, protection against UV damage	CS, SCGs, coffee husks, immature/defective coffee beans	Chiari et al. (2014); Kitagawa et al. (2011); Marto et al. (2016); Mazzafera (1999); Ramalakshmi et al. (2008)
Caffeoylquinic acids/feruloylquinic acids/p-coumaroylquinic acids	Antiaging activity, protection against UV damage, antimicrobial activity, antiinflammatory activity	CS, SCGs, coffee husks, immature/defective coffee beans	Kitagawa et al. (2011); Rodrigues et al. (2015a); Rodrigues et al. (2015b)
Fatty acids • Linoleic acid • Palmitic acid • Oleic acid • Stearic acid	Emollient properties	SCGs, immature/defective coffee beans	Campos-Vega et al. (2015); Franca and Oliveira (2008); Ribeiro et al. (2013)
Minerals	Emollient properties, hydration	CS, SCGs, coffee husks	Ballesteros et al. (2014); Esquivel and Jiménez (2012); Jiménez-Zamora et al. (2015); Murthy and Naidu (2012); Mussatto et al. (2011b)

CS, Coffee silverskin; SCGs, spent coffee grounds.

but according to different authors it is mainly composed by potassium and calcium (Clarke and Walker, 1974; de Assuncao et al., 2012). Napolitano et al. (2007) have reported that the caffeine content ranges between 0.81% and 1.37%. Other bioactive compounds such as CGA, caffeic acid, quinic acid, and melanoidins, among others with putative health benefits, have been reported (Bresciani et al., 2014; Rodrigues et al., 2015b; Sato et al., 2011). Different research groups have also verified the antioxidant content of CS (Borrelli et al., 2004; Bresciani et al., 2014; Costa et al., 2014; Jiménez-Zamora et al., 2015; Narita and Inouye, 2012; Rodrigues et al., 2015a). Recently, Iriondo-DeHond et al. (2016) have shown positive outcomes using CS extract for skin-care protection against oxidative stress and accelerated aging induced by UV radiation (Iriondo-DeHond et al., 2016).

7.2.2 SPENT COFFEE GROUNDS

SCGs, the residues obtained after the treatment of coffee with hot water or steam for extracting flavor substances, are traditionally used in industrial applications such as high quality biodiesel production (Couto et al., 2009). However, this coffee by-product contains high amounts of bioactive compounds, such as fatty acids, amino acids, phenolic compounds, minerals, and polysaccharides, as extensively reported by different research groups (Al-Hamamre et al., 2012; Ballesteros et al., 2014; Campos-Vega et al., 2015; Couto et al., 2009; Mussatto et al., 2011a). Based on composition, SCGs are considered a coffee by-product with added value. According to Campos-Vega et al. (2015), the lipid fraction can reach approximately 29% of the organic composition of SCGs. The lipids that remain in coffee brews prepared by different methods are mainly composed of triacylglycerols (84.4%), diterpene alcohol esters (12.3%), sterols (1.9%), and sterol esters (0.1%) (Campos-Vega et al., 2015; Ratnayake et al., 1993). Thus, due to the high content of lipids, particularly fatty acids, SCGs might also find a suitable application in cosmetic products, where they can be used as the main components of the oil phase. On the other hand, SCGs contain high amounts of CGA and related compounds (caffeoylquinic acids, feruloylquinic acids, p-coumaroylquinic acids, and mixed diesters of caffeic and ferulic acids with quinic acid). These compounds are well described as powerful antioxidants. Moreover, in vitro antitumor, antiinflammatory, and antiallergenic activities have also been evaluated, with promising antioxidant properties, probably due to the presence of phenolics and CGAs in appreciable amounts.

7.2.3 COFFEE HUSKS

According to Murthy and Naidu (2012), husks represent about 12% of the cherry on a dry-weight basis. This coffee by-product is mainly composed by carbohydrates (58%–85%), followed by reducing sugars (14%), protein (8%–11%), tannins (~5%), minerals (3%–7%), lipids (0.5%–3%), and caffeine (~1%) (Franca and Oliveira, 2009). Different authors have reported the richness of secondary metabolites, such as caffeine and polyphenols, in husks (Esquivel and Jiménez, 2012; Murthy and Naidu, 2012; Pandey et al., 2000). In fact, 5-O-caffeoylquinic acid is the major phenolic, followed by quercetin-3-O-rutinoside, quercetin-3-O-glucoside, quercetin-3-O-galactoside, (+)-catechin, and (−)-epicatechin, procyanidin dimers, trimmers, and tetramers (Mullen et al., 2013).

7.2.4 DEFECTIVE/IMMATURE COFFEE BEANS

Currently, defective beans make up about 20% of total coffee production; these are separated from nondefective beans prior to sale. In defective coffee beans, the oil yield ranged from 10% to 12% (dry weight) (Oliveira et al., 2008). The fatty acids composition of oil is not significantly different than that of mature coffee beans, with linoleic and palmitic acids being the predominant ones (with averages of 44% and 34%, respectively) while miristic and palmitoleic acids are present in trace amounts (Oliveira et al., 2006). These coffee by-products can also be a good source of caffeine. Low-grade coffee beans, which are defective coffee beans obtained after grading, have approximately 1.7% caffeine, while coffee from arabica plants and robusta cherries have 1.6% and 2.4%, respectively (Mazzafera, 1999). Moreover, low-grade green coffee presents high total phenolic compound content and CGAs (Ramalakshmi et al., 2009). These beans are separated and excluded from regular coffee processing since they can confer an undesirable flavor to the beverage. Thus, a cosmetic application of the oil extracted from the defective beans is an alternative to be considered in order to improve the added value of this by-product (Oliveira et al., 2006).

7.3 POSSIBLE COSMETIC APPLICATION OF COFFEE BY-PRODUCTS

As reported in the previous section, several bioactive compounds can be obtained from coffee wastes, but few of them have been studied and incorporated into cosmetic formulations. Fig. 7.2 summarizes the most important compounds and their potential effects on cosmetic products.

(A)

Antiaging
- Coffee silverskin

Protection against UV damage
- Coffee silverskin
- Spent coffee grounds oil fraction
- Defective/immature coffee beans

Emollient properties
- Coffee silverskin
- Spent coffee grounds oil fraction
- Defective/immature coffee beans

Antimicrobial activity
- Coffee silverskin
- Spent coffee grounds

Antiinflammatory activity
- Spent coffee grounds

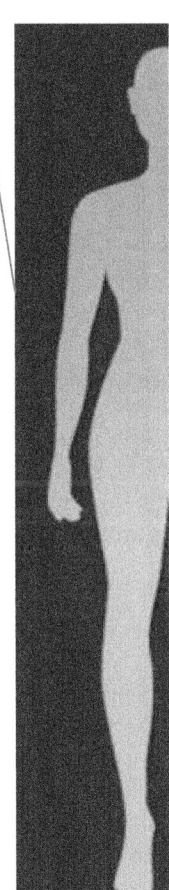

(B)

Antiaging
- Chlorogenic acid
- Caffeic acid
- Acidic polysaccharides

Protection against UV damage
- Caffeine
- Phenolic compounds
Flavonoids
Quinic acid
Ferulic acid
Chlorogenic acid

Anticellulite activity
- Caffeine

Emollient properties
- Fatty acids
  - Linoleic acid
  - Palmitic acid
  - Oleic acid
  - Stearic acid
- Minerals
  - Potassium
  - Calcium

Antimicrobial activity
- Coffee melanoidins

Antiinflammatory activity
- Chlorogenic acid
- Caffeic acid
- Gallic acid
- Trigonelline
- Protocatechuic acid
- Caffeine

FIGURE 7.2

Potential application in cosmetic products of ingredients (A) and bioactive compounds (B) derived from coffee by-products.

7.3.1 PROTECTION AGAINST UV DAMAGE

As it is well known, UV radiation contributes to skin-related disorders when overexposure occurs. These skin disorders, most of time caused by chronic UV exposure, can be observed as erythema, edema, hyperpigmentation, photoaging, and, in the worst scenario, skin cancer (Chiari et al., 2014). Nevertheless, these situations can be prevented if a proper UV protection is used daily. UV radiation, depending on its wavelength, can be divided into short wavelengths, or UVC (200–280 nm), medium wavelengths, or UVB (280–320 nm), and long wavelengths, or UVA (320–400 nm). UVA radiation penetrates deeply into the epidermidis and dermis and induces the formation of reactive oxygen species (ROS) (Soto et al., 2015). UVC radiation is completely absorbed by the earth's atmosphere while UVB radiation constitutes only about 4%–5% of high-energy photons of total UV radiation. Despite this proportion, UVB radiation has a high skin sunburn capacity, and it is commonly known as "burning rays." In fact, UVB can penetrate into the skin's epidermis layer, causing skin injuries. Under ideal circumstances, skin melanin absorbs UV radiation, as the first line of an extremely efficient human defense system against the deleterious effects of UV exposure. If not, a situation of oxidative stress characterized by an excess of ROS is generated. Major protective systems include the natural pigment melanin, enzymatic antioxidants (glutathione peroxidase, superoxide dismutase, glucose-6-phosphate dehydrogenase, and catalase), and nonenzymatic antioxidants (vitamin C, glutathione, vitamin E, coenzyme Q10, and alpha lipoic acid) (Zillich et al., 2015). However, the body's natural defenses are not sufficient to protect the skin, resulting in the development of sunburn (Farris, 2007; Saewan and Jimtaisong, 2015). When the skin is exposed to internal and external factors, such as UV light, free radicals are generated, causing oxidative stress. Antioxidants have the capacity to bind to the free radicals caused by oxidative stress and prevent and/or slow the skin aging process. Nowadays, the application of sunscreen products is a matter of public health. Moreover, consumer awareness of environmental factors and synthetic ingredients has led to increased attention about the utilization of natural ingredients and their advantages (potential safeness, multiple biological actions, and lower cost).

Antioxidants can provide protection by quenching UV-induced ROS. Matsui et al. (2009) demonstrated in human studies that combining antioxidants with sunscreen has a protective effect. Two sunscreens were tested, one combined with antioxidants and the other one without. The added compounds were caffeine, vitamin E, vitamin C, *Echinacea pallida* extract, gorgonian extract, and chamomile essential oil. Results suggested that the combined sunscreen decreased UV-induced damage compared with sunscreen alone (Matsui et al., 2009). Kitagawa et al. (2011) evaluated skin protection against UV-induced damage using CGA incorporated in an oil-in-water microemulsion. The formulation was applied on excised guinea pig dorsal skin and Yucatan micro pigskin (Kitagawa et al., 2011). Results showed positive outcomes and a potential use of hydrophilic CGA with oil-in-water microemulsions as a vehicle to protect skin against UV-induced oxidative damage. The protective effect could be related to the reduction of ROS by CGA (Kitagawa et al., 2011).

Sun protection factor (SPF) measures the capacity of a compound or product to absorb UV radiation. Overall, SPF indicates how well a sunscreen will protect skin from UVB radiation (Wagemaker et al., 2011). Wagemaker et al. (2011) studied the lipid fraction and determined the SPF of 10 species of *Coffea*, obtaining very interesting results. The study showed a significant variability in wax, oil, and unsaponifiable matter content among species. Fatty acids, such as linoleic (C18:2) and oleic (C18:1) acids, were also found in various proportions. The highest SPF values were found in *Coffea eugenioides*, *C. salvatrix*, and *C. stenophylla* while the lowest were detected in *C. kapakata*, *C. liberica* var. *liberica* "Passipagore," *C. liberica* var. *dewevrei abeokutae*, and *C. canephora*. Nevertheless, the predominant coffee species that grows around the world is *C. arabica*. This species has a high SPF (1.5) when compared with the other species evaluated. Thus, these results foresee green coffee oil as an interesting natural ingredient for sunscreens (Wagemaker et al., 2011).

Recently, the biological effect of the oil fraction of SCGs was evaluated by Marto et al. (2016). Producers only select beans with high quality for coffee production. A large amount (about 15%) presents low quality, and are not entered into coffee processing. Therefore, they can be a rich source of green coffee oil with several applications in cosmetic products, namely as sunscreens (Chiari et al., 2014). The oil fraction of SCGs extracted with supercritical CO_2 and green coffee oil was evaluated and incorporated into a new generation of sunscreens with improved sun protection performance (Marto et al., 2016). Results demonstrated the effectiveness of the final formulation in terms of UVB/A protection, biological activity, and better tolerability, showing that the reuse of the lipidic fraction of spent coffee is an excellent opportunity to add value to this coffee by-product (Marto et al., 2016).

Couto et al. (2009) also reported that the lipid fraction of SCGs extracted with supercritical CO_2 can be used in the development of new sunscreens. On average, one-fifth of Brazilian coffee production consists of defective beans, and several studies have been developed to find an alternative use for such products, including use of the extracted oil in cosmetic applications. The use of SGC oil in the cosmetic industry seems to be a suitable approach to recycle and valorize wastes from the coffee industry. Moreover the coffee oil presents promising characteristics for the goal of improvement of sunscreen performance.

Another natural compound present in coffee with UV-absorption properties is caffeine. In 1988, L'Oreal patented the use of coffee-bean oil as a sun filter that allowed a natural browning while at the same time protect skin against solar radiation and development of skin burns. The composition of coffee-bean oil essentially consists of about 75% triglycerides and 20% fatty acid monoesters of two diterpene alcohols: caffestol and kahweol. Grollier and Plessis (1988) have also reported the presence of phosphatides, sterol esters, caffestol, kahweol, and free fatty acids.

Green coffee oil is generally obtained by mechanical cold pressing. Considering that green coffee oil in cosmetics is a source of caffeine, sterols, terpenes, and tocopherols, the fractionation of these compounds is a way to get valuable bioactive compounds to add to cosmetic formulations. Moreover, raw green coffee oil presents a cloudy aspect and a dark green color due to the presence of different constituents, such as wax, chlorophyll, and terpenes, among others, which can be a constraint to the utilization of this oil in cosmetic products. Therefore, clean technologies, such as supercritical CO_2 extraction, have been proposed to refine the coffee oil obtained by mechanical pressing (de Azevedo et al., 2008). Supercritical fluid extraction has been used for the extraction of essential oils from plants based on its capacity to prevent the degradation of bioactive compounds. In fact, supercritical fluid extraction is a sustainable green technology that presents main advantages as, safeness, absence of chemical solvent use, and low associated costs (Couto et al., 2009; de Melo et al., 2014).

Green coffee oil, similar to other oils, is susceptible to lipid oxidation, which leads to a loss of its beneficial chemical properties and the formation of unpleasant odors. Different technologies, such as microencapsulation, have been successfully used to avoid the degradation processes by transforming the oil into a stable powder (Frascareli et al., 2012). However, further research for industrial applications is needed. Microencapsulation, acting as a functional barrier, aims to protect the encapsulated bioactive ingredients and keep them safe from the deterioration caused by different factors, such as temperature, oxygen, light, water, and microorganisms. Additionally, it allows the release of encapsulated bioactive compounds under particular or determined circumstances (Bakry et al., 2016).

Microencapsulation, the most widely used encapsulation method, allows for green coffee oil incorporation into, for instance, powder cosmetics as make-up products (Silva et al., 2014b). Also, when green coffee oil is encapsulated, direct contact with human skin does not occur, which reduces the potential for allergenic effects caused by cinnamic acid (Carvalho et al., 2014). The improvement of oxidative stability and the antioxidant profile using the microencapsulation method has been studied. Microencapsulation by spray-drying of green coffee oil using lecithin–chitosan and lecithin emulsion stabilizers showed positive results, demonstrating an improvement in oxidative stability and maintaining the SPF of green coffee oil (Carvalho et al., 2014). Also, coffee-oil encapsulation by spray-drying, using gum arabic as an encapsulating agent was studied by Ray et al. (2016) with very good results. Overall, microencapsulation is an effective functional methodology to improve the protection of compounds' bioactivity. Particularly, the application of this technique in oils for cosmetic formulations purposes is a great improvement and should be promoted. Additionally, other advantages have emerged, such as the maintenance of the quality and physicochemical properties of cosmetic products in terms of shelf life and the controlled release of active ingredients (Bakry et al., 2016; Ray et al., 2016).

Caffeine has been shown to prevent UVB-induced skin-cancer development in mouse models. Lu et al. (2002) have studied the effect of topical applications of caffeine on SKH-1 hairless mice. Animals were irradiated with UVB twice a week over 20 weeks. This type of irradiation induces a high risk of developing skin tumors over the following several months. Afterward, the animals were treated topically with caffeine once a day (5 days per week) for 18 weeks in the absence of further treatment with UVB. Results showed that topical applications of caffeine decreased the number of nonmalignant and malignant skin tumors per mouse by 44% and 72%, respectively, (Lu et al., 2002). Other authors have also studied the possible effects of topical caffeine application after UVB irradiation in animal models (Koo et al., 2007). In this study, hairless mice were treated with UVB followed immediately by a topical application of caffeine 3 times per week for 11 weeks. The topical application of caffeine in mouse skin after UV irradiation promoted the deletion of DNA-damaged keratinocytes and also partially diminished photo damage, as well as photocarcinogenesis (Koo et al., 2007). Also, the topical application of caffeine 30 minutes before a high dose of UVB radiation demonstrated a protective effect against sunburn lesions (Lu et al., 2006).

7.3.2 ANTICELLULITE ACTIVITY

Coffee by-products have several classes of health-related chemicals, such as phenolic compounds, mellanoidins, and methylxanthines (Magalhães et al., 2016). Caffeine is the major methylxanthine recovered and can be applied in cosmetics that are intended to be used for cellulite treatment. Nevertheless, the recovery and application of caffeine in innovative sustainable cosmetic products is a challenging task that involves several coffee value-chain actors. The first line is the coffee producers. This is not an easy task, considering that the several coffee-producing countries have different levels of environmental policies. At the end of the line are coffee beverage consumers and cosmetic consumers, or, most of the time, both. In fact, awareness of the sustainable management of coffee by-products needs to be initiated with household coffee consumers (it is known that, nowadays, a great portion of the coffee is consumed at home) and also should be directed to consumers who are looking for ecofriendly products with a low environmental impact. Cosmetic products such as those that are intended for cellulite treatment can benefit from caffeine extracted from coffee by-products. Cellulite is a skin disorder developed because of multiple factors that affect many postpubertal human females. It is estimated that between 85% and 98% of women present some degree of cellulite (Silva et al., 2014a). Nevertheless, this is not a condition specific to overweight women. The body areas more susceptible to cellulite are the upper outer thighs, posterior thighs, and buttocks. However, it can be found in any area of the body that contains subcutaneous adipose tissue. The common aspect of cellulite is consistent with an "orange peel" or "cottage cheese" like appearance caused by the herniation of subcutaneous fat within fibrous connective tissue (Alizadeh et al., 2016). According to Rawlings (2006), cellulite is a multifactorial and complex condition that involves not only subcutaneous fat but also microcirculation and the lymphatic system. Currently, a great number of products are available on the market focusing on the treatment of cellulite using caffeine as an active ingredient, despite some controversy. It has been proposed that caffeine penetrates the skin barrier and reaches the dermis where fat protrusion occurs and breaks it down through stimulation of the lipolysis process. Therefore, fat-cell deposits can be reduced

(Lupi et al., 2007; Velasco et al., 2008). Mechanistically, caffeine exerts lipolytic effects via induction of cyclic adenosine monophosphate (cAMP) and inhibition of phosphodiesterase (PDE) in adipocytes. An increase in cAMP levels stimulates the protein kinase A to phosphorylate. In that way, hormone-sensitive lipase is activated. Phosphorylated hormone-sensitive lipase hydrolyzes triglycerides into diglycerides, monoglycerides, free fatty acids, and glycerol (Byun et al., 2015; Diepvens et al., 2007).

Caffeine is a hydrophilic compound that can easily penetrate the skin barrier and reach the dermis (Silva et al., 2014a). Advanced formulas have been developed to improve caffeine efficiency in cellulite treatment (Hamishehkar et al., 2015; Silva et al., 2014a; Vyas et al., 2013). A topical formula using caffeine-loaded solid lipid nanoparticles was studied in rat skin, showing that solid lipid nanoparticles can ameliorate caffeine efficiency. Comparing drug accumulation in the skin with caffeine-solid lipid nanoparticles hydrogel and caffeine hydrogel, the first one presented a higher drug accumulation (Hamishehkar et al., 2015). Nevertheless, studies are not always in agreement. Rodrigues et al. (2016a) encapsulated a CS extract in nanostructured lipid carriers (NLCs) and demonstrated that encapsulation did not conduce to a higher permeability. In another study, biocellulose membranes were investigated as caffeine topical delivery systems using human epidermal membranes and Franz cells (Silva et al., 2014a). The permeation of caffeine from biocellulose membranes was compared between aqueous and gel formulations, showing promising outcomes, mainly considering a prolonged and predictable caffeine release over time. In fact this system is already in the market in cosmetic masks (BioCellulose and NanoMasque), in which biocellulose membranes are impregnated with active compounds that have moisturizing properties (Silva et al., 2014b). Innovative products are emerging in the market whose purpose is a continuous and controlled release of caffeine throughout the day (Silva et al., 2014a).

Notwithstanding, controversial scientific discussions related to caffeine penetration into the epidermis, several studies have been developed to increase caffeine permeation ability (Mujica Ascencio et al., 2016). Since several coffee by-products have a noteworthy amount of caffeine, it is of utmost importance to consider them as added-value products for the cosmetic industry. Among the coffee by-products, such as coffee pulp, coffee husks, CS, and spent coffee, the major sources of caffeine are coffee pulp and coffee husks. These by-products have 1.5% and 1.0% caffeine, respectively. CS and spent coffee present moderate caffeine content (0.03% and 0.02%, respectively) (Murthy and Naidu, 2012).

7.3.3 EMOLLIENT PROPERTIES

The skin barrier protects the body from water loss, as well as from the permeation of harmful and allergenic substances. A healthy *stratum corneum* is imperative to an effective permeability barrier. As previously described, the *stratum corneum* is composed of corneocytes surrounded by a continuous matrix of lipids. The intercellular lipids comprise a mixture of ceramides, cholesterol, and free fatty acids, organized into tightly packed lamellar formations—the lipid lamellae (Moncrieff et al., 2013). It has been suggested that lipids content affects the skin-barrier function (Halvarsson and Lodén, 2007).

The corneocytes contain the natural moisturizing factor (NMF) that is responsible for maintaining proper moisture levels in the stratum corneum, and the appropriate barrier homeostasis, desquamation, and plasticity (Moncrieff et al., 2013). In a normal status of hydration they contribute to a healthy skin aspect and sensorial surface properties. NMF compounds are present in elevated concentrations within the corneocytes and represent up to 20%–30% (dry weight) of the *stratum corneum*, consisting of a mixture of amino acids and their derivatives (pyrrolidone carboxylic acid and urocanic acid), lactates, and urea. Minerals as chloride, sodium, potassium, calcium, and magnesium are also present (Verdier-Sévrain and Bonté, 2007).

Emollients are moisturizers with skin-softening properties. They may contain a variety of compounds (for example, lipids) that have multiple actions as agents of water retention and improvement of the skin-barrier function. These products can be presented in several formulations, such as creams, lotions, bath products, and ointments. Reduced levels of NMF and variations in the lipid composition of the *stratum corneum* lead to dry skin, which is closely associated with age-related skin conditions (Correa and Nebus, 2012). Thus, the emollient utilization aims to improve skin hydration and sebum levels (Ribeiro et al., 2013). Green coffee oil has been used in cosmetic formulations based on its emollient properties provided by fatty acids. The major fatty acids present are linoleic (C18:2) and palmitic (C16:0) acids, followed by oleic (C18:1) and stearic (C18:0) acids. Also, cafestol and kahweol, natural diterpenes only found in coffee, are present in the unsaponifiable lipid fraction of green coffee (Carvalho et al., 2014). Different authors have studied the emollient properties of SCGs. For example, Campos-Vega et al. (2015) reported that SCGs present an oil fraction that range between 10% and 29% of it organic composition (dry weight). Ribeiro et al. (2013), using supercritical CO_2 methodology, extracted the lipid fraction of SCGs to assess the possibility of using the lipid fraction in the development of cosmetic formulations (oil-in-water creams) and evaluate the effect on skin hydration and sebum capacity. Green coffee oil was used to compare both formulations. Results showed that in the SCG and green coffee lipidic fractions, palmitic (C16:0) and linoleic (C18:2) acids were the major fatty acids quantified. These two fatty acids comprised almost 80% of the total fatty acids. Both creams increased skin hydration and sebum levels, suggesting that skin-barrier properties were improved. Also, the SGC oil formulation presented a good acceptance by consumers when compared with green coffee oil and placebo formulations (Ribeiro et al., 2013). According to Franca et al. (2005), linoleic (C18:2) and palmitic (C16:0) acids are the major fatty acids present in defective coffee beans composition, while oleic (C18:1) and stearic (C18:0) acids are quantified in moderate amounts. Fatty acids have been used in cosmetics due to their softening, smoothing, and protective properties. Unsaturated fatty acids [such as palmitoleic (16:1), oleic (18:1), linoleic (18:2), and α-linolenic (18:3) acids] play an important role in cosmetology. For instance, linoleic acid (C18:2) is a natural component of sebum. In acneic skin it is possible to observe a decrease in linoleic acid content in sebum. This condition leads to the obstruction of pores, and consequently formation of comedos and eczemas. The use of linoleic (C18:2) acid for oily skin results in an improvement of the activity of sebaceous glands, unblocking of pores, and decrease of comedos (Zielinska and Nowak, 2014). Moreover, fatty acids, such as linoleic (C18:2), linolenic (C18:3), myristic (C14:0), oleic (C18:1), palmitic (C16:0), and stearic (C18:0) acids have been reported in the literature as skin penetration enhancers (also called absorption promoters) for topical and transdermal formulations (Lane, 2013). Those agents may increase skin permeation by disrupting the *stratum corneum*, forming solvated complexes, increasing compound diffusivity, and improving partitioning through the skin barrier. As skin penetration enhancers, monounsaturated oleic (C18:1) acid, polyunsaturated linoleic (C18:2) acid, and linolenic (C18:3) acid, as well as the saturated lauric acid (C12:0) have been shown to be effective for many compounds (Santoyo and Ygartua, 2000; Williams and Barry, 2012).

In hydrated skin, the water-content in the *stratum corneum* ranges between 15% and 20%, and the NMF presents high amounts of lactate and ions, such as potassium, sodium, and calcium (Williams and Barry, 2012). Different authors have demonstrated that lactate and potassium are important compounds for the maintenance of physical properties of the *stratum corneum* (Nakagawa et al., 2004; Sugawara et al., 2012). Potassium is a particularly important mineral component of the NMF that has been associated with the state of skin surface hydration and *stratum corneum* stiffness. As coffee by-products are rich in minerals (particularly potassium), the recovery and purification of these compounds can be strategic to reuse and valorize these wastes. Further research is needed, considering the scarce information about the mineral profile of several coffee by-products, despite some studies that have already evaluated their composition as added-value ingredients for the cosmetic and food industries. In fact, coffee by-products, such as coffee husks, skin, and pulp can present a mineral content of about 11%. SCGs and CS present between 1% and 5%, respectively, of minerals (Ballesteros et al., 2014; Esquivel and Jiménez, 2012; Jiménez-Zamora et al., 2015; Murthy and Naidu, 2012; Mussatto et al., 2011b). Mussatto et al. (2011b) quantified the amount of different minerals in SCGs: potassium (3549.0 mg/kg), phosphorus (1475.1 mg/kg), magnesium (1293.3 mg/kg), calcium (777.4 mg/kg), iron (118.7 mg/kg), manganese (40.1 mg/kg), copper (32.3 mg/kg), and zinc (15.1 mg/kg). Ballesteros et al. (2014) also have determined the mineral profile of SCGs and CS. According to the authors, potassium is the major mineral present in both by-products, followed by magnesium and phosphorus in SCGs, and by calcium and magnesium in CS. The mineral compounds are associated with NMF, which in turn is correlated with the state of hydration, stiffness, and pH of *stratum corneum* (Rawlings and Matts, 2005).

7.3.4 ANTIMICROBIAL ACTIVITY

Coffee by-products have shown antimicrobial activity against different microorganisms, which could be a promising application for the prevention and treatment of skin diseases or even as a preservative for final cosmetic formulations (Rodrigues et al., 2015a; Rufián-Henares and de la Cueva, 2009). Rodrigues et al. demonstrated the antimicrobial activity of CS extracts against *Staphylococcus aureus* (ATCC

6538 and MRSA), *Staphylococcus epidermidis*, *Escherichia coli* (ATCC 1576 and MRSA), and *Klebsiella pneumoniae* (ATCC 4352). However, the authors did not report antimicrobial activity against *Candida albicans* and *Pseudomona aeruginosa*, probably due to the existence of the yeast wall and extracellular matrix, respectively. In another study, Jiménez-Zamora et al. (2015) demonstrated that SCGs and CS without coffee melanoidins did not present an antimicrobial effect against *S. aureus* and *E. coli*, whereas SCGs with coffee melanoidins shown antimicrobial activity against both microorganisms. Moreover, the antimicrobial activity of coffee melanoidins was higher when the compounds were isolated, suggesting an interference of SCGs. Therefore, these studies suggest that: (1) antimicrobial activity is probably related to melanoidins present in coffee by-products and (2) to attain the most relevant antimicrobial activity, coffee melanoidins must be removed from the by-product matrix.

In fact, the Maillard reaction is one of the main chemical reactions that occur during coffee roasting. In this reaction the reducing sugars react with the amino groups, leading to the formation of end compounds, the melanoidins (Bekedam et al., 2008). Depending on their concentration, coffee melanoidins can have bacteriostatic or bactericide actions. The antimicrobial activity of coffee melanoidins against different pathogenic bacteria has been studied by different research groups (Rodrigues et al., 2015a; Rufián-Henares and de la Cueva, 2009). It has been proposed that such activity is related to those compounds that present metal-chelating properties (Rufián-Henares and de la Cueva, 2009). The inhibitory activity of melanoidins against microorganisms has been previously demonstrated against urease-gastric mucin adhesion of *Helicobacter pylori* in the stomach (Hiramoto et al., 2004). Also, its protective effect in the regulation of dental plaque formation possibly by the minimization of *Streptococcus mutans* on the tooth surface has been reported by Daglia et al. (2002). Martinez-Saez et al. (2014) demonstrated that CS can be an important source of melanoidins and that the utilization of CS extracts can be chosen instead of raw CS as a source of melanoidins due to the higher concentration of the first. Melanoidins are high molecular-weight nitrogenous, brown-colored compounds, despite their heterogeneous structure, and present high solubility in water (Bekedam et al., 2008). However, more studies are needed to identify the compounds responsible for the antimicrobial activity of coffee by-products.

7.3.5 ANTIAGING ACTIVITY

Antioxidants are used topically to prevent and treat photoaged and chronologically aged skin. As it is well known, free radicals promote oxidation of nucleic acids, proteins, and lipids and can damage intracellular structures, including DNA, leading to aging. Free radicals can upregulate transcription factors, such as activator protein 1 (AP-1) and nuclear transcription factor-kappa B (NF-κB) that are involved in the aging mechanism (Meyer et al., 1994). Dense collagen-rich fibrils, which provide structural and mechanical support, are the major components of dermis in human skin while fibroblasts are the most important cells responsible for collagen production (Qin et al., 2014).

During aging, the collagen fibrils and other extracellular matrix proteins that comprise the dermal connective tissue fragment (Quan et al., 2009). AP-1 is the transcription factor responsible for production of metalloproteinases that break down existing collagen (Fisher and Voorhees, 1998). This process is normally initiated by matrix metalloproteinase 1 (MMP-1) (Qin et al., 2014). This transcription factor induces the expression of matrix metalloproteinases collagenase and stromelysin, which degrade collagen and other proteins that comprise the dermal extracellular matrix (Fisher and Voorhees, 1998). Another important transcriptor factor is NF-κB. This factor upregulates transcription of proinflammatory mediators, such as interleukin-1 (IL-1), IL-6, and IL-8, and tumor necrosis factor-α (Salminen and Kaarniranta, 2009). These proinflammatory mediators further activate AP-1 and NF-κB, resulting in more damage. Also, transforming growth factor (TGF)β, a cytokine that promotes collagen production, is involved in the regulation of collagen production (Kang et al., 1997).

Under ideal circumstances, the two major groups of the human antioxidant system, enzymatic and nonenzymatic antioxidants, are used by skin to protect itself from free radicals (Carocho and Ferreira, 2013). Enzymatic antioxidants are divided into primary and secondary enzymatic defenses. The primary defense is composed of three important enzymes that prevent the formation of or neutralize free radicals: glutathione peroxidase, catalase, and superoxide dismutase. The secondary enzymatic defense includes glutathione reductase and glucose-6-phosphate dehydrogenase. The nonenzymatic antioxidants include cofactors (such as coenzyme Q10), flavonoids, minerals, phenolic acids, vitamins and derivates, carotenoids, and nitrogen nonprotein compounds (Carocho and Ferreira, 2013). Thus, without adequate antioxidant protection, the free radicals generated are left unchecked, resulting in skin aging. As previously reported, coffee by-products are rich in antioxidants, such as CGAs and caffeic, and these are promising ingredients for antiwrinkle products. In this vein, Furusawa et al. (2011) investigated the inhibitory effect of CS on hyaluronidase enzymes. According to the authors, the higher molecular-weight substances present in CS extracts could contribute to hyaluronidase inhibition while acidic polysaccharides, mainly composed of uronic acid, played a major role in hyaluronidase inhibition. This observation is in line with the results of an in vivo evaluation, performed in human volunteers, by Rodrigues et al. (2016c). In this study, 20 volunteers used a cream containing CS extract twice a day over 28 days. After this period an improvement in skin hydration and firmness was observed. A decrease in wrinkle depth, volume of cavities, and roughness was also observed, but without statistical differences. The same research group had previous reported the skin compatibility and safety of CS extract (Rodrigues et al., 2015b). An evaluation of CS extract with in vitro skin and ocular irritation assays, using reconstructed human epidermis, EpiSkin, and the human corneal epithelial model, SkinEthics HCE, respectively, was performed by Rodrigues et al. (2015b). The in vitro results demonstrated that the extract was not classified as irritant and the histological analysis proved that both models' structures were not affected after exposure. An in vivo assay was also described by the same research group aiming to guarantee the extract's safety. A patch test was performed using 20 volunteers over 48 h, with the most promising extract (hydroalcoholic) showing that, with respect to irritant effects, it can be regarded as safe for topical application (Rodrigues et al., 2015b). Rodrigues et al. (2016b,d) also developed and described a hand cream and a body cream containing CS extract. For both, antioxidant activity and consumer acceptability was high, and no decrease in cell viability in keratinocytes and fibroblasts was observed. Nevertheless, and as reported by Toschi et al. (2014), CS could also contain undesirable compounds, such as ochratoxin A and phytosterol oxidation products, whose presence should be analyzed when the extract is intended to be used in cosmetics.

7.3.6 ANTIINFLAMMATORY ACTIVITY

Inflammation can be defined as a bodily response to an injury, infection, or destruction, normally characterized by heat, redness, pain, swelling, and disturbed physiological functions. ROS are well known to be involved in human cutaneous aging and in dermal inflammatory response (Kammeyer and Luiten, 2015). The process is simple: ROS and reactive nitrogen species (RNS) combined with increased secretion of inflammatory mediators and enzyme expressions (such as collagenase and/or elastase) are responsible for inflammation, decreasing the tensile strength and elasticity of the skin. As previously reported, ROS lead to the activation of AP-1, which suppresses TGFβ receptors, which in turn blocks procollagen synthesis, reducing collagen levels. At the same time, AP-1 activation leads to NF-κB production, which is a major activator of the inflammatory process (Kammeyer and Luiten, 2015; Natarajan et al., 2014; Peng et al., 2015). Normally, to control inflammation, nonsteroidal antiinflammatory drugs are used, which have several adverse effects, especially gastric irritation.

The inflammatory response conduces to cytokines and interleukins production, which amplify the inflammatory process, inducing keratinocyte proliferation and monocyte recruitment to the site of injury (Menezes et al., 2016). In recent studies, the extracts from the residue of brewed coffee exhibited antiinflammatory activity, mainly due to the presence of phenolic compounds, such as CGA, caffeine, caffeic acid, trigonelline, and protocatechuic acid (Borrelli et al., 2004; Yen et al., 2005). Shin et al. (2015) evaluated the effects of CGA and its metabolites on IL-8 production in human intestinal Caco-2 cells induced by combined stimulation with tumor necrosis factor alpha (TNFα) and H_2O_2. Results demonstrated that CGA and caffeic acid inhibited TNFα- and H_2O_2-induced IL-8 production. Recently, López-Barrera et al. (2016) reported on the ability of SCGs to prevent inflammation. According to the authors, SCGs exert antiinflammatory activity, mediated by short-chain fatty acid produced from their dietary fiber, by reducing the release of inflammatory mediators, such as IL-10, CCL-17, CXCL9, IL-1β, and IL-5 cytokines. Also, this matrix is rich in CGA and gallic acid, which probably explain its activity. Nevertheless, these results are not in agreement with those obtained by Ramalakshmi et al. (2009), who reported that spent coffee showed limited antiinflammatory and antiallergic activities. Further in vitro studies, using skin cells, such as fibroblasts and keratinocytes, are needed to draw conclusions about the potentialities of coffee by-products as antiinflammatory ingredients.

7.3.7 OTHER POTENTIAL APPLICATIONS OF COFFEE BY-PRODUCTS IN COSMETICS

Despite the previously mentioned coffee by-product applications, novel approaches have been developed to add value to these raw materials (Murthy and Naidu, 2012). Currently, a noticeable preference by consumers for natural food additives has been reported by the media. On the other hand, industries need to find suitable alternatives to the chemical synthesis of aroma compounds and their direct extraction from plants, since the processes present economical and ecological disadvantages.

Many fungi and yeasts have the capacity to produce new compounds or transform in an added substrate/precursor compound (Vandamme, 2003). The use of coffee by-products as substrates in biotechnological processes seems to be a valuable alternative. Bonilla-Hermosa et al. (2014) tested coffee residues for the production of aroma/flavor compounds that could be interesting for the food, cosmetic, and perfume industries. Therefore, the selected yeasts strains were inoculated in a fermentation medium containing coffee pulp and coffee wastewater. The volatile compounds were analyzed and 35 compounds corresponding to 6 groups of volatile compounds were identified: higher alcohols, acetates, ethyl esters, aldehydes, terpenes, and volatile acids. Among the compounds, 2-phenylethanol has a particular interest for industry due to its rose-like odor. Relative to ethyl esters, ethyl butyrate was produced by two strains. Characteristic pleasant fruity notes are attributed to acetates and ethyl esters. Only one strain produced the highest concentration of terpenes. Terpenes are considered positive aromas due to their attributed floral and fruity notes (Bonilla-Hermosa et al., 2014; Salgado et al., 2012). This study showed a promising utilization of coffee residues as aroma compounds, despite the fact that further research is necessary to expand its potential.

7.4 FUTURE PERSPECTIVES

Beyond the therapeutic and cosmetic benefits of bioactive compounds identified in coffee wastes, the safety, stability, and quality of the topical formulations developed have to be ensured. For example, the incorporation of phenolic compounds in topical products represents a real challenge once their functional activity and aesthetic properties have to be guaranteed. Due to the sensitivity of natural compounds to environmental factors, such as light or oxygen, novel technologies (for example, encapsulation) can be used to improve their stability, avoid incompatibilities, reduce the odor of compounds, and control their release (Jung et al., 2016). The compounds' active function especially must be assured (Soto et al., 2015). The field of sunscreens, continuously growing, is of particular interest for the cosmetics industry. These formulations are generally based on synthetic chemicals. However, consumers are increasingly demanding natural ingredients, and most of the time the sunscreens used are in nanoscale. Therefore, the development of "green" formulations is of huge interest. However, there are always two questions that are obligatory to answer: (1) whether the "green" formulation is better when compared to others, and (2) whether the "green" formulation is efficient in terms of photoprotection activity (Chiari et al., 2014). Until now, data have been scarce concerning clinical research, functional ingredients used, and combined activity of natural antioxidants and synthetic sunscreens (Matsui et al., 2009). Unlike sunscreens, and to the best of our knowledge, the antiwrinkle formulations using coffee wastes have demonstrated their potential. In the coming years the focus of the cosmetic industry could be the anticellulite activity field, based on the richness of coffee by-products in caffeine.

7.5 CONCLUSIONS

According to the Food and Agriculture Organization (FAO) of the United Nations, one-third of the edible parts of food produced globally for human consumption is lost or wasted. The field of coffee production is not an exception. Different coffee by-products are originated during coffee processing, including not only defective/immature coffee beans, but also coffee husks, CS, and SCGs. All these wastes are extremely rich in bioactive compounds, such as antioxidants, caffeine, and fatty acids, being a challenge to discover new final applications. In particular, the field of cosmetics may benefit from these remaining materials, as those bioactive compounds can fulfill a real skin function and activity. However, issues such as safety, skin permeation, and efficacy delivery, are concerns in the development of new cosmetic products using these agro-industrial by-products. The data analyzed through this chapter supports the growing knowledge that bioactive compounds can be added to cosmetic formulations, such as sunscreens or antiaging formulations, to optimize, respectively, the photoprotection and antiwrinkles effects of the formulations. However, it is necessary to assure the presence and biological activity of compounds in the final products. Therefore, compounds stability in the formulations should first be evaluated. Also, another requirement for antioxidant functionality is the permeation in the *stratum corneum* and maintenance of adequate concentrations in the *epidermis* and *dermis* to have positive outcomes (Chen et al., 2012). Nevertheless, as opposed to other food by-products, in the field of coffee by-products extremely important progress has been made up to now, as different research groups are developing cosmetic formulations with these wastes as active ingredients, and evaluating their in vivo safety and efficacy. As the main conclusion, the biggest challenges can now be summarized:

1. To create an in-depth collaboration between coffee producers and the cosmetic industry.
2. To raise awareness among household coffee consumers about their importance in this chain, since they are the major producers of SCGs.
3. To convince consumers of these ingredients' efficacy and safety.
4. To alert the world population about environmental issues such as this one, stimulating awareness and a proactive attitude.

ACKNOWLEDGMENTS

The authors are grateful for the financial support from the project Operação NORTE-01-0145-FEDER-00001, titled Qualidade e Segurança Alimentar—uma abordagem (nano)tecnológica. This work was also supported by the project UID/QUI/50006/2013–POCI/01/0145/FEDER/007265 with financial support from FCT/MEC through national funds and cofinanced by FEDER.

Francisca Rodrigues and M. Antónia Nunes are grateful for the research grants from project UID/QUI/50006/2013.

REFERENCES

Al-Hamamre, Z., Foerster, S., Hartmann, F., Kröger, M., Kaltschmitt, M., 2012. Oil extracted from spent coffee grounds as a renewable source for fatty acid methyl ester manufacturing. Fuel 96, 70–76.

Alizadeh, Z., Halabchi, F., Mazaheri, R., Abolhasani, M., Tabesh, M., 2016. Review of the mechanisms and effects of noninvasive body contouring devices on cellulite and subcutaneous fat. Int. J. Endocrinol. Metabol. 14 (4), e36727.

Bakry, A.M., Abbas, S., Ali, B., Majeed, H., Abouelwafa, M.Y., Mousa, A., Liang, L., 2016. Microencapsulation of oils: a comprehensive review of benefits, techniques and applications. Compr. Rev. Food Sci. Food Safety 15, 143–182.

Ballesteros, L.F., Teixeira, J.A., Mussatto, S.I., 2014. Chemical, functional, and structural properties of spent coffee grounds and coffee silverskin. Food Bioprocess Technol. 7, 3493–3503.

Barbulova, A., Colucci, G., Apone, F., 2015. New trends in cosmetics: by-products of plant origin and their potential use as cosmetic active ingredients. Cosmetics 2, 82.

Bekedam, E.K., Schols, H.A., Van Boekel, M.A.J.S., Smit, G., 2008. Incorporation of chlorogenic acids in coffee brew melanoidins. J. Agricult. Food Chem. 56, 2055–2063.

Bonilla-Hermosa, V.A., Duarte, W.F., Schwan, R.F., 2014. Utilization of coffee by-products obtained from semi-washed process for production of value-added compounds. Biores. Technol. 166, 142–150.

Borrelli, R.C., Esposito, F., Napolitano, A., Ritieni, A., Fogliano, V., 2004. Characterization of a new potential functional ingredient: coffee silverskin. J. Agricult. Food Chem. 52, 1338–1343.

Bresciani, L., Calani, L., Bruni, R., Brighenti, F., Del Rio, D., 2014. Phenolic composition, caffeine content and antioxidant capacity of coffee silverskin. Food Res. Int. 61, 196–201.

Byun, S.-Y., Kwon, S.-H., Heo, S.-H., Shim, J.-S., Du, M.-H., Na, J.-I., 2015. Efficacy of slimming cream containing 35% water-soluble caffeine and xanthenes for the treatment of cellulite: clinical study and literature review. Ann. Dermatol. 27, 243–249.

Campos-Vega, R., Loarca-Piña, G., Vergara-Castañeda, H.A., Oomah, B.D., 2015. Spent coffee grounds: a review on current research and future prospects. Trends Food Sci. Technol. 45, 24–36.

Carocho, M., Ferreira, I.C., 2013. A review on antioxidants, prooxidants and related controversy: natural and synthetic compounds, screening and analysis methodologies and future perspectives. Food Chem. Toxicol. 51, 15–25.

Carvalho, A.G.S., Silva, V.M., Hubinger, M.D., 2014. Microencapsulation by spray drying of emulsified green coffee oil with two-layered membranes. Food Res. Int. 61, 236–245.

Chen, L., Hu, J.Y., Wang, S.Q., 2012. The role of antioxidants in photoprotection: a critical review. J. Am. Acad. Dermatol. 67, 1013–1024.

Chiari, B.G., Trovatti, E., Pecoraro, É., Corrêa, M.A., Cicarelli, R.M.B., Ribeiro, S.J.L., Isaac, V.L.B., 2014. Synergistic effect of green coffee oil and synthetic sunscreen for health care application. Ind. Crops Prod. 52, 389–393.

Clarke, R.J., Walker, L.J., 1974. Potassium and other mineral contents of green, roasted and instant coffees. J. Sci. Food Agricul. 25, 1389–1404.

Correa, M.C.M., Nebus, J., 2012. Management of patients with atopic dermatitis: the role of emollient therapy. Dermatol. Res. Pract. 2012, 15.

Costa, A.S.G., Alves, R.C., Vinha, A.F., Barreira, S.V.P., Nunes, M.A., Cunha, L.M., Oliveira, M.B.P.P., 2014. Optimization of antioxidants extraction from coffee silverskin, a roasting by-product, having in view a sustainable process. Ind. Crops Prod. 53, 350–357.

Couto, R.M., Fernandes, J., da Silva, M.D.R.G., Simões, P.C., 2009. Supercritical fluid extraction of lipids from spent coffee grounds. J. Supercrit. Fluid. 51, 159–166.

Daglia, M., Tarsi, R., Papetti, A., Grisoli, P., Dacarro, C., Pruzzo, C., Gazzani, G., 2002. Antiadhesive effect of green and roasted coffee on *Streptococcus mutans* adhesive properties on saliva-coated hydroxyapatite beads. J. Agricult. Food Chem. 50, 1225–1229.

de Assuncao, L.S., da Luz, J.M., da Silva Mde, C., Vieira, P.A., Bazzolli, D.M., Vanetti, M.C., Kasuya, M.C., 2012. Enrichment of mushrooms: an interesting strategy for the acquisition of lithium. Food Chem. 134, 1123–1127.

de Azevedo, A.B.A., Kieckbush, T.G., Tashima, A.K., Mohamed, R.S., Mazzafera, P., Vieira de Melo, S.A.B., 2008. Extraction of green coffee oil using supercritical carbon dioxide. J. Supercrit. Fluid. 44, 186–192.

de Melo, M.M.R., Barbosa, H.M.A., Passos, C.P., Silva, C.M., 2014. Supercritical fluid extraction of spent coffee grounds: measurement of extraction curves, oil characterization and economic analysis. J. Supercrit. Fluid. 86, 150–159.

Diepvens, K., Westerterp, K.R., Westerterp-Plantenga, M.S., 2007. Obesity and thermogenesis related to the consumption of caffeine, ephedrine, capsaicin, and green tea. Am. J. Physiol. Regul. Integr. Comp. Physiol. 292, R77–R85.

European Commission, 2003. Directive 2003/15/EC, Relating to Cosmetic Products. OJEU L66, 26–35.

European Commission, 2009. Regulation (EC) No 1223/2009, on Cosmetic Products. OJEU L342, 59–209.

European Commission Scientific Committee on Consumer Safety (SCCS), 2012. Notes of Guidance for the Testing of Cosmetic Substances and Their Safety Evaluation. SCCS Note of Guidance. European Commission, Health & Consumers Directorate D: Health Systems and Products, Brussels, p. 133. Available from: http://www.sftox.com/actualites/sccs_s_006.pdf

Esquivel, P., Jiménez, V.M., 2012. Functional properties of coffee and coffee by-products. Food Res. Int. 46, 488–495.

Farris, P., 2007. Idebenone, green tea, and Coffeeberry® extract: new and innovative antioxidants. Dermatol. Ther. 20, 322–329.

Fisher, G.J., Voorhees, J.J., 1998. Molecular mechanisms of photoaging and its prevention by retinoic acid: ultraviolet irradiation induces MAP kinase signal transduction cascades that induce Ap-1-regulated matrix metalloproteinases that degrade human skin in vivo. J. Invest. Dermatol. 3, 61–68.

Franca, A., Oliveira, L., 2008. Chemistry of defective coffee beans. In: Koeffer, E.N. (Ed.), Food Chemistry Research Developments. Nova Science Publishers, New York, NY, pp. 105–138.

Franca, A.S., Oliveira, L.S., 2009. Coffee processing solid wastes: current uses and future perspectives. In: Ashworth, G.S., Azevedo, P. (Eds.), Agricultural Wastes. Nova Science Publishers, New York, NY.

Franca, A.S., Oliveira, L.S., Mendonça, J.C.F., Silva, X.A., 2005. Physical and chemical attributes of defective crude and roasted coffee beans. Food Chem. 90, 89–94.

Frascareli, E.C., Silva, V.M., Tonon, R.V., Hubinger, M.D., 2012. Effect of process conditions on the microencapsulation of coffee oil by spray drying. Food Bioprod. Process. 90, 413–424.

Furusawa, M., Narita, Y., Iwai, K., Fukunaga, T., Nakagiri, O., 2011. Inhibitory effect of a hot water extract of coffee "silverskin" on hyaluronidase. Biosci. Biotechnol. Biochem. 75 (6), 1205–1207.

Grollier, J.F., Plessis, S., 1988. Use of coffee bean oil as a sun filter. Patent number US4793990A. Google Patents.

Halvarsson, K., Lodén, M., 2007. Increasing quality of life by improving the quality of skin in patients with atopic dermatitis. Int. J. Cosmet. Sci. 29, 69–83.

Hamishehkar, H., Shokri, J., Fallahi, S., Jahangiri, A., Ghanbarzadeh, S., Kouhsoltani, M., 2015. Histopathological evaluation of caffeine-loaded solid lipid nanoparticles in efficient treatment of cellulite. Drug Dev. Ind. Pharm. 41, 1640–1646.

Hiramoto, S., Itoh, K., Shizuuchi, S., Kawachi, Y., Morishita, Y., Nagase, M., Suzuki, Y., Nobuta, Y., Sudou, Y., Nakamura, O., 2004. Melanoidin, a food protein-derived advanced maillard reaction product, suppresses *Helicobacter pylori* in vitro and in vivo. Helicobacter 9, 429–435.

Iriondo-DeHond, A., Martorell, P., Genovés, S., Ramón, D., Stamatakis, K., Fresno, M., Molina, A., del Castillo, M., 2016. Coffee silverskin extract protects against accelerated aging caused by oxidative agents. Molecules 21, 721.

Jiménez-Zamora, A., Pastoriza, S., Rufián-Henares, J.A., 2015. Revalorization of coffee by-products. Prebiotic, antimicrobial and antioxidant properties. LWT-Food Sci. Technol. 61, 12–18.

Jung, K., Everson, R.J.C., Joshi, B., Bulsara, P.A., Upasani, R., Clarke, M.J., 2016. Structural- functional relationship of phenolic antioxidants in topical skin health products. Int. J. Cosmet. Sci 39 (2), 217–223.

Kammeyer, A., Luiten, R.M., 2015. Oxidation events and skin aging. Ageing Res. Rev. 21, 16–29.

Kang, S., Fisher, G.J., Voorhees, J.J., 1997. Photoaging and topical tretinoin: therapy, pathogenesis, and prevention. Arch. Dermatol. 133, 1280–1284.

Kitagawa, S., Yoshii, K., Morita, S.-y., Teraoka, R., 2011. Efficient topical delivery of chlorogenic acid by an oil-in-water microemulsion to protect skin against UV-induced damage. Chem. Pharm. Bull. 59, 793–796.

Koo, S.W., Hirakawa, S., Fujii, S., Kawasumi, M., Nghiem, P., 2007. Protection from photodamage by topical application of caffeine after ultraviolet irradiation. Br. J. Dermatol. 156, 957–964.

Lane, M.E., 2013. Skin penetration enhancers. Int. J. Pharm. 447, 12–21.

López-Barrera, D.M., Vázquez-Sánchez, K., Loarca-Piña, M.G.F., Campos-Vega, R., 2016. Spent coffee grounds, an innovative source of colonic fermentable compounds, inhibit inflammatory mediators in vitro. Food Chem. 212, 282–290.

Lu, Y.-P., Lou, Y.-R., Xie, J.-G., Peng, Q.-Y., Liao, J., Yang, C.S., Huang, M.-T., Conney, A.H., 2002. Topical applications of caffeine or (−)-epigallocatechin gallate (EGCG) inhibit carcinogenesis and selectively increase apoptosis in UVB-induced skin tumors in mice. PNAS 99, 12455–12460.

Lu, Y.-P., Lou, Y.-R., Xie, J.-G., Peng, Q.-Y., Zhou, S., Lin, Y., Shih, W.J., Conney, A.H., 2006. Caffeine and caffeine sodium benzoate have a sunscreen effect, enhance UVB-induced apoptosis, and inhibit UVB-induced skin carcinogenesis in SKH-1 mice. Carcinogenesis 28, 199–206.

Lupi, O., Semenovitch, I.J., Treu, C., Bottino, D., Bouskela, E., 2007. Evaluation of the effects of caffeine in the microcirculation and edema on thighs and buttocks using the orthogonal polarization spectral imaging and clinical parameters. J. Cosmet. Dermatol. 6, 102–107.

Magalhães, L.M., Machado, S., Segundo, M.A., Lopes, J.A., Páscoa, R.N.M.J., 2016. Rapid assessment of bioactive phenolics and methylxanthines in spent coffee grounds by FT-NIR spectroscopy. Talanta 147, 460–467.

Martinez-Saez, N., Ullate, M., Martín-Cabrejas, M.A., Martorell, P., Genovés, S., Ramón, D., Castillo, M.D.d., 2014. A novel antioxidant beverage for body weight control based on coffee silverskin. Food Chem. 150, 227–234.

Marto, J., Gouveia, L.F., Chiari, B.G., Paiva, A., Isaac, V., Pinto, P., Simões, P., Almeida, A.J., Ribeiro, H.M., 2016. The green generation of sunscreens: using coffee industrial sub-products. Ind. Crops Prod. 80, 93–100.

Matsui, M.S., Hsia, A., Miller, J.D., Hanneman, K., Scull, H., Cooper, K.D., Baron, E., 2009. Non-sunscreen photoprotection: antioxidants add value to a sunscreen. J. Invest. Dermatol. 14, 56–59.

Mazzafera, P., 1999. Chemical composition of defective coffee beans. Food Chem. 64, 547–554.

Menezes, A.C., Campos, P.M., Euletério, C., Simões, S., Praça, F.S.G., Bentley, M.V.L.B., Ascenso, A., 2016. Development and characterization of novel 1-(1-Naphthyl)piperazine-loaded lipid vesicles for prevention of UV-induced skin inflammation. Eur. J. Pharmaceut. Biopharmaceut. 104, 101–109.

Menon, G.K., Cleary, G.W., Lane, M.E., 2012. The structure and function of the stratum corneum. Int. J. Pharm. 435, 3–9.

Meyer, M., Pahl, H.L., Baeuerle, P.A., 1994. Regulation of the transcription factors NF-kappa B and AP-1 by redox changes. Chem. Biol. Interact. 91, 91–100.

Moncrieff, G., Cork, M., Lawton, S., Kokiet, S., Daly, C., Clark, C., 2013. Use of emollients in dry-skin conditions: consensus statement. Clin. Exp. Dermatol. 38, 231–238.

Mujica Ascencio, S., Choe, C., Meinke, M.C., Müller, R.H., Maksimov, G.V., Wigger-Alberti, W., Lademann, J., Darvin, M.E., 2016. Confocal Raman microscopy and multivariate statistical analysis for determination of different penetration abilities of caffeine and propylene glycol applied simultaneously in a mixture on porcine skin ex vivo. Eur. J. Pharmaceut. Biopharmaceut. 104, 51–58.

Mullen, W., Nemzer, B., Stalmach, A., Ali, S., Combet, E., 2013. Polyphenolic and hydroxycinnamate contents of whole coffee fruits from China, India, and Mexico. J. Agricult. Food Chem. 61, 5298–5309.

Murthy, P.S., Naidu, M.M., 2012. Sustainable management of coffee industry by-products and value addition—a review. Resour. Conserv. Recy. 66, 45–58.

Mussatto, S.I., Ballesteros, L.F., Martins, S., Teixeira, J.A., 2011a. Extraction of antioxidant phenolic compounds from spent coffee grounds. Sep. Purif. Technol. 83, 173–179.

Mussatto, S.I., Carneiro, L.M., Silva, J.P.A., Roberto, I.C., Teixeira, J.A., 2011b. A study on chemical constituents and sugars extraction from spent coffee grounds. Carbohydr. Polymers 83, 368–374.

Nakagawa, N., Sakai, S., Matsumoto, M., Yamada, K., Nagano, M., Yuki, T., Sumida, Y., Uchiwa, H., 2004. Relationship between NMF (lactate and potassium) content and the physical properties of the *stratum corneum* in healthy subjects. J. Invest. Dermatol. 122, 755–763.

Napolitano, A., Fogliano, V., Tafuri, A., Ritieni, A., 2007. Natural occurrence of Ochratoxin A and antioxidant activities of green and roasted coffees and corresponding byproducts. J. Agricult. Food Chem. 55, 10499–10504.

Narita, Y., Inouye, K., 2012. High antioxidant activity of coffee silverskin extracts obtained by the treatment of coffee silverskin with subcritical water. Food Chem. 135, 943–949.

Narita, Y., Inouye, K., 2014. Review on utilization and composition of coffee silverskin. Food Res. Int. 61, 16–22.

Natarajan, V.T., Ganju, P., Ramkumar, A., Grover, R., Gokhale, R.S., 2014. Multifaceted pathways protect human skin from UV radiation. Nat. Chem. Biol. 10, 542–551.

Oliveira, L.S., Franca, A.S., Mendonça, J.C.F., Barros-Júnior, M.C., 2006. Proximate composition and fatty acids profile of green and roasted defective coffee beans. LWT-Food Sci. Technol. 39, 235–239.

Oliveira, L.S., Franca, A.S., Camargos, R.R.S., Ferraz, V.P., 2008. Coffee oil as a potential feedstock for biodiesel production. Biores. Technol. 99, 3244–3250.

Pandey, A., Soccol, C.R., Nigam, P., Brand, D., Mohan, R., Roussos, S., 2000. Biotechnological potential of coffee pulp and coffee husk for bioprocesses. Biochem. Eng. J. 6, 153–162.

Park, Y., Kim, K.S., Chung, M., Sung, J.H., Kim, B., 2016. Fabrication and characterization of dissolving microneedle arrays for improving skin permeability of cosmetic ingredients. J. Ind. Eng. Chem. 39, 121–126.

Peng, Y., Xuan, M., Leung, V.Y., Cheng, B., 2015. Stem cells and aberrant signaling of molecular systems in skin aging. Ageing Res. Rev. 19, 8–21.

Pourfarzad, A., Mahdavian-Mehr, H., Sedaghat, N., 2013. Coffee silverskin as a source of dietary fiber in bread-making: optimization of chemical treatment using response surface methodology. LWT-Food Sci. Technol. 50, 599–606.

Prow, T.W., Grice, J.E., Lin, L.L., Faye, R., Butler, M., Becker, W., Wurm, E.M.T., Yoong, C., Robertson, T.A., Soyer, H.P., Roberts, M.S., 2011. Nanoparticles and microparticles for skin drug delivery. Adv. Drug Deliv. Rev. 63, 470–491.

Qin, Z., Voorhees, J.J., Fisher, G.J., Quan, T., 2014. Age-associated reduction of cellular spreading/mechanical force up-regulates matrix metalloproteinase-1 expression and collagen fibril fragmentation via c-Jun/AP-1 in human dermal fibroblasts. Aging Cell 13, 1028–1037.

Quan, T., Qin, Z., Xia, W., Shao, Y., Voorhees, J.J., Fisher, G.J., 2009. Matrix-degrading metalloproteinases in photoaging. J. Invest. Dermatol. 14, 20–24.

Ramalakshmi, K., Rahath Kubra, I., Jagan Mohan Rao, L., 2008. Antioxidant potential of low-grade coffee beans. Food Res. Int. 41, 96–103.

Ramalakshmi, K., Rao, L.J.M., Takano-Ishikawa, Y., Goto, M., 2009. Bioactivities of low-grade green coffee and spent coffee in different in vitro model systems. Food Chem. 115, 79–85.

Ratnayake, W.M.N., Hollywood, R., O'Grady, E., Stavric, B., 1993. Lipid content and composition of coffee brews prepared by different methods. Food Chem. Toxicol. 31, 263–269.

Rawlings, A.V., 2006. Cellulite and its treatment. Int. J. Cosmet. Sci. 28, 175–190.

Rawlings, A.V., Matts, P.J., 2005. Stratum corneum moisturization at the molecular level: an update in relation to the dry skin cycle. J. Invest. Dermatol. 124, 1099–1110.

Ray, S., Raychaudhuri, U., Chakraborty, R., 2016. An overview of encapsulation of active compounds used in food products by drying technology. Food Biosci. 13, 76–83.

Ribeiro, H., Marto, J., Raposo, S., Agapito, M., Isaac, V., Chiari, B.G., Lisboa, P.F., Paiva, A., Barreiros, S., Simões, P., 2013. From coffee industry waste materials to skin-friendly products with improved skin fat levels. Eur. J. Lipid Sci. Technol. 115, 330–336.

Rodrigues, F., Palmeira-de-Oliveira, A., das Neves, J., Sarmento, B., Amaral, M.H., Oliveira, M.B., 2015a. Coffee silverskin: a possible valuable cosmetic ingredient. Pharm. Biol. 53, 386–394.

Rodrigues, F., Pereira, C., Pimentel, F.B., Alves, R.C., Ferreira, M., Sarmento, B., Amaral, M.H., Oliveira, M.B.P.P., 2015b. Are coffee silverskin extracts safe for topical use? An in vitro and in vivo approach. Ind. Crops Prod. 63, 167–174.

Rodrigues, F., Alves, A.C., Nunes, C., Sarmento, B., Amaral, M.H., Reis, S., Oliveira, M.B.P.P., 2016a. Permeation of topically applied caffeine from a food by-product in cosmetic formulations: is nanoscale in vitro approach an option? Int. J. Pharm. 513, 496–503.

Rodrigues, F., Gaspar, C., Palmeira-de-Oliveira, A., Sarmento, B., Helena Amaral, M. P.P., Oliveira, M.B., 2016b. Application of coffee silverskin in cosmetic formulations: physical/antioxidant stability studies and cytotoxicity effects. Drug Dev. Ind. Pharm. 42, 99–106.

Rodrigues, F., Matias, R., Ferreira, M., Amaral, M.H., Oliveira, M.B.P.P., 2016c. In vitro and in vivo comparative study of cosmetic ingredients coffee silverskin and hyaluronic acid. Exp. Dermatol. 25, 572–574.

Rodrigues, F., Sarmento, B., Amaral, M.H., Oliveira, M.B.P.P., 2016d. Exploring the antioxidant potentiality of two food by-products into a topical cream: stability, in vitro and in vivo evaluation. Drug Dev. Ind. Pharm. 42, 880–889.

Rufián-Henares, J.A., de la Cueva, S.P., 2009. Antimicrobial activity of coffee melanoidins— a study of their metal-chelating properties. J. Agricult. Food Chem. 57, 432–438.

Saewan, N., Jimtaisong, A., 2015. Natural products as photoprotection. J. Cosmet. Dermatol. 14, 47–63.

Salgado, J.M., González-Barreiro, C., Rodríguez-Solana, R., Simal-Gándara, J., Domínguez, J.M., Cortés, S., 2012. Study of the volatile compounds produced by *Debaryomyces hansenii* NRRL Y-7426 during the fermentation of detoxified concentrated distilled grape marc hemicellulosic hydrolysates. World J. Microbiol. Biotechnol. 28, 3123–3134.

Salminen, A., Kaarniranta, K., 2009. NF-κB signaling in the aging process. J. Clin. Immunol. 29, 397–405.

Santoyo, S., Ygartua, P., 2000. Effect of skin pretreatment with fatty acids on percutaneous absorption and skin retention of piroxicam after its topical application. Eur. J. Pharmaceut. Biopharmaceut. 50, 245–250.

Sato, Y., Itagaki, S., Kurokawa, T., Ogura, J., Kobayashi, M., Hirano, T., Sugawara, M., Iseki, K., 2011. In vitro and in vivo antioxidant properties of chlorogenic acid and caffeic acid. Int. J. Pharm. 403, 136–138.

Schurer, N.Y., Elias, P.M., 1991. The biochemistry and function of stratum corneum lipids. Adv. Lipid Res. 24, 27–56.

Shin, H.S., Satsu, H., Bae, M.-J., Zhao, Z., Ogiwara, H., Totsuka, M., Shimizu, M., 2015. Antiinflammatory effect of chlorogenic acid on the IL-8 production in Caco-2 cells and the dextran sulphate sodium-induced colitis symptoms in C57BL/6 mice. Food Chem. 168, 167–175.

Silva, N.H.C.S., Drumond, I., Almeida, I.F., Costa, P., Rosado, C.F., Neto, C.P., Freire, C.S.R., Silvestre, A.J.D., 2014a. Topical caffeine delivery using biocellulose membranes: a potential innovative system for cellulite treatment. Cellulose 21, 665–674.

Silva, V.M., Vieira, G.S., Hubinger, M.D., 2014b. Influence of different combinations of wall materials and homogenisation pressure on the microencapsulation of green coffee oil by spray drying. Food Res. Int. 61, 132–143.

Soto, M., Falqué, E., Domínguez, H., 2015. Relevance of natural phenolics from grape and derivative products in the formulation of cosmetics. Cosmetics 2, 259.

Sugawara, T., Kikuchi, K., Tagami, H., Aiba, S., Sakai, S., 2012. Decreased lactate and potassium levels in natural moisturizing factor from the stratum corneum of mild atopic dermatitis patients are involved with the reduced hydration state. J. Dermatol. Sci. 66, 154–159.

Toschi, T.G., Cardenia, V., Bonaga, G., Mandrioli, M., Rodriguez-Estrada, M.T., 2014. Coffee silverskin: characterization, possible uses, and safety aspects. J. Agricult. Food Chem. 62, 10836–10844.

Vandamme, E.J., 2003. Bioflavours and fragrances via fungi and their enzymes. Fungal Divers. 13, 153–166.

Velasco, M.V.R., Tano, C.T.N., Machado-Santelli, G.M., Consiglieri, V.O., Kaneko, T.M., Baby, A.R., 2008. Effects of caffeine and siloxanetriol alginate caffeine, as anticellulite agents, on fatty tissue: histological evaluation. J. Cosmet. Dermatol. 7, 23–29.

Verdier-Sévrain, S., Bonté, F., 2007. Skin hydration: a review on its molecular mechanisms. J. Cosmet. Dermatol. 6, 75–82.

Vyas, L.K., Tapar, K.K., Nema, R.K., Parashar, A.K., 2013. Development and characterization of topical liposomal gel formulation for anticellulite activity. Int. J. Pharm. Pharmaceut. Sci. 5, 512–516.

Wagemaker, T.A.L., Carvalho, C.R.L., Maia, N.B., Baggio, S.R., Guerreiro Filho, O., 2011. Sun protection factor, content and composition of lipid fraction of green coffee beans. Ind. Crops Prod. 33, 469–473.

Williams, A.C., Barry, B.W., 2012. Penetration enhancers. Adv. Drug Deliv. Rev., 128–137.

Yen, W.-J., Wang, B.-S., Chang, L.-W., Duh, P.-D., 2005. Antioxidant properties of roasted coffee residues. J. Agricult. Food Chem. 53, 2658–2663.

Zielinska, A., Nowak, I., 2014. Fatty acids in vegetable oils and their importance in cosmetic industry. CHEMIK 68, 103–110.

Zillich, O.V., Schweiggert-Weisz, U., Eisner, P., Kerscher, M., 2015. Polyphenols as active ingredients for cosmetic products. Int. J. Cosmet. Sci. 37, 455–464.

Biotechnological applications of coffee processing by-products

Suzana E. Hikichi, Rafaela P. Andrade, Eustáquio S. Dias, Whasley F. Duarte

University of Lavras (UFLA), Lavras, Minas Gerais, Brazil

ABSTRACT

The search for new substrates to be used in biotechnological processes has grown in recent years, especially, for the bioprocesses in which microorganisms are used to produce value-added products. In this scenario, the coffee processing by-products, wastewater, pulp, mucilage, husk, silverskin, and spent coffee ground, appear as interesting alternative substrate to be used in bacteria, filamentous fungi, and yeast cultivation. The composition of the coffee processing by-products including water-soluble sugars, proteins, lignin, cellulose, hemicellulose, oils, etc., allows its microbial conversion into products, such as ethanol, acids (lactic, citric, gibberellic, and gallic), aromatic volatile compounds (terpenes, higher alcohols, and esters), pigments (carotenoids), microbial biomass (edible mushroom), and enzymes. In this context, the current chapter presents several examples of coffee processing by-products used as substrate for the cultivation of microorganisms with an ultimate goal to generate value-added products.

Keywords: microbial biotechnology; coffee wastewater; coffee pulp; coffee husk

8.1 COFFEE AND ITS BY-PRODUCTS

Coffee is the world's second largest commodity in trading volume being only surpassed by oil. The *Coffea arabica* and *C. robusta* are the two varieties produced worldwide for commercial purposes (Pandey et al., 2000). According to ICO (2016), the overall coffee production is estimated for 2015/2016 in 144.8 million bags of coffee (60 kg), an increase of 1.6% compared to the 2014/2015. Also according to ICO (2016), the world coffee consumption during the year of 2015 was 152.1 million bags. The demand is high, and tends to increase with the growth of the world population.

The large volume of coffee produced and processed by industry results in the generation of a range of waste and by-products, which implies in the contamination

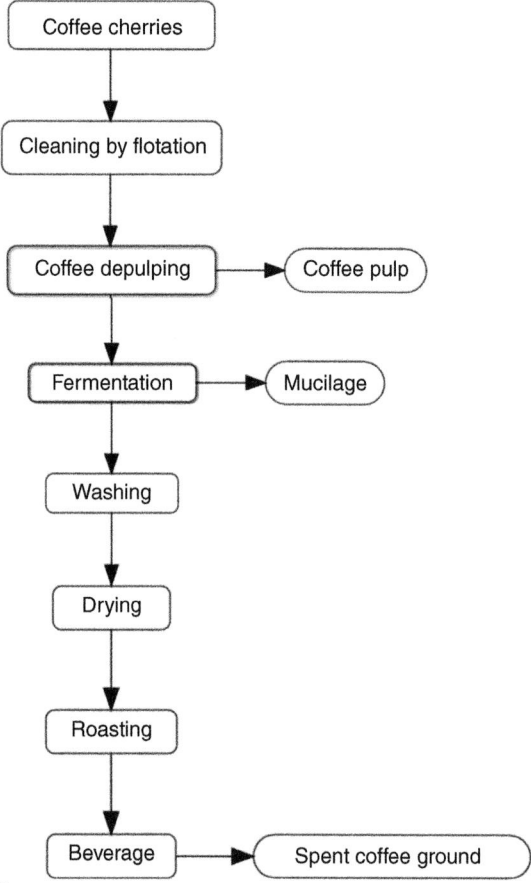

FIGURE 8.1 Coffee Wet Processing and Its By-Products

of water bodies and lands around production units, representing a serious environmental problem for producing countries (Murthy and Naidu, 2012a).

In coffee processing, the main objective is the removal of the pulp, mucilage, parchment, and husk surrounding the bean. After harvesting the coffee is commonly processed in two different ways: by wet or dry process (Oliveira and Franca, 2015; Schwan and Wheals, 2003). A simplified scheme of the coffee wet processing and generated coffee by-products is shown in Fig. 8.1.

The dry process, which results in so-called unwashed or natural coffee, is the oldest and simple coffee-processing method. After harvest, the fruit is dried to reach moisture content around 10%–11% being ready after about 12–15 days of natural drying in dry weather, which also prevents the formation of molds (Murthy and Naidu, 2012a). This method involves complete grain fermentation that is scattered on the ground, concrete, or asphalt platform in layers that are collected overnight and respread every day to dryness. At the end of processing, husk is the main produced by-product and is removed mechanically.

In the wet-processing method the ripe beans are depulped mechanically after harvest and then fermented for approximately 24–48 h to remove the mucilage layer. After fermentation, the beans are washed, immersed in pure water for 12 h, aiming to provide better quality of the product by the removal of terpenes and polyphenols, and dried to reach moisture content similar to the grain processed via dry method (Murthy and Naidu, 2012a). The main by-products produced in the wet process are pulp and wastewater.

In a third type of processing, called semidry or semiwashed, the husk and mucilage are removed mechanically using large amounts of water, producing at the end of the process, husk, pulp and wastewater as abundant by-products.

8.2 COFFEE BY-PRODUCTS WITH POTENTIAL USES IN FERMENTATION BIOTECHNOLOGY

The by-products from coffee production chain are husk, mucilage, pulp, silverskin, wastewater and spent coffee ground. The amount and type of by-product generated is dependent on the type of coffee processing. Taking into account the world's coffee production and the fact that, according to Esquivel and Jiménez (2012), over 50% of the fresh fruit does not compose the marketed product, high amount of coffee by-products is generated from coffee processing.

The coffee husk is obtained when coffee beans are processed via dry method. This by-product is rich in organic compounds, nutrients, contains compounds, such as tannins, caffeine, and polyphenols (Pandey et al., 2000). The coffee pulp, the main by-product from wet processing, corresponds to about 30% of coffee bean dry weight, containing more than 17% cellulose (Bhoite et al., 2013a). The coffee wastewater is the water used for depulping and demucilage of cherries and has high concentration of organic pollutants (Haddis and Devi, 2008). Coffee silverskin is the coffee tegument obtained from coffee roasting process (Toschi et al., 2014). According to Tehrani et al. (2015), the largest portion of by-product generated by the coffee is produced during consumption after filtration in filter paper or by percolation process in coffee machines, the spent coffee ground.

The composition (sugar, proteins, lipids, etc.) of the aforementioned coffee by-products, allow their use in biotechnological processes with microorganisms, such as yeasts, filamentous fungi, and bacteria, generating value-added products with consequent reduction of risks caused by the by-products.

In this context, many studies (Table 8.1) have been developed in recent years aimed at the conversion of coffee by-products, such as ethanol, volatile aromatic compounds, enzymes, edible mushrooms, gallic acid, lactic acid, carotenoids, polyhydroxyalkanoate, and lipids.

The current chapter presents some examples of coffee by-products used as substrate for the production of value-added products via fermentation processes (some not yet fully exploited) illustrating the potential of coffee by-products and microorganisms in the discussed context.

Table 8.1 Coffee By-Products Used in Biotechnological Process for Production of Value-Added Products From Fermentation

Value-Added Products	Coffee By-Products	References
Volatile compounds and ethanol	Wastewater and coffee pulp	Bonilla-Hermosa et al. (2014)
Alcoholic beverage	Spent coffee grounds	Sampaio et al. (2013)
Gallic acid	Coffee pulp	Bhoite et al. (2013a)
Lactic acid	Coffee pulp and mucilage	Pleissner et al. (2016), Neu et al. (2016)
Citric acid	Coffee husks	Shankaranand and Lonsane (1994), Vandenberghe et al. (2000)
Gibberellic acid	Coffee husks	Machado et al. (2002)
Carotenoids	Spent coffee grounds	Obruca et al. (2015), Petrik et al. (2014)
Polyhydroxyalkanoates	Oil from spent coffee grounds	Cruz et al. (2014), Obruca et al. (2014)
Ethanol	Spent coffee grounds	Choi et al. (2012), Mussatto et al. (2012), Tehrani et al. (2015)
Ethanol	Coffee pulp	Bhoite et al. (2013b), Menezes et al. (2013), Shenoy et al. (2011)
Ethanol	Coffee silverskin	Mussatto et al. (2012)
Ethanol	Coffee husks	Gouvea et al. (2009)
Edible mushroom	Coffee husks	Dias et al. (2003), Fan et al. (2006)
Exoglucanase	Coffee husks	Navya et al. (2012a)
Endoglucanase	Coffee husks	Navya and Murthy (2013)
β-Glucosidase	Coffee husks	Navya et al. (2012b)
Protease	Coffee husks, coffee pulp, parchment, silverskin, and spent coffee grounds	Murthy and Naidu (2010)
Tannase	Coffee husks, coffee pulp	Battestin and Macedo (2007), Bhoite et al. (2013c), Sabu et al. (2006)
Xylanase	Coffee pulp, coffee husks, silverskin, and spent coffee grounds	Murthy and Naidu (2012b)
α-Amylase	Coffee pulp, coffee husks, silverskin, and spent coffee grounds	Murthy et al. (2009)
Fructosyl transferase	Coffee pulp, coffee husk, and spent coffee grounds	Sangeetha et al. (2004)
Pectinase	—	Antier et al. (1993), Boccas et al. (1994)

8.3 PRODUCTION OF BIOETHANOL

In the current global scenario, bioethanol is the main environmentally sustainable energy alternative, being a promising substitute of traditional fossil fuels such as gasoline in the transportation sector. Nowadays, bioethanol leads the renewable energy market and offers a reduction of over 80% in carbon dioxide emissions, compared to nonrenewable fuels (Bhoite et al., 2013a). Despite representing a viable alternative to traditional fuels, the success of the bioethanol market is still limited, for example, by the cost of raw materials used in its production. The sugarcane and corn account for the majority of biomass used in the generation of this biofuel, which has led to a competition between transport and food sectors, resulting in an increase of these crops prices (Gurram et al., 2016). This fact has directly affected the cost of the production process in both sectors.

The bioethanol consumed around the world is currently produced from alcoholic fermentation process of simple sugars, such as glucose, by microorganisms, usually yeast. However, it is expected that agricultural and forestry waste will become the future sources of lignocellulosic biomass for the production of this biofuel (Tehrani et al., 2015). The lignocellulosic substrates are characterized as promising materials for the production of ethanol mainly due to their low cost and abundance (Bonilla-Hermosa et al., 2014; Choi et al., 2012). However, according to Bhoite et al. (2013a), the efficient use of raw materials rich in lignin requires an initial chemical, physical, or physicochemical pretreatment. This step aims to convert lignocellulosic biomass into simple monomers and making them accessible carbohydrates for fermentation and consequent production of bioethanol (Tehrani et al., 2015).

It is interesting to note that in most studies related to the use of coffee by-products in bioethanol generation, there are several examples of pretreatments application (for biomass containing lignin), which show the importance of this stage for the production process. Despite being recent, the literature about the application of coffee by-products has many successful examples of bioethanol production, especially for the use of spent coffee ground. The methods used usually involve an initial step of hydrolysis leading to the use of two different fermentation techniques: separated saccharification and fermentation (SHF), simultaneous saccharification and fermentation (SSF), wherein the latter has a higher efficiency due to reduction in operation time (Choi et al., 2012). The importance of enzymatic pretreatment of coffee lignocellulosic biomass for subsequent synthesis of bioethanol has been verified by Choi et al. (2012). In particular, these researchers treated spent coffee grounds samples acquired from Starbucks (Gwangju, Korea) from a pooping machine by applying different pressures in the reactor: 0.49, 0.98, 1.47, and 1.96 MPa, for 10 min reaction time. The best performance was obtained on the pressure 1.47 MPa. Hydrolysis was performed by cellulase and pectinase that were added to the spent coffee ground samples in concentrations of 1.2–18.3 and 1.1–17.4 mg protein/g of residue, respectively. Enzymatic hydrolysis was performed in 1% material (v/v) in phosphate citrate buffer (pH 4.8) at 180 rpm for 48 h at 37°C. Using the highest cellulase concentration of 18.3 mg/g, hydrolysis process resulted in a high content of fermentable sugars, such as glucose, mannose, and galactose.

Among others possible applications, bioethanol production from spent coffee grounds has also been evaluated by Tehrani et al. (2015). Initially, the biomass was dried for 24 h at 120°C and then subjected to saccharification by direct hydrolysis. The solid and liquid phases served as medium for *S. cerevisiae*, being tested in different fermentation conditions, including temperature, substrate flow, yeast inoculum, and the fermentation time. The best ethanol yield was obtained at 30°C using a yeast concentration of 30 mg/g of substrate combining the aqueous and solid phases. In another approach, Bhoite et al. (2013a) promoted simultaneous saccharification and fermentation of coffee pulp, which was additionally used as a carbon source for the biosynthesis of β-glucosidase by *Penicillium verrucosum*. This enzyme breaks down the β-1,4 glycosidic bonds, liberating fermentable sugars and enabling ethanol production. The saccharification step was performed by adding 10 mL of partially purified crude extract of *P. verrucosum* to the medium containing 10 g/L yeast extract, 10 g/L peptone, and 100 g of autoclaved coffee pulp. In the fermentation step, 1% (v/v) of *Saccharomyces cerevisiae* was added to the flasks and incubated under agitation (120 rpm) at 37°C. After 24 h of fermentation under these conditions (simultaneous saccharification and fermentation of pretreated coffee pulp) was obtained a maximum ethanol yield of 3.3%.

Similar to Bhoite et al. (2013a), Bonilla-Hermosa et al. (2014) produced bioethanol using coffee pulp and wastewater, another coffee by-product, which is frequently associated with water and soil contamination. This choice was justified by the fact that the pulp and the wastewater from wet processing of coffee, presented fermentable sugars and other nutrients sources (Table 8.2), which made them suitable substrates

Table 8.2 Composition of Coffee Pulp and Coffee Wastewater reported by Bonilla-Hermosa et al. (2014)

Parameter (g/100 g)	Coffee Pulp (dry matter)	Coffee Wastewater
Water content	82.44	97.56
Total sugars	9.7	1.21
Reducing sugars	9.63	1.13
Nonreducing sugars	0.07	0.08
Protein	14.79	4.26
Lipids	1.2	—
Cellulose	20.7	—
Hemicellulose	3.6	—
Lignin	14.3	—
Total pectin	11.37	—
Soluble pectin	0.7	—
Total phenolic	2.62	0.022
Chemical oxygen demand	—	10.26
Biochemical oxygen demand	—	6.5
pH	—	5.25

for the growth of microorganisms and the production of value-added compounds. Satisfactory results were also obtained by fermenting a mixture of these two by-products by *Hanseniapora uvarum* UFLA CAF76. In this case, coffee pulp was dried at a temperature of 60°C for 48 h until constant weight, then ground and stored until the fermentation step. In a screening step, the yeasts *S. cerevisiae* CA11, *Pichia anomala* UFLA CAF70, *Kluyveromyces marxianus* UFLA CH1-1, *Candida tropicalis* UFLA CES-Y573, *Pichia guilliermondi* UFLA CAF725, *H. uvarum* UFLA CAF76, *Torulaspora delbruekii* UFLA CAF58, and *Pichia stipitis* NCYC 1541 (National Collection of Yeast Cultures, Norwich, UK) were used at a concentration of 0.3 g/L for the fermentation of a medium composed of 10 g coffee pulp and 100 mL of coffee wastewater (previously sterilized by autoclaving). The fermentation experiment was kept in agitation at 105 rpm and 27°C, with the fermentation time being determined by the maximum sugars consumption analyzed by HPLC.

The yeast *H. uvarum* UFLA CAF76 produced 14.67 g/L (1.86% v/v) of ethanol and was selected for the fermentation of an optimized medium composed by 12 g coffee pulp, 100 mL coffee pulp, and 1.0 g/L yeast extract, resulting in an ethanol concentration of 20.39 g/L. Taking into account the state of Minas Gerais (Brazil) coffee production in 2013 (about 1.44 ton of coffee per hectare) and the estimated generation of 0.58 tons of coffee pulp, 125.67 L/ha of bioethanol could be produced using coffee pulp and coffee wastewater as substrate for fermentation.

Coffee pulp has been used as a source of carbohydrate for bioethanol production by Menezes et al. (2013), too. Thereby, a coffee pulp sample from wet processing was collected and stored at −18 ± 2°C until usage. Four different methods for the extraction of soluble fraction of the pulp (manual pressure, mechanical treatment, thermomechanical with heating in a water bath, and thermomechanical with heating in an autoclave) were tested, and after the selection of the best one, the fermentations were performed in the culture medium containing coffee pulp extract, sugarcane juice and/or molasses. Menezes et al. (2013) used an inoculum of 10% (v/v) *S. cerevisiae* CA11 in Erlenmeyer flask with 700 mL of medium fermentation. The flasks were incubated for 24 h at 30°C, and samples were collected at 0, 2, 4, 8, 10, 12, 14, and 24 h. The highest ethanol concentration reported by the authors was 70 g/L and best conditions were grinding the coffee pulp at room temperature, manual press, and use coffee pulp extract with sugarcane juice or molasses for bioethanol fermentation. The production of bioethanol from coffee pulp has also been reported by Shenoy et al. (2011). Specifically, the coffee bean pulp, obtained from wet and dry processing, were pretreated with 2% sulfuric acid (w/v) and autoclaved at 120°C for 10 min. The pH was adjusted to 6.0 and the medium was inoculated with *S. cerevisae* at a concentration of 5 g/L. The fermentation was carried out at 30°C, 120 rpm for 48 h. At this case, the bioethanol yield was 0.46 g/g for dry and wet coffee pulp.

The coffee silverskin and spent coffee ground were the by-products used by Mussatto et al. (2012) for the production of bioethanol. The samples were dried at 60°C before hydrolysis in sulfuric acid solution under the following conditions: liquid/solid ratio of 10 g/g at 163°C/45 min for hydrolyses of spent coffee ground; ratio 14 g/g, 170°C/45 min for coffee silverskin. The hydrolysates were centrifuged and pH adjusted to 5.5 by adding NaOH. Fermentation media were inoculated with yeast in a concentration of 1 g/L and the flasks were incubated in a shaker (200 rpm) at 30°C for 48 h. In this work three different yeasts, *P. stipitis*, *S. cerevisiae*, and *K. fragilis* were evaluated, and the best ethanol yield was obtained in the fermentation with the hydrolyzate of spent coffee ground fermented by *S. cerevisiae*.

Finally, Gouvea et al. (2009) performed experiments using whole and ground husk, as well as the aqueous extract of ground coffee husk generated from the dry processing method. The fermentation media consisted of 13 g of each aforementioned substrates in 100 mL of distilled water inoculated with *S. cerevisiae* in concentrations of 3, 4, and 5 g/L. The fermentations were conducted under stirring at 100 ppm and at three temperatures: 25, 30, and 35°C. The best result was obtained using whole husk coffee as substrate, yeast concentration of 3 g/L, and 30°C, producing 13.6 g/L of ethanol.

8.4 PRODUCTION OF AROMATIC VOLATILE COMPOUNDS, ACIDS, CAROTENOIDS, AND POLYHYDROXYALKANOATES

8.4.1 AROMATIC VOLATILE COMPOUNDS

The low molecular weight compounds that are volatile at room temperature, produce olfactory effects, and are widely applied in industry. Aromas, such as "floral, leafy green, fruity, spicy, among others, are frequently associated with the volatile compounds of economic relevance in cosmetics and food industry. Although the major source of these compounds are plants, according Dubal et al. (2008) factors such as cost and dependence on climate conditions are great disadvantages in the extraction from plants. In recent years, the search for microorganisms and fermentation substrates for low cost production of volatile aromatic compounds in large-scale is a great challenge.

Microorganisms, such as bacteria, filamentous fungi, and yeasts have been used in the production of volatile aromatic compounds using synthetic substrates and those alternatives substrates as agro industrial residues. In this context, the work of Bonilla-Hermosa et al. (2014) was one of the first that has explored the potential of coffee pulp and coffee wastewater for the production of these compounds. They used yeasts, such as *S. cerevisiae*, *P. anomala*, *P. guilliermondii*, *Torulaspora delbrueckii*, *K. marxianus*, and *H. uvarum* in a screening step for the fermentation of a medium containing 10% coffee pulp in coffee wastewater. Thereafter, *H. uvarum* was selected for the fermentation of an optimized medium consisting of 12% of coffee pulp (dry weight) in wastewater added of 1 g/L of yeast extract. The incubation was performed in a shaker at 27°C and 105 rpm. After 24–48 h fermentation, many different volatile aromatic compounds were identified by gas chromatography, including higher alcohols (11), acetates (6), terpenes (6), volatile fatty acids (5), aldehydes (4), and ethyl ester (1).

The yeast *P. guilliermondii* produced 712.87 µg/L of higher alcohols, whereas *P. anomala* resulted in the production of 28.39 µg/L of 2-phelylethanol, one of the most important aromatic compounds of the industry. This yeast also produced, among terpenes found, β-citronellol and linalool in concentrations of 9.8 and 22.73 µg/L. They also reported the production of economically important compounds in the fragrance industry, such as linalool, α-terpeniol, β-citronellol, and geraniol.

Sampaio et al. (2013) investigated the production of volatile aromatic compounds using thermal extraction of spent coffee ground before the fermentation with *S. cerevisiae*. In the developed process, spent coffee grounds (1 g/10 mL of water) were heat treated for 45 min at 163°C. After heat treatment, the obtained extract was added of 180 g/L of sucrose and 175 mg/L of potassium sulfate and inoculated with 1 g/L of *S. ceresiviae*. After the fermentation at 150 rpm/30°C and distillation of the fermented broth, the authors identified and quantified volatile aromatic compounds. Among them, the most abundant were 3-methyl-1-butanol and 2-methyl-1-propanol found in concentrations of 810 and 269 mg/L, respectively. The authors also reported the presence in the distillate of esters, such as ethyl butanoate, ethyl hexanoate, ethyl octanoate, and phenylethyl acetate. It is noteworthy that all the volatile compounds produced from the fermentation of spent coffee ground extract reported by Sampaio et al. (2013) are known to be impactful in the aroma and flavor of beverages, leading to the conclusion that there is a real feasibility of using spent coffee ground extract to produce a distilled beverage containing volatile aromatic compounds with aroma descriptors, such as "coffee, frankly, and elegance."

8.4.2 ORGANIC ACIDS

The production of acids from the fermentation of different coffee by-products has been carried out using solid-state and submerged fermentation, e.g., the production of gallic acid from coffee pulp (Bhoite et al., 2013b) or lactic acid from coffee pulp and mucilage (Neu et al., 2016; Pleissner et al., 2016), citric acid (Shankaranand and Lonsane, 1994; Vandenberghe et al., 2000), and gibberellic acid from coffee husk (Machado et al., 2002).

In the latest study, coffee husk was used to produce an extract by mixing 200 g of husk in 1 L of water, after which the solid material was separated and used for submerged fermentation. The solid residue obtained after extraction, was used as a substrate for solid-state fermentation after treatment with KOH in different concentrations and extraction times. The fermentations of these substrates were performed using different strains of *Gibberella fujikuroi* and one strain of *Fusarium moniliforme*. The gibberellic acid production by the strain *G. fujikuroi* LPB-6 was approximately 30 mg/L and 38.4 mg/kg for submerged fermentation and solid-state fermentation (solid extraction residue), respectively. These values were higher than 18.3 mg/kg obtained from the fermentation of coffee husk without prior treatment with KOH. The obtained concentration of gibberellic acid was considerably enlarged when the coffee husk was treated with KOH, resulting in the production of 99.7 and

112.6 mg/kg when solid residue was treated with 5 and 2.5 g/L of KOH, respectively. At the optimized conditions—70% moisture, pH 5.0, incubation temperature of 29°C and addition of bagasse cassava—492.4 mg/kg of gibberellic acid were produced, demonstrating the feasibility of use of coffee husk and filamentous fungi for the production of gibberellic acid.

Citric acid (an important acid in the food industry due to its preservative function) is mainly produced using *Aspergillus niger* via solid state fermentation. The use of coffee husk as a substrate for citric acid production was assessed by Vandenberghe et al. (2000) and Shankaranand and Lonsane (1994). When coffee husk with 65% moisture was inoculated with 10^7 spores/g (dry substrate) from *A. niger* NRRL 2001, the production of citric acid was 12.7 g/kg. This value was lower compared to the yield produced from cassava residues (Vandenberghe et al., 2000). However, the production reached by Shankaranand and Lonsane (1994) after 120 h of fermentation with *A. niger* was 150 g citric acid for each kg of dry coffee husk (moistened with 0.075 M NaOH solution) supplemented with 1 0.2 and 0.1 ppm of iron, copper and zinc, respectively. This value [about 12-fold higher than that found by Vandenberghe et al. (2000)], may be a function of the physicochemical characteristics of the substrate, the *A. niger* strain used and enrichment or treatment with NaOH, zinc, iron, and copper. The discrepancy between the values obtained in both works, indicate that there is still need for further studies to explore the coffee husk potential for citric acid production.

More recently, the use of coffee pulp and mucilage for lactic acid production was investigated by Pleissner et al. (2016) and Neu et al. (2016), respectively. In both cases, unlike the aforementioned use of fungi for the production of citric acid and gibberellic acid, the production of lactic acid from the coffee by-products have been studied using *Bacillus coagulans*. For the production of lactic acid, Pleissner et al. (2016) obtained the coffee pulp from dried fruits, grinding it into particles of 1 mm for subsequent hydrolysis in two steps. In the first step, the coffee pulp was treated with sulfuric acid 0.18 mol, 121°C for 30 min. This pretreated material was then subjected to a second hydrolytic step with enzyme mixture for hydrolyzing lignocellulose, using 0.3 mL of the enzyme mixture Accellerase 1500 for each gram of coffee pulp. The hydrolysate was used for fermentation in a pilot scale with 45 L of hydrolysate supplemented with yeast extract at a concentration of 10 g/L. The fermentation was carried out using 5% *B. coagulans* inoculum, 400 rpm and pH control initially set to 6. The results for *B. coagulans* coffee pulp hydrolysate were 937 g/L lactic acid L(+) with purity of 99.7% and productivity of 2.4 g/L/h. The potential for the conversion of coffee pulp into lactic acid was evidenced in this study by the possibility of producing 200 to 300 kg lactic acid for every 1000 kg of dry coffee pulp.

The use of coffee mucilage, as well as coffee pulp for the production of lactic acid has also shown promising results. Neu et al. (2016) characterized coffee pulp as containing 6.5%–7.2% dry matter, 21.5–21.7 g/L glucose, 27.8–28.2 g/L galactose + fructose + xylose, and 3.1–10.8 g/L of sucrose. This coffee pulp when added of 5 g/L of yeast extract, inoculated with *B. coagulans* and fermented at 52°C, pH 6.0 and 400 ppm, resulted in the production 40 g/L of lactic acid with productivity of 4–5 g/L/h. In the fermented broth, lactic acid was detected at a concentration of 43.4 g/L, whereas, in the permeate after microfiltration, the concentration of lactic acid concentration was 43.1 g/L. In the following steps, the permeate of the nanofiltration, which showed a concentration of 29.2 g/L lactic acid was subjected to the mono and bipolar electro-dialysis, anion and cation exchange resin and distillation, resulting in production of 0.8 L of lactic acid solution 930.0 g/L with optical purity of 99.8%. The works of Pleissner et al. (2016) and Neu et al. (2016) showed the real potential of coffee pulp and coffee mucilage for the production of lactic acid, due to the fact that these experiments were carried out in a pilot scale mode and obtained the product in a pure form.

Future works should aim at the use of other microorganisms, nutritional supplementation of substrates, and the evaluation of relevant parameters in fermentative processes, such as agitation, inoculum size, and temperature.

8.4.3 CAROTENOIDS

The carotenoids are pigments associated with many important functions including the protection against some diseases (Vachali et al., 2012). These functions make carotenoids an important group of molecules synthesized by plants and microorganisms. The production of these pigments with microbial fermentation has received increasing attention in recent years, mainly due to the use of alternative and low cost substrates, including coffee processing by-products. Among the used microorganism for carotenoids production, filamentous fungi, and yeasts has been more often exploited when compared to bacteria and microalgae.

Two examples of using spent coffee grounds to produce carotenoids are the work of Obruca et al. (2015) and Petrik et al. (2014). In both studies, the authors cite the use of *Sporobolomyces roseus*, one "red" yeast recognized as a producer of pigments, mainly β-carotene, torulene torularhodin (Davoli and Weber, 2002; Davoli et al. 2004). Petrik et al. (2014) used the spent coffee grounds with or without oil extraction for subsequent acidic hydrolysis, when treated with cellulase and/or protease in the production of carotenoids. The culture medium composed of spent coffee grounds was supplemented with 4 g/L $(NH_4)_2SO_4$, 4 g/L of KH_2PO_4, and 0.34 g/L of $MgSO_4$ and used for cultivation of carotenogenic yeasts *Rhodotorula mucilaginosa*, *R. glutinis, Cystofilobasidium capitatum*, and *S. roseus*. The production of both carotenoids and β-carotene was higher when *S. roseus* was used for fermentation of spent-coffee-ground based media prepared with oil extraction, acid hydrolysis, and treatment with cellulases and proteases producing 12.59 mg/L of total carotenoids and 7.81 mg/L of β-carotene in flasks experiments. When the fermentation was performed in bioreactor operated in fed batch, the best yields were 29.9–15.62 mg/L of total carotenoids and β-carotene, respectively. When compared to previously reported values in the literature, the obtained results reinforce that spent coffee grounds are an economically viable substrate for use in the commercial scale production of carotenoids by *S. roseus*. The predicted production, according to the authors, assuming the use of 100 m^3 reactors, can be around 174 kg of carotenoids and 109 kg of β-carotene annually. It is important to mention that in this study, the potential of spent coffee grounds was exploited beyond the production of pigments, since the best yields were obtained from the spent coffee grounds from which oil content was previously extracted resulting in approximately 15% of oil. This oil is also value-added product and can be used as substrates for fermentation to produce other products, such as poly(3-hydroxybutyrate) Obruca et al. (2014) and 3-hydroxybutyrate (Cruz et al., 2014), what we will discuss next.

8.4.4 POLYHYDROXYALKANOATES

The polyhydroxyalkanoates (PHA) are produced by microorganisms acting as storage compounds in the cells. Its degradability and biocompatibility make them interesting as an alternative to petroleum (Anjum et al., 2016). Several microorganisms and substrates are used for the production of PHAs; however, bacteria are most frequently used when compared to other microorganisms. The coffee by-products reported as used for PHA production are spent coffee grounds and their oil.

The extraction of oil from spent coffee grounds for the production of PHA has been performed reported by Cruz et al. (2014) using supercritical fluid extraction in a semi continuous high-pressure system equipped with four extractors. Meanwhile, Obruca et al. (2014), used Soxhlet extraction for 70 g spent coffee ground and 250 mL of *n*-hexane, with subsequent separation of the solvent by distillation. The extracted oil was composed of palmitic and linoleic acids in the proportions of 35.7% and 43.7%, respectively. These two acids are preferably used by *Cupriavidus necator* as a carbon source in the production of PHA as reported by Cruz et al. (2014).

The bacteria *C. necator* grown in spent-coffee-ground oil, resulting in the production of 10 g/L poly (3-hydroxybutyrate), which is higher than the values found for substrates in waste frying palm oil and waste frying sunflower oil. The estimated annual production of poly (3-hydroxybutyrate) in a 100 m^3 bioreactor, operated in fed-batch, using oil extracted from 5573.3 tons of spent coffee grounds, would be around 668.8 tons. This data reinforces the feasibility of the spent-coffee-ground oil used to produce poly (3-hydroxybutyrate) (Obruca et al., 2014).

8.5 MUSHROOM CULTIVATION ON COFFEE BY-PRODUCTS

The use of coffee to grow mushrooms is not a new approach, as they have been used for decades in countries like Colombia and Mexico. This approach is not yet common in Brazil, but a few studies have been published since Brazil is the biggest coffee producer in the world. In Brazil, eucalyptus sawdust and sugarcane bagasse are the main residues used for growing shiitake and oyster mushrooms, respectively. Sawdust and sugarcane bagasse need supplementation to be a good mushroom substrate, whereas other substrates, like different kinds of grasses, could be used without supplementation to get high mushroom production. In spite of this, the former are available at low cost and in large quantities in different regions of Brazil, but to use grasses, it is necessary to cultivate them, so their obtainment is time consuming and demands some cost. That is why sawdust and sugarcane bagasse have been the preferential substrates for mushroom growing in Brazil. However, coffee residues would be a good alternative, too; for example, (1) coffee is cultivated in different regions in Brazil, so their residues are just as or more available than sawdust and sugarcane bagasse; (2) arabica coffee is cultivated in regions of cool weather where some species of mushrooms, like shiitake and shimeji, could be cultivated without the necessity of high costly mushroom houses; (3) in the last years, sugarcane bagasse has been used more and more as an energy source by the alcohol industry, so its utilization as mushroom substrate tends to become more and more limited; (4) has higher nitrogen content than both sawdust and sugarcane bagasse, so the costs with supplementation will be lower.

On the other hand, the published studies about the utilization of coffee in Brazil have used the axenic cultivation system, which is not the most appropriate for the small mushroom growers with low investment capacity. For this reason, the use of coffee depends on the strategy of treatment to control contaminants, pests, and diseases. Short composting followed by steam pasteurization is a very appropriate strategy since it results in a stable compost, which, after steam pasteurization, is very selective to allow the colonization by the mushroom species and avoid the growing of contaminants. Besides, this approach is just as appropriate for mushroom compost production in high scale as in low scale, requiring only an adjustment in the dimensions of the compost facility. However, although having a short composting time, the process is somewhat time consuming, besides being unpleasant, dirty, and smelly. Another approach, is the treatment of the substrate with lime. This approach seems to be an interesting one because there is no need for a compost facility; it is not time consuming, not smelly, and less dirty. Despite these advantages, there are no detailed studies published on this matter. There are some questions about the lime concentration and time of treatment in function of the substrate. Another important point is that pulp, husks, and ground coffee have a property of absorbing a lot of water. When using the axenic system, this characteristic is not a problem because we may add water in the exact quantity, but to use the lime treatment, it is necessary to merge the substrate in the lime solution resulting in an overly wet substrate that is difficult to drain. Considering all the aforementioned questions and difficulties in using coffee as substrate for mushroom cultivation, we will describe next some experiences about the preparation of coffee-husk-based substrates including those treated with lime.

8.5.1 AUTOCLAVED SUBSTRATE FOR AXENIC CULTIVATION

The autoclaved substrate is used for axenic cultivation to have only the inoculated mushroom species growing on the substrate. The conventional procedure to prepare the substrate is to add water to have a substrate with around 60% of moisture, but another possibility is to merge the substrate in water and then drain it to remove the excess water. As discussed earlier, coffee husk absorbs too much water, so the draining process needs to be long (12–24 h) to have the moisture around 60%. To reduce draining time, it is possible to spread the wet substrate and dry it under the sun for a period of 4–6 h. The wet substrate must be put in autoclavable bags (polypropylene or high density polyethylene) with 2 kg capacity, which contributes to cost increase of production in the axenic cultivation system.

According to the obtained experiences, coffee-husk substrate presents a big problem of contamination even when "sterilized" in an autoclave in which many bags may be lost in function due to a high level of contamination. It is known autoclaving is not a 100% efficient process because the elimination of contaminants depends on the microbial population present in the substrate. Naturally, coffee beans present a rich microbiota on their surface, which corresponds to the coffee husks after de-hulling and may increase in function of the moisture and temperature during the coffee beans fermentation step when dry coffee processing is the method chosen to process coffee. During this process, sporulating fungi colonize the grain surface producing millions of spores, which will be present on the coffee. Considering that there is too much powder present in this substrate, it may be recommended to wash the coffee husk to decrease the microbial population. However, the substrate will absorb too much water, as reported earlier, requiring a judicious process of draining so that the substrate may have just the moisture the mushroom species will require for growing. Previous studies have shown that the moisture of coffee husks must be 60% at the maximum. It is important to emphasize that this strategy will be recommended only if the level of contamination is lower than the conventional process. If so, water will be added just to get the necessary moisture before autoclaving. Additionally, excess water is not the only problem when using coffee husks. Some studies have reported an inhibiting effect of caffeine and tannin present in this by-product (Fan et al., 2000), which are suggested to be a determinant factor in stopping the mycelial growth after an apparently normal initial spawning run. According to Fan et al. (2000) the use of untreated coffee husk to cultivate shiitake resulted in mycelial regression and increase of moisture. Dias et al. (2003) reported a similar problem for the utilization of autoclaved coffee husk to cultivate *Pleurotus sajor-caju*, but considering that the heat treatment is sufficient to eliminate the inhibiting effect of caffeine, that effect was attributed to excess moisture and consequently the anaerobic conditions created. It seems that coffee husk and pulp have a high population of anaerobic microorganisms, which ferment the substrate and contribute to increasing the moisture in function of the anaerobic degradation of the substrate. When we use the heat treatment, these microorganisms are totally or partially eliminated, avoiding anaerobic fermentation, but if the treatment is not sufficient to do that, we will have problems of increasing moisture and inhibiting effect of the mushroom mycelial growth.

The *Pleurotus* species are considered to be able to degrade or to reduce caffeine and tannin present in coffee husk (Fan et al., 2006). Despite this, the same authors reported some unsuccessful studies to produce *Pleurotus* on coffee husk, where the mycelial growth was initially vigorous but was interrupted some days after. Therefore, different strains or species of *Pleurotus* show, probably, different levels of tolerance or ability to degrade caffeine and tannin. Fan and Soccol (2005) reported a hot-water treatment to eliminate the problem of caffeine and tannin toxicity; however, it is possible that this treatment has not only a degradation effect but a washing effect, too. Therefore, even for autoclaved substrate coffee husk and pulp, washing may be important to prevent problems of inhibition of mycelial growth. In this case we still face the problem of environmental pollution and wastewater, so appropriate management of the water used in the coffee residues washing is required.

8.5.2 TREATMENT OF SUBSTRATE WITH LIME SOLUTION

This treatment involves the immersion of the substrate in the lime solution with a subsequent absorption of water. Therefore, after lime treatment, it is necessary to drain the excess water until the moisture decreases to 60%. The most practical approach is to spread the wet substrate under the sun until the desired moisture is obtained. This strategy works well during sunny days but during cold and rainy days it could be a problem, probably requiring more draining time. If the substrate moisture is not appropriate, the excess of water will result in anaerobic conditions and bacteria will develop and then inhibit the mushroom mycelial growing.

The Oyster mushrooms may be cultivated in this kind of substrate because they are very tolerant to alkaline pH, which can reach values up to 14 using 2% lime treatment. Few microorganisms may support such alkaline pH, but some species may survive, especially some bacteria that grow if the substrate is too wet. Therefore, the control of moisture after lime treatment is critical for the success of the process. A good alternative to drain excess water is to press the substrate using an appropriate machine with a sufficient pressure to remove all excessive water, but it will add costs to the process.

Considering the control of the moisture is a critical point in the mushroom substrate preparation, the mushroom grower must determine the minimal time of draining to have around 60% or less of moisture, both for sunny or rainy days. A sample must be taken from the drained substrate and put to dry under the sun for 2 days and then calculate its moisture.

The production of mushroom substrate using lime treatment is conceptually much simpler than short composting because there is no need for a mushroom compost facility and it is less time consuming compared to the composting process followed by steam pasteurization. The same type of bags and the spawning process used for compost may be used for the substrate treated by lime. Therefore, this strategy may be stated as the ideal for small mushroom growers, especially for small-scale production. However, it is important to emphasize that the quality of the substrate may not be the same compared to the compost obtained by short composting, since the last one is more stable and selective and does not normally have problems with excess water. In our experience, any excess water will favor anaerobic fermentation and more water will be "produced" during the substrate degradation, increasing even more the anaerobic conditions and producing more toxic compounds that will inhibit the mycelia growth until stopping it.

The use of lime treatment has the advantage of allowing the inoculation process before putting the compost in the bags, so the utilization of a high spawn rate is extremely beneficial to avoid contamination during spawn-run and allowing a faster colonization of the substrate.

8.5.3 SHORT COMPOSTING AND STEAM PASTEURIZATION

The composting process is an interesting way to have stable and selective compost as that desired in mushroom cultivation. For *Agaricus* mushrooms, the composting is longer and more complex, but for oyster mushrooms, a short composting is usually sufficient to have a compost of good quality. Some variations may occur, but the short composting usually lasts for 5–7 days, after which the compost must be pasteurized to kill pests, pathogens, and contaminants. Contrary to what occurs with the axenic system, the inoculation of the compost does not need a laminar flow hood to prevent contaminants, but just a clean room. Another advantage of this system, as reported for the substrate treated with lime, is that the spawn can be mixed throughout the compost before it is packaged in bags, favoring a fast colonization, shortening the spawning-run step. Therefore, the production of the mushroom substrate using short composting and steam pasteurization has the advantage of being cheap, simple, and accessible for small growers, which may scale out the facility according to their needs.

8.5.4 GENERAL CONSIDERATIONS ABOUT THE PREPARATION OF MUSHROOM SUBSTRATES

The *Pleurotus ostreatus* and *Pleurotus pulmonarius* are able to colonize the unpasteurized coffee-husk and pulp-based substrate, but, besides the emergence of larvae, we did observe an increasing water content and anaerobic fermentation. In this case, we did not have problems with excess water, but the presence of a high population of anaerobic bacteria in the unpasteurized substrate was sufficient enough to promote the anaerobic fermentation. Therefore, we may suggest that the anaerobic fermentation may occur in different conditions. In autoclaved substrates, when the moisture is too high, a few survival anaerobic bacteria may grow and establish an anaerobic condition by producing a large quantity of gases. However, for autoclaved substrates, no problems occurred when the moisture of the substrate is 65%. In lime-treated substrate, it is necessary to face problems with moisture and a high survival population of anaerobic bacteria. In this case, 65% moisture is too high to prevent anaerobic fermentation, making it necessary to reduce it to less than 60%. An interesting strategy is to perforate the bags to guarantee continuous aerobic conditions, but it is necessary to have good conditions of air humidity and asepsis. Finally, the substrate obtained by short composting only needs to be pasteurized and to have a moisture of 65% to avoid any problems of insects, contaminants, or anaerobic fermentation. Although composting is dirty, smelly, and time-consuming, there is no doubt that it is very appropriate for oyster-mushroom production. For shiitake production, this system is not appropriate because there is no colonization, unfortunately. Therefore, the heat treatment is currently the most appropriate strategy to prepare the mushroom substrate.

8.5.5 THE MUSHROOM YIELD ON COFFEE-HUSK-BASED SUBSTRATES

The mushroom yield may be given in productivity (P) or biological efficiency (BE) both in percentage (%). Productivity is calculated in function of the reason between fresh mushroom mass and wet substrate mass [(P = fresh mushroom mass/wet substrate mass) $\times$ 100], whereas BE is calculated in function of the reason between fresh mushroom mass and dry substrate mass [(EB = fresh mushroom mass/dry substrate mass) $\times$ 100]. Productivity is not precise to define the efficiency of a substrate to convert its mass in mushrooms because it will depend on the substrate moisture. Biological efficiency is more adequate when comparing different substrates. However, productivity is a more direct measure to evaluate the mushroom yield and calculate the production costs. Therefore, biological efficiency may be more appropriate for scientific communication, but productivity (%) is more appropriate for the mushroom growers. It is common to refer to mushroom yield simply as the production of fresh mushroom/mass of wet substrate (e.g., g/kg) or per square meter (e.g., kg/m^2).

Oyster mushrooms are known to be of very high productivity on a large number of types of substrates, some of them needing no kind of supplementation (Dias et al., 2003; Gonçalves et al., 2010; Siqueira et al., 2012). Therefore, it is possible to cultivate oyster mushroom on, theoretically, any kind of agricultural residue, with or without supplementation. As discussed earlier, coffee residues are superior substrates when compared to many others, making it possible to cultivate oyster mushrooms without the need for supplementation. The use of only coffee husk and pulp as substrate has resulted in BE of more than 90% for oyster mushroom production (Fan et al., 2000; Fan and Soccol, 2005).

Coffee by-products can be used to cultivate shiitake, too, but, according to Fan and Soccol (2005) *Lentinula edodes* needs an adequate compaction of the substrate to promote the ideal conditions for fruiting, and the particles size of coffee husks is not appropriate to allow that. Therefore, Fan and Soccol (2005) suggest combining the coffee husk and pulp with another residue like spent coffee grounds or sawdust and supplement it with a source of nitrogen like wheat or rice bran. However, these results are contradictory to those obtained by Fan et al. (2000), which used coffee husk and spent coffee ground for shiitake production. When used separately, coffee husk and spent coffee ground provided higher biological efficiency (85.5% and 88.6%, respectively) to that obtained from the mixture of the both residues (78.4%). Jaramillo (2005) reported a successful project of cultivating shiitake in Colombia using coffee residues but did not give any details about substrate formulation or give it in publications that are not widely available. Therefore, we may conclude about the potential for cultivating shiitake using coffee residues but some questions remain about the need of supplementation with a source of nitrogen like wheat or rice bran.

This type of supplementation is very important when using sawdust considering that this residue is too poor in nitrogen.

Zied et al. (2016) compared the production of shiitake in different substrate formulations using the traditional substrate as control (80% sawdust + 20% wheat bran). The substrate containing a diversity of supplements (90% sawdust, 1% wheat bran, 1% cottonseed meal, 1% coarse cornmeal, 6.8% rice bran, and 0.2% $CaCO_3$) provided higher BE than the traditional substrate. The increase of EB varied according to the strain so that the greatest increase in EB was 37%. Use of the substrate with twice the concentration resulted in an even higher BE for most strains, but the highest increase was 15.8%. Therefore, the increase of BE was not proportional to the increase in the supplements concentrations for the substrate with a diversity of supplements. It is important to consider that coffee husks and pulp are much richer substrates, so supplementation is probably not necessary in such proportion as reported for sawdust. However, supplementation is a remaining question for future studies.

8.6 ENZYMES PRODUCTION FROM COFFEE BY-PRODUCTS AS SUBSTRATES

Microorganisms have been widely exploited for their ability to produce enzymes and still are an interesting source of many economically important enzymes. According to Adrio and Demain (2014), microbial enzymes are applied to several markets, such as food and beverages, detergents and textiles, pulp and paper, animal feed and personal care, leather, pharmaceuticals, chemical and biofuels.

The search for new strains and species of fungi and bacteria, improvement of production processes, and the need for new substrates are still facing challenges in the production of microbial enzymes. In this context, the potential of coffee processing by-products as substrates for the cultivation of bacteria and fungi for the production of enzymes, such as pectinases, exoglucanase, endoglucanase, β-glucosidase, protease, tannase, xylanase, α-amylase, and fructosyl transferase, has been shown in several studies in recent years. Coffee pulp and coffee husks are the main by products used for microbial enzymes products as discussed next.

Coffee husks were used Navya et al. (2012a) as substrate in solid-state fermentation for the cultivation of *Rhizopus stolonifer* to produce exoglucanase. To optimize the enzyme production, pH, moisture (%), and time of fermentation were evaluated ranging from 3.32 to 6.68, 43.18% to 76.82%, and 79.64 to 160.36 h, respectively. Before the inoculation with *R. stolonifer,* the coffee husk was pretreated with steam during 15 min at 121°C. The highest production of exoglucanase, 390.3 U/mL was achieved after 99.8 h of fermentation at pH 6.7 and coffee husk moisture of 43.2%. This result allowed the authors to conclude that under the optimized conditions coffee husk is an interesting substrate for the commercial production of exoglucanase by *R. stolonifer*, once the obtained value was higher than that reported by others fungi and bacteria know as good producers of this enzyme.

The combination of coffee husks and *R. stolonifer* for the production of endoglucanase were also investigated by Navya and Murthy (2013). Similar to the earlier-cited work, endoglucanase production was optimized by studying pH, moisture (%), and time of fermentation. After 96 h of fermentation with pH 5.43 and moisture of 50%, *R. stolonifer* produced 21,082 U/gds of endoglucanase. The produced enzyme was used for coffee husk saccharification, which resulted, after fermentation with *S. cerevisiae*, in the production of 65.5 g/L of ethanol. Additionally, the enzyme potential as detergent additive was showed. The approach presented by Navya and Murthy (2013) for the use of coffee husks, reinforces the feasibility of this coffee by-product as an interesting substrate to cultivate both filamentous and unicellular fungi producing two very important added value microbial metabolites.

The three fermentation parameters studied by Navya et al. (2012a) and Navya and Murthy (2013), was also evaluated by Navya et al. (2012b) to optimize the production of β-glucosidase from the fermentation of coffee husk by *R. stonolonifer*. Under the conditions pH 5.2, moisture 53%, and fermentation time of 112 h, the highest (448.92 U/mL) enzyme yield was produced. The prepurified enzyme presented maximum activity at 60°C and pH 6, showing a Km of 59 mM and V_{max} of 78 μmol/min/mg cellobiose.

Coffee by-products have also been used for solid state fermentation with *Aspergillus oryzae* to produce enzymes, such as fructosyl transferase (Sangeetha et al., 2004) and protease (Murthy and Naidu, 2010). Among coffee pulp, coffee husks, and spent coffee grounds, the last by-product allowed the higher production of fructosyl transferase when added of yeast extract (Sangeetha et al., 2004). When coffee husks, coffee pulp, parchment, silverskin, and spent coffee grounds were used to produced protease by *A. oryzae*, 12,236 U/g was the best result obtained (Murthy and Naidu, 2010), showing that among studied coffee by-products, coffee husks pretreated with steam (121°C/15 min) had the potential to be used as substrate for cultivation of *A. oryzae* to produce protease.

Xylanase is another enzyme that can be produced from coffee by-products fermentation. Murthy and Naidu (2012b) showed that among coffee pulp, parchment, coffee husks, silverskin, and spent coffee grounds, only coffee parchment did not allow the production of xylanase by *Penicillium* sp. On the other hand, coffee pulp resulted in the highest enzyme activity of 9475 U/g. This valued was increased to 23,494 U/g when the used conditions were as follows: pretreatment with steam (121°C/30 min), particle size 1.5 mm, moisture 50%, xylose carbon source, peptone as nitrogen source, inoculum size 20%, pH 5.0, temperature 30°C, and 5 days of fermentation.

Pectinase is an enzyme commonly used in processes involving degradation of plant materials, such as in the production of wine and beer. This enzyme is widely explored from fungi.

Antier et al. (1993) evaluated strains of fungi isolated from coffee plants or from the soil of coffee plantations, for pectinase production in solid-state fermentation of coffee pulp. Screening of 248 strains, the authors selected a strain of *Aspergillus niger* as the best pectinase producer. The dried coffee pulp was added to KH_2PO_4,

$(NH_4)_2SO_4$, and urea at concentrations of 4.9%, 8%, and 4.3%, respectively. The coffee pulp was then sterilized at 10 psi for 30 min, the initial pH was adjusted to 5.5 by addition of H_3PO_4 0.1 M or 0.1 M NaOH. The fermentations with coffee pulp of 60% moisture were kept aerated with fixed rate of 60 mL/min saturated air. After inoculated with 2×10^7 spores/g of dry pulp, the cylinders were kept at 25°C in a water bath. To quantify the production of pectinase, the medium fermented by *A. niger* was pressed, the extract obtained was centrifuged, and the supernatant was used for enzyme assays. A total of 27.7 U/mL of pectinase was found in fermented extract corresponding to 138 U/g of dry coffee pulp.

Boccas et al. (1994) performed a screening with 248 fungi strains isolated from Mexico coffee growing areas to select fungi with pectinase production potential fermenting coffee pulp by solid state fermentation. From this screening, a strain of *A. niger* showed capacity to produce fourfold more pectinase in coffee pulp when compared to the used reference strain. For fermentation, 34 g of dried coffee pulp was added of 0.8 g urea, 3.3 g $(NH_4)_2SO_4$, and 30 mL of distilled water and sterilized by autoclaving for 20 min at 121° C, followed by inoculation with 2×10^7 spores/g dry coffee pulp. The incubation conditions were similar to those described by Antier et al. (1993). For the enzyme extraction, 30 g of fermented material was mixed with 30 mL of water and pressed. The extract obtained after the pressing was centrifuged and the supernatant was used for enzyme assays. Both authors, Antier et al. (1993) and Boccas et al. (1994) found that the fungus *A. niger* showed high efficiency in the production of pectinase from the solid state fermentation of coffee pulp. Another interesting fact is that both authors, with similar methods, showed a high production of pectinase, reinforcing the feasibility of pectinase production by coffee pulp solid-state fermentation using *A. niger*.

Amylase is another enzyme also reported as produced from fermentation of coffee by-products. This enzyme may be derived from plants, animals, and microorganisms, and is used in fermentation, textile, and paper industries. Murthy et al. (2009) evaluated the capacity of *Neurospora crassa* for the fermentation of coffee pulp, coffee husk, coffee silverskin, spent coffee grounds, and the mixture of these by-products to produce amylase. Each coffee by-product, as well as a combination of 10 g of all coffee by-products, was crushed into particles of 1 mm. The substrates were inoculated with the *N. crassa* CFR 308 and incubated for 5 days at room temperature for fermentation. Several parameters that affect enzyme production were optimized during fermentation, and different treatments were tested on coffee pulp and by-products mixture to improve the enzyme activity. The substrates were treated with steam, enzyme hydrolysis (cellulase containing 50 μg/mL of Lowry protein), 1% hydrogen peroxide, 1.5% methanol, and 1% sodium hydroxide. Among the different coffee by-products used, the coffee pulp and the by-products mixture showed higher enzymatic activities.

Some factors, such as temperature, moisture, pH, fermentation time, inoculum size, and particle size can influence the production of the enzyme amylase. The optimum conditions for the production of amylase were 60% moisture, temperature 27°C, pH 4.5, and 5 days of fermentation. The inoculum of 10^7 spores showed higher efficiency in the production of amylase when compared to the inocula of 10^8 and 10^6 spores.

Battestin and Macedo (2007) and Sabu et al. (2006), have also used coffee residues for enzyme production. Battestin and Macedo (2007) evaluated the production of tannase from the fermentation of wheat bran with coffee husk by the fungus *Paecilomyces variotii*. The effect of some variables, such as temperature, percentage of by-productin f tannic acid, saline solutions and 3, 5, and 7 days of fermentation were evaluated for the production of tannase. The best fermentation conditions were 29–34°C, 8.5%–14% tannic acid, 50:50 wheat bran and coffee husk, and 5 days of fermentation. Under the optimized fermentation conditions it was observed an increase of 8.6 times in the production of the enzyme.

Sabu et al. (2006), unlike most authors that use fungi for the production of enzymes, reported the use of *Lactobacillus* sp. ASR S1 for tannase production. This author tested various by-products for the production of tannase in solid-state fermentation including coffee husks. Among the used by-products, coffee husks allowed a maximum productin of extra-cellular tannase. For fermentation, 5 g of coffee husks wetted in a saline solution were added to 0.5% NH_4NO_3, 0.1% $MgSO_4.7H_2O$, and 0.1% of NaCl. The fermentation medium pH 5 was inoculated with 1 mL of cell suspension, grown for 18–20 h, and then incubated at 30°C for 48 h. Some parameters were evaluated aiming an increase in the production of tannase. The best conditions for coffee husk fermentation and better tannase production were supplementation with 0.6% tannic acid, 50% moisture, inoculation with 1 mL of cell suspension incubated at 33°C for 72 h, resulting in the production of 0.85 U/g of tannase. Although Battestin and Macedo (2007) and Sabu et al. (2006) have conducted studies with different microorganisms, fungi and bacteria, both noted that it is possible to produce tannase from coffee husk fermentation.

8.7 CONCLUSIONS

Considering the aforementioned examples, we find that all coffee by-products present potential to be use in microbial biotechnology. Besides generating added value microbial products, this use also allows a possible reduction of risks, considering that some coffee by-products, such as wastewater and coffee pulp have polluting potential. Thus, the biotechnological application of coffee by-products is in line with the growing worldwide need for the sustainable production. Although several success examples of the coffee by-products use have already been reported, further studies are needed to maximize these potential uses and even for finding new alternatives to the exploitation of coffee by-products.

REFERENCES

Adrio, J.L., Demain, A.L., 2014. Microbial enzymes: tools for biotechnological processes. Biomol. 4, 117–139.

Anjum, A., Zuber, M., Zia, K.M., Noreen, A., Anjum, M.N., Tabasum, S., 2016. Microbial production of polyhydroxyalkanoates (PHAs) and its copolymers: A review of recent advancements. Int. J. Biol. Macromol. 89, 161–174.

Antier, P., Minjares, A., Roussos, S., Raimbault, M., Viniegra-Gonzalez, G., 1993. Pectinase-hyperproducing mutants of *Aspergillus niger* C28B25 for solid-state fermentation of coffee pulp. Enzyme Microb. Technol. 15, 254–260.

Battestin, V., Macedo, G.A., 2007. Tannase production by *Paecilomyces variotii*. Bioresour. Technol. 98, 1832–1837.

Bhoite, R.N., Navya, P.N., Murthy, P.S., 2013a. Statistical optimization of bioprocess parameters for enhanced gallic acid production from coffee pulp tannins by *Penicillium verrucosum*. Prep. Biochem. Biotechnol. 43, 350–363.

Bhoite, R.N., Navya, P.N., Murthy, P.S., 2013b. Statistical optimization, partial purification, and characterization of coffee pulp β-glucosidase and its application in ethanol production. Food Sci. Biotechnol. 22, 205–212.

Bhoite, R.N., Navya, P.N., Murthy, P.S., 2013c. Purification and characterisation of a coffee pulp tannase produced by *Penicillium verrucosum*. J. Food Sci. Engineer. 3, 323–331.

Boccas, F., Roussos, S., Gutierrez, M., Serrano, L., Viniegra, G.G., 1994. Production of pectinase from coffee pulp in solid-state fermentation system - selection of wild fungal isolate of high potency by a simple 3-step screening technique. J. Food Sci. Technol. 31, 22–26.

Bonilla-Hermosa, V.A., Duarte, W.F., Schwan, R.F., 2014. Utilization of coffee by-products obtained from semi-washed process for production of value-added compounds. Bioresour. Technol. 166, 142–150.

Choi, I.S., Wi, S.G., Kim, S.B., Bae, H.J., 2012. Conversion of coffee residue waste into bioethanol with using popping pretreatment. Bioresour. Technol. 125, 132–137.

Cruz, M.V., Paiva, A., Lisboa, P., Freitas, F., Alves, V.D., Simões, P., Barreiros, S., Reis, M.A.M., 2014. Production of polyhydroxyalkanoates from spent coffee grounds oil obtained by supercritical fluid extraction technology. Bioresour. Technol. 157, 360–363.

Davoli, P., Weber, R.W.S., 2002. Carotenoid pigments from the red mirror yeast. Sporobolomyces roseus. Mycologist. 16, 102–108.

Davoli, P., Mierau, V., Weber, R.W.S., 2004. Carotenoids and fatty acids in red yeasts *Sporobolomyces roseus* and *Rhodotorula glutinis*. Appl. Biochem. Microbiol. 40, 392–397.

Dias, E.S., Koshikumo, E.M.S., Schwan, R.F., Silva, R.D.A., 2003. Cultivation of the mushroom *Pleurotus sajor*-caju in different agricultural residues. Cienc. e Agrotecnologia. 27, 1363–1369.

Dubal, S.A., Tilkari, Y.P., Momin, S.A., Borkar, I.V., 2008. Biotechnological routes in flavour industries. Advanced Biotech. 6, 20–31.

Esquivel, P., Jimenez, V.M., 2012. Functional properties of coffee and coffee by-products. Food Res. Int. 46, 488–495.

Fan, L., Soccol, C.R., 2005. Coffee Residues for Shiitake Cultivation. Mushroom Grower's Handabook 2 - Shitake Cultivation. Mushworld, Seoul, (pp. 92-95).

Fan, L., Pandey, A., Soccol, C.R., 2000. Solid state cultivation-an efficient method to use toxic agro-industrial residues. J. Basic Microbiol. 40, 187–197.

Fan, L., Soccol, A.T., Pandey, A., Porto, L., Vandenberghe, D.S., Soccol, C.R., 2006. Effect of caffeine and tannins on cultivation and fructification of *Pleurotus* on Coffee Husks. Braz. J. Microbiol. 37, 420–424.

Gonçalves, C.C. de M., Paiva, P.C. de A., Dias, E.S., de Siqueira, F.G., Henrique, F., 2010. Avaliação do cultivo de *Pleurotus sajor*-caju (fries) sing. sobre o resíduo de algodão da industria têxtil para a produção de cogumelos e para alimentação animal. Cienc. e Agrotecnologia. 34, 220–225.

Gouvea, B.M., Torres, C., Franca, A.S., Oliveira, L.S., Oliveira, E.S., 2009. Feasibility of ethanol production from coffee husks. Biotechnol. Lett. 31, 1315–1319.

Gurram, R., Al-Shannag, M., Knapp, S., Das, T., Singsaas, E., Alkasrawi, M., 2016. Technical possibilities of bioethanol production from coffee pulp: A renewable feedstock. Clean Technol. Environ. Policy 18, 269–278.

Haddis, A., Devi, R., 2008. Effect of effluent generated from coffee processing plant on the water bodies and human health in its vicinity. J. Hazard. Mater. 152, 259–262.

ICO, 2016. International Coffee Organization. Available from: http://www.ico.org/

Jaramillo, C.L., 2005. Mushroom growing project in Colombia. Mushroom Grower's Handabook 2 - Shitake cultivation. Mushworld, Seoul, (pp. 234–243).

Machado, C.M.M., Soccol, C.R., de Oliveira, B.H., Pandey, A., 2002. Gibberellic acid production by solid-state fermentation in coffee husk. Appl Biochem Biotechnol. 102, 179–191.

Menezes, E.G.T., Do Carmo, J.R., Menezes, A.G.T., Alves, J.G.L.F., Pimenta, C.J., Queiroz, F., 2013. Use of different extracts of coffee pulp for the production of bioethanol. Appl. Biochem. Biotechnol. 169, 673–687.

Murthy, P.S., Naidu, M.M., 2010. Proteases production by *Aspergillus oryzae* in solid state fermentation utilizing coffee by products. World Appl. Sci. J. 8, 199–205.

Murthy, P.S., Naidu, M.M., 2012a. Sustainable management of coffee industry by-products and value addition - A review. Resour. Conserv. Recycl. 66, 45–58.

Murthy, P.S., Naidu, M.M., 2012b. Production and application of xylanase from *Penicillium* sp. utilizing coffee by-products. Food Bioprocess Technol. 5, 657–664.

Murthy, P.S., Naidu, M.M., Srinivas, P., 2009. Production of α-amylase under solid-state fermentation utilizing coffee waste. J. Chem. Technol. Biotechnol. 84, 1246–1249.

Mussatto, S.I., Machado, E.M.S., Carneiro, L.M., Teixeira, J.A., 2012. Sugars metabolism and ethanol production by different yeast strains from coffee industry wastes hydrolysates. Appl. Energy 92, 763–768.

Navya, P.N., Murthy, P.S., 2013. Production, statistical optimization and application of endoglucanase from *Rhizopus stolonifer* utilizing coffee husk. Bioprocess Biosyst. Eng. 36, 1115–1123.

Navya, P.N., Bhoite, R.N., Murthy, P.S., 2012a. Bioconversion of coffee husk cellulose and statistical optimization of process for production of exoglucanase by *Rhizopus stolonifer*. World Appl. Sci. J. 20, 781–789.

Navya, P.N., Bhoite, R.N., Murthy, P.S., 2012b. Improved β-glucosidase production from *Rhizopus stolonifer* utilizing coffee husk. Int. J. Curr. Res. 4, 123–129.

Neu, A.K., Pleissner, D., Mehlmann, K., Schneider, R., Puerta-Quintero, G.I., Venus, J., 2016. Fermentative utilization of coffee mucilage using *Bacillus coagulans* and investigation of down-stream processing of fermentation broth for optically pure l(+)-lactic acid production. Bioresour. Technol. 211, 398–405.

Obruca, S., Benesova, P., Petrik, S., Oborna, J., Prikryl, R., Marova, I., 2014. Production of polyhydroxyalkanoates using hydrolysate of spent coffee grounds. Process Biochem. 49, 1409–1414.

Obruca, S., Benesova, P., Kucera, D., Petrik, S., Marova, I., 2015. Biotechnological conversion of spent coffee grounds into polyhydroxyalkanoates and carotenoids. N. Biotechnol. 32, 569–574.

Oliveira, L.S., Franca, A.S., 2015. An overview of the potential uses for coffee husks. In: Preedy, V.R. (Ed.), Coffee in Health and Disease Prevention. Academic Press, San Diego, CA, (Chapter 31).

Pandey, A., Soccol, C.R., Nigam, P., Brand, D., Mohan, R., Roussos, S., 2000. Biotechnological potential of coffee pulp and coffee husk for bioprocesses. Biochem. Eng. J. 6, 153–162.

Petrik, S., Obruča, S., Benešová, P., Márová, I., 2014. Bioconversion of spent coffee grounds into carotenoids and other valuable metabolites by selected red yeast strains. Biochem. Eng. J. 90, 307–315.

Pleissner, D., Neu, A.-K., Mehlmann, K., Schneider, R., Puerta-Quintero, G.I., Venus, J., 2016. Fermentative lactic acid production from coffee pulp hydrolysate using *Bacillus coagulans* at laboratory and pilot scales. Bioresour. Technol. 218, 167–173.

Sabu, A., Augur, C., Swati, C., Pandey, A., 2006. Tannase production by *Lactobacillus* sp. ASR-S1 under solid-state fermentation. Process Biochem. 41, 575–580.

Sampaio, A., Dragone, G., Vilanova, M., Oliveira, J.M., Teixeira, J.A., Mussatto, S.I., 2013. Production, chemical characterization, and sensory profile of a novel spirit elaborated from spent coffee ground. LWT - Food Sci. Technol. 54, 557–563.

Sangeetha, P.T., Ramesh, M.N., Prapulla, S.G., 2004. Production of fructosyl transferase by *Aspergillus oryzae* CFR 202 in solid-state fermentation using agricultural by-products. Appl. Microbiol. Biotechnol. 65, 530–537.

Schwan, R.F., Wheals, A.E., 2003. Mixed microbial fermentations of chocolate and coffee. In: Boekhout, T., Robert, V. (Eds.), Yeasts in Food. Behr's Verlag, Hamburg, pp. 426–459.

Shankaranand, V.S., Lonsane, B.K., 1994. Coffee Husk: an inexpensive substrate for production of citric acid by *Aspergillus Niger* in a solid state fermentation system. World J. Microbiol. Biotechnol. 10, 165–168.

Shenoy, D., Pai, A., Vikas, R.K., Neeraja, H.S., Deeksha, J.S., Nayak, C., Rao, C.V., 2011. A study on bioethanol production from cashew apple pulp and coffee pulp waste. Biomass and Bioenergy 35, 4107–4111.

Siqueira, F.G., Maciel, W.P., Martos, E.T., Duarte, G.C., Miller, R.N.G., Silva, R., Dias, E.S., 2012. Cultivation of *Pleurotus* mushrooms in substrates obtained by short composting and steam pasteurization. Afr. J. Biotechnol. 11, 11630–11635.

Tehrani, N.F., Aznar, J.S., Kiros, Y., 2015. Coffee extract residue for production of ethanol and activated carbons. J. Clean. Prod. 91, 64–70.

Toschi, T.G., Cardenia, V., Bonaga, G., Mandrioli, M., Rodriguez-Estrada, M.T., 2014. Coffee Silver skin: characterization, possible uses, and safety aspects. J. Agric. Food Chem. 62, 10836–10844.

Vachali, P., Bhosale, P., Bernstein, P.S., 2012. Microbial carotenoids. In: Barredo, J.-L. (Ed.), Microbial carotenoids from fungi: Methods and Protocols. Humana Press, Totowa, NJ, pp. 41–59.

Vandenberghe, L.P.S., Soccol, C.R., Pandey, A., Lebeault, J.M., 2000. Solid-state fermentation for the synthesis of citric acid by *Aspergillus niger*. Bioresour. Technol. 74, 175–178.

Zied, D.C., Maciel, W.P., Marques, S.C., da Silveira e Santos, D.M., Rinker, D.L., Dias, E.S., 2016. Selection of strains for shiitake production in axenic substrate. World J. Microbiol. Biotechnol. 32, 168.

Environmental applications of coffee processing by-products

Mejdi Jeguirim*, Lionel Limousy*, Madona Labaki**

*Institute of Materials Science of Mulhouse, Mulhouse, France; **Lebanese University, Fanar, Jdeidet, Lebanon

ABSTRACT

This chapter explores the possibilities of using coffee processing by-products (e.g., coffee husks, coffee grounds and coffee beans) for environmental applications (e.g., adsorption of heavy metals, dye removal etc.). The elaboration of char from coffee processing by-products is presented and some applications devoted to the removal of cations and dyes from wastewater. The preparation of activated carbon (AC) is detailed. Chemical and physical activation methods are described as well as the characteristics of the AC obtained (textural, structural properties and surface chemistry). The role of the activation process on the adsorption of cations, anions, and organic molecules (pharmaceutical products, pesticides, micropollutants) contained in water is discussed. The use of AC for gas treatment is detailed, especially for acid pollutants (H_2S, NO_2) and CO_2 coming from postcombustion. The last part of this chapter is devoted to the preparation of AC for catalytic application (decomposition of organic molecules). For this specific application, AC has to be functionalized (sulfonation of the surface, or impregnation of K and Fe) in order to obtain catalytic properties.
Keywords: biosorbent; biochar; activated carbon; adsorption; catalysis

9.1 INTRODUCTION

This chapter examines the recovery of coffee processing by-products for environmental applications. In particular, the use of raw and modified coffee residues for the removal of pollutants from aqueous and gaseous phases is addressed. The modification of coffee residues includes chars, activated carbon (AC), and catalyst support production.

This chapter is divided into four main parts. The first part is devoted to the direct use of raw coffee-processing residues as biosorbents for the removal of heavy metals and organic dyes from aqueous solution. This section enumerates the main results obtained in literature including the adsorption capacities, as well as kinetics,

Handbook of Coffee Processing By-Products. http://dx.doi.org/10.1016/B978-0-12-811290-8.00009-8

equilibrium, and thermodynamic parameters. The second part presents the few attempts for chars elaboration from coffee-processing residues and their performance in wastewater treatment. The third part is dedicated to the preparation of ACs using coffee-processing residues as precursor through different activation protocols. The morphological, textural, and surface chemistry of the prepared ACs are analysed using different analytical techniques. Then, the applications of these ACs for the removal of pollutants from aqueous and gaseous effluents are presented. The use of AC as a catalyst support for the elimination of several organic compounds is presented in the fourth section.

9.2 THE USE OF COFFEE-PROCESSING RESIDUES AS BIOSORBENTS FOR POLLUTANTS REMOVAL FROM AQUEOUS SOLUTION

The interest on low-cost adsorbents derived from agriculture and agrifood-processing residues is growing. Numerous works have evaluated the performance of these cheap adsorbents for removal of pollutants from aqueous solution (Belala et al., 2011; Chouchène et al., 2014). These adsorbents are available in large amounts and request low processing costs compared to other adsorbents materials, such as ACs and zeolites. Several investigations have examined the application of coffee processing residues in the removal of organic compounds, dyes, and heavy metals from aqueous effluent. These studies focused basically on the determination of the biosorbents adsorption capacities at laboratory scale using wastewater solution models. The experimental results were used to extract the kinetics, thermodynamic, and equilibrium parameters for pollutants adsorption.

9.2.1 APPLICATION FOR THE REMOVAL OF HEAVY METALS

The adsorption behaviors of several heavy metals ions including cadmium (Cd), lead (Pb), copper (Cu), chromium (Cr), and Zinc (Zn) on the different coffee processing residues were examined in the literature. The adsorption capacities were evaluated at various operating conditions, such as temperature, pH, contact time (Ct), adsorbent dosage, and heavy metals concentrations (C_i). Table 9.1 summarizes the adsorption capacities for coffee processing by-products available in literature for different metals ions.

9.2.1.1 Coffee bean
The performance of coffee-bean residue after extraction with hot water was evaluated first as a heavy metals adsorbent in aqueous solution. Coffee-bean residues generated from four Arabica species treated at five roasting temperatures and times were tested for the removal of cadmium (II) and copper (II) from aqueous solution (Minamisawa et al., 2005). The adsorption tests were performed in batch mode by immersing 2.5 g of adsorbent in 500 mL of sample solution containing 5 mg/L of heavy metals. The

Table 9.1 Coffee-Processing Residues Capacity for Heavy Metal Removal in Aqueous Solution

Coffee Residues	Metal	Operating Conditions	Adsorption Capacity (mg/g)	References
Coffee grounds	Cd (II)	pH = 7, T = 20°C, Ct = 120 min, C_i = 100 mg/L	15.65	Azouaou et al. (2010)
Coffee grounds	Pb (II)	T = 25°C, Ct = 24 h, C_i = 200 µg/L	0.70	Tokimoto et al. (2005)
Degreased coffee beans	Cd (II)	pH = 8.8, T = 30°C, Ct = 24 h, C_i = 5–500 mg/L	6.72	Kaikake et al. (2007)
Coffee husk	Cr (IV)	pH = 4, T = 25°C, Ct = 72 h, C_i = 100 mg/L	7.00	Oliveira et al. (2008b)
	Cu (II)		7.50	
	Zn (II)		5.60	
	Cd (II)		6.90	
Coffee beans	Cu (II)	pH = 6.5–6.7, T = 20°C, Ct = 180 min, C_i = 5 mg/L	1.82–2.00	Minamisawa et al. (2005)
	Cd (II)		1.82–1.98	

contact time was 180 min and the pH was adjusted between 6.5 and 6.7. The authors found that Cd(II) and Cu(II) are removed very quickly and the adsorption capacities ranged between 1.8 and 2.0 mg/g for all the examined samples. These capacities were similar to those obtained for zeolithe and AC in similar operating conditions. However, it is difficult to point out the efficiency of coffee-bean residues due to the low initial concentrations (5 mg/L) of heavy metals used in the aqueous solution.

Kaikake et al. (2007) have also evaluated the capacity of coffee beans for the removal of cadmium ion from aqueous solution. In this investigation, coffee beans were firstly degreased using ethanol soxhlet extraction. Authors observed through FTIR and SEM analysis that coffee ingredients were removed after oil extraction leading to a macroporous structure (Fig. 9.1) with a specific surface area of 1.2 m²/g. The remaining substrate was used as an adsorbent in an aqueous solution containing cadmium concentrations ranging from 5 to 500 mg/L at 30°C and pH equal to 8.8. Under these operating conditions, the authors found an adsorption capacity of 6.72 mg/g and identified an ion exchange reaction between metal ions and the adsorbent as a reaction mechanism. Furthermore, Kaikake et al. (2007) have found that a solution of hydrochloric acid (1 mol/L) is effective for the total cadmium desorption.

9.2.1.2 Coffee husk

Oliveira et al. (2008b) have evaluated the performance of coffee husks for the removal of cadmium, copper, zinc, and chromium from aqueous solution. The adsorption tests were realized in the following operating conditions: initial concentrations: 50–100 mg/L, pH: 4–7, T: 25°C. They have characterized the adsorbent before and

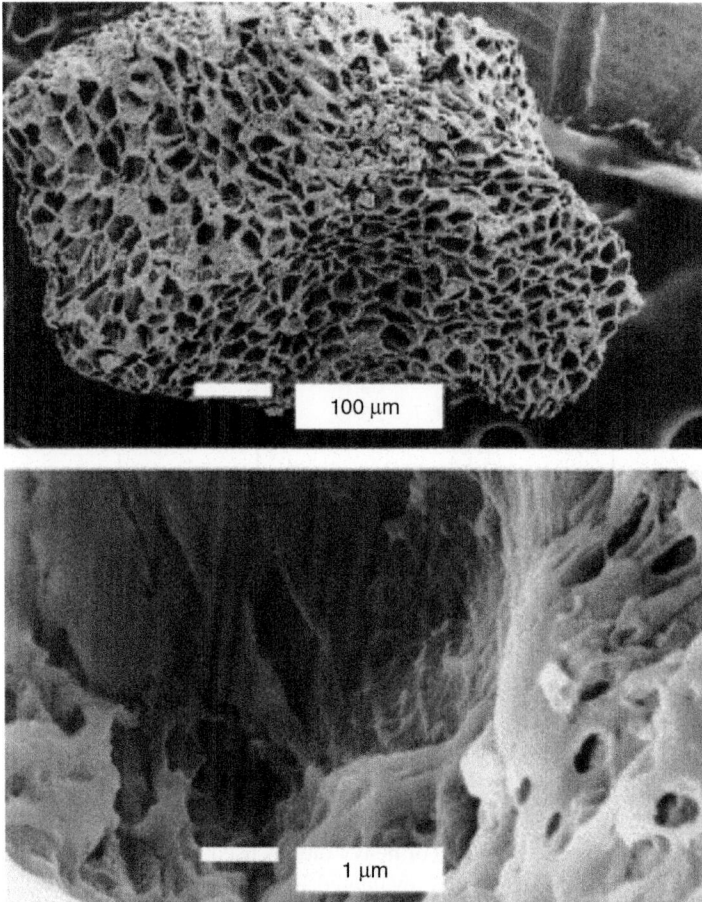

FIGURE 9.1 Scanning Electron Micrograph of Degreased Coffee Bean

Reprinted with permission from Kaikake, K., Hoaki, K., Sunada, H., Dhaka, L R.P., Baba, Y., 2007. Removal characteristics of metal ions using degreased coffee beans: adsorption equilibrium of cadmium(II). Biores. Technol. 98, 2787–2791. Copyright 2016 Elsevier.

after adsorption tests in order to identify the reaction mechanism. Oliveira et al. (2008b) noted that the biosorption capacity for copper ions was higher than the ones for the other metals at pH = 4. This tendency changed when the pH value increased from 4 to 7. In fact, the pH has an effect on the presence of the negative charge on the biosorbent surface, as well as on the heavy metals form, which affect their interactions with the surface of the biomass. Moreover, the characterization of the biosorbent before and after adsorption tests showed different morphological aspects between the coffee husk sample before (Fig. 9.2A) and after adsorption of copper (Fig. 9.2B) and chromium (Fig. 9.2B). In particular, the coffee husk surface becomes

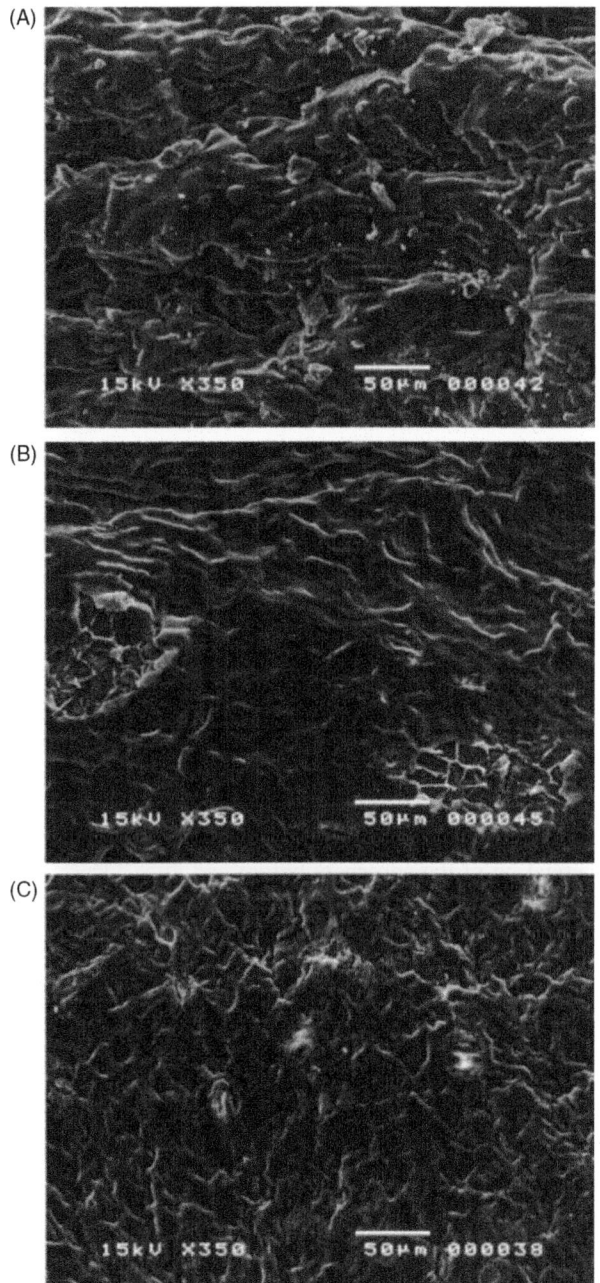

FIGURE 9.2 SEM Images Showing Surface Morphologies of the Coffee Husk

(A) Before and after biosorption of (B) copper and (C) chromium ions.

Reprinted with permission from Oliveira, W.E., Franca, A.S., Oliveira, L.S., Rocha, S.D., 2008b.
Untreated coffee husks as biosorbents for the removal of heavy metals from aqueous solutions.
J. Haz. Mater. 152, 1073–1081. Copyright 2016 Elsevier.

smoother after copper adsorption, whereas the morphology looks like a network of interconnected veins after chromium adsorption. The authors attributed this difference in ionic species present in solution for each metal ion—that is, divalent copper, zinc, and cadmium ions in comparison to oxygenated anionic species for chromium, and the respective sorption mechanisms—that is, simple cation exchange for divalent metal ions or direct and indirect reduction for chromium oxyanions. Furthermore, the analysis of the biosorbent group using Boehm titration showed the presence of phenolic (2.24 mmol/g$_{sorbent}$), followed by lactonic (1.05 mmol/g$_{sorbent}$), carboxylic (0.60 mmol/g$_{sorbent}$), and basic (0.49 mmol/g$_{sorbent}$) groups. These groups were depleted after adsorption tests in various quantities depending on the heavy metals that indicate the implication of surface groups on the adsorption mechanism. The study of the kinetics and equilibrium models showed that the second pseudo order described well the biosorption kinetics and Langmuir model fitted well the sorption isotherm.

9.2.1.3 Coffee ground

Azouaou et al. (2010) have conducted batch kinetic and equilibrium experiments to evaluate the performance of coffee ground for the elimination of cadmium ions from aqueous solution. The authors have determined firstly the chemical and the physical characteristics of the coffee ground. Then, they have studied the effects of the initial concentration (10–700 mg/L), contact time (0–240 min), adsorbent dose (3–24 g), pH (2–8), and temperature (20–50°C). Azouaou et al. (2010) showed that the optimal operating conditions were: ambient temperature (20°C), adsorbent dose = 9 g, and pH = 7 (Fig. 9.3). Under these experimental parameters, authors obtained the highest adsorption capacity in literature for heavy-metals removal using coffee processing residues. The maximum adsorption capacity of cadmium ion by coffee ground was equal to 15.65 mg/g. This capacity was also higher than several adsorbents available in literature, such as olive waste, olive cake, and bagasse fly ash.

The characterization of the adsorbent before and after adsorption tests showed, as mentioned previously for coffee husks, that the principal functional sites, such as carboxyl and hydroxyl groups, participated in the sorption process. The fit of the experimental data with the different available kinetic and equilibrium models showed that the Langmuir model described well the sorption isotherm compared to Freundlich and Dubinin–Radushkevich models while the pseudo second order model are in agreement with the adsorption kinetics. The thermodynamic study showed that the adsorption process was exothermic, favorable, and of physical nature. The thermodynamic parameters were: enthalpy change $\Delta H° = -11.884$ kJ/mol, entropy change $\Delta S = -47.036$ J/mol/K, and free enthalpy change $\Delta G° = 1.896$ kJ/mol at 293 K.

9.2.2 APPLICATION TO THE DYES REMOVAL

Several studies have examined the feasibility of the use of coffee processing residues as dyes adsorbent in textile industry wastewater. The adsorbents performance was evaluated under synthetic and real wastewater for various operating conditions. These various conditions were used to extract the isotherm and kinetics parameters.

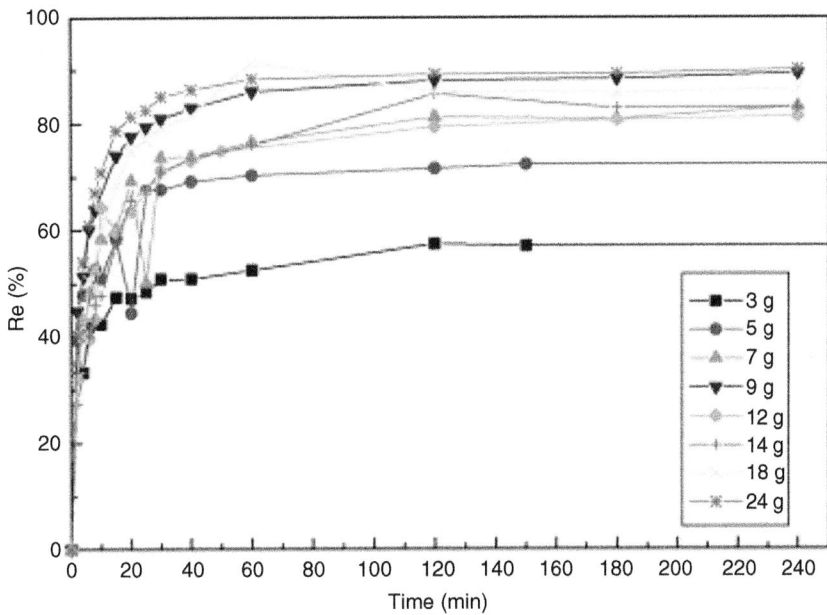

FIGURE 9.3 Effect of Contact Time on the Rate of Adsorption of Cadmium for Different Mass of Adsorbent

pH = 7, $(Cd)_0$ = 100 mg/L, dp = 0.63–0.85 mm, T = 20°C.

Reprinted with permission from Azouaou, N., Sadaoui, Z., Djaafri, A., Mokaddem, H., 2010. Adsorption of cadmium from aqueous solution onto untreated coffee grounds: Equilibrium, kinetics and thermodynamics. J. Hazard. Mater. 184, 126–134. Copyright 2016 Elsevier.

Basically, the Langmuir model was used to describe the sorption isotherm, and the pseudo-second-order model was effective to describe the biosorption kinetics. The thermodynamic data has no significant tendency except for the endothermic character of the adsorption process. Therefore, the used methods are not adapted for the thermodynamic parameters extraction. The main results available in the literature for the dyes removal using coffee processing residues are presented in Table 9.2.

9.2.2.1 Coffee husk

Oliveira et al. (2008a,b) have examined the removal of methylene blue (MB) from aqueous solution using untreated coffee husks (CH). The authors have examined the effect of different operating conditions, such as adsorbent dose (2–15 g/L), initial dye concentration (50–500 mg/L), contact time (15 min–12 h), pH (3–11), and temperature (30–50°C). The authors found that a contact time of 12 h is sufficient to attain the equilibrium for initial MB concentrations below 400 mg/L. They have noted faster adsorption of methylene blue at lower concentrations indicating that the adsorption of MB occurs mainly on the adsorbent surface. When the MB concentration increases, the adsorption process occurs in two steps, the first one at the adsorbent surface

Table 9.2 Coffee-Processing Residues Capacity for Dyes Removal From Wastewater

Coffee Residues	Dyes	Operating Conditions	Langmuir Model	Pseudo-Second Order Model	Thermodynamic Data	References
Coffee husk	Methylene blue	pH = 8, T = 30°C, Ct = 12 h, C_i = 50–400 mg/L	Qm = 90.1 (mg/g); K_L = 0.023 (L/mg)	Ci = 100 mg/L; k_2 = 0.1673	ΔH = 17.7 kJ/mol; ΔS = − 33.1 J/mol/K ΔG = 8.06 kJ/mol	Oliveira et al. (2008a)
Spent coffee grounds	Methylene blue	pH = 5, T = 25°C, Ct = 24 h, C_i = 50–350 mg/L	Qm = 18.73 (mg/g); K_L = 0.269 (L/mg)	Ci = 200 mg/L k_2 = 0.1624; q_e = 14.23	ΔG = − 9.506 kJ/mol	Franca et al. (2008)
Greek coffee grounds	Ramazol Red, yellow, and blue	pH = 10, T = 25°C, Ct = 24 h, C_i = 5–1000 mg/L	Qm = 175 (mg/g); K_L = 0.009 (L/mg)	C_i = 700 mg/L; k_2 = 0.026	ΔH = 1.68 kJ/mol; ΔS = 746 J/mol/K ΔG = 2.05 kJ/mol	Kyzas (2012)
Degreased coffee beans	Malachite green	pH = 4, T = 25°C, Ct = 60 min, C_i = 25–100 mg/L	Qm = 55.3 (mg/g); K_L = 0.094 (L/mg)	Ci = 100 mg/L; k_2 = 0.00375; q_e = 20.26	ΔH = 27.6 kJ/mol; ΔS = 27.6 J/mol[-1]/K ΔG = − 8.19 kJ/mol	Baek et al. (2010)

(faster) and the second one (slower) in the adsorbent pores. Furthermore, Oliveira et al. (2008a) have found that the adsorption process is effective in the pH range 6–11. Such behavior is linked to the coffee husk pZc value (4.3–4.5) and, therefore, above this value the adsorbent surface became predominantly negatively charged, enhancing the electrostatic attraction between the surface and MB cations. This optimization study showed, as expected, that Langmuir model describes the sorption isotherm while pseudo-second order model describes the biosorption kinetics. The maximum adsorption capacity obtained with Langmuir model was 90.5 mg/g which is higher than various adsorbents available in literature. They have found that during MB adsorption, the rate limiting step may be chemisorption promoted by either valence forces, through sharing of electrons between biosorbent and sorbate, or covalent forces, through the exchange of electrons between the parties involved. Thermodynamic data indicated that adsorption process is endothermic and spontaneous.

9.2.2.2 Spent coffee ground

Similar investigation was performed by the same research group (Franca et al., 2008) using spent coffee ground (SCG) instead of coffee husks for the removal of methylene blue from aqueous solution. The same operating conditions of contact time, initial concentration, adsorbent dosage, pH, and temperature were evaluated. Under these parameters, the authors found a low adsorption capacity for SCG (18.5 mg/g) comparing to CH (90.5 mg/g). They have also described, as usual, the adsorption isotherm by the Langmuir model, and the adsorption kinetic was described by the pseudo-second-order model and intraparticle diffusion model. The low adsorption capacity of SCG compared to CH was not explained by Franca et al. (2008). However, the analysis of both biosorbents characteristics showed the higher surface groups amount of coffee husk (phenolic: 2.24 mmol/g, lactonic: 1.05 mmol/g, carboxylic: 0.60 mmol/g, basic: 0.49 mmol/g) comparing to spent coffee ground (phenolic: 0.12 mmol/g, lactonic: 0.07 mmol/g, carboxylic: 0.80 mmol/g, basic: 0.78 mmol/g). In particular, CH has a higher phenolic and lactonic groups amount that may play a significant role in the interaction of the cationic dye with the adsorbent surface.

The performance of spent coffee grounds was also evaluated in the literature for the elimination of three reactive dyes from synthetic and real wastewater—namely, Remazol Red (RR), Remazol Yellow (RY), and Remazol Blue (RB) (Kyzas, 2012). The tested adsorbent was a special variety of coffee—namely, "Greek coffee" from cafeterias, whereas the aqueous effluent was a synthetic and real textile wastewater containing 700 mg/L of dyes concentration (RR = 197 mg/L, RY = 223 mg/L, RB = 280 mg/L). The other operating conditions were: pH = 2–12, $T = 25°C$, adsorbent dosage = 1 g/L^{-1}, contact time: 24 h. Before the adsorption test, Kyzas (2012) has characterized the textural properties and the surface chemistry of the adsorbent. He has found that the functional groups at the surface of coffee grounds were phenolic: 0.14 mmol/g, carboxylic: 0.97 mmol/g, lactonic: 0.11 mmol/g, and basic 0.93 mmol/g. The PZC values were in the range of 3.3–3.5, whereas the specific surface area calculated from BET was equal to 2.3 m^2/g. During the adsorption tests, the author found that the pH has not a significant effect on the adsorption capacity (a decrease of 5% between 2 and 12). He has attributed this result to the complexity of the phenomena occurring at the adsorbent surface. In fact, in presence of a complex adsorbent with a heterogeneous surface chemistry and in a wide range of chemical structures (three different dye molecules were used), pH, and salt concentrations, the interaction phenomena are complicated. Some of the reported interactions may include: (1) ion-exchange, (2) complexation, (3) coordination/chelation, (4) electrostatic interactions, (5) acid-base interactions, (6) hydrogen bonding, (7) hydrophobic interactions, (8) physical adsorption, (9) precipitation. The comparison of the experimental data with available equilibrium and kinetic models showed that the best agreement for sorption isotherm was obtained for the Langmuir-Freundlich model. The calculated maximum adsorption capacities (Q_{max}) for total dye removal at 25°C was 241 mg/g (pH = 2) and 179 mg/g (pH = 10). Kinetic data were well fitted to the pseudo-first-order model. Thermodynamic data indicated that the adsorption process is spontaneous having an endothermic nature.

9.2.2.3 Coffee bean

The feasibility of the use of degreased coffee bean (DCB) for the removal of malachite green (MG) from aqueous solution was studied by Baek et al. (2010). In particular, the authors have examined the effect of degreasing process intensity with NaOH on the adsorbent performance at different operating conditions: initial dye concentration (25–100 mg/L), pH (2–10), and temperature (25–45°C). They have observed during SEM, nitrogen adsorption, and FTIR analyses that the increase of NaOH concentration led to an increase of the amount of oil extraction, which liberates the pores and induces an increase of the porosity. The specific surface area increased from 120 m²/g for raw coffee husk to 173 m²/g for coffee husk degreased with a 5 mol/L NaOH solution. During the adsorption tests, the authors have noted that the adsorption capacity was higher for degreased sample and linked to the textural properties evolution. However, the effect of NaOH treatment on the surface chemistry was not analyzed in their investigation. The modification of the surface chemistry may play also a role on the adsorption mechanism. The authors have also examined the effect of the initial concentration and the contact time on the adsorption capacity (Fig. 9.4). As expected, the adsorption capacity at equilibrium condition increased as MG concentration increased. It could be said that the higher the adsorbate concentration, the more diffusion would occur from the adsorbent surface into the micropores.

Different kinetics and equilibrium models were used to determine sorption isotherm and kinetic parameters of the adsorption process. The authors found that the adsorption kinetics may have followed the pseudo second-order kinetic model. In addition, they observed that the sorption isotherm obeyed both the Langmuir and Freundlich models. The Langmuir adsorption capacity was 55.3 mg/g, whereas the Freundlich parameters indicate a chemisorption process. The adsorption capacity decreased with temperature showing that the adsorption process is endothermic. The authors recommended the use of degreased coffee beans as low cost adsorbent.

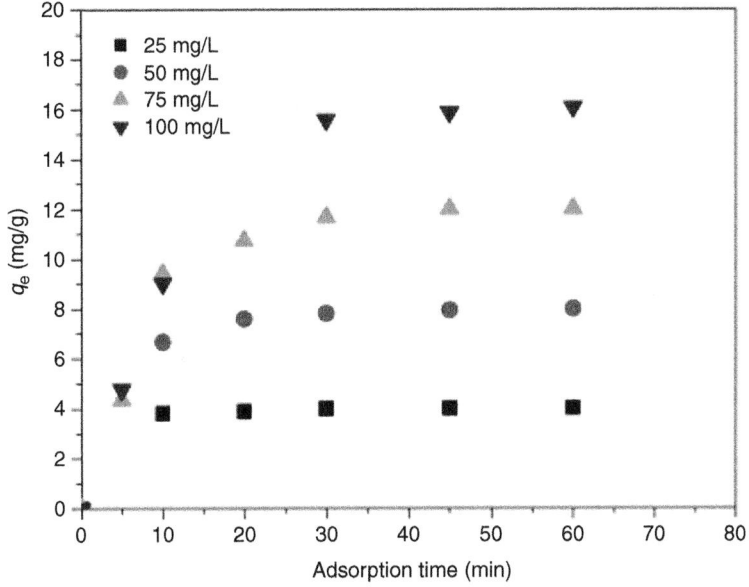

FIGURE 9.4

Equilibrium sorbed amount of MG onto DCB according to adsorption time at various initial MG concentrations (adsorbent dosage: 6 g/L, pH: 4 ± 0.1, temperature: 25 ± 0.1°C).

Reprinted with permission from Baek, M.H., Ijagbemi, C.O., O, S.J., Kim, D.S., 2010. Removal of

Malachite Green from aqueous solution using degreased coffee bean. J. Hazard. Mater. 176, 820–828.

Copyright 2016 Elsevier.

9.3 ELABORATION OF CHAR FROM COFFEE PROCESSING RESIDUES FOR WASTEWATER TREATMENT

In order to improve the adsorption capacity of coffee processing residues, some researchers have carbonized these raw materials and used the obtained char for the removal of pollutants from wastewater.

Boonamnuayvitaya et al. (2004) have tested the effect of carbonization temperature of coffee residue on the adsorption capacity of copper ion in aqueous solution. The authors have prepared chars at different temperatures ranging between 300 and 800°C. The adsorption capacity was evaluated in solution containing 200 ppm Cu^{2+} ion solution at 30°C for contact time between 1 and 30 min. Its value capacity was not significantly increased comparing to the raw material. In fact, the highest adsorption capacity 18 mg/g was obtained for the char prepared at 700°C, whereas the raw material had a capacity of 16.4 mg/g. The authors proposed a mixture of chars and clays to improve the adsorption capacity of several metals in aqueous solution. In particular, the authors performed an optimization study and selected a pyrolysis temperature of 500°C, weight ratio of pyrolyzed coffee residue to clay of 80:20, and

Table 9.3 Maximum Adsorption Capacity V_m (mg/g) of the Adsorbent for Five Heavy Metal Ions Determined by Langmuir Isotherms

Metal	Cd^{2+}	Cu^{2+}	Ni^{2+}	Pb^{2+}	Zn^{2+}
V_m (mg/g)	39.5	31.2	11.0	19.5	13.4

Reprinted with permission from Boonamnuayvitaya, V., Chaiya, C., Tanthapanichakoon, W., Jarudilokkul, S., 2004. Removal of heavy metals by adsorbent prepared from pyrolyzed coffee residues and clay. Sep. Purif. Technol. 35., 11–22. Copyright 2016, Elsevier.

a diameter of 4 mm for the granular adsorbent. The maximum values of adsorption capacity (V_m) of the char–clay adsorbents for Cd^{2+}, Cu^{2+}, Ni^{2+}, Pb^{2+}, and Zn^{2+} are shown in Table 9.3.

Nakamura et al. (2003) have prepared three charcoals from coffee ground carbonized at 800, 1000, and 1200°C. As expected, the surface specific area and the pore volume increased with the carbonization temperature from 0.17 m²/g and 0.001 cm³/g at 800°C to 61.71 m²/g and 0.041 cm³/g at 1200°C. However, these values are lower compared to those found for other adsorbents, such as AC. These charcoals were used by the authors for the removal of acid orange 7 from aqueous solution at 25°C during 20 days. The initial dye concentrations were varied between 50 and 1000 mg/L. Nakamura et al. (2003) obtained an adsorption capacity of 3.09, 13.38, and 14.51 mg/g for the charcoal prepared at 800, 1000, and 1200°C, respectively. Moreover, they have identified the intraparticle diffusion of the dye onto pores of adsorbents as the rate-limiting step in the adsorption process.

Oh and Seo (2016) have also pyrolyzed coffee grounds at 550°C during 4 h under N_2 flow. The obtained char had a surface specific area of 18 m²/g and PZC of 8.13. The performance of the prepared char was evaluated for the removal of nine halogenated phenols and two pharmaceuticals (triclosan and ibuprofen) from waste water using a series of batch experiments at 25°C, pH = 7, and initial organic compound concentrations in the range 50–500 mg/L. This attempt was not successful since the adsorption capacity was below to 7 mg/g lower than those found for AC and chars prepared from other biomasses.

The use of chars prepared from coffee processing by-products for the removal of pollutants from wastewater seems not to be interesting. Although, a slight increase of surface specific area and pore volume, the adsorption capacities are very low. Such behavior may be attributed to the removal of surface groups during the carbonization step. These surface groups are requested for the interaction between the organic pollutants and the adsorbent surface. Therefore, it is necessary to add an activation step after the carbonization one in order to improve the textural properties and the surface chemistry of the adsorbent materials. The AC synthesis, characterization, and applications are examined in details in the following sections.

9.4 SYNTHESIS, CHARACTERIZATION AND APPLICATIONS OF ACTIVATED CARBONS

ACs is the most effective adsorbent used in gas and water purification, metal extraction, sewage treatment, and many other applications. The adsorbents can be obtained after the activation of chars obtained during the carbonization at elevated temperature of precursors, such as coal, date pits (Belhachemi et al., 2014), olive waste (Ghouma et al., 2015), palm shell, cardboard, plastics, municipal waste (Aboud et al., 2015), and many other biomass sources. This activation step could be physical through char treatment under CO_2 and/or steam flow or chemical through immersing the char in activation agents, such as phosphoric acid, potassium hydroxide, sulfuric or citric acid, and zinc chloride. The obtained ACs have generally a high surface area (>1000 m^2/g) and a pore volume (>0.5 cm^3/g).

Numerous investigations have examined the preparation and application of ACs from coffee processing by-products. In particular, fifty papers were identified on this subject in the bibliographic database scopus. This section presents the synthesis methods of AC derived from coffee residues, as well as their morphological, textural, and surface chemistry characteristics. Then, the environmental application of these carbonaceous materials including gas storage, pollutants elimination from gaseous and aqueous effluents is addressed.

9.4.1 SYNTHESIS AND CHARACTERIZATION

The most coffee processing by-product used for the AC in literature is spent coffee grounds. Its physical properties, such as low particle size and homogeneity have motivated several researchers to test various protocols to elaborate efficient ACs. In general, the production of ACs includes carbonization and activation steps. The activation step could be physical, chemical, or a mixture of both methods. Chemical activation is the most common method applied for AC synthesis using coffee grounds as the precursor. The major activated agents used during the chemical activation step are zinc chloride ($ZnCl_2$), potassium hydroxide (KOH), and phosphoric acid (H_3PO_4) for various precursor/agent ratios and temperature. Few attempts have examined the use of other chemical activated agents, such as iron chloride ($FeCl_3$) (Oliveira et al., 2009), hydrogen peroxide (H_2O_2) (Pavlovic et al., 2015), sodium hydroxide (NaOH) (Pechyen et al., 2010), sulfuric acid (H_2SO_4) (Ching et al., 2011) or citric acid ($C_6H_8O_7$) (Cerino-Córdova et al., 2013). Furthermore, the physical activation is not commonly used for the production of coffee-based AC. Few investigations have examined the physical activation using carbon dioxide (CO_2) and steam (H_2O) as activation agents and the mixed activation using CO_2 or H_2O with a chemical agent (Boonamnuayvitaya et al., 2005).

9.4.1.1 Production of activated carbons

Chemical activation protocols are mainly used for the production of ACs from spent coffee grounds. The major investigations have used potassium hydroxide, zinc chloride, and phosphoric acid as activation agents. The production of the chemically AC includes a carbonization step and activation step. The step order changes between the different investigations. In general, the impregnation of spent coffee ground with the chemical activation agent is performed first and followed by a carbonization at low heating rate. However, recent investigations have started the AC production by the carbonization step followed by the chemical activation one. These protocols lead to different AC yields and characteristics.

Boudrahem et al. (2011) have examined the production of AC from coffee grounds using phosphoric acid and zinc chloride as activated agent. The precursor was impregnated with different impregnation ratios (mass of chemical agent/mass of precursor) ranging from 0 to 100% at 85°C for 7 h to ensure the access of activating agents to the interior of the precursor. After a drying step, the impregnated samples were pyrolyzed during 1 h at 600°C in a furnace heated from room temperature to 600°C at 10°C/min. Similar procedure was used by Ma et Ouyang (2013) and Kante et al. (2012) for the elaboration of AC from spent coffee grounds. The differences between the different investigations are mainly the impregnation ratio or carbonization temperature. Oliveira et al. (2009) has also applied similar protocol for the production of AC from coffee husk. The authors have impregnated the raw material in $ZnCl_2$ solution at 1:1 impregnation ratio then carbonized the mixture under nitrogen flow at 550°C for 3 h.

Recently, the use of potassium hydroxide (KOH) as an activating agent for the elaboration of coffee-based AC has received particular attention. The KOH impregnation was used before or after the carbonization step. Kemp et al. (2015) have mixed 100 g of spent coffee grounds and 100 mL of KOH solution (7 mol/L) at 65°C during 24 h. Then, the solid product was heated under argon atmosphere from room temperature to various temperatures (700–900°C) at 7°C/min heating rate. In contrast, Nowicki et al. (2013) and Travis et al. (2015) have started their AC production by the carbonization step. Nowicki et al. (2013) have selected 400 and 700°C as carbonization temperature while Travis et al. (2015) have chosen 500 and 800°C. Both research teams have performed their chemical activation step with potassium hydroxide at 700°C.

Some attempts have examined the use of two chemical activating agents for the production of AC using coffee processing residues as precursor. Oliveira et al. (2009) have used a mixture of zinc chloride and iron chloride for the impregnation of coffee husks. The obtained solid was pyrolyzed at 280°C. Bouchenafa-Saib et al. (2014) have prepared an AC through mixing 100 mL of 50% $ZnCl_2$ and 50% H_3PO_4 solution with 50 g of coffee grounds. The obtained mixture was refluxed at 80°C for 3 h and carbonized at 600°C for 1 h under nitrogen atmosphere.

Physical activation and mixture activation (chemical and physical activations) were also examined for the elaboration of coffee-based carbons. Boonamnuayvitaya et al. (2005) have prepared different ACs by applying physical and mixture activations. For the physically AC, the coffee residues were heated under nitrogen flow from room temperature to 600°C at 10°C/min and kept at this temperature during

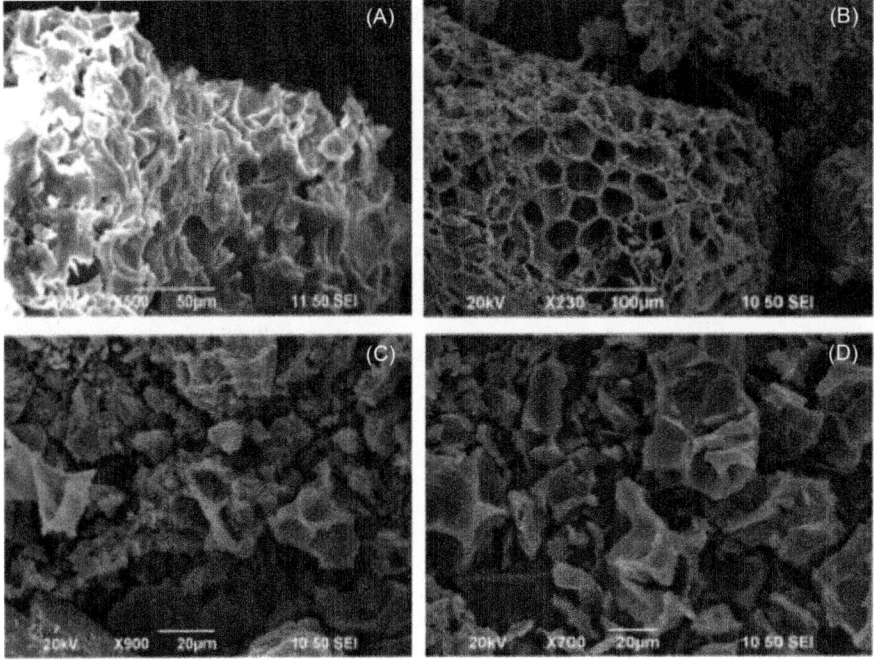

FIGURE 9.5

Scanning electron micrographs (SEM) for (A) coffee residue, (B) carbonized coffee residue, (C) impregnated with H_3PO_4, (D) impregnated with KOH.

Reprinted with permission from Yeung, P.T., Chung, P.Y., Tsang, H.C., Tang, J.C.O, Cheng, G.Y.M., Gambari, R., Chui, C.H., Lam, K.H., 2014. Preparation and characterization of bio-safe activated charcoal derived from coffee waste residue and their application for removal of lead and copper ions. RSC Adv. 4, 38839–38847. Copyright 2016 Royal Society of Chemistry.

4 h. Then, the activation step was performed under carbon dioxide or steam flow. During the mixture activation, coffee residues were impregnated with zinc chloride (3:1) and then followed the same procedure as the physically AC. Castro et al. (2011) have prepared an AC using water vapor as an activating agent. During this investigation, spent coffee grounds were heated under N_2 at a rate of 10°C/min up to 550°C and kept at this temperature for 3 h then activated with water vapor at 800°C for 2 h.

9.4.1.2 Morphological properties

Scanning electronic microscopy (SEM) was usually applied to analyze the surface morphology of ACs. In particular, the evolution of morphological properties was observed after each step (carbonization step, chemical or physical step). Yeung et al. (2014) have compared the SEM micrographs of raw coffee ground (Fig. 9.5A), carbonized coffee residue (Fig. 9.5B) activated with H_3PO_4 (Fig. 9.5C) and activated with KOH (Fig. 9.5D). The authors have observed rough and highly folded surface

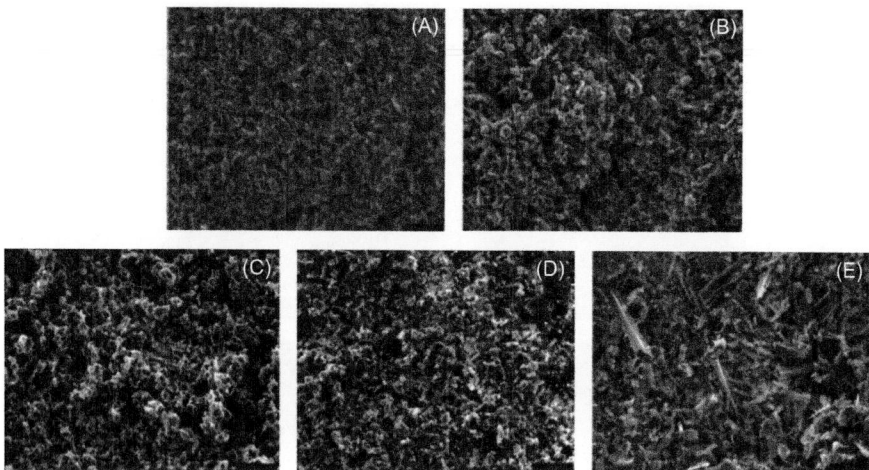

FIGURE 9.6

Scanning electron micrographs (SEM) of (A) nonactivated (0%), (B) AC 25%, (C) AC 50%, (D) AC 75%, and (E) AC 100%.

Reprinted with permission from Khenniche, L., Aissani, F., 2010b. Preparation and characterization of carbons from coffee residue: adsorption of salicylic acid on the prepared carbons. J. Chem. Eng. Data 2010, 728–734. Copyright 2016 American Chemical Society.

for the raw biomass. After carbonization step, a well-developed porosity was observed. After chemical activation, disrupted structure was observed for both samples.

Khenniche and Aissani (2010b) have analyzed the SEM micrographs of ACs (AC) prepared by $ZnCl_2$ activation at different impregnation ratios (Fig. 9.6). The authors observed a well-developed porous structure containing different pore sizes and shapes resulting from the activation process. In particular, the authors observed pores enlargement with increasing $ZnCl_2$ impregnation ratio. The authors attributed $ZnCl_2$ action to its intercalation in carbon matrix. During carbonization, an interaction between the zinc compounds and the carbon atoms leads to carbon layers widening and pores formation in the carbon matrix.

9.4.1.3 Textural properties

The different protocols used for the production of ACs lead to different textural properties. Table 9.4 summarizes the textural properties of ACs obtained for the different activation techniques. Coffee-residue carbonization followed by chemical activation using potassium hydroxide seems the most interesting protocols. In fact, the specific surface area (S_{BET}) and total pore volume (V_t) reached 2785 m^2/g and 1.36 cm^3/g during coffee ground (CG) carbonization at 400°C followed by KOH activation at 4:1 impregnation weight ratio (Travis et al., 2015). In addition, the use of higher impregnation ratio led to more hierarchical pore architecture through a high degree of microporosity in combination with a developed narrow mesoporosity as observed in Fig. 9.7 from the wider knee in the adsorption isotherm. In contrast, the use of a

Table 9.4 Textural Properties of the Different Activated Carbons Prepared From Coffee-Processing Residues

Precursor	Carbon	Activation Method: Agent/Char Ratio	S_{BET} (m²/g)	V_{mic} (cm³/g)	V_t (cm³/g)	Pore Width (nm)	Reference
Coffee grounds	AMC-700	KOH 1:1 (65°C), pyrolysis at 700, 800, and 900°C (1 h)	536.5	0.286	0.320	N.C	Kemp et al. (2015)
	AMC-800		815.2	0.437	0.498	N.C	
	AMC-900		1040.3	0.574	0.635	N.C	
Coffee grounds	CG-AC	H_3PO_4 1.5, calcined at 450°C (1 h)	1110	N.C	N.C		Ma and Ouyang (2013)
Coffee waste	CP5Ac	Pyrolysis (1 h) at 500/300°C, KOH 2:1 (700°C)	2076	1.06	1.25	2.42	Nowicki et al. (2013)
	CP8Ac		1553	0.76	1.06	2.64	
Coffee grounds	COFAC0.5	$ZnCl_2$ 0.5:1, 1:1, 2:1 (30°C), pyrolysis at 300°C (1 h)	645	0.326	0.328	N.C	Kante et al. (2012)
	COFAC1		905	0.399	0.492	N.C	
	COFAC2		1121	0.445	0.954	N.C	
Coffee grounds	CG 400 2-1	Pyrolysis (1 h) at 400/700°C, KOH 2:1, 4:1 (700°C)	2073	0.731	0.869	0.926	Travis et al. (2015)
	CG 400 4-1		2785	0.716	1.360	1.051	
	CG 700 2-1		1624	0.589	0.662	0.966	
	CG 700 4-1		2620	0.793	1.225	0.926	
Coffee husk	AC-Zn	$ZnCl_2$ 1:1, pyrolysis at 550°C FeCl₃, $ZnCl_2$/FeCl₃ 1:1, pyrolysis at 280°C	1522	0.600	0.75	0.90	Oliveira et al. (2009)
	AC-Fe		965	0.40	0.51	0.63	
	AC-Zn-Fe		1374	0.53	0.65	0.77	
Coffee residue	CN_2CO_2	Pyrolysis, CO_2 or H_2O at 600°C (4 h)	11	0.00	0.015	5.43	Boonamnuayvitaya et al. (2005)
	CN_2ST	$ZnCl_2$ 3:1, pyrolysis 600°C, CO_2/H_2O (600°C)	469	0.170	0.36	3.09	
	$CZnN_2$		470	0.090	0.454	3.87	
	$CZnN_2CO_2$		914	0.084	1.010	4.42	
	$CZnN_2ST$		305	0.073	0.275	3.60	

(Continued)

Table 9.4 Textural Properties of the Different Activated Carbons Prepared From Coffee-Processing Residues (*cont.*)

Precursor	Carbon	Activation Method: Agent/Char Ratio	S_{BET} (m²/g)	V_{mic} (cm³/g)	V_t (cm³/g)	Pore Width (nm)	Reference
Coffee grounds	$ZnCl_2$	$ZnCl_2$ or H_3PO_4 or mixture 2:1 (80°C), pyrolysis at 600°C	1020	0.74	1.23	N.C	Bouchenafa-Saib et al. (2014)
	H_3PO_4		910	0.32	0.98	N.C	
	$ZnCl_2$-H_3PO_4		960	0.41	1.02	N.C	
Coffee grounds	ACK	K_2CO_3 1:1 (80°C), pyrolysis (800°C)	950	0.38	0.45	N.C	Castro et al. (2011)
	ACW	Pyrolysis 550°C (3 h), H_2O 800°C.	620	0.26	0.36	N.C	
Coffee residue	$ZnCl_2$	Agent 1:1 (85°C), pyrolysis (600°C)	889	0.418	0.765	3.44	Boudrahem et al. (2011)
	H_3PO_4		1003	0.423	0.618	2.46	

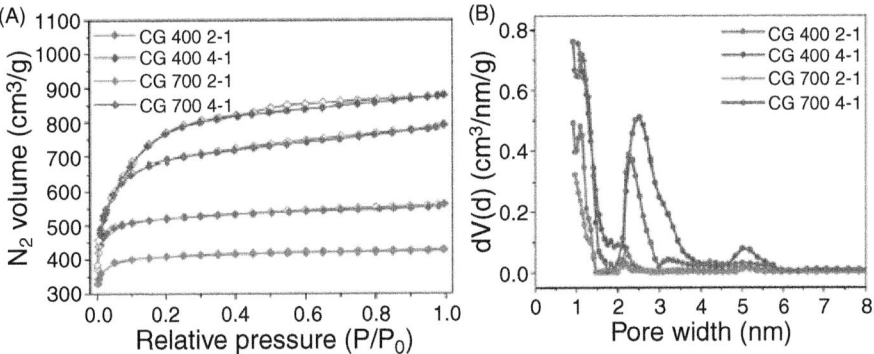

FIGURE 9.7

(A) Nitrogen adsorption at 77 K (solid) and desorption (open) isotherm for KOH activated CGs; (B) corresponding pore size distribution plots.

Reprinted with permission from Travis, W., Gadipelli, S., Guo, Z., 2015. Superior CO₂ adsorption from waste coffee ground derived carbons. RSC Adv. 5, 29558–29562. Copyright 2016 Royal Society of Chemistry.

2:1 impregnation ratio showed highly microporous structures with type-I adsorption curves. Nowicki et al. (2013) also observed such behavior during close protocol applied for AC preparation (carbonization temperature: 500°C). The authors obtained a microporous carbon with the highest microporous volume (V_{mic} = 1.06 cm³/g) available in literature for coffee-based ACs. The developed narrow mesoporosity for samples with higher impregnation ratio can be attributed to the higher degree of carbon burn-off upon the potassium reduction that creates larger porous architectures. In fact, the sample yield was 11 wt.% for higher KOH concentration and temperature carbonization.

Otherwise, it seems that the KOH chemical activation at higher temperature after the carbonization step leads to better textural properties than to KOH chemical activation before the carbonization step. In fact, a microporous AC with low specific surface area (S_{BET} = 536.5 m²/g) and microporous volume (V_{mic} = 0.286 cm³/g) were obtained by Kemp et al. (2015) during the preparation of AC by KOH impregnation (1:1) at 65°C followed by carbonization at 700°C. The increase of the carbonization temperature to 900°C was requested to increase the textural properties values.

Among the various chemical activation protocols, the use of zinc chloride as activation agent leads also to suitable textural properties. In the literature, zinc chloride activation at 80°C was performed before a carbonization step in the 550–800°C temperature range. The use of low impregnation ratio (<1:1) and low carbonization temperature leads to a microporous AC. In fact, Kante et al. (2012) obtained a specific surface area of 645 m²/g and a microporous volume of 0.326 cm³/g representing 99% of the total porous volume. In contrast, the use of high impregnation ratio and carbonization temperature develops larger porosity (Kante et al., 2012). These pores are formed when a large quantity of zinc is removed from the hot carbon matrix (boiling point of zinc chloride is 732°C).

Several investigations have examined the textural properties during the AC preparation under further protocols, such as phosphoric acid activation, physical activation, and mixture activation. The analysis of the AC characterization shows that these available protocols are less interesting in terms of textural properties comparing to KOH and $ZnCl_2$ chemical activation. In particular, the phosphoric acid accelerates cleavage of bonds between biopolymer (principally cellulose and lignin), which is followed by a recombination reaction in a rigid cross-linked structure. This chemical activation leads to low microporous volume (0.32 cm³/g) and more mesoporous with a ratio of approximately 63% (Bouchenafa-Saib et al., 2014). The physical activation under carbon dioxide was not favorable for the development of porous structure (Boonamnuayvitaya et al., 2005; Nowicki et al., 2013) and poor textural parameters were obtained. This behavior is attributed to the high content of inorganic elements in coffee residues, which can be deposited in the pores and block the access of the adsorbate to smaller pores. The increase of activation temperature led to the increase of carbon consumption under CO_2 atmosphere in presence of minerals and, therefore, the porous structure destruction (Bouraoui et al., 2015; Bouraoui et al., 2016). The physical activation under vapor steam was more interesting than the carbon dioxide one. However, low microporous volume ranging between 0.17 and 0.26 cm³/g (Boonamnuayvitaya et al., 2005; Castro et al., 2011) was obtained in literature compared to the textural parameters of the chemically AC. The difference between the textural parameters during the physical activation under carbon dioxide and water vapor was attributed to different pathways for both activation agents (Guizani et al., 2016). Furthermore, the use of mixture activation (chemical and physical activations), as well as the use of two chemical activation agents did not improve significantly the textural properties. An intermediate product is generally obtained arising from the competition between the activating agents.

9.4.1.4 Surface chemistry properties

Different techniques were used in the literature to characterize the surface chemistry of the ACs prepared from coffee processing residues. Boehm titration method was applied to identify the amount of the acidic and basic groups on the AC surface. This characterization technique indicated that the chemical activation using KOH, $ZnCl_2$, and H_3PO_4 as activating agents lead to acidic AC. The total amount of the acidic surface groups ranged between 1.5 and 3.5 mmol/g depending on the activation agent and temperature. Khenniche and Aissani (2010a,b) have compared the total amount of oxygen surface groups for different impregnation ratios (0.25–1) during the activation of coffee grounds with zinc chloride at 85°C followed by carbonization at 600°C. The authors have observed a decrease in the amount of basic groups on the surface, whereas the acidic groups remained constant (3.5 mmol/g). Similar behavior was observed on the impregnation ratio effect during the elaboration of chemically AC from coffee grounds using phosphoric acid (Reffas et al., 2010). The authors found that the total acidic groups remained more or less constant to about

(1.5 mmol/g). This amount is less than the one found during zinc chloride activation. This difference may be attributed to the low temperature used by Reffas et al. (2010) during phosphoric acid activation. In fact, Bouchenafa-Saib et al. (2014) have compared the surface chemistry of ACs prepared by zinc chloride and phosphoric acid activation under similar conditions (impregnation at 80°C for 3 h, carbonization at 600°C for 1 h). The authors showed that the maximum acidity was obtained for H_3PO_4 (3.2 mmol/g) comparing to $ZnCl_2$ (1.44 mmol/g). In addition, the carbonization temperature affected the amount of the acidic groups. In fact, Nowicki et al. (2013) showed that the amount of the acidic groups increased from 1.95 to 2.35 mmol/g when the carbonization temperature increased from 500 to 800°C and followed by KOH activation at 700°C. These authors showed that KOH activation generated a significant amount of basic groups (0.90 mmol/g). Contrary to the chemical activation, physical activation leads to the formation of basic ACs. Nowicki et al. (2013) detected an amount of 10.25 mmol/g of basic groups during the activation of coffee grounds under carbon dioxide at 500°C. The amount of basic groups increased with increasing activation temperature.

Analytical techniques were also used to assess the surface chemistry of ACs prepared from coffee-processing residues. Khenniche and Aissani (2010b) analyzed the FTIR spectra of the carbonized coffee grounds (without activation) and the ACs prepared with different $ZnCl_2$ impregnation ratios (25%, 50%, 75%, and 100%). The authors identified different bands in the FTIR spectra. These bands were associated to different surface groups including: OH group of the phenol function (3399 cm), hydroxyls vibration fixed on the carbon surface and water chemisorbed on carbon (3100 to 3600 cm), —CH_2 groups (2906 cm), —O—CH_3 groups (2847 cm), the CO stretch of the carbonyl group in a quinone representing the γ-pyrone structure with strong vibrations from a combination of CO and CC (1565 cm), the C—O stretch or O—H deformation in carboxylic acids (1424 cm), ketones, alcohols, pyrones, and aromatic C—H in-plane deformations (1249 cm), the C—O single bond in carboxylic acids, alcohols, phenols, and esters (1000 to 1350 cm). The authors indicated that the spectra of the prepared AC were similar and the main difference was in the intensity of peaks. The presence of hydroxyl groups of phenolic and carboxylic character brings acidic surface properties, whereas carbonyl and quinone groups cause surface basicity.

X-Ray photoelectron spectroscopy (XPS) analysis was also applied to assess the surface chemistry of the elaborated AC. Travis et al. (2015) shows that the ACs prepared by KOH impregnation (impregnation weight ratios: 2:1 and 4:1) and carbonization temperature (400–700°C) contained only carbon and oxygen atoms. The authors have observed that no nitrogen was detected. The main peaks at binding energies (BE) of 284.5 and 533.0 eV characterize C 1s and O 1s environments (Fig. 9.8A–B). The C 1s peaks at 284.5 eV shows that the carbon species are predominately in graphitic-type C=C systems. The O 1s broad peaks at 532.5–533.5 eV can be attributed to the phenolic type OH groups. This fact indicates that the KOH activated samples possess higher oxygen percentage than the chars before activation. Such difference could be observed in the C 1s spectra for the activated species where features

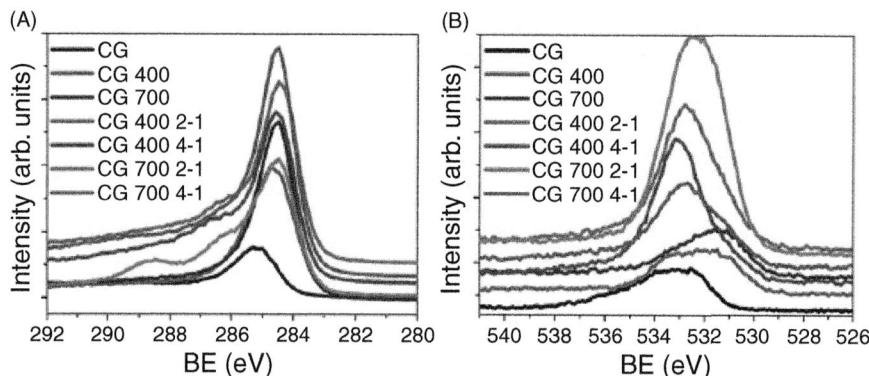

FIGURE 9.8 XPS Analysis of Coffee Grounds Activated at Different Temperatures and With Different KOH Ratios

(A) C 1s spectra and (B) O 1s spectra.

Reprinted with permission from Yeung, P.T., Chung, P.Y., Tsang, H.C., Tang, J.C.O, Cheng, G.Y.M., Gambari, R., Chui, C.H., Lam, K.H., 2014. Preparation and characterization of bio-safe activated charcoal derived from coffee waste residue and their application for removal of lead and copper ions. RSC Adv. 4, 38839–38847. Copyright 2016 Royal Society of Chemistry.

at 288–286 eV are indicative of carbon in an oxygen bound environment. The O 1s peak of CG 700 at 531.5 eV characterizes C=O functionality; this peak shifts to 532.5–533.5 eV, which may be attributed to phenolic–OH functionality resulting from KOH treatment or adsorbed surface H_2O.

Similar results were observed by Gonçalves et al. (2013b) during the analysis of the surface chemistry of AC prepared from coffee pulp using disodium hydrogen phosphate (Na_2HPO_4) at different impregnation ratios. The authors have fitted the O 1s AC spectrum to three peaks:

- Peak I 530.7–531.2 eV, attributed to double-bond oxygen in quinones, carbonyl groups or O=P.
- Peak II 532.5 eV, attributed to C—OH of phenol and alcohol groups.
- Peak III 534.2 eV, corresponds to the binding energy of carboxylic groups.

Rufford et al. (2008) have used XPS analysis to characterize the AC obtained with zinc chloride at 1:1 activation ratio. They indicated that the carbon enclosed 6.8% oxygen and 1.5% nitrogen. The oxygen was in phenols/ether (58.8%), chemisorbed O_2 (21.5%), and carbonyl–quinone (19.7%) configurations. In contrast, the nitrogen was in pyridinic (35.9%), quaternary nitrogen (30.6%), pyrrolic/pyridine (20.5%), and pyridine-*N*-oxide (13%) forms.

X-ray diffraction (XRD) was used in some studies for the characterization of AC. Djilani et al. (2012) have analyzed coffee grounds and the corresponding AC prepared by carbonization followed by chemical activation using phosphoric acid and nitric acid solution. The XRD patterns are shown in the Fig. 9.9. The authors noted that the peaks at 6.5°, 20°, 43°, 81.5°, and 98° indicated the presence of silicon

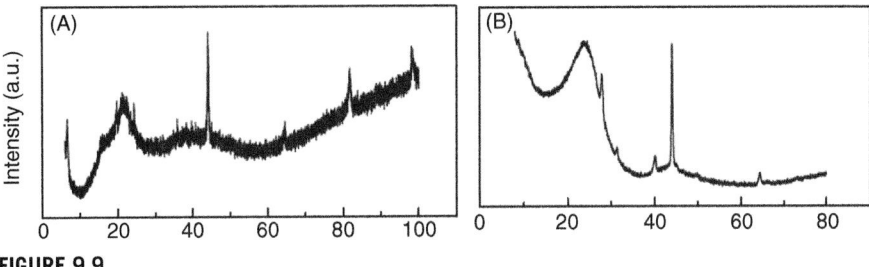

FIGURE 9.9

XRD pattern of (A) coffee grounds and (B) activated carbon.

Reprinted with permission from Djilani, C., Zaghdoudi, R., Modarressi, A., Rogalski, M., Djazi, F., Lallam, A.,
2012. Elimination of organic micropollutants by adsorption on activated carbon prepared from agricultural
waste. Chem. Eng. J. 189–190, 203–212. Copyright 2016 Elsevier.

dioxide, native cellulose, hemicellulose dehydrate, and calcium carbonate, whereas peaks at 24° and 23.5° indicated the presence of carbon and graphite.

The morphological, textural, and surface chemistry properties of coffee-based ACs are well characterized in the literature. These characteristics are essential for the investigation of pollutants interaction with AC in aqueous and gaseous effluents.

9.4.2 APPLICANTS IN AQUEOUS EFFLUENT TREATMENTS

AC obtained from coffee residues can be used for the removal of organic molecules (dyes, acids, pesticides, antibiotics…) and also of cationic and anionic species (lead, chromium, copper, fluoride…). Depending on the preparation routes but also on the experimental conditions, the performances can be strongly affected. In the following part, we will first describe the use of AC for the removal of ions present in solution, and then the adsorption of organic wastes.

9.4.3 REMOVAL OF MINERALS PRESENT IN WATER

The removal of ions from water can be achieved with ACs in different ways. The first one corresponds to adsorption, which is the main phenomenon described in literature, and the other one corresponds to ionic exchange. The latter is often forgotten at the benefit of adsorption by lack of characterization. Different works were done on the removal of copper ions (Cu^{2+}) using ACs prepared from spent coffee grounds (SCG) (Cerino-Córdova et al., 2013; Jutakridsada et al., 2016; Yeung et al., 2014). Experiments were carried out in batch reactor with low concentrations of copper (from 5 to 300 ppm), with ACs obtained by different chemical activations (KOH, H_3PO_4, citric acid, or $ZnCl_2$). Jutakridsada et al. (2016) observed a low efficiency of copper removal (18 wt.%), they explained that the activation with $ZnCl_2$ was responsible for this result because of the surface charge that was generated. They modeled the adsorption of copper by a Freundlich isotherm, but the value of the 1/n parameter, which is a function of the adsorption strength, was below a normal value, indicating a different mechanism of adsorption.

Table 9.5 Adsorption Capacities Obtained in the Presence of Pb^{2+} and Cu^{2+} Ions for the Different AC Obtained by Alkaline or Acid Treatment

Biosorbent Activation Method	pH	Lead (II) q_{max} (mg/g)	Copper (II)
Acid citric	4 (Pb^{2+})/5 (Cu^{2+})	159.5	97.2
H_3PO_4	5	95.2	38.2
KOH	5	45.4	21.2

The removal of copper ions from water strongly depends on the pH value that is used. At a pH higher than 5, copper hydroxyls will form and precipitate. At pH lower than 5, copper ions will compete with proton because they can adsorb on the same sites. Cerino-Córdova et al. (2013) and Yeung et al. (2014) investigated the removal of copper and lead ions from water at pH = 5 and pH = 4, respectively, with ACs obtained after a treatment with citric acid or KOH and H_3PO_4 respectively. The acid treatments lead to a decrease of the pKa and pHpzc values (from pH = 5.5 to 2.8), which may be explained by the increase of acid surface groups (carboxylic) on the biosorbent surface. The removal of copper and lead ions was higher with the AC treated with an acid than with KOH (Table 9.5). The Langmuir model was used to model the adsorption of isotherms, and the adsorption kinetics fitted well with the linear pseudo-second-order model for copper and lead ions. The main difference corresponds to the mechanism described by Cerino-Córdova et al. (2013). They suggested that the removal of copper and lead from the solution should be attributed to a calcium exchange mechanism, and not only by a proto-exchange mechanism. This result was supported by the absence of carboxylic groups on the surface of the nonactivated SCG, and it adsorbed considerable amounts of copper and lead ions.

Boudrahem et al. (2009) have also studied the adsorption of lead in SCG activated by $ZnCl_2$. Adsorption tests were carried out in a batch reactor with increasing concentrations of Pb^{2+} ions (from 10 to 90 mg/L). The ACs were prepared with different ratios of $ZnCl_2$/SCG (from 0 to 100 wt.%). The best adsorption performances were obtained with a ratio of 75 wt.%, where the AC presented a surface mainly covered by acid sites. In this specific case, adsorption took place very rapidly in the presence of lead. The adsorption equilibrium was reached after 15 min; the adsorption capacity was approximately 63 mg/g for an initial concentration of 90 mg/L (1 g of AC in 1 liter solution, adsorption yield of 70%). Boudrahem et al. (2009) studied the effect of ionic strength (presence of NaCl) on the adsorption capacity of the AC in the presence of lead ions. The results indicated that an optimal concentration of 5.10^{-3} mol/L NaCl induced the removal of 100% of lead ions (10 mg/L, at 25°C). Above this concentration, the electrostatic double layer of the AC is compressed, which induced a decrease of the adsorption capacity. No effect of the temperature was observed in the range of 25–60°C, which may be due to low sorptive forces between lead ions and the surface of the AC. The adsorption kinetic was modeled by a pseudo-second-order equation, and the adsorption isotherm was best fitted by the Freundlich isotherm. In a more recent work, Boudrahem et al. (2011) compared the influence of the activation treatment ($ZnCl_2$ or H_3PO_4) on the adsorption of lead. The activation with H_3PO_4 induced the increase of the adsorption capacity of the AC, which was correlated to the higher concentration of acid groups at the surface of the AC (both carboxylic and carbonylic groups). The improvement of the adsorption capacity was not only attributed to the increase of the surface area or to the microporosity of the AC, but to a combined effect of these parameters with the availability of the acid surface groups. The adsorption kinetics were modeled by a pseudo-second-order model, and the adsorption isotherm was fitted with the Langmuir model. In this work, Boudrahem et al. (2011) observed that cadmium behaved similarly to lead, but the adsorption efficiency of cadmium was lower. This was attributed to higher interferences of cadmium with the positive charges present at the surface of the AC.

The removal of other cations present in water was studied by Ching et al. (2011) and Giraldo and Moreno-Pirajan (2012). The adsorption of mercury and zinc was performed on ACs obtained by chemical ($ZnCl_2$ or KOH) and physical activations, at pH = 5.1 (Giraldo and Moreno-Pirajan, 2012). The adsorption capacities were higher with the AC obtained with KOH than with $ZnCl_2$ (0.38 and 0.33 mmol/g for Hg, 0.30 and 0.274 mmol/g for Zn, respectively). These differences were not directly proportional to the textural properties of the different AC. Hg^{2+} cations presented a better affinity with the surface of the AC than Zn^{2+}, which was explained by a higher diffusion of Hg^{2+} ions (diffusion coefficient of 8.47×10^{-10} m^2/s in pure water) compared to Zn^{2+} ones (diffusion coefficient of 7.03×10^{-10} m^2/s in pure water), and also to a steric overcrowding at the surface of the AC in the presence of Zn^{2+} due to the difference of ionic radius (ionic radius of 1.16 Å for Hg^{2+} and 1.34 Å for Zn^{2+}). The adsorption isotherms of Zn^{2+} and Hg^{2+} were modeled using the Freundlich model. The competitive adsorption of Hg^{2+} and Zn^{2+} was also studied. In particular, Giraldo and Moreno-Pirajan (2012) showed that the adsorption capacity of the AC was reduced when Zn^{2+} and Hg^{2+} were present simultaneously in the water solution. Ching et al. (2011) have investigated the adsorption of total iron (Fe^{2+} and Fe^{3+}) on the surface of a carbon activated with sulfuric acid. Low concentration of total iron was studied (4.57 mg/L). The main difference with the previous works corresponds to the optimal pH where an optimal adsorption takes place. The authors have found that at pH higher than 11, the removal of iron from the solution was complete (yield of 100%) in the presence of the AC. The adsorption isotherm was modeled by the Freundlich isotherm, and the adsorption kinetics followed a pseudo-second-order model.

The removal of anions from water was studied by Ching et al. (2011) and Getachew et al. (2015). They have carried out experiments in the presence of phosphate and fluoride ions, respectively. The adsorption of phosphate seems to be possible with AC, but the results given by Ching et al. (2011) are not consistent. The adsorption of phosphate did not follow a specific trend, and the adsorption isotherm and adsorption kinetic were not correctly modeled. These results must be considered carefully. The adsorption of fluoride was performed with ACs (activation made with sulfuric acid) prepared from coffee husk (Getachew et al., 2015). This study was focused on the removal of fluoride from the water of a lake (Hawassa, Ethiopia), in order to find a simple solution to recover values into the permissible limit of 0.5–1.5 mg/L. The experimental tests were carried out with synthetic solutions containing from 5 to 20 mg/L of fluoride ions. The efficiency of the removal was higher at low pH values, especially at pH = 2 (optimal value close to 0.14 mg/g). Getachew et al. (2015) obtained a removal higher than 80% after 3 h of contact in the presence of 18 g of AC (pH = 2, initial concentration of fluoride lower than 20 mg/L). The adsorption of fluoride was attributed to strong coulombic force of attraction between the fluoride anions and the positively charged surface of AC. The adsorption isotherm and the adsorption kinetic were modeled using the Langmuir model and the pseudo-second-order equation, respectively.

These results show that the adsorption of cations and anions on AC obtained from coffee residue is a promising route to treat water. The performances of adsorption depend on different parameters: the activation mode, the pH, and the initial concentration of minerals. The modeling of adsorption was found to follow Freundlich or Langmuir isotherm, but the adsorption kinetic corresponded each time to a pseudo second order model.

9.4.4 REMOVAL OF ORGANIC MOLECULES PRESENT IN WATER

Among the different applications, adsorption of organic molecules on ACs was widely studied, especially for ACs obtained from coffee residues. The removal of dyes was particularly studied, as well as phenol and other pollutants (e.g., antibiotics).

9.4.4.1 Adsorption of dyes

Dyes come from different industries, mainly the textile industry. They represent a problem for the entire ecosystem since they are not biodegradable but very stable in the environment. The removal of such pollutant by AC has been widely studied. Several works reported the adsorption of different dyes on coffee ACs prepared with NaOH or KOH as activating agents (Ahmad and Rahman, 2011; Jung et al., 2016; Pechyen et al., 2010). Experimental conditions were adapted according to the nature of the dye (anionic or cationic), in order to obtain the best adsorption performances. As an example, Jung et al. (2016) have adjusted the pH value at 3 for the adsorption of acid orange 7 (AO7: anionic dye) and at 11 for the adsorption of methylene blue (MB: cationic dye). This result is not surprising because the pH_{PZC} of the AC was 7.6, which explains the affinity of the AC surface with the anionic and the cationic dyes at pHs below and above the pH_{PZC} value, respectively. The same result was reported by Ahmad and Rahman (2011) for the adsorption of Remazol brilliant Orange 3R (RBO3R: anionic dye). The adsorption of these dyes was performed at 10, 20, and 30°C. The adsorption capacities increased with temperature (from 460.5 to 634.6 mg/g for AO7, and from 465.5 to 710.2 mg/g for MB), which indicates an endothermic adsorption mechanism. Ahmad and Rahman (2011) obtained similar results for the adsorption of RBO3R at temperatures ranging from 30 to 60°C. The adsorption of these different dyes followed different models. For anionic dyes, the adsorption of RBO3R was correlated with the Langmuir model for Ahmad and Rahman (2011), whereas the Langmuir and Sips models represented the adsorption of AO7 for Jung et al. (2016). For MB, the adsorption isotherm was well described by the Sips model. There are three different potential mechanisms to describe the adsorption process: particle or film diffusion and adsorption in the adsorbent pores. The authors concluded that the adsorption mechanism of RBO3R was controlled by film diffusion, and by pore diffusion for AO7 and MB. Jung et al. (2016) have also studied the influence of ionic strength on the adsorption capacity of AO7 and MB. They observed that the presence of NaCl (from 0.5 to 3 g/L) did not influence the adsorption of the dyes. It means that the adsorption of these dyes take place at the internal layer of the AC (ionic bonds), rather than at the external layer (Van der Waals forces), but the modification of the electrostatic forces due to the presence of NaCl did not modify the adsorption capacities.

The adsorption of dyes was studied by Reffas et al. (2010) on coffee-ACs prepared with H_3PO_4 and Na_2HPO_4, respectively. The adsorption capacities at 25°C of the different AC are lower than the previous ones obtained by alkaline treatment. Reffas et al. (2010) have obtained adsorption capacities of 175.5 mg/g for MB at pH = 6 and of 350 mg/g for Nylosan Red N – 2RBL (NR: anionic dye) at pH = 4, while et al. (2013) have observed adsorption capacities of 150 mg/g for MB at pH = 11 and of 110 mg/g for direct red (DR: anionic dye) on their AC. The adsorption isotherms were of type I, which were suggesting a monolayer adsorption. Reffas et al. (2010) have found that the adsorption isotherm could be well described by the Langmuir model for both MB and NR. They have also suggested that the adsorption of MB could not only be due to electrostatic interactions because the adsorption was carried out at a pH lower than the pH_{pzc}. The difference of adsorption capacities between MB and NR was attributed to the better affinity of NR with the surface of the AC. The adsorption capacity of NR was also proportional to the mesoporous volume of the AC, which indicates that both surface and textural properties of the AC are responsible for the adsorption performances of the AC.

Oliveira et al. (2009) have prepared AC using iron chloride as activated agent. They found that this route for preparing AC generated specific material with a more acidic surface than other ones, activated as an example by zinc chloride. The adsorption of MB at pH = 5.7 has shown a poor adsorption capacity (close to 50 mg/g) at 25°C. This result compared with the adsorption of MB on the same AC prepared with $ZnCl_2$ (263 mg/g) was explained by the formation of smaller pores with iron chloride (0.63 nm) than with zinc chloride (0.9 nm), which avoids the diffusion and the accessibility of MB (Stokes radii of 0.486 nm) to the adsorption sites inside the porous material.

The adsorption of MB was studied by Nunes et al. (2009) on a physically AC. The main difference they observed corresponds to the adsorption efficiency, which remains high even for pH values below the pH_{PZC}. This behavior was explained by the π–π dispersion interactions between MB and the surface of the AC. In fact, the physical activation leads to different surface adsorption centers, and low surface charges. The decrease of the particle size was also responsible for higher adsorption capacity,

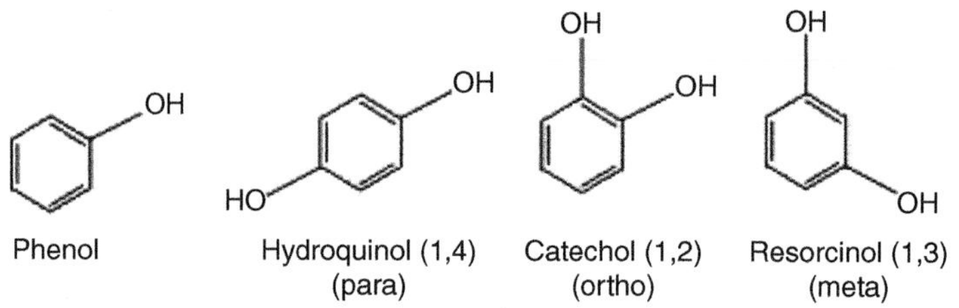

FIGURE 9.10

Phenol (C_6H_5OH) and disubstituted phenols ($C_6H_4(OH)2$) (para, ortho, meta) used for the evaluation of the AC adsorption capacity.

Reprinted with permission from Namane, A., Mekarzia, A., Benrachedi, K., Belhaneche-Bensemra, N., Hellal, A., 2005. Determination of the adsorption capacity of activated carbon made from coffee grounds by chemical activation with ZnCl₂ and H₃PO₄. J. Hazard. Mater. B 119, 189–194. Copyright 2016 Elsevier.

indicating the importance of surface area, as well as the pore accessibility. The adsorption of MB was important during the first 30 minutes, which was attributed to the fact that adsorption mainly occurs at the surface of the AC. However, the adsorption capacity was very low (18.4 mg/g) in comparison to the other AC presented earlier. The adsorption isotherm was described by the Freundlich model, which indicates a heterogeneous and multilayer adsorption. The mechanism of adsorption corresponds to a chemisorption process as the adsorption kinetic was perfectly represented by a pseudo-second-order model.

9.4.4.2 Adsorption of phenol

Phenol is a common contaminant of water (carcinogenic). Its concentration in wastewater is limited in several countries (1 mg/L in the United States). This product is used in several industrial applications, such as the production of resins. The use of AC for the removal of this molecule is interesting because it presents interesting characteristics (hydroxyl function and aromatic ring), which are able to interact with the surface of AC. Coffee biochar activated with H_3PO_4 and Na_2HPO_4 were used to remove phenol from solutions. To identify the importance of the hydroxyl group during the adsorption process, Namane et al. (2005) have compared the adsorption of phenol and three other organic molecules (catechol, resorcinol, and hydroquinol), which present the same formula as phenol but have a second hydroxyl group oriented in ortho (catechol), meta (resorcinol), and para (hydroquinol) (Fig. 9.10). A chemical activation of the biochar was made with $ZnCl_2$ and H_3PO_4. The results indicate that hydroquinol and phenol were more adsorbed than the other compounds. The authors explained that the orientation of the -OH radicals in position para limits the cross sectional of the molecule, and then allows a better adsorption on the AC. The adsorption capacity of phenol was limited (3.2 mg/g), and the adsorption isotherm followed a Langmuir model, which indicates that the adsorption sites are energetically homogeneous and form a saturated monolayer. The same result was reported by Lamine et al. (2014), with a higher adsorption capacity for the AC (55.5 mg/g).

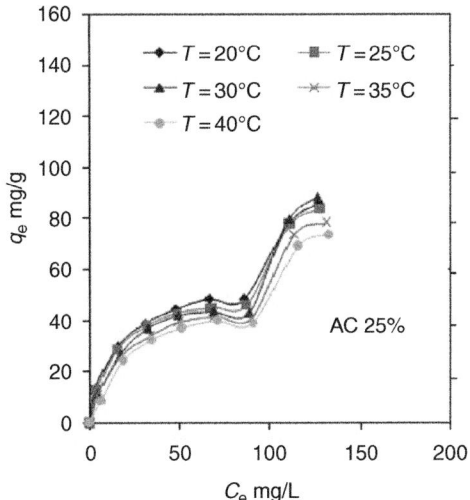

FIGURE 9.11 Adsorption Isotherms of Phenol Onto AC 25%

Conditions: pH = 3, AC 25%, and dosage = 1 g.

Reprinted with permission from Khenniche, L., Aissani, F., 2010a. Adsorptive removal of phenol by coffee residue activated carbon and commercial activated carbon: equilibrium, kinetics and thermodynamics. J. Chem. Eng. Data 55, 4677–4686. Copyright 2016 American Chemical Society.

In another recent work, Gonçalves et al. (2013b) obtained higher adsorption capacity for an AC prepared from Na_2HPO_4 at pH = 2 and a temperature of 25°C (105 mg/g).

The adsorption of phenol was studied using a highly microporous AC (Khenniche and Aissani, 2010a), activated with $ZnCl_2$. The adsorption was performed at pH = 3 and at different temperatures (from 20 to 40°C). The adsorption process was found to be exothermic (decrease of the adsorption capacity with temperature), purely physical (adsorption enthalpy <40 kJ/mol) and followed a pseudo-second-order kinetic. Contrary to the different results presented previously (L-type isotherms), the adsorption isotherms obtained by Khenniche and Aissani (2010a) are of type IV (Fig. 9.11). About 50% of the total adsorption capacity is obtained at the first plateau and the second plateau is attributed to the displacement of water molecules adsorbed at the AC surface by phenol ones. The total adsorption capacity reaches a value of 90 mg/g.

Castro et al. (2011) have used two different activation modes to prepare their AC: water vapor and K_2CO_3. The first difference observed with the different AC corresponds to their PZC: it corresponds to pH = 8 when steam is used as the activating agent and pH = 3 for K_2CO_3. Their adsorption capacities correspond to 149 and 159 mg/g, respectively (at pH = 7 and 25°C). During this work, the authors observed that the maximal adsorption capacity was obtained with the AC presenting the lowest surface area. It means that the affinity of the AC with phenol mainly depends on the functional groups, which are present at the surface of the AC, more than the surface area. They have also found that the adsorption capacity was a function of the porous volume (Fig. 9.12). This observation has not been reported by other authors.

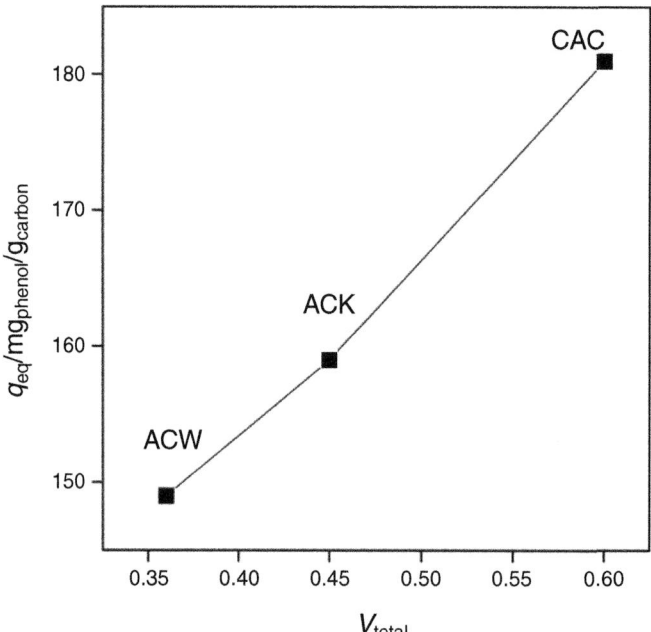

FIGURE 9.12 Relationship Between the Phenol Adsorption Capacity (q_{eq}) and the Total Pore Volume (V_{total}) of the ACs

9.4.4.3 Adsorption of pharmaceutical products, pesticides, and other micropollutants

In parallel to the previous studies, some works have focused on the adsorption of salicylic acid by AC. Khenniche and Aissani (2009) produced AC after a treatment with different ratios of $ZnCl_2$. The best adsorption yield was obtained for a ratio of 25 wt.% $ZnCl_2$, the adsorption capacity reached 128 mg/g (pH = 3, T = 25°C). This result was attributed to the basicity of the surface and to the microporosity of the AC (96% of the total pore volume). The adsorption process was found to be exothermic, the adsorption isotherm was modeled by both Langmuir and Freundlich isotherms (L-type isotherm according to Giles' classification) and the kinetic by a pseudo second order equation. Khenniche and Aissani (2009) have investigated the adsorption competition between salicylic acid and phenol molecules at the surface of the AC. The adsorption of phenol was more affected (−30%) than the adsorption of salicylic acid (−5%). This result can be explained by the stronger affinity of the carboxylic function of salicylic acid than the hydroxyl function of phenol.

The adsorption of malathion (carcinogenic pesticide, Fig. 9.13) on an AC (pH = 6, T = 30°C) was studied by Bouchenafa-Saib et al. (2014). The biochar was activated with $ZnCl_2$, H_3PO_4, or both of them to compare the performance of adsorption. The best results were obtained with the AC prepared from H_3PO_4 activation. Bouchenafa-Saib et al. (2014) have attributed this result to the stronger acidity character obtained at the surface and also to the larger mesoporous volume of the AC prepared with H_3PO_4. As malathion

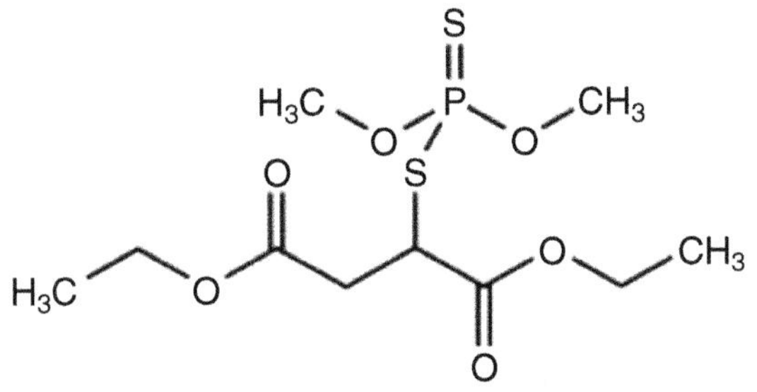

FIGURE 9.13 Chemical Structure of Malathion

is a large molecule, the diffusion inside the micropore of the AC may be limited, whereas the presence of larger pores enables the adsorption in the porosity of the AC.

The presence of pharmaceutical products in wastewater, as well as in rivers has become of great importance for the environment. These compounds are persistent pollutants, which are present at low concentrations in water. Flores-Cano et al. (2016) have studied the adsorption of metronidazole, dimetridazole, and diatrizoate on AC (chemically activated with H_3PO_4) prepared from coffee residues. The adsorption was described to be optimal due to the presence of phenolic groups at the surface of the AC, which promote dispersive interactions between π electrons of the graphene planes and the π electrons of the aromatic ring of the different molecules. The adsorption mechanism for the three pollutants was found to mainly correspond to surface diffusion, which controls the intraparticle diffusion. Another work reported by Djilani et al. (2012) focused on the adsorption of two micropollutants (o-nitrophenol and p-nitrophenol) on an AC prepared from H_3PO_4 and HNO_3 activation. The best results were obtained for p-nitrotoluene, where adsorption was complete even for an initial concentration of 20 mg/L. The authors have explained this result by the hydrophobic character of p-nitrotoluene in regard to o-nitrotoluene. The decrease of the cross-sectional of the molecule was not used to explain this result, at the difference of Namane et al. (2005) for similar molecules.

To conclude this part devoted to the adsorption of molecules or ions at the surface of AC prepared from coffee residue, we can say that the adsorption process is strongly dependent on different parameters:

- nature of the activation agent (physical or chemical),
- textural properties of the AC (ratio of the micro-mesoporosity),
- nature and size of the organic molecule,
- charge of the ions and molecules (acid, base),
- pH_{PZC} of the AC.

Then, the properties of the AC must be tailored in regard to the solute to be adsorbed, to improve the efficiency of the process, but also to avoid adsorption competition between different solutes or with the solvent.

Table 9.6 The Maximum Adsorption Volumes of Ethylene and *n*-Butane Onto Commercial Activated Carbon (AC), Activated Carbon Prepared From Spent Coffee Grounds (CG-AC), and Activated Carbon Prepared From Pomelo Skin (PS-AC)

Sample	Maximum Adsorption of Ethylene (cm^3/g)	Maximum Adsorption of *n*-Butane (cm^3/g)
AC	70	75
CG-AC	51	84
PS-AC	48	129

Reprinted with permission from Ma, X., Ouyang, F., 2013. Adsorption properties of biomass-based activated carbon prepared with spent coffee grounds and pomelo skin by phosphoric acid activation. Appl. Surf. Sci. 268, 566–570. Copyright 2016 Elsevier.

9.4.5 GAS- AND VAPOR-PHASE ADSORPTION

ACs are interesting adsorbent materials. They can efficiently adsorb many gases, because of their porous structure and high surface area and also chemical surface composition. For economical reasons, some researchers proposed to prepare AC adsorbents from wastes and biomasses, such as coffee grounds. The following section reports the use of adsorbents prepared from coffee wastes to store energy gases or to remove pollutants, such as volatile organic compounds (VOC), acid gases, and carbon dioxide.

9.4.5.1 Storage of energy gases

AC materials derived from waste coffee grounds were studied as adsorbents for methane and hydrogen by Kemp et al. (2015). Storing methane in compressed gas cylinders is dangerous (safety problems) and heavy. For this aim, low-weight adsorptive materials, such as AC, to adsorb and release this valuable gas at lower pressures, are required. Furthermore, hydrogen, a gas whose combustion does not lead to the greenhouse gas effect CO_2, is also dangerous to be stored in cylinders, and thus the search of materials that could store it (then to store energy) is a necessity. In this context, Kemp et al. (2015) activated coffee grounds by KOH and further subsequent annealing under argon at three different temperatures—700, 800, and 900°C—to lead, respectively, to materials designated by AMC700, AMC800, and AMC900. The highest surface area (about 1040 m^2/g), total pore volume (about 0.64 cm^3/g), and microporous volume (about 0.57 cm^3/g) were obtained for AMC900. The authors deduced that treating activated coffee grounds at higher temperature led to the increase of the surface area. Indeed, the authors presented a temperature-dependent mechanism for AC formation from coffee grounds. Accordingly, the highest adsorption capacity of CH_4, at 0°C, and H_2 at −196°C, under a pressure of 100 kP_a, were noted for AMC900. In addition, the CH_4 storage capacity of AMC900 was found to increase with increasing pressure to 3000 kP_a and temperature to 15–35°C (Table 9.6), where an adsorption capacity of about 4.2 mmol/g was obtained. It is also to be noted that reversible and stable adsorption capacity was observed on this material at 15–35°C. The storage capacities obtained for methane (4.2 mmol/g) and for hydrogen (1.75 wt.%) on

AMC900 were among the highest obtained in literature for methane and hydrogen adsorption on carbon materials with high specific surface areas (Kemp et al., 2015).

9.4.5.2 Adsorption of volatile organic compounds pollutants

Ma and Ouyang (2013) investigated the adsorption, at room temperature, of ethylene and *n*-butane on AC from spent coffee grounds. Ethylene and *n*-butane were regarded as VOCs pollutants that could be eliminated by adsorption. Spent coffee grounds were treated by phosphoric acid and then heated under air at 500°C. The sample was designated by CG-AC. The specific surface area was about 1110 m^2/g, higher than many types of commercial ACs. The authors compared the adsorption behavior of CG–AC with a carbon issued from pomelo skin and activated in the same way (PS-AC) and a commercial AC (AC). The isotherms were performed at 34°C (temperature close to room one), under a pressure range of 0–101 kP_a. The equilibrium was not reached with ethylene whereas for *n*-butane the equilibrium was reached at a pressure of about 30 kP_a, because ethylene has a higher value of saturation vapor pressure than *n*-butane, at room temperature. The isotherms fitted well Langmuir equation and are of type I according to the IUPAC classification. After micropore filling, the amount of gas adsorbed increased slowly. The maximum adsorption capacities are reported in Table 9.6. All the samples adsorbed more *n*-butane than ethylene. PS-AC and CG-AC showed better adsorption capacity for *n*-butane than AC. However, AC adsorbs more ethylene than PS-AC and CG-AC. The authors (Ma and Ouyang, 2013) stated that the smaller the pore, the easier will be the adsorption of ethylene. The isosteric enthalpy of adsorption, $\Delta_{ads}H$, was also determined at different coverages, from the Clausius–Clapeyron relation:

$$\Delta_{ads}H = \frac{R(LnP)}{1/T}, \text{ where } R \text{ is the gas constant, } P \text{ the pressure, and } T \text{ the adsorption}$$

temperature. The results are plotted in Fig. 9.14. The isosteric heats of adsorption were quite large at low coverages, due to the great adsorption potential in the micropores. At low coverages, the isosteric heats of *n*-butane were higher than those of ethylene, indicating a stronger interaction between *n*-butane and carbon adsorbent than between ethylene and carbon adsorbent. Such a result is linked to the fill of the micropores by *n*-butane rather than by ethylene. Furthermore, the heats of ethylene adsorption were higher on AC than on the other adsorbents, revealing a stronger interaction between ethylene and AC, in accordance with the higher ethylene adsorption capacities obtained for AC. In addition, Ma and Ouyang (2013) tested the same sample of each adsorbent for more than forty times and found stable adsorption capacities. Hence, CG-AC and PS-AC are stable adsorbents, as well as AC.

The removal of formaldehyde, a recognized VOC pollutant, by adsorption on AC prepared from coffee residue was studied by Boonamnuayvitaya et al. (2005). These researchers compared untreated coffee residues or residues treated with $ZnCl_2$ and activated by pyrolysis, at 600 °C, under either nitrogen flow (CN_2 or $CZn-N_2$), carbon dioxide flow (CCO_2 or $CZn-CO_2$), or nitrogen flow followed by steam activation (CN_2-ST or $CZn-N_2$-ST). In the total, six samples were prepared from coffee residues and compared to each other and with a commercial AC

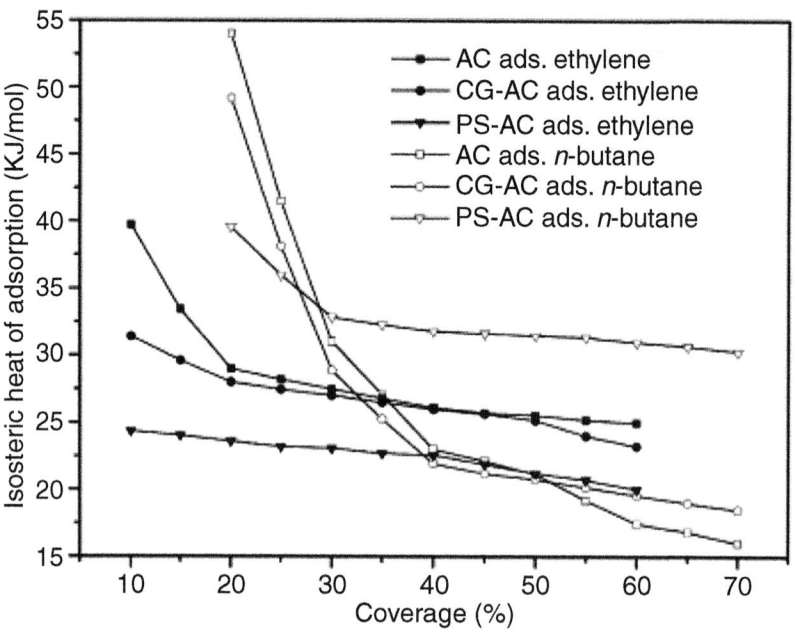

FIGURE 9.14

Isosteric heats of ethylene and *n*-butane adsorptions, at different coverages, onto commercial activated carbon (AC), AC prepared from spent coffee grounds (CG-AC), and AC prepared from pomelo skin (PS-AC).

Reprinted with permission from Ma, X., Ouyang, F., 2013. Adsorption properties of biomass-based activated carbon prepared with spent coffee grounds and pomelo skin by phosphoric acid activation. Appl. Surf. Sci. 268, 566–570. Copyright 2016 Elsevier.

(CH-I1000). The specific surface areas of carbons activated by $ZnCl_2$ were higher than those of the untreated with $ZnCl_2$. Indeed, during the mixing of carbon with $ZnCl_2$, the latter could be intercalated in the carbon matrix. Adsorption of formaldehyde was carried out on the studied samples at 30°C, with different formaldehyde concentrations (Fig. 9.15).

The order of formaldehyde adsorption is as follows: $CZn-N_2$ > CH-I1000 > $CZn-N_2$-ST > $CZn-N_2$-CO_2 > CN_2-ST > CN_2-CO_2 > CN_2. It is known that the adsorption capacity could be influenced by the textural properties and also by the surface chemistry. The authors (Boonamnuayvitaya et al., 2005) demonstrated that the order of adsorption in their case was rather linked to the surface chemistry, since $CZn-N_2$, which exhibits the highest adsorption capacity shows lower specific surface area and pore volume than some other studied carbon adsorbents. The authors stated that the functional hydrophilic groups, O—H, C=O, and C—O, that are present in higher amounts on the adsorbents with higher adsorption capacities, significantly affect the adsorption of formaldehyde. For example, the strongest peak of O—H group was observed for $CZn-N_2$, which also exhibits the highest adsorption capacity toward

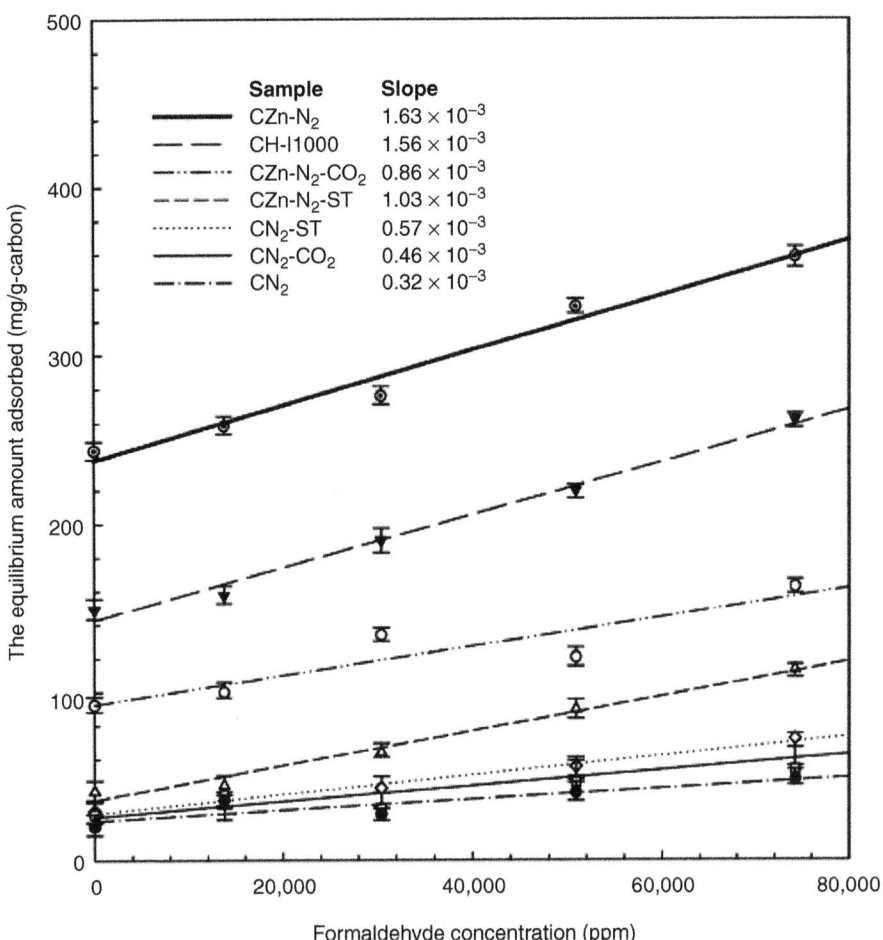

FIGURE 9.15

The equilibrium amount adsorbed by tested ACs prepared by Boonamnuayvitaya et al. (2005) at various concentrations of formaldehyde and a temperature of 30°C.

Reprinted with permission from Boonamnuayvitaya, V., Sae-Ung, S., Tanthapanichakoon, W., 2005.
Preparation of activated carbons from coffee residue for the adsorption of formaldehyde. Sep. Purif. Technol.
42, 159–168. Copyright 2016 Elsevier.

formaldehyde. Moreover, CN_2-ST and CZn-N_2 have similar surface areas, but the adsorption capacity of CZn-N_2 is higher. This observation is explained by the presence on the surface of CN_2-ST of higher quantity of hydrophobic groups, C—H, C=C, and CC, that not favor formaldehyde adsorption. The samples CN_2-CO_2 and CN_2, which have the lowest adsorption capacities, have the lowest amounts of hydrophilic groups along with the lowest surface areas and pore volumes. The sample CH-I1000 has good capacity of formaldehyde adsorption since it combines high surface area and pore volume to the presence of hydrophilic groups. CZn-N_2-CO_2, which showed lower adsorption capacity than CZn-N_2, despite its much higher surface area and pore volume, was found to have higher amount of hydrophobic groups on its surface.

9.4.5.3 Adsorption of acid gases pollutants

Removing the pollutants gases with acidic character, such as nitrogen dioxide (NO_2) and hydrogen sulfide (H_2S), by ACs derived from coffee wastes was investigated by Kante et al. (2012) and Nowicki et al. (2013, 2016).

Kante et al. (2012) activated ground coffee with $ZnCl_2$. Three different ratios of dry coffee grounds: $ZnCl_2$ were used: 1:0.5 (COFAC-0.5); 1:1 (COFAC-1), and 1:2 (COFAC-2). The increase of the amount of $ZnCl_2$ led to an increase in the specific surface area (from 645 m^2/g for COFAC-0.5 to 1121 m^2/g for COFAC-2). Pores with large volumes, 10–30 Å, were formed. Such a size was reported (Kante et al., 2012) to be optimal for water and hydrogen sulfide adsorption. The different adsorbents prepared were subjected, at room temperature, to a flow of H_2S diluted with moist air (1000 ppm H_2S, 80% humidity). The following H_2S adsorption capacities were found: COFAC-0.5: 81.3 mg/g, COFAC-1: 127.0 mg/g, and COFAC-2: 18.3 mg/g. The value obtained for COFAC-1 was higher than that found on some other carbon materials. For example, the catalytic carbon, Centaur, is able to adsorb 104 mg/g of H_2S in the same experimental conditions. Washing the carbon materials with concentrated HCl acid, to remove inorganic matter, led to decreased H_2S adsorption capacities (respectively, 26, 58, and 20.7 mg/g for COFAC-0.5, COFAC-1, and COFAC-2). This result suggests that some inorganic matter (ash) soluble in HCl play a role in H_2S adsorption. However, even if the removal of ash decreased the adsorption capacity, it is not the only factor influencing the adsorption behavior, since there is a difference in H_2S adsorption capacities between the studied samples. The authors proved that it is not only zinc species or alkali and alkaline metals that are responsible for the good adsorption capacity of COFAC-1 but also the nitrogen functional groups located in the pores where water can be adsorbed. Indeed, these groups could be incorporated in the carbon matrix and are not removed by HCl washing. It was demonstrated, that carbons containing high volumes of small pores with incorporated nitrogen species, such as Centaur, gave high selectivity for sulfuric acid because of their basicity and their ability to activate oxygen. Nitrogen contained in the pores comes from the caffeine, $C_8H_{10}N_4O_2$. Nitrogen functional groups play a catalytic role in hydrogen sulfide oxidation.

Hence, the factors playing a role in H_2S adsorption are the porous surface and pore volumes of the AC, the residual inorganic matter that forms a salt with H_2S, and nitrogen species, which play a catalytic role in the oxidation of H_2S.

Nowicki et al. (2013) investigated the effect of pyrolysis temperature (500 and 800°C, samples CP5 and CP8, respectively) and the nature of the activating agent on the NO_2 adsorption performances of the ACs issued from coffee wastes. Coffee-industry waste was activated chemically by KOH (CP5Ac and CP8Ac) or physically by CO_2 (CP5Ap and CP8Ap). The sample CAd was obtained by direct activation process by CO_2 (without pyrolysis). Activation by CO_2 does not lead to the development of the porous structure of carbon; the specific surface areas and pore volume

Table 9.7 NO_2 Adsorption Capacities (mg/g) of the Activated Carbons

Sample	Dry Conditions	Wet Conditions
CP5	6.3	21.0
CP8	9.0	84.1
CAd	3.4	31.8
CP5Ac	13.0	12.6
CP5Ap	36.4	65.8
CP8Ac	44.5	28.8
CP8Ad	31.9	76.0

Prepared by Nowicki, P., Skibiszewska, P., Pietrzak, R., 2013. NO_2 removal on adsorbents prepared from coffee industry waste materials. Adsorption 19, 521–528.

Table 9.8 Adsorption Capacities of NO_2 and H_2S (mg/g) on Coffee Waste Activated by KOH or CO_2 in Dry and Wet Conditions (Nowicki et al., 2016)

Sample	NO$_2$ Dry	NO$_2$ Wet	H$_2$S Dry	H$_2$S Wet
CP7PA	23.1	54.7	11.8 (52.9)*	149.6 (118.8)*
CP7CA	22.4	13.4	4.7 (11.1)*	7.2 (9.2)*

The bed was subjected to moisture before adsorption of H_2S.

obtained for the sample activated by CO_2 are very low—5 to 10 m^2/g—whereas those obtained for coffee wastes activated with KOH are in the range 1550–2100 m^2/g with predominance of micropores. NO_2 was adsorbed on the samples put in the form of granules of 0.75–1.5 mm in diameter. Dry or wet (70% humidity) air with 0.1% NO_2 was passed through the carbon matrix. The adsorption capacities toward NO_2 are listed in Table 9.7. The adsorption capacity depends on the temperature of pyrolysis, the activating agent, and also the adsorption conditions (dry or wet). The carbons activated by CO_2 exhibit better adsorption capacities in wet conditions than in dry ones. Furthermore, the authors noted that despite the high specific surface area and pore volumes of the samples activated by KOH, their NO_2 adsorption capacities are lower than those activated by CO_2. The authors suggested that the adsorbed nitrogen dioxide was strongly bonded by the functional groups (basic oxygen functional groups) present on the surface of ACs or by the metal ions (minerals) in their structure, revealing a chemisorption of the NO_2. The interaction of NO_2 with the oxygen groups and the minerals is enhanced in water presence.

Nowicki et al. (2016) investigated, in H_2S and NO_2 removals, the same series of coffee wastes pyrolyzed at 700°C (CP7) and activated either chemically by KOH (CP7CA) or physically by CO_2 (CP7PA). The samples were sieved to a particle size between 0.75 and 1.5 mm and subjected to a dry or wet (70% humidity) air with 0.1% H_2S or NO_2. For H_2S adsorption, the effect of wetting the carbon bed before adsorption experiment was studied. Adsorption capacities are listed in Table 9.8. The dependence of the adsorption capacity on the nature of the activating agent and the presence/absence of humidity are the same as in the previous work of these authors (Nowicki et al., 2013). The decrease of adsorption capacity with the presence of humidity is observed for samples activated with KOH; these samples contain lower amounts of minerals. The authors concluded that the presence of minerals is conducive to effective adsorption of H_2S. When the bed was moistured before H_2S adsorption, higher adsorption capacities were obtained. In fact, a generation of a thin film of water on the surface seems to be conducive to H_2S bonding. When, in addition to wetting the bed previous to the adsorption, H_2S is adsorbed in wet conditions, the adsorption capacities are generally enhanced.

The authors concluded that high contents of inorganic were crucial for efficient removal of NO_2 and H_2S.

9.4.5.4 Adsorption of carbon dioxide from postcombustion

Adsorption of carbon dioxide CO_2 issued from combustion was also studied on AC prepared from coffee residues. Indeed, CO_2 is emitted in large amounts by combustion and, being a greenhouse gas effect, is responsible for climate change. Thus, CO_2 capture and storage (CCS) is a process that could contribute to the decrease of its emissions in atmosphere.

CCS was proposed as alternative for CO_2 absorption with aqueous alkanolamines solutions. Absorption by solutions is less easy to use since it requires high amounts of solvents to deal with, difficulty in separation, solvent regeneration, high energy to heat the solvent, with the risk of corrosion, toxicity, amines losses, and so on. Adsorption is less expensive and necessitates a lower size for the setup and does not require heating and cooling large amounts of water. The use of adsorbents coming from biomass, helps to reduce the cost of the adsorption process.

Plaza et al. (2012) investigated carbon materials derived from coffee grounds chemically activated by KOH or physically activated by CO_2. All the prepared samples were essentially microporous. This is desirable since it favors CO_2 adsorption at atmospheric pressure. CO_2 adsorption experiments were performed at atmospheric pressure of 101 kP_a. At 0°C, the CO_2 adsorption capacities were in the range of 3.2–4.9 mmol/g for the samples activated by KOH and of 3.4–3.6 mmol/g for those activated by CO_2. At 25°C, these capacities decreased to, respectively, 2.7–3.0 mmol/g for the former series and to 2.2–2.4 mmol/g for the latter one. Furthermore, raising the temperature slowly from 25 to 100°C under CO_2 flow shows a decrease of the CO_2 adsorbed amount from 2–3 to 0.6 mmol/g, the physisorption being an exothermic process. This decrease in the amount of CO_2 was less in the case of the samples activated with CO_2 than those activated with KOH. Thus, a higher interaction adsorbate-adsorbent is supposed to exist with the samples activated with CO_2, due to their narrower porosity and greater surface basicity.

These latter features were also found to be responsible for the higher CO_2/N_2 selectivity of the samples activated with CO_2 compared to those activated by KOH. N_2 was studied because it is the most abundant gas present with CO_2 in postcombustion emissions. The authors recommended the use of physical activation because it associates lower environmental constraints to a higher carbon yield for low burn off degrees. In a work from the same research group (González et al., 2013), the authors presented the

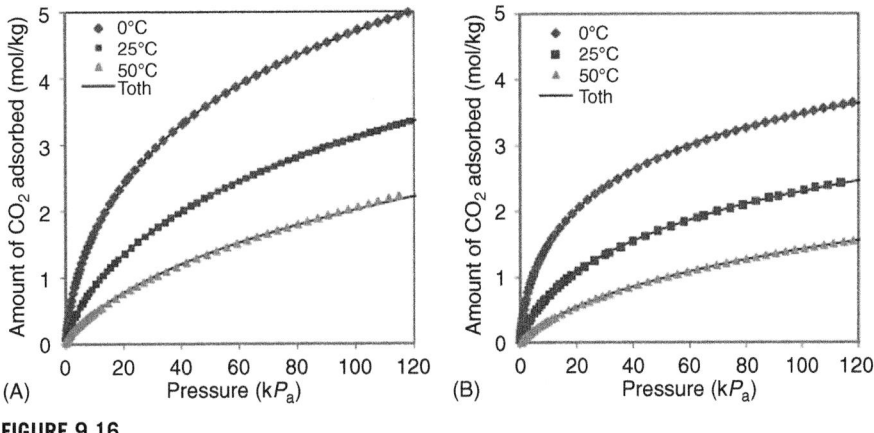

FIGURE 9.16

CO_2 adsorption isotherms, at 0, 25, and 50°C, of AC prepared from coffee grounds activated by (A) KOH (NCLK3) and (B) CO_2 (NCHA29).

Reprinted with permission from González, A.S., Plaza, M.G., Pis, J.J., Rubiera, F., Pevida, C., 2013.
Post-combustion CO_2 capture adsorbents from spent coffee grounds. Energy Procedia 37, 134–141.
Copyright 2016 Elsevier.

adsorption isotherms of CO_2 at 0, 25, and 50°C (Fig. 9.16). The adsorption was completely reversible by lowering the pressure (desorption part not shown), as predicted for physisorption process. The sample activated by KOH (NLCK3) showed higher CO_2 uptake than its homologous activated by CO_2 (NCHA29), due to its higher ultramicropore volume (0.36 versus 0.25 cm^3/g). The authors also proved the higher selectivity of CO_2 compared to N_2, obtained over their materials. The samples were easily regenerated by reducing the pressure of the system. The isotherms fitted the Toth model (Fig. 9.16) for CO_2 adsorption and Langmuir one for N_2 adsorption.

The isosteric heats of CO_2 and N_2 adsorption were calculated by the Clausius-Clapeyron equation. For CO_2, they were 27.19 and 36.42 kJ/mol, respectively for NLCK3 and NCHA29. The isosteric heats of N_2 adsorption were found to be the half of those of CO_2, 15.70 and 16.04 kJ/mol, respectively for NLCK3 and NCHA29. The interaction between CO_2 and the adsorbent is higher than that between N_2 and the adsorbent. The isosteric heats of adsorption depend mainly on the pore size distribution and to a less extent on the surface chemistry. The values found on the coffee-ground based samples were in the range of different types of commerical carbons but lower than that of zeolite 13X (40 kJ/mol) (because of the higher electrostatic interaction between the CO_2 molecule and the polar surface of zeolite). The higher value obtained for the CO_2 activated sample is due to its narrower microporosity. The binary adsorption isotherm of N_2 and CO_2 was below that of pure N_2 and close to that of pure CO_2, indicating that the presence of N_2 (weak adsorptive) affects little the adsorption of CO_2 (strong adsorptive), whereas the presence of CO_2 reduces significantly the N_2 adsorption in a N_2/CO_2 binary mixture. Similar behavior was reproted over zeolite 13X.

The working capacity (difference between the amount of CO_2 adsorbed at the high-pressure adsorption stage and the low-pressure evacuation step) was evaluated. When the partial pressure of the CO_2 in the feed gas is close to the evacuation pressure, the work capacity tends to zero, and vice versa—the working capacity increases with the partial pressure of CO_2 in the gas phase. NCLK3 presents working capacity higher than that of zeolite 13X. Furthermore, heat effects lead to the reduction of the working capacity. This reduction will be smaller for NCLK3 because of its lower isosteric heat of adsorption.

NCLK3 showed higher CO_2 adsorption capacities, higher selectivity toward CO_2, and lower CO_2 adsorption heat than NCHA29. Coffee-based materials showed better selectivtiy and working capacity compared to zeolite 13X along with lower cost, higher stability, and easier regeneration.

In a study from the same research group in 2015 (Plaza et al., 2015), the researchers investigated the same type of material in CO_2 adsorption, but with the use of the adsorbent under the pellet form instead of the fine powder form. The use of pellets allows using the adsorbents at a high scale in a fixed bed unit. The spent coffee grounds were shaped into pellets with a diameter of 4 mm and a height of 3 mm, without a binder addition. The cylindrical pellets obtained were activated by CO_2. The activation conditions were optimized to maximize the micropore volume. The activation process did not alter the shape of the pellets (cylinders with 2.67 ± 0.05 mm diameter and 1.88 ± 0.13 mm height). The sample will be designated by PPC. PPC surface was rich in nitrogen and oxygen species, coming from the caffeine. PPC was totally microporous with pore volume and surface area values comparable to those of the nonpelletized sample (González et al., 2013; Plaza et al., 2012). The micropores presented a pore-size diameter of 3.5–6.5 Å, in favor of CO_2 adsorption. The bulk density of PPC—421 kg/m^{-3}—was higher than those of many granular ACs derived from biomass. This property is important since it implies a more reduced volume of the adsorber.

Fig. 9.17 shows different isotherms for CO_2 and N_2 adsorption. The CO_2 adsorption capacities of PPC (0.7 mmol/g or 337 mol/m^3 at 50 °C and 1.2 mmol/g or 509 mol/m^3 at 25 °C) were higher than those of different carbon materials.

Similar results to those of the previous works of the same group were obtained on the PPC pellet:

- Higher selectivity toward CO_2 than N_2, the presence of surface N and O donor groups enhancing the interactions between CO_2 and the PPC, leading to higher CO_2 capacity and selectivity than for N_2;
- The selectivity toward CO_2 higher than that toward N_2, thus the material could be successfully used to separate these two gases;
- CO_2 adsorption fits the Toth model, N_2 adsorption fits the Langmuir model (Langmuir being a particular case of Toth);
- The isosteric heat of adsorption was in the same range of magnitude (around 30 kJ/mol) as for the nonpelletized materials.

One of the advantages of carbonaceous materials compared to other adsorbents is their easy regeneration. The shape of the CO_2 adsorption isotherms (not excessively steep) reveals this fact. Hence, the adsorbent could be regenerated by pressure reduction (vacuum swing adsorption VSA). Other adsorbents, such as zeolithe 13X,

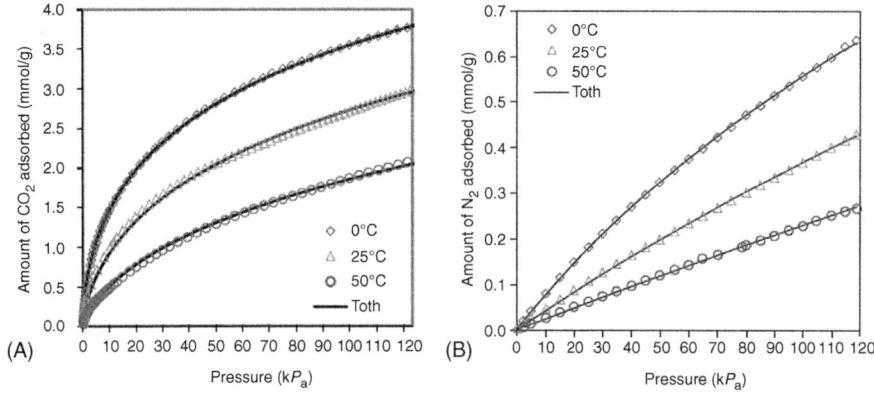

FIGURE 9.17

Adsorption isotherms at 0, 25, and 50°C on pelletized coffee grounds activated by carbon dioxide (PPC) of (A) CO_2 and (B) N_2.

Reprinted with permission from Plaza, M. G., González, A. S., Pevida, C., Rubiera, F., 2015. Green coffee based CO_2 adsorbent with high performance in postcombustion conditions. Fuel 140, 633–648. Copyright 2016 Elsevier.

possess higher adsorption capacities but steeper isotherms indicating the need of higher pressures to achieve high working capacities. For a moderate pressure ratio (desirable from an economical point of view), similar CO_2 working capacities could be obtained on carbon and zeolites.

Different cycles of adsorption, desorption by reducing pressure and/or increasing temperature, were performed over the PPC sample. The results obtained led the authors to conclude that although the purity of CO_2 of the cycles is not enough for transport and storage applications, CO_2 recovery and productivity are large because the minimum pressure reached within the cycle is relatively high (10 kP_a). This would contribute to reduce the cost of operating the adsorption unit. Furthermore, no adsorbent deactivation was obtained during the different cycles, revealing the good stability of the PPC material.

In a recent article from the same research group (Plaza et al. 2016), the researchers investigated water adsorption on PPC material and its influence on CO_2 materials. Water is the third (N_2 being the first and CO_2 the second) most abundant compound present in postcombustion flue gases. The adsorption and desorption isotherms of H_2O at 12.5, 25, 50, 70, and 85°C were performed under relative pressures in the range 0–0.95. The adsorption isotherms display two zones: the first one, takes place, at low relative pressures, and is attributed to the adsorption of water molecules on the oxygen and nitrogen present on the surface of PPC. This adsorption occurs by hydrogen bonding between water and the functional groups containing oxygen and nitrogen. Thus, clusters of water will be formed and will increase in size. When the size of the clusters becomes critical, their dispersive force will be enough to penetrate into the micropores leading to the second zone, with higher water uptake. Micropore filling occurs at relative pressures between 0.1 and 0.4. This range of relative pressure is lower on PPC than on other carbons materials, due to the narrow pore size of PPC and the functional groups existing on its surface. Moreover, the hysteresis loops between adsorption and desorption branches of the isotherm are narrow, due to the narrow pore size of PPC. If the isotherms are depicted as the amount of H_2O adsorbed versus relative pressure, they will practically overlap, meaning that the water vapor adsorption on PPC is rather not affected by temperature. The maximum water adsorption capacity on PPC (about 12 mmol/g) is lower than that on other carbon materials and on zeolithe 13X. This is regarded as advantageous for postcombustion applications since the competition between CO_2 and H_2O on the PPC sites will be low. The adsorption/desorption isotherms of water fit the DJD (Do, Junpirom, and Do) model, in the experimental conditions studied. The average isosteric heat of water adsorption was found to be of 46 kJ/mol. The difference between the adsorption energy and the evaporation energy of water (43 kJ/mol) is weak (3 kJ/mol), suggesting a physisorption process. H_2O hinders CO_2 adsorption on PPC only after long adsorption times. Thus, by selecting appropriate adsorption time, the adsorption of CO_2 could be optimized. In addition, PPC is fully regenerated even after long exposure to water vapor. The advantage of using carbon materials is that water interaction with hydrophobic carbon materials is weak and thus, the carbon adsorbent will be easily regenerated by vacuum swing adsorption (VSA) or temperature-aided vacuum-swing adsorption (VTSA). Hence, there is no need to a dehydration unit and thus the operating cost of the installation will be reduced.

The lower H_2O uptake and its lower heat of adsorption obtained on PPC compared to other inorganic materials are advantageous for postcombustion applications. PPC could be easily regenerated by simple VSA or VTSA. Travis et al. (2015) prepared microporous carbons by activating spent coffee grounds by KOH. Coffee grounds were heated under N_2 at 400 and 700°C (designated by CG 400 and CG 700, respectively) and then activated by KOH according to two KOH:CG weight ratios (2:1 and 4:1). The surface area of the materials ranged between 1624 m^2/g for CG 700 2-1 and 2785 m^2/g for CG 400 4-1, and the pore volume between 0.662 cm^3/g for CG 700 2-1 and 1.36 cm^3/g for CG 400 4-1. The pore width was in the range 0.926–1.051 nm.

CO_2 adsorption on these materials was performed at 0, 25, and 50°C, and under 1 bar and 10 bars. The adsorption capacities obtained are listed in Table 9.9. The highest adsorption capacities were obtained for the samples, which are totally microporous, with narrow microporosity, the micropores being mainly responsible for CO_2 adsorption at low pressures. The samples that presented larger micropore volumes exhibited lower CO_2 uptake at low pressure. The desorption curves of the isotherms did not show hysteresis, revealing that the adsorption is fully reversible. The adsorption curves at 0, 25, and 50°C fit with the Hill-Langmuir model. The isosteric heats of adsorption were of 22–26 kJ/mol, indicating a physisorption. The authors observed no loss of performance of the materials studied over 10 cycles of the temperature-swing process under 1 bar CO_2 at 25°C. The materials are fully regenerated at relatively low regeneration temperature (100°C).

Travis et al. (2015) reported that for adsorption technology to compete with liquid amine technology for CO_2 removal, adsorbent materials have to reach a CO_2 uptake of about 3 mmol/g at temperatures ≥ 50°C and ambient flue pressures. To achieve such capacities, N-doping of the adsorbents has been frequently used. The advantage

Table 9.9 CO_2 Capacities for Coffee Grounds (CG) Derived Carbons

Material	CO_2 Uptake at 1 bar (mmol/g)			CO_2 Uptake at 10 bar (mmol/g)	
	0°C	**25°C**	**50°C**	**0°C**	**25°C**
CG 400 2-1	7.17	4.21	2.47	16.44	12.74
CG 400 4-1	5.09	2.81	1.52	23.27	16.05
CG 700 2-1	7.55	4.42	2.86	13.34	11.59
CG 700 4-1	6.89	4.00	2.38	23.26	16.81

Reprinted with permission from Travis, W., Gadipelli, S., Guo, Z., 2015. Superior CO_2 adsorption from waste coffee ground derived carbons. RSC Adv. 5, 29558–29562. Copyright 2016, Royal Society of Chemistry.

of carbon derived from coffee waste is that it contains nitrogen originating from caffeine as evidenced by Plaza et al. (2012, 2015, 2016) and González et al. (2013). The adsorption capacity reached by carbons derived from coffee wastes did not reach values greater than 3 mmol/g at 50°C and pressures of 1 bar or 100 kP_a (atmospheric pressure). Even though, these materials are good candidates in such applications, because of their low cost, high regenerability and stability, high-uptake values.... Some little optimization is needed to further increase their adsorption capacities. Coffee wastes conveniently chemically and physically treated, have shown interesting adsorption/desorption capacities for methane (CH_4), hydrogen (H_2), ethylene (C_2H_4), n-butane (C_4H_{10}), hydrogen sulfide (H_2S), nitrogen dioxide (NO_2), mainly due to:

- their high specific surface area,
- the high volume of micropores,
- the narrow size of the micropores, and
- the surface chemistry of the adsorbent.

Furthermore, these low cost materials are easy to generate.

9.5 CATALYST PREPARATION FOR POLLUTANTS OXIDATION (DECOMPOSITION)

AC derived from coffee products was used as catalyst or catalyst support in various reactions for pollutants removal or biodiesel production—that is, for environmental goals and waste valorization. Indeed, the characterization of the carbon materials obtained from coffee products evidenced their high surfaces areas and porosities, rich surface chemistry that confers to them interesting properties for adsorption and catalysis. Furthermore, their chemical inertness, good thermal and chemical stability, as well as their availability from different wastes, make them good candidates for the use as catalyst-based materials.

Table 9.10 groups the different catalysts based on coffee products used and the reactions studied.

Table 9.10 The Different Catalysts Based on Coffee Products Used in Reactions of Environmental Interest: Preparation, Reactions Studied, and References

Catalyst	Activating Agent and Conditions	Reaction Studied	References
Iron oxide supported on activated carbon produced by coffee grounds	• Spent coffee grounds were impregnated with K_2CO_3, with a mass ratio (waste/K_2CO_3) of 1/1 in a 100 mL aqueous solution • Heating at 80°C until complete drying • Drying 24 h at 110°C • Heating 2 h at 800°C • Washing with deionized water to lead to AC (ACK) • Addition of iron: mixing ACK with iron nitrate $Fe(NO_3)_3/9H_2O$ salt in a 100 mL aqueous solution with a weight ratio iron/ACK of 5% • Heating at 80°C until complete drying • Calcination 3 h at 300°C to produce ACK/Fe • A subsequent pretreatment of ACK/Fe during 1 h under hydrogen at 400°C was performed to give ACK/Fe-H_2	Methylene blue oxidation by H_2O_2, according to a Fenton-like process	Castro et al. (2009a,b)
Iron oxide (goethite, Gt) supported on activated coffee husk waste 5, 10, 20 wt.% Gt	• Impregnation of $ZnCl_2$ on the coffee husk at 100°C • with a mass ratio (Waste/$ZnCl_2$) of 1/1 • Activation under N_2 at 500°C for 1 h • Washing by HCl and distilled water • Drying at 100°C • Addition of iron precursor	Methylene blue, caffeine, and phenol oxidation by H_2O_2 according to a Fenton-like process	Gonçalves et al. (2013a)

Catalyst	Preparation	Application	Reference
Sulfonated incompletely carbonized coffee residue (SCAC)	• Dried coffee residue mixed with $ZnCl_2$ with a mass ratio (waste/$ZnCl_2$) of 1/3 • Drying at 110°C for 12 h • Treatment under CO_2 during 4 h at 600°C • Washing with deionized water • Drying in air at 110°C for 12 h to get the CAC (carbonized coffee residue) • Sulfonation with concentrated sulfuric acid (H_2SO_4) (1 g CAC for 20 mL H_2SO_4) at 140, 160, 180, and 200°C to give sulfonated CAC (SCAC) SCAC-140, SCAC-160, SCAC-180, and SCAC-200 • Washing with deionized water • Drying in air at 110°C for 12 h	Esterification of caprylic acid by methanol	Ngaosuwan et al. (2016)
Coffee ground treated with sulfuric acid and fuming sulfuric acid	• Pyrolysis of coffee ground under 100 mL/min N_2 at 400°C during 4 h to lead to black carbon (BCC) • Treating BCC with sulfuric acid (5 g of BCC with 50 mL of acid) under reflux at 180°C for different times (2, 5, and 10 h) • Washing with deionized water • Drying 24 h at 120°C to lead to BCC-S2h, BCC-S5h, BCC-S10h • BCC was also treated with fuming sulfuric acid, about 20% SO_3, (50 mL of acid with 5 g BCC) under reflux at 70°C for 2, 5, or 10 h • Washing with water and drying to lead to BCC-SF2h, BCC-SF5h, and BCC-SF10h	Glycerol etherification with tert-butyl alcohol (TBA)	Gonçalves et al. (2016)

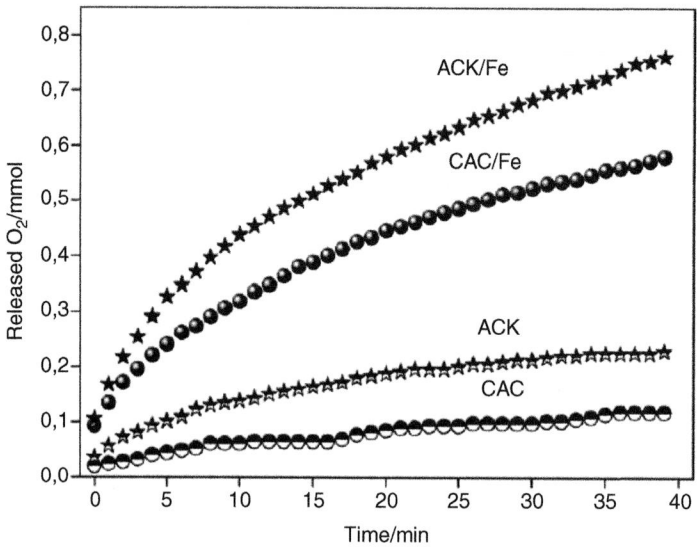

FIGURE 9.18

Hydrogen peroxide decomposition in the presence of commercial activated carbon (CAC) and AC from coffee grounds (ACK), before and after iron addition (CAC/Fe and ACK/Fe, respectively).

Reprinted with permission from Castro, C.S., Guerreiro, M.C., Oliveira, L.C.A., Gonçalves, M., Anastácio, A.S., Nazzarro, M., 2009a. Iron oxide dispersed over activated carbon: support influence on the oxidation of the model molecule methylene blue. Appl. Catal. A General 367, 53–58. Copyright 2016 Elsevier.

Castro et al. (2009a,b) compared the performance, in methylene blue oxidation by hydrogen peroxide H_2O_2, of iron supported on two carbon materials, one issued from coffee grounds (ACK/Fe) and one from commercial AC. ACK/Fe showed higher activity (Fig. 9.18). The physico-chemical characterization evidenced high surface area and iron dispersion in both cases. The iron presence increased the removal of methylene blue (Fig. 9.18). The methylene blue oxidation occurred via a radical mechanism. However, the activity was lost after four cycles of use (Castro et al., 2009a). A decrease of activity of 45% was observed between the first and the second cycle. No leaching of iron was put into evidence, revealing the high stability of the prepared materials in the reaction conditions. Castro et al. (2009b) evidenced that treating ACK/Fe by hydrogen results in an increase in the activity. Indeed, hydrogen led to partial reduction of iron species resulting in higher activity toward hydrogen peroxide decomposition and oxidation of organic compounds. Hydrogen treatment resulted in materials with textural properties similar to the solids before hydrogen treatment, without significant reduction in surface area. The kinetic analysis of hydrogen peroxide decomposition suggests a pseudo-first-order rate with respect to H_2O_2 at the beginning of the reaction (Castro et al., 2009b).

Gonçalves et al. (2013a) evidenced that iron dispersion was primordial for better catalytic activity in methylene blue oxidation by hydrogen peroxide; the highest activity was obtained over 10 wt.% goethite supported on AC derived from coffee wastes, which presents the smallest particle size compared to 5 and 20 wt.%. The catalytic systems were also tested in many cycles in methylene blue oxidation and showed some resistance to deactivation. No significant iron leaching was noted, revealing the good stability of the prepared materials. Such systems were also performant in oxidation of caffeine and atrazine. In absence of iron, the AC plays only the role of a sorbent whereas in presence of iron, the pollutant removal (methylene blue) occurred mainly by catalytic oxidation (Castro et al., 2009a,b; Gonçalves et al., 2013a). Both adsorption and catalytic properties confer to iron species supported on AC derived from coffee wastes interesting properties for wastewater treatment.

Ngaosuwan et al. (2016) investigated the esterification of caprylic acid by methanol on sulfonated carbonized coffee residue (SCAC) catalysts. It is to be reminded that esterification reaction is crucial in organic synthesis (medicine, plastics, cosmetics, food…) and in the pretreatment step of biodiesel production. The authors demonstrated that the sulfonation temperature plays a role in determining the density of $-SO_3H$ sites in the catalyst and that the sulfonation process tends to increase the specific surface area and the pore volume of the carbonized coffee residue (CAC). Better activities in caprylic acid esterification were obtained on SCAC compared to the typical (and expensive) solid acid catalyst for esterification, the resin Amberlyst-15. In addition, SCAC catalysts exhibited high stability. The used SCAC could be fully regenerated by washing with methanol. Adsorption of water produced by esterification was a reason for deactivation. The most active SCAC was that sulfonated at 200°C (SCAC-200), because of the balanced number of strong acid sites $-SO_3H$ and weak acid ones (carboxylic and phenolic, which constitute the hydrophilic surface of carbon black materials). On SCAC-200, the caprylic acid conversion to methyl caprylate was of 71.5% at 4 h and the initial turnover frequency (TOF), based on $-SO_3H$, was of 0.79 min. SCAC-200 also exhibits greater resistance to deactivation than the other SCAC catalysts (treated at other sulfonation temperatures).

Gonçalves et al. (2016) investigated carbonized coffee ground (BCC) sulfonated by sulfuric (BCC-S) and fuming sulfuric acid (BCC-SF) as catalysts in the reaction of glycerol etherification by tert-butyl alcohol (TBA). This reaction was considered attractive from many points of view: valorizing two wastes available in high amounts in Brazil, glycerol (by-product of biodiesel production) and coffee ground, to produce fuel additives (mono-tert-butyl glycerol MTBG, di-tert-butyl glycerol DTBG, and tri-tert-butyl-glycerol TTBG). As in the case of Ngaosuwan et al. (2016), the sulfonation process was efficient to insert sulfur groups ($-SO_3H$) on the surface. The sulfonation extent was higher with fuming sulfuric acid (7.5–8.4% of sulfur groups) compared to sulfuric acid (1.5–3.5% of sulfur groups). BCC-S and BCC-SF catalysts produced higher amounts of oxygenated additives than the commercial Amberlyst-15. The yield of MTBG was about 42%. The sulfur presence on the surface (concentration and strength) plays the main role in enhancing the catalytic performance; the activity of nonsulfonated BCC was negligible. Typically, the hydrophilic surface groups, such as –COOH and –SO$_3$H, act as anchoring sites for polar reactants, such as TBA, which leads to carbocation formation and subsequent MTBG production. The time of treating the BCC with either acid has a nonsignificant influence on the glycerol conversion. Furthermore, the sulfur groups exhibiting high stability, the catalysts could be used in many consecutive cycles. Washing the catalyst with hot water removes the site blockage by adsorbed reactants and lead to reproducible activities during the different reaction cycles.

All the aforementioned studies put the accent on the economy that could be realized by the use of coffee products to get the carbon matrix: availability in high amounts, valorisation of waste, no need for a size reduction step or pretreatment step, and so on. Moreover, for Fenton-like process, the alternative to substitute homogeneous catalysis with a heterogeneous one with coffee-based catalytic systems was highlighted (Castro et al., 2009a,b; Gonçalves et al., 2013a). Indeed, iron species were more dispersed on high surface area of AC support leading to higher performances. Also, the replacement of H$_2$SO$_4$ in homogeneous esterification processes by sulfonated AC (obtained from coffee waste) solid catalysts was also proposed to realize esterification and etherification with heterogeneous catalysis (Gonçalves et al., 2016; Ngaosuwan et al., 2016). Indeed, the limits of homogeneous catalysis are the corrosion (acid solutions), the environmental pollution, the side reactions, pH of solutions, harder separation processes, less possibility to recover the catalyst, sensitivity to water, and so on. All these disadvantages could be overcome by the use of solid catalysts.

Deactivation of catalysts could be due to either leaching of the supported species or adsorption of reactants. In all the studied cases, no leach of iron or sulfur was detected, revealing the stability of the prepared catalyst systems. Deactivation was mainly due to reactants adsorption. In some cases, a full recovery of the catalyst activity could be obtained by washing with methanol or hot water.

9.6 CONCLUSIONS

The recovery of raw and modified coffee processing by-products for the removal of pollutants from aqueous and gaseous phases is examined. The different raw residues were used as biosorbents for the removal of various heavy metals. Among the various residues, spent coffee grounds were the most efficient for the removal for cadmium ions. Furthermore, the raw materials were applied for dyes removal from aqueous solutions and showed interesting adsorption capacities.

In order to improve the adsorption capacity, several attempts were applied for chars elaboration from coffee processing residues. However, the performance in wastewater treatment was not significantly improved comparing the raw materials. Therefore, a physical and/or a chemical activation steps are required to improve the textural and surface chemistry of the elaborated adsorbents.

Different ACs were prepared using coffee processing residues as a precursor through various protocols of carbonization and activation. These ACs were applied for the removal of pollutants from aqueous and gaseous effluents. The physico-chemical characterization of AC issued from coffee waste evidenced its developed contact surface and porosity, interesting properties in adsorption and catalysis. Furthermore, such materials are environmentally friendly. Moreover, it is economically interesting to use AC prepared from coffee waste as catalysts or sorbents for different pollutants removal; they are issued from the valorization of a waste available in high amounts.

The good adsorption capacities of pollutants in aqueous and gaseous effluents, of AC derived from coffee wastes, are mainly due to the microporosity and to the surface chemistry of these materials. AC materials prepared from coffee grounds exhibit an ease of regeneration and a high stability under the adsorption experimental conditions. The adsorption properties of such ecologically and economically efficient materials should be optimized.

The use of AC as catalyst support was investigated in the literature. Iron supported on AC derived from coffee grounds is promising material in hydrogen peroxide decomposition and organic compounds oxidation; thus, they are useful for wastewater treatment by heterogeneous catalysis. Sulfonated AC issued from coffee wastes are good candidates for esterification and etherification reactions, leading to many interesting intermediates that could be successfully obtained by heterogeneous catalysis rather than homogeneous catalysis.

REFERENCES

Aboud, M.F.A., Alothman, Z.A., Habila, M.A., Zlotea, C., Latroche, M., Cuevas, F., 2015. Hydrogen storage in pristine and d10-block metal-anchored activated carbon made from local wastes. Energies 8, 3578–3590.

Ahmad, M.A., Rahman, N.K., 2011. Equilibrium, kinetics and thermodynamic of remazol brilliant orange 3R dye adsorption on coffee husk-based activated carbon. Chem. Eng. J. 170, 154–161.

Azouaou, N., Sadaoui, Z., Djaafri, A., Mokaddem, H., 2010. Adsorption of cadmium from aqueous solution onto untreated coffee grounds: Equilibrium, kinetics and thermodynamics. J.Haz. Mater. 184, 126–134.

Baek, M.H., Ijagbemi, C.O., O, S.J., Kim, D.S., 2010. Removal of Malachite Green from aqueous solution using degreased coffee bean. J. Hazard. Mater. 176, 820–828.

Belala, Z., Jeguirim, M., Belhachemi, M., Addoun, F., Trouvé, G., 2011. Biosorption of copper from aqueous solutions by date stones and palm-trees waste. Environ. Chem. Lett. 9, 65–69.

Belhachemi, .M., Jeguirim, M., Limousy, L., Addoun, F., 2014. Comparison of NO_2 removal using date pits activated carbon and modified commercialized activated carbon via different preparation methods: effect of porosity and surface chemistry. Chem. Eng. J. 253, 121–129.

Boonamnuayvitaya, V., Chaiya, C., Tanthapanichakoon, W., Jarudilokkul, S., 2004. Removal of heavy metals by adsorbent prepared from pyrolyzed coffee residues and clay. Sep. Purif. Technol. 35, 11–22.

Boonamnuayvitaya, V., Sae-Ung, S., Tanthapanichakoon, W., 2005. Preparation of activated carbons from coffee residue for the adsorption of formaldehyde. Sep. Purif. Technol. 42, 159–168.

Bouchenafa-Saib, N., Mekarzia, A., Bouzid, B., Mohammedi, O., Khelifa, A., Benrachedi, K., Belhanech, N., 2014. Removal of malathion from polluted water by adsorption onto chemically activated carbons produced from coffee grounds. Desalin. Water Treat. 52, 4920–4927.

Boudrahem, F., Aissani-Benissad, F., Aït-Amar, H., 2009. Batch sorption dynamics and equilibrium for the removal of lead ions from aqueous phase using activated carbon developed from coffee residue activated with zinc chloride. J.Environ. Manag. 90, 3031–3039.

Boudrahem, F., Soualah, A., Aissani-Benissad, F., 2011. Pb(II) and Cd(II) removal from aqueous solutions using activated carbon developed from coffee residue activated with phosphoric acid and zinc chloride. J.Chem. Eng. Data 56, 1946–1955.

Bouraoui, Z., Jeguirim, M., Guizani, C., Limousy, L., Dupont, C., Gadiou, R., 2015. Thermogravimetric study on the influence of structural, textural and chemical properties of biomass chars on CO_2 gasification reactivity. Energy 88, 703–710.

Bouraoui, Z., Dupont, C., Jeguirim, M., Limousy, L., Gadiou, R., 2016. CO_2 gasification. of woody biomass chars: the influence of K and Si on char reactivity. Comptes Rendus Chim. 19, 457–465.

Castro, C.S., Guerreiro, M.C., Oliveira, L.C.A., Gonçalves, M., Anastácio, A.S., Nazzarro, M., 2009a. Iron oxide dispersed over activated carbon: support influence on the oxidation of the model molecule methylene blue. Appl. Catal. A General 367, 53–58.

Castro, C.S., Oliveira, L.C.A., Guerreiro, M.C., 2009b. Effect of hydrogen treatment on the catalytic activity of iron oxide based materials dispersed over activated carbon: investigations toward hydrogen peroxide decomposition. Catalysis Lett. 133, 41–48.

Castro, C.S., Abreu, A.L., Silva, C.L.T., Guerreiro, M.C., 2011. Phenol adsorption by activated carbon produced from spent coffee grounds. Water Sci. Technol. 64, 2059–2065.

Cerino-Córdova, F.J., Díaz-Flores, P.E., García-Reyes, R.B., Soto-Regalado, E., Gómez-González, R., Garza-González, M.T., Bustamante-Alcántara, E., 2013. Biosorption of Cu(II) and Pb(II) from aqueous solutions by chemically modified spent coffee grains. Int.J.Environ. Sci. Technol. 10, 611–622.

Ching, S.L., Yusoff, M.S., Abdul Aziz, H., Umar, M., 2011. Influence of impregnation ratio on coffee ground activated carbon as landfill leachate adsorbent for removal of total iron and orthophosphate. Desalination 279, 225–234.

Chouchène, A., Jeguirim, M., Trouvé, G., 2014. Biosorption performance, combustion behavior, and leaching characteristics of olive solid waste during the removal of copper and nickel from aqueous solutions. Clean Tech. Environ. Policy 16, 979–986.

Djilani, C., Zaghdoudi, R., Modarressi, A., Rogalski, M., Djazi, F., Lallam, A., 2012. Elimination of organic micropollutants by adsorption on activated carbon prepared from agricultural waste. Chem. Eng. J. 189–190, 203–212.

Flores-Cano, J.V., Sanchez-Polo, M., Messoud, J., Velo-Gala, I., Ocampo-Perez, R., Rivera-Utrilla, J., 2016. Overall adsorption rate of metronidazole, dimetridazole and diatrizoate on activated carbons prepared from coffee residues and almond shells. J.Environ. Manag. 169, 116–125.

Franca, A.S., Oliveira, L.S., Ferreira, M.E., 2008. Kinetics and equilibrium studies of methylene blue adsorption by spent coffee grounds. J. Haz. Mater. 155, 507–512.

Getachew, T., Hussen, A., Rao, V.M., 2015. Defluoridation of water by activated carbon prepared from banana (*Musa paradisiaca*) peel and coffee (*Coffea arabica*) husk. Int. J. Environ. Sci. Tech. 12, 1857–1866.

Ghouma, I., Jeguirim, M., Dorge, S., Limousy, L., Matei Ghimbeu, C., Ouederni, A., 2015. Activated carbon prepared by physical activation of olive stones for the removal of NO_2 at ambient temperature. Comptes Rendus Chim. 18, 63–74.

Giraldo, L., Moreno-Pirajan, J.C., 2012. Synthesis of activated carbon mesoporous from coffee waste and its application in adsorption zinc and mercury ions from aqueous solution. E-J. Chem. 9, 938–948.

Gonçalves, M., Guerreiro, M.C., Oliveira, L.C.A., Castro, C.S., 2013a. A friendly environmental material: iron oxide dispersed over activated carbon from coffee husk for organic pollutants removal. J.Environ. Manag. 127, 206–211.

Gonçalves, M., Guerreiro, M.C, Ramos, P.H, Alves De Oliveira, L.C., Sapag, K., 2013b. Activated carbon prepared from coffee pulp: potential adsorbent of organic contaminants in aqueous solution. Water Sci. Tech. 68, 1085–1090.

Gonçalves, M., Soler, F.C., Isoda, N., Carvalho, W.A., Mandelli, D., Sepúlveda, J., 2016. Glycerol conversion into value-added products in presence of a green recyclable catalyst: acid black carbon obtained from coffee ground wastes. J. Taiwan Inst. Chem. Eng. 60, 294–301.

González, A.S., Plaza, M.G., Pis, J.J., Rubiera, F., Pevida, C., 2013. Post-combustion CO_2 capture adsorbents from spent coffee grounds. Energy Procedia 37, 134–141.

Guizani, C., Jeguirim, M., Esceduro Sainz, J., Gadiou, R., Salvador, S., 2016. Biomass char gasification by H_2O, CO_2 and their mixture: evolution of chemical, textural and structural properties of the chars. Energy 112, 133–145.

Jung, K.W., Choi, B.H., Hwang, M.J., Jeong, T.U., Ahn, K.H., 2016. Fabrication of granular activated carbons derived from spent coffee grounds by entrapment in calcium alginate beads for adsorption of acid orange 7 and methylene blue. Biores. Tech. 219, 185–195.

Jutakridsada, P., Prajaksud, C., Kuboonya-Aruk, L., Theerakulpisut, S., Kamwilaisak, K., 2016. Adsorption characteristics of activated carbon prepared from spent ground coffee. Clean Tech. Environ. Policy 18, 639–645.

Kaikake, K., Hoaki, K., Sunada, H., Dhaka, l, R.P., Baba, Y., 2007. Removal characteristics of metal ions using degreased coffee beans: Adsorption equilibrium of cadmium(II). Biores. Tech. 98, 2787–2791.

Kante, K., Nieto-Delgado, C., Rangel-Mendez, J.R., Bandosz, T.J., 2012. Spent coffee-based activated carbon: specific surface features and their importance for H_2S separation process. J. Hazard. Mater. 201–202, 141–147.

Kemp, K.C., Baek, S.B., Lee, W.-G., Meyyappan, M., Kim, K.S., 2015. Activated carbon derived from waste coffee grounds for stable methane storage. Nanotech. 26, 385602–385610.

Khenniche, L., Aissani, F., 2009. Characterization and utilization of activated carbons prepared from coffee residue for adsorptive removal of salicylic acid and phenol: kinetic and isotherm study. Desalin. Water Treat. 11, 192–203.

Khenniche, L., Aissani, F., 2010a. Adsorptive removal of phenol by coffee residue activated carbon and commercial activated carbon: equilibrium, kinetics and thermodynamics. J. Chem. Eng. Data 55, 4677–4686.

Khenniche, L., Aissani, F., 2010b. Preparation and characterization of carbons from coffee residue: adsorption of salicylic acid on the prepared carbons. J. Chem. Eng. Data 2010, 728–734.

Kyzas, G.Z., 2012. A decolorization technique with spent "Greek Coffee" grounds as zero-cost adsorbents for industrial textile wastewaters. Materials 5, 2069–2087.

Lamine, S.M., Ridha, C., Mahfoud, H.M., Mouad, C., Lofti, B., Al-Dujaili, A.H., 2014. Chemical activation of an activated carbon prepared from coffee residue. Energy Procedia 50, 393–400.

Ma, X., Ouyang, F., 2013. Adsorption properties of biomass-based activated carbon prepared with spent coffee grounds and pomelo skin by phosphoric acid activation. Appl. Surf. Sci. 268, 566–570.

Minamisawa, M., Nakajima, S., Minamisawa, H., Yoshida, S., Takai, N., 2005. Removal of Copper(II) and Cadmium(II) from Water Using Roasted Coffee Beans. In: Lichtfouse, E., Schwarzbauer, J., Robert, D. (Eds.), Environmental Chemistry: Green Chemistry and Pollutants in Ecosystems. Springer-Verlag, Berlin Heidelberg, Germany, pp. 259–265.

Nakamura, T., Tokimoto, T., Tamura, T., Kawasaki, N., tanada, S., 2003. Decolorization of acidic dye by charcoal from coffee grounds. J. Health Sci. 49, 520–523.

Namane, A., Mekarzia, A., Benrachedi, K., Belhaneche-Bensemra, N., Hellal, A., 2005. Determination of the adsorption capacity of activated carbon made from coffee grounds by chemical activation with $ZnCl_2$ and H_3PO_4. J.Haz. Mater. B 119, 189–194.

Ngaosuwan, K., Goodwin, J.G., Prasertdham, P., 2016. A green sulfonated carbon-based catalyst derived from coffee residue for esterification. Renew. Energy 86, 262–269.

Nowicki, P., Skibiszewska, P., Pietrzak, R., 2013. NO_2 removal on adsorbents prepared from coffee industry waste materials. Adsorption 19, 521–528.

Nowicki, P., Kazmierczak-Razna, J., Skibiszewska, P., Wiśniewska, M., Nosal-Wiercińska, A., Pietrzak, R., 2016. Production of activated carbons from biodegradable waste materials as an alternative way of their utilisation. Adsorption 22, 489–502.

Nunes, A.A., Franca, A.S., Oliveira, L.S., 2009. Activated carbons from waste biomass: an alternative use for biodiesel production solid residues. Biores. Tech. 100, 1786–1792.

Oh, S.Y., Seo, Y.D., 2016. Sorption of halogenated phenols and pharmaceuticals to biochar: affecting factors and mechanisms. Environ. Sci. Pollut. Res. 23, 951–961.

Oliveira, L.S., Franca, A.S., Alves, T.M., Rocha, S.D.F., 2008a. Evaluation of untreated coffee husks as potential biosorbents for treatment of dye contaminated waters. J.Hazard. Mater. 155, 507–512.

Oliveira, W.E., Franca, A.S., Oliveira, L.S., Rocha, S.D., 2008b. Untreated coffee husks as biosorbents for the removal of heavy metals from aqueous solutions. J.Haz. Mater. 152, 1073–1081.

Oliveira, L.C.A., Pereira, E., Guimaraes, I.R., Vallone, A., Pereira, M., Mesquita, J.P., Sapag, K., 2009. Preparation of activated carbons from coffee husks utilizing $FeCl_3$ and $ZnCl_2$ as activating agents. J.Haz. Mater. 165, 87–94.

Pavlovic, M.D., Buntic, A.V., Šiler-Marinkovic, S.S., Antonovic, D.G., Dimitrijević-Brankovic, S.I., 2015. Recovery of (−)-epigallocatechingallate (EGCG) from aqueous solution by selective adsorption onto spent coffee grounds. Eur. Food Res. Technol. 241, 399–412.

Pechyen, C., Aht-Ong, D., Sricharoenchaikul, V., Atong, D., 2010. Optimization of manufacturing conditions for activated carbon from coffee (*Coffea Arabica* L.) bean waste by chemical activation. Mater. Sci. Forum 658, 113–116.

Plaza, M.G., González, A.S., Pevida, C., Pis, J.J., Rubiera, F., 2012. Valorisation of spent coffee grounds as CO_2 adsorbents for postcombustion capture applications. Appl. Energy 99, 272–279.

Plaza, M.G., González, A.S., Pevida, C., Rubiera, F., 2015. Green coffee based CO_2 adsorbent with high performance in postcombustion conditions. Fuel 140, 633–648.

Plaza, M.G., González, A.S., Rubiera, F., Pevida, C., 2016. Water vapour adsorption by a coffee-based microporous carbon: effect on CO_2 capture. J. Chem. Tech. Biotechnol. 90, 1592–1600.

Reffas, A., Bernardet, V., David, B., Reinert, L., Bencheikh Lehocine, M., Dubois, M., Batisse, N., Duclaux, L., 2010. Carbons prepared from coffee grounds by H_3PO_4 activation: characterization and adsorption of methylene blue and Nylosan Red N-2RBL. J. Haz. Mater. 175, 779–788.

Rufford, T.E., Hulicova-Jurcakova, D., Zhu, Z., Lu, G.Q., 2008. Nanoporous carbon electrode from waste coffee beans for high performance supercapacitors. Electrochem. Commun. 10, 1594–1597.

Tokimoto, T., Kawasaki, N., Nakamura, T., Akutagawa, J., Tanada, S., 2005. Removal of lead ions in drinking water by coffee grounds as vegetable biomass. J. Colloid Interf. Sci. 281, 56–61.

Travis, W., Gadipelli, S., Guo, Z., 2015. Superior CO_2 adsorption from waste coffee ground derived carbons. RSC Adv. 5, 29558–29562.

Yeung, P.T., Chung, P.Y., Tsang, H.C., Tang, J.C.O., Cheng, G.Y.M., Gambari, R., Chui, C.H., Lam, K.H., 2014. Preparation and characterization of bio-safe activated charcoal derived from coffee waste residue and their application for removal of lead and copper ions. RSC Adv. 4, 38839–38847.

The potential of pyrolysing exhausted coffee residue for the production of biochar

10

Wen-Tien Tsai

National Pingtung University of Science and Technology, Pingtung, Taiwan

ABSTRACT

Coffee may be the most popular beverage brewed from roasted bean. However, its producing generates various by-products, such as pulp, husk, sliver silk, mucilage, and spent coffee grounds (SCG). Among them, SCG (known as exhausted coffee residue) is significantly generated from soluble coffee factories and commercial coffee shops. This chapter explores the possibilities of utilizing SCG as a precursor for the production of biochar via pyrolysis. In addition, its potential agricultural and fabric applications are also reviewed taking into account that it is a carbon-rich and nutrient-enriched material. Initially, the thermochemical conversion of biomass into various energy forms and the biochar characterization are described. The exhausted coffee residue and its potential utilization for the production of various energy forms were then addressed. The pyrolysis process for the sustainable management of exhausted coffee residue is further reviewed. Finally, the thermochemical properties of exhausted coffee residue and its resulting biochar are summarized.

Keywords: exhausted coffee residue; pyrolysis; biochar; thermochemical property; agricultural use

10.1 INTRODUCTION

The reuse of agricultural residues and food processing by-products as valuable materials, energy sources or alternative fuels for the replacement of fossil-based feedstocks has received much attention since the 1990s (Basu, 2013). This transition from nonrenewable resources to renewable resources could be connected with the economic consideration (e.g., cost down), energy supply (e.g., biogas and biodiesel), and environmental and agricultural issues (e.g., global warming) (Lehmann and

Handbook of Coffee Processing By-Products. http://dx.doi.org/10.1016/B978-0-12-811290-8.00010-4

Joseph, 2015). The energy supply from domestic biomass resources not only enhances fuel diversification, but also reduces the air pollution because the biomass resource contains relatively low contents of sulfur and heavy metals in comparison with fossil fuels, such as coal (Klass, 1998). These bioresources may contain large amounts of lignocellulosic constituents (i.e., cellulose, hemicellulose, and lignin) or organic elements (i.e., carbon, hydrogen, oxygen, and nitrogen), thus featuring the characteristics of carbon-rich, high-energy, and low-ash substrates. Therefore, biomass materials contain available carbon and hydrogen sources, which can be utilized for the production of biomass energy based on benefits of both energy utilization and resource recycling. For these reasons, there is an increasing interest in cofiring biomass with coal or oil in the traditional power plants and industrial utility boilers (Van Loo and Koppejan, 2008).

Coffee is one of the most important agricultural commodities in the world because it is a popular beverage brewed from roasted bean. Coffee trees generally are planted in tropical regions where the climate is abundant in rainfall and sunshine. Regarding the cross-structure of a coffee cherry (Acchar and Dultra, 2015; Bressani, 1979; Franca and Oliveira, 2009; Murthy and Naidu, 2012; Narita and Inouye, 2014), green coffee beans exist inward and are sequentially covered by a thin seed skin (silverskin), an endocarp (parchment or coffee hull), a pectic adhesive layer (mucilage), pulp, and epicarp (outer skin). After harvesting, the coffee berries are transported to the processing plant. For the purification and separation purposes, two methods (i.e., wet method and dry method) are generally used to obtain the commercial coffee beans depending on the operation complexity and the quality of the resultant raw coffee (Bressani, 1979; Murthy and Naidu, 2012). These processing by-products consist of about 50% of coffee cherry by weight. Due to the contents of functional ingredients in these resources, they might be valuable materials for several purposes (Bonilla-Hermosa et al., 2014; Esquivel and Jimenez, 2012; Franca and Oliveira, 2009; Kim et al., 2014; Murthy and Naidu, 2012; Narita and Inouye, 2014; Taylor and Antonio, 2006), including livestock feed, silage, fuel, fermented products (e.g., enzyme, citric acid, and flavoring substances), feedstocks for the production of biogas and mushroom, biosorbents, soil amendments (organic fertilizers), and other value-added compounds. Fig. 10.1 shows the flow sheet for the production of coffee bean and its resulting by-products, including pulp, husk, sliver silk, mucilage, and spent coffee grounds (SCGs) (Acchar and Dultra, 2015; Campos-Vega et al., 2015; Murthy and Naidu, 2012). In addition, green coffee grounds (also called exhausted coffee residue) are further processed during the brewing process, where SCGs are jointly generated. On average, about 2 kg of wet SCGs are obtained to each kilogram of soluble coffee produced (Murthy and Naidu, 2012).

As mentioned earlier, coffee-derived industries generate significant amounts of processing by-products that can be considered as valuable bioresources due to their abundance in the contents of carbohydrates (e.g., cellulose and hemicellulose), functional components (e.g., proteins and pectins), and bioactive compounds like polyphenols. With high coffee beverage consumption, there is an urgent need for the development of sustainable management of coffee by-products and applications.

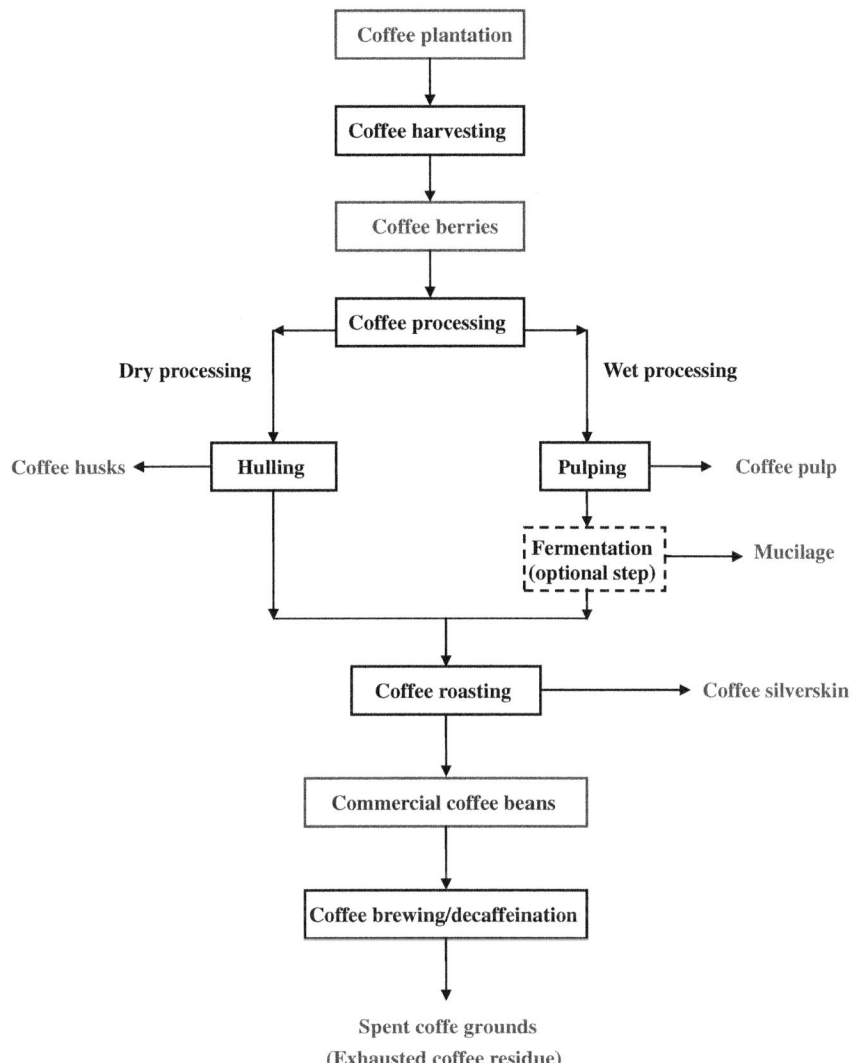

FIGURE 10.1 Flow Sheet for Illustrating the Production of Various By-Products From Coffee Processing

Therefore, several researches have focused on the recovery of fine chemicals and the production of precious metabolites via chemical and biotechnological processes (Campos-Vega et al., 2015; Murthy and Naidu, 2012), but not commercially operated. The current chapter shows the practical possibilities of utilizing exhausted coffee residue as a precursor for the production of biochar via pyrolysis. In addition, its potential agricultural applications are reviewed, taking into account that it is a carbon-rich and nutrient-enriched material.

10.2 BRIEF DESCRIPTION OF BIOMASS CONVERSION INTO BIOCHAR BY PYROLYSIS

10.2.1 COMPOSITION OF BIOMASS

Biomass generally refers to any organic materials (living organisms and their residues) that are derived from microbes, plants (including algae), or animals. According to the definition by the United Nations Framework Convention on Climate Change (UNFCCC), it includes products (e.g., food crops and energy crops), by-products, residues and waste from agriculture (e.g., rice straw, bagasse, corn stalk, and fruit tree trimmings), forestry (e.g., wood, sawdust, and timber), livestock (e.g., manure from cattle, poultry, and hogs), and related industries, as well as the nonfossilized and biodegradable organic fractions of industrial and nonindustrial (municipal) wastes, such as sewage sludge, yard clippings, food waste, waste paper, and food-processing residues. Due to its lignocellulosic compositions, it has become a promising feedstock for a variety of renewable chemicals and fuels. It is well known that lignocellulosic materials mainly contain a mixture of biopolymers and others, which are simply described next (Basu, 2013; Klass, 1998):

- Cellulose: It is a long-chain linear polymer composed of glucose, making up the primary structural component in the cell wall of biomass, and providing the skeletal structure of most terrestrial biomass.
- Hemicellulose: It is another constituent involved in the cell walls of a plant, which is a group of heterogeneous polysaccharides containing both 6-carbon and 5-carbon sugars.
- Lignin: It is a complex and highly branched polymer with phenylpropane units linked in a three-dimensional matrix, ranking the third important constituent of the cell walls of woody biomass.
- Others: They include extractives and ash. The former refers to protein, oil (or fat), starch, and sugar, which can be separated from organisms' tissue. The latter is the inorganic components of the biomass. The primary ingredients of ash are silica, aluminum, and oxides of iron, calcium, magnesium, potassium, titanium, and sodium.

10.2.2 FEATURES OF BIOMASS

The reuse of biomass as renewable chemicals, energy sources, or alternative fuels in replacement of fossil-based feedstocks has received much attention since the early 1990s. This transition from nonrenewables to renewables could be attributed to the energy crisis and environmental issues, such as global warming and ambient air pollution. For example, the biomass contains relatively low contents of sulfur and heavy metals in comparison with fossil fuels, such as coal (Klass, 1998). In this respect, agricultural wastes or residues contain large amounts of lignocellulosic constituents (i.e., cellulose, hemicellulose, and lignin), thus possessing high heating values when dry. Therefore, there is an increasing interest in cofiring biomass with coal in the power plants and industrial utility boilers (Demirbas, 2005), which have been well designed to control the emissions of toxic air pollutants from the flue stack. However, biomass inherently has the following features or disadvantages (Basu, 2013; Chen, 2015):

- High moisture: The moisture content of biomass generally ranges from 30 to 60 wt.%, depending to its type, location source, time of harvest, period of storage after harvest, and methods of processing or pretreatment.
- Low heating value: It is due to the high moisture and low bulk density as compared to fossil fuels, such as coal and petrol oil. Therefore, its volume may be large or different in forms.
- High hygroscopicity: This is inherently derived from its lignocellulosic constituents because moisture can be absorbed into the biomass due to the hydroxyl groups of cell wall components.
- High heterogeneity: It means that the specific properties of biomass can vary significantly due to its different sources (e.g., species, location, and plantation).

As a result, biomass cannot be handled economically, stored, or transported easily, forming the barriers to transition from fossil to biomass resources. Therefore, a number of pretreatment and conversion methods have been developed to upgrade its energy density and handle, or conveniently store it during the use of biomass for the production of chemicals, fuels, or energy.

10.2.3 PRETREATMENT OF BIOMASS

Due to the heterogeneity and complexity of the lignocellulosic biomass in nature, biomass has to be pretreated to make it easy for enzyme action or higher hydrophobicity to directly combust. These methods include a variety of physical (e.g., comminution and grinding), chemical (e.g., acid, alkali, ozone, and ionic liquid), physicochemical (e.g., steam explosion and ammonia explosion), and biological techniques for the disruption of the naturally resistant biopolymer shield that limits the accessibility of enzymes to cellulosic fibers (Binod and Pandey, 2015). However, the latter is a mild pyrolysis process (i.e., torrefaction process) that adopts relatively low temperature (200–300°C) to produce solid fuels with higher hydrophobicity and lower oxygen content as compared to the feedstock biomass. It should be noted that the torrefied biomass has been used as an auxiliary fuel in replacement of solid fuels (e.g., coal and densified biomass pellets or briquettes) in the boilers and utilities because of its calorific value ranging from 5500 to 6500 kcal/kg (Basu, 2013; Chen, 2015).

10.2.4 CONVERSION OF BIOMASS

To make the biomass more energy dense and convenient during handling and storage, its conversion has to be achieved through one of two major routes (Basu, 2013):

1. Biochemical conversion

   Traditionally, biomass has been biochemically converted into biogas (rich in methane) for local energy use by anaerobic microbial digestion of animal manures. In the modern society, bioethanol or wine is produced from starch-/sugar-rich feedstocks (e.g., rice, corn, and sugarcane) using fermentation processes. In biochemical conversion, biomass macromolecules will be broken down smaller molecules (e.g., hexoses, pentoses, acetic acid, ethanol, and methane) by bacteria (e.g., yeast) or enzymes. Although this process is much slower than the thermochemical conversion processes, it does not require much external energy due to its operation at normal conditions. There are three principle routes for biochemical conversion: digestion, fermentation, and enzymatic (or acid hydrolysis).

2. Thermochemical conversion

As described earlier, biomass is a complex heterogeneous material mainly composed of cellulose, hemicellulose, lignin, and other organics (primarily carbon C, hydrogen H, nitrogen N, oxygen O, and sulfur S). The energy involved in biomass and stored in plants is originally from solar energy via photosynthesis. As a consequence, biomass can be thermochemically converted into heat (steam or power) via direct combustion or cofiring, and a variety of carbon-rich solids and gases, which are then converted into the desired chemicals. For example, the Fischer–Tropsch (FT) synthesis process has been applied to convert syngas (mainly composed of hydrogen and carbon monoxide) into liquid transport fuels (e.g., ethanol and methanol). Generally, there are five principle routes for thermochemical conversion: combustion, carbonization/torrefaction, pyrolysis, gasification, and liquefaction (Basu, 2013; Klass, 1998).

10.2.5 PYROLYSIS OF BIOMASS

As mentioned earlier, pyrolysis involves thermal decomposition of biomass or other feedstocks in the absence of air or oxygen at an optimal temperature for a specified holding time that will be converted into three main products:

- Noncondensable gases: They are mainly composed of lower molecular weight gases, such as carbon dioxide (CO_2), carbon monoxide (CO), methane (CH_4), acetylene (C_2H_2), ethylene (C_2H_4), and benzene (C_6H_6).
- Solid char: It is a carbon-rich material, also called charcoal or biochar. Aside from the carbon content, it also contains some oxygen, hydrogen, and residual minerals (inorganic ash). The calorific value of biomass-derived char may be close to or higher than that of coal. In addition, it may feature high specific surface area (SSA) due to the devolatilization and condensation reactions during the pyrolysis process.
- Condensable liquid products.

The liquid product, also known as tar, biooil, or biocrude, is a black slurry mainly composing of hydrocarbons and water. The yield and composition of a biooil product depends on the physical and chemical properties of biomass feedstock, and several operating parameters including heating rate, pyrolysis temperature, residence time in the reaction zone, and other factors (e.g., system pressure, oxygen concentration in the reaction zone, and the presence of catalysts).

Regarding the effect of biomass composition on the pyrolysis yield, cellulose and hemicellulose are the main sources of resulting volatiles, which are the primary sources of condensable and noncondensable vapors, respectively. Due to its aromatic structure, however, lignin degrades more slowly over a broader temperature range (250–500°C), thus making a major contribution to the char yield. On the other hand, the physical structure of the biomass also has some influence on the composition and yields of pyrolysis products. For example, fine biomass particles offer less diffusion resistance to the release of condensable gases from intraparticles, thus resulting in higher yield of biooil. Large biomass particles, on the other hand, can facilitate secondary cracking reaction due to the higher diffusion resistance. In a tentative pyrolysis design, there are three operation types for maximizing product yield:

- Adopt the operation mode with a slow heating rate ($<2°C/s$), a low pyrolysis (final) temperature and a long gas (vapor) residence time for maximizing char yield.
- Adopt the operation mode with a high heating rate, a moderate pyrolysis (final) temperature (450–600°C), and a short gas (vapor) residence time for maximizing biooil yield.
- Adopt the operation mode with a slow heating rate, a high pyrolysis (final) temperature (700–900°C), and a long gas (vapor) residence time for maximizing gas yield.

In general, the complex compounds found in the condensable liquid product, which is a mixture of complex hydrocarbons with large fractions of water and organic oxygen, can be categorized into the following types: hydroxyaldehydes, hydroxyketones, carboxylic acids (e.g., formic acid and acetic acid), phenolic compounds (e.g., phenol and benoquinone), alcohols, and polycyclic aromatic compounds.

Pyrolysis process, depending on the heating rate, can generally be classified as slow and fast types. The former is usually used for the production of biocoal, char, biochar, or charcoal at slower heating rates (usually less than 50°C/min). So, the residence time of resulting vapor in the pyrolysis reaction zone must hold on the order of minutes or longer. As described in Section 10.2.4, it can be further grouped into two types: torrefaction and carbonization. Torrefaction is a pretreatment process performed at a relatively low and narrow temperature range (200–300°C), whereas carbonization takes place at a higher and broader temperature range (400–800°C). In the torrefaction process, its main features can be briefly summarized as follows (Basu, 2013):

- To increase the O/C ratio of the torrefied biomass, which can improve its energy efficiency when the torrefied biomass is used as a feedstock in the gasification process.
- To reduce power requirements for size reduction, thus improving its handling.
- To provide a cleaner fuel with little acid in the smoke when the torrefied biomass is used as a feedstock in the combustion process.
- To reduce the moisture adsorbed in the torrefied biomass when it is stored.

- To obtain a fuel gas with higher heating value (HHV) when the torrefied biomass is used as a feedstock in the gasification process.
- To produce high-quality solid fuel because the torrefied biomass has a higher volumetric energy density.

Carbonization process often provides a longer time for converting the condensable vapor into char and noncondensable gases, thus upgrading the yield of char as a result of secondary cracking reaction. On the other hand, the residence time of resulting vapor in the fast pyrolysis is on the order of seconds or less because it is intended for the production of biooil and gas. Herein, the heating rate may be as high as 1000°C/s, but the peak pyrolysis temperature can be either below 650°C for the production of target biooil, or up to 1000°C for the production of target gas. In brief, there are four significant features in the fast pyrolysis process that increase the biooil yield:

- Very high heating rate.
- Moderate reaction temperature range (425–600°C).
- Short residence time of vapor in the pyrolysis reactor (<3 s).
- Rapid quenching of the product gas using cryogenic condensation.

10.3 BIOCHAR CHARACTERIZATION

According to the International Biochar Initiative (http://www.biochar-international.org/), biochar is defined as "A solid material obtained from thermochemical conversion of biomass in an oxygen-limited environment." As mentioned earlier, biochar is produced by pyrolysis process, where it is a solid by-product or residue. Also, its characterization is dependent on the biomass type and operation parameters, including final temperature, heating rate, residence time, system pressure, and presence or absence of oxygen. Regarding the characterization of biochar used as a soil amendment, the International Biochar Initiative (IBI) has published a manual (Standardized Product Definition and Product Testing Guidelines for Biochar) in 2015 that included a working definition of biochar and a set of standardized methods for biochar analysis. These methods can be used as initial guidelines for analysis of physical and chemical properties of biochar.

10.3.1 PHYSICAL PROPERTIES OF BIOCHAR

10.3.1.1 Specific surface area, pore volume, and pore size distribution
The physical properties of biochar generally refer to the pore properties, including SSA, pore volume, pore size distribution, and average pore diameter (or width) (Lowell et al., 2006). Among them, SSA (total surface area per gram; unit: m^2/g) should be the most important parameter. The SSA of biochar is linked to its pore volume and pore size distribution. According to the definition by the International Union of Pure and Applied Chemistry (IUPAC), there are three main pore size ranges, given below:

- Micropores: Pore diameter or width <2.0 nm.
- Mesopores: Pore diameter or width 2.0–50.0 nm.
- Macropores: Pore diameter or width >50.0 nm.

In fact, the SSA measurement of porous particle (adsorbent) involves the theory of physical adsorption of gas, where inert gas nitrogen (N_2) is usually used as a probe molecule (adsorbate) in its equilibrium coverage (also called adsorption isotherm). Due to the van der Waals force increased with the pressure of adsorbate (measured by relative pressure P/P_0, where P is the adsorbate pressure, and P_0 is the saturation pressure at measurement temperature 77K), the volume adsorbed will be quickly increased from monolayer coverage to multilayers at equilibrium.

- Specific (BET) surface area: In the standard method for measuring SSA, the amount adsorbed by monomolecular coverage must be identified in order to determine the SSA. A theoretical model first developed by Brunauer, Emmett, and Teller has been used to form the famous BET equation, which is valid up to about $P/P_0 = 0.35$.
- Total pore volume: To calculate the total pore volume, it is necessary to measure the total adsorbed volume at a relative pressure close to unity (i.e., $P/P_0 = 1.0$). If W_s g of nitrogen (its liquid density is ρ_L, 0.808 g/cm^3 at 77K) is adsorbed at $P/P_0 = 0.99$, its total pore volume (V_t) can be given by the ratio of W_s to ρ_L.
- Average pore diameter: Herein, all the pores are assumed to be straight, cylindrical, not interconnected, and have the same diameter D_a and length L. The average pore diameter of particles, if there are n pores per particle with the mass m_p, can be roughly calculated by writing equations for its total SSA (i.e., BET surface area, S_{BET}) and total pore volume (V_t) (Smith, 1981):

$$m_p \times S_{BET} = (\pi \times D_a \times L) \times n$$

$$m_p \times V_t = (1/4 \times \pi \times D_a^2 \times L) \times n$$

Dividing the two equations, average pore diameter (D_a) can be given as:

$$D_a = 4 \times (V_t / S_{BET})$$

- Pore size distribution: The most common methods for determining the pore size distribution are based on nitrogen adsorption–desorption isotherm. In the nitrogen adsorption–desorption method, all pores are filled with adsorbed and condensed nitrogen at $P/P_0 \rightarrow 1.0$. As the vapor pressure of liquid nitrogen evaporating from a capillary depends on the size of the capillary, the data on the desorption isotherm (i.e., volume adsorbed vs. pore size) can be plotted to give the pore size distribution. It should be noted that the nitrogen adsorption–desorption method is not suitable for pores larger than 20.0 nm because the Kelvin equation is not applied effectively (Smith, 1981).

10.3.1.2 Densities

It is well known that density is defined as the ratio of mass to volume. However, the volume of porous particles may be confused with the space it occupies. In general, two types of density are used to describe biochars: true density and bulk density. The bulk density of the biochar and its porosity could be decreased at very high pyrolysis temperature due to the collapse of the resulting pores. Typical true densities for biochars are around 2.0 g/cm^3, depending on both biomass feedstock and pyrolysis conditions. In general, true density will increase with the increase of pyrolysis temperature and longer residence time at a specified pyrolysis temperature (Chia et al., 2015). This should be associated with the structural conversion of low-density disordered carbon present in the biomass feedstock to turbostratic carbon in the resulting biochar. This result has been reported by Tsai et al. (2012), where the values of the true densities (i.e., 1.520 ± 0.009, 1.720 ± 0.019, 1.843 ± 0.023, and 2.030 ± 0.038 for 10 measurements) proportionally increased with increase in the pyrolysis temperature (i.e., 673, 773, 873, and 973K, respectively) in the case of biochars resulting from the pyrolysis of the exhausted coffee residue. More significantly, a higher true density of biochar is generally consistent with its stronger mechanical property because of more organized structure and higher concentrations of crystalline phases in the biochars. It should be noted that the data on true densities for biochars are normally measured with helium pycnometry.

> Particle size: Particle size (including average particle size and particle size distribution) is also a factor in determining the bulk density because the volume in its measurement includes the void volume between particles. For these reasons, standard screens (i.e., US sieve size) are often used in determining the particle size distribution of a granular material or powder in the size range up to 400 mesh (37 μm or 0.0015 in.). The higher the mesh number, the smaller the screen opening. Practically, if the particle size of a powder is noted with a mesh number range, such as 10/30, its means that its size range is smaller than number10 mesh no. (opening size 2 mm) and larger than number 30 mesh no. (opening size 0.6 mm). Thus, the notation 10/30 means "through number10 mesh no. and on number 30 mesh no." As a result, its average particle size is about 1.3 mm; that is, $(2 + 0.6)/2 = 1.3$ mm.

10.3.2 CHEMICAL PROPERTIES OF BIOCHAR

10.3.2.1 Ultimate analysis (elemental analysis)

Herein, the ultimate analysis of biomass or its resulting biochar is basically expressed in terms of its organic elements except for its moisture and inorganic constituents (derived from ash). In general, carbon (C), hydrogen (H), oxygen (O), nitrogen (N), and sulfur (S) are involved in determining their mass percentages on a dry, ash-free basis. Sometimes, organic chlorine (Cl) is also involved in the elemental analysis, but often it is ignored because there are only trace amount in most cases. In fact, these organic elements contribute to the compositions of volatile matter and fixed carbon in biomass or biochar. Proximate analysis, another chemical property of biochar not discussed here, gives its composition in terms of moisture, volatile matter, ash, and fixed carbon. It should be noted that not all biochars contain all of these organic elements. For example, the vast majority of wood-derived biochar may not contain sulfur (S). In this respect, biochar, in addition to being neutral in carbon dioxide (CO_2) emission, will reduce the emission of sulfur oxides (SO_x) from the combustion of biochar derived from wood biomass. Due to the analytical difficulty in the determination of O element, its content (mass percentage) can be calculated by the difference.

- Heating value: The HHV (also called gross calorific value) of common biomass generally ranges from 15 to 20 MJ/kg, which is lower than that of coal in the range of 25–35 MJ/kg. This value refers to the heat released from the sample combustion at an adiabatic oxygen bomb calorimeter with the original and generated water in a condensed state. However, the measurement of heating value poses a slightly complicated and time-consuming process, suggesting that many attempts have been studied to calculate it with correlating the data of proximate analysis and ultimate analysis (Sheng and Azevedo, 2005). It was found that the correlations based on the ultimate analysis seemed to be the most accurate. For example, the formula for calculating HHV with more than 90% predictions in the range of ±5% error, is based on the contents of main organic elements (in wt.%) C, H, and O (Sheng and Azevedo, 2005):

$$HHV\,(MJ/kg) = -1.3675 + 0.3137\,C + 0.7009\,H + 0.0318\,O$$

Herein, O is obtained by difference, which is the sum of the contents of oxygen and other elements in the sample.
- Inorganic (mineral) elemental analysis: In general, increasing pyrolysis temperature tends to increase the pH of biochar due to the increase in ash (or inorganic constituents), but decrease cation exchange capacity (CEC). As a result, biochar has been extensively reused as a soil improvement for upgrading crop productivity through nutrients available. Except for carbon (C), nitrogen (N), and sulfur (S), these nutrients in biochar include potassium (K), phosphorus (P), sodium (Na), calcium (Ca), magnesium (Mg), cobalt (Co), iron (Fe), silicon (Si), aluminum (Al), and so on (Ippolito et al., 2015). It should be noted that some toxic heavy metals (i.e., lead, nickel, cadmium, chromium, zinc, and copper) and inorganic elements [i.e., arsenic (As), mercury (Hg), and selenium (Se)] are of concern because of their hazards in the soil environment. In the measurement of inorganic elements, an inductively coupled plasma–optical emission spectrometer (ICP-OES) was commonly used to determine simultaneously the contents of relevant elements, including nutrients and toxic inorganic elements.
- pH: It has been shown that most biochars are alkaline (i.e., pH > 7.0). This could be attributed to the fact that biochar produced at high pyrolysis temperature will lose more acidic functional groups (e.g., quinone, chromene, or diketone groups) and remain the ash content increased, thus causing it to be more basic. More consistently, increasing pyrolysis temperature will result in

increased nutrients (ash) in mineral salt forms with oxide, carbonate, hydroxide, or chloride (such as KOH, NaOH, $CaCO_3$, and $MgCO_3$) and elevated pH values. Due to its basicity, biochar has been used to mitigate acidic soil conditions, raising the pH of acidic soil due to liming effect and also affecting the mobility of cation in the soil. As a consequence, biochar can be used as a liming agent, which is commonly applied to the soil according to its calcium carbonate equivalency. More significantly, raising the pH value of soil will result in increased base saturation and, thus, microbial activity is increased.

- Cation exchange capacity: CEC is a quantitative measure of exchangeable cations (e.g., Ca^{2+}, Mg^{2+}, K^+, and Na^+), representing the ability to electrostatically adsorb or attract cations in soil water. Therefore, it is indicative of the soil quality and productivity. In general, higher CEC values in the manure-based biochars were observed as compared to the plant-based biochars (Lee et al., 2010). Also, increasing pyrolysis temperature will tend to induce a decrease in the CEC of biochar due to the greater removal of organic functional groups at higher processing temperature. However, temperature effects are not consistent, suggesting that different mechanisms have been proposed for explaining the two contradicting trends (Mukome and Parikh, 2016). Due to the negatively charged surface of biochar, the addition of biochar to soil as an amendment agent has been shown to raise the soil's CEC values, meaning greater retention of the soil cations through electrostatic interactions.

- Electrical conductivity: Electrical conductivity (EC) is a measure for describing the ability of a material to conduct an electrical current. As a consequence, the EC value of biochar is highly related to its ash content and elemental compositions. In general, the greater EC values are associated with higher amounts of soluble salts present in the manure-derived biochars due to their significant ash contents.

In summary, biochars exhibit highly complex and heterogeneous structures composed of mineral (oxide, carbonates, chlorides, silicates, etc.) and organic fractions (aromatic C, carboxyl C, carbonyl C, etc.). The chemical properties of these biochars, including pH, EC, CEC, functional groups, and elemental composition, mainly depend on the biomass feedstock type and the production conditions during the pyrolysis process.

10.4 EXHAUSTED COFFEE RESIDUE AND ITS REUSE AS AN ENERGY SOURCE

Exhausted coffee residue (sometimes called SCGs), is inevitably generated from soluble coffee production during the extraction (brewing) process, in which the roasted and grinding coffee is introduced into the process to extract the flavors (soluble materials) in the soluble or instant coffee-manufacturing factories and coffee shops. The solid final by-product is thus obtained, and the insoluble residue (a slurry containing

Table 10.1 Imported Amounts of Coffee Beans in Taiwan During the Period of 2003–2015 (in Metric Tons)

Year	Coffee				
	Not Roasted/ Not Decaffeinated	Not Roasted/ Decaffeinated	Roasted/Not Decaffeinated	Roasted/ Decaffeinated	Sum
2003	7,602	5	1,025	49	8,681
2004	9,589	21	1,195	73	10,878
2005	10,179	12	1,220	66	11,477
2006	9,423	29	1,231	82	10,765
2007	12,298	67	1,416	92	13,873
2008	9,761	66	1,673	99	11,599
2009	11,585	24	1,686	100	13,395
2010	15,879	48	1,843	117	17,887
2011	15,211	88	2,258	129	17,686
2012	15,802	56	2,507	77	18,442
2013	18,782	108	2,846	65	21,801
2014	20,597	103	2,997	76	23,773
2015	23,782	199	4,433	128	28,542

Data from Customs Administration (Ministry of Finance, Taiwan).

SCGs) is screw pressed. The residue weighs approximately 50% of the total input mass of coffee feedstock, representing a significant bioresource because it contains large amounts of organic compounds, such as cellulose, hemicellulose, lignin, fatty acid, and other polysaccharides (Campos-Vega et al., 2015). In Taiwan, for example, the annual generation of exhausted coffee residue from food processors was estimated to be over 10,000 MT based on about 24,000 MT of imported coffee (not roasted and not decaffeinated) in 2015 (Table 10.1).

The data in Table 10.2 indicate that exhausted coffee residue is chemically characterized by high contents of carbon and hydrogen (i.e., >50% and around 7.0 wt.%, respectively), and low content of ash. Therefore, its heating value is relatively high, up to be over 20 MJ/kg-dried basis as shown in Table 10.2, suggesting that exhausted coffee residue was suitable as a starting feedstock for the production of various energy forms, such as stream, heat, and fuels (i.e., biogas, syngas, biooils, and solid fuel). Fig. 10.2 shows a variety of conversion methods (i.e., thermochemical, biological, and physical/chemical processes) for biomass utilization, as well as their alternative fuels. These processes have been used to study the conversion of exhausted coffee residue into various energy forms. They include torrefaction (Chen et al., 2012; Tsai and Liu, 2013), pyrolysis (Bok et al., 2012; Cho et al., 2015; Fischer et al., 2015; Jeguirim et al., 2014; Li et al., 2014; Tsai et al., 2012; Vardon et al., 2013; Yang et al., 2014), combustion (Silva, 1998), cofiring (Cao et al., 2008; Limousy et al., 2013, 2015; Oliveira et al., 2015), gasification (Masek et al., 2008),

Table 10.2 Thermochemical Properties of Spent Coffee Grounds (SCGs) Reported in the Literature

References	Approximate Analysis (wt.%)			EA (wt.%)						HCV (MJ/kg)	Comments
	VM	FC	Ash	M	C	H	O	N	S	Cl	
Masek et al. (2008)	—[a]	—	0.8	6.7	54.2	7.3	35.6[b]	2.0	<0.1	—	—	Dry basis (ash/EA)
Skreibert et al. (2011)	76.67	16.75	6.58	10.7	51.33	6.79	38.60	3.02	0.21	0.055	19.82	Dry ash free basis (EA)
Bok et al. (2012)	77.51	19.83	1.35	1.31	54.61	6.59	34.83[b]	3.97	0	—	22.74	Dry ash free basis (EA)
Chen et al. (2012)	68.8	14.94	1.76	14.50	—	—	—	—	—	—	—	Dry basis (EA)
Tsai et al., 2012	79.52	8.23	0.73	11.52	52.54	6.95	34.82	3.46	0.10	—	23.5	Dry basis (ash/EA)
Limousy et al. (2013)	—	—	1.94	—	61.13	8.99	26.60[b]	2.91	0.37	0.0113	—	Dry basis[c] (ash/EA)
Vardon et al. (2013)	—	—	2.4	—	51.8	6.3	38.8	2.8	0.17	—	20.1	Dry basis[d] (ash/EA)
Vardon et al. (2013)	—	—	1.8	—	56.1	7.2	34.0	2.4	0.14	—	23.4	Dry basis (EA)
Zuorro and Lavecchia (2012)	—	—	—	—	—	—	—	—	—	—	23.72 24.07	Dry basis (EA)
Li et al. (2014)	82.0	16.3	1.7	8.1	54.5	7.1	34.2	2.4	0.1	—	23.2	Dry basis (EA)
Fischer et al. (2015)	—	—	—	—	55.6	7.0	35.5[b]	1.8	—	—	—	

EA, Elemental analysis; FC, fixed carbon; HCV, higher calorific value; M, moisture; VM, volatile matter.
[a]*Not determined.*
[b]*By difference.*
[c]*SCGs derived from commercial coffee beverage production.*
[d]*Defatted coffee grounds produced from SCGs after lipid extraction.*

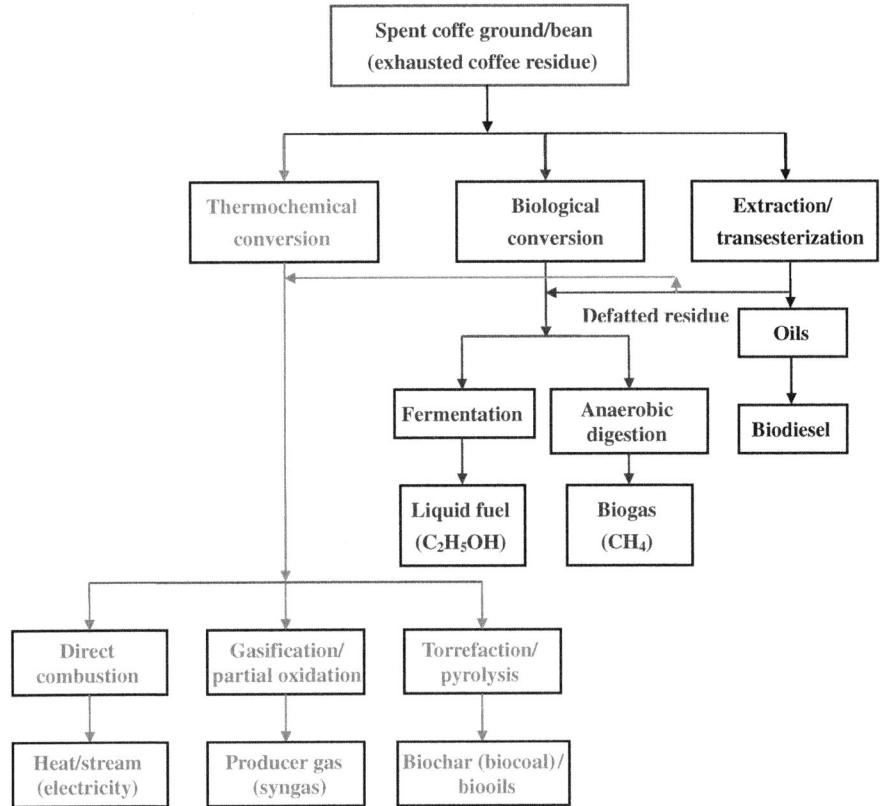

FIGURE 10.2 Conversion of Exhausted Coffee Residue to Alternative Fuels and Other Energy

transesterification after extraction (Al-Hamamre et al., 2012; Caetano et al., 2014; Kondamudi et al., 2008; Kwon et al., 2013; Ponte et al., 2014; Vardon et al., 2013), fermentation (Choi et al., 2012; Kwon et al., 2013; Ponte et al., 2014), and anaerobic digestion (Neves et al., 2006). Of a variety of thermochemical conversion processes of exhausted coffee residue, this chapter will focus on pyrolysis due to its technological significance in the desired production of solid char (biochar) with relatively high yield (Meyer et al., 2011; Manya, 2012).

10.5 PYROLYSIS FOR THE PRODUCTION OF BIOCHAR FROM EXHAUSTED COFFEE RESIDUE

Although biomass characterizes its high content of carbon source, the reuse of raw biomass residue as a solid fuel or feedstock for the production of fuels and chemicals often encounters practical limitations, which must take into account its applications in the agricultural and industrial fields. These limiting factors include large bulk volume, high moisture content, low heating value and energy density, hygroscopic nature, disagreeable odor, biodegradable nature, and smoke during combustion, making it low in combustion efficiency (Pimchuai et al., 2010). In view of overcoming some of the aforementioned limitations of biomass residue, thermochemical conversion is a promising route to convert lignocellulosic biomass to fuel, chemical, and renewable power (Klass, 1998). Pyrolysis is generally described as the thermochemical decomposition of the organic components (long polysaccharides) in biomass in an inert atmosphere at mediate temperature (e.g., 400–800°C) to yield gas, liquid (biooil), and char (biochar, charcoal, or biocoal) (Basu, 2013; Uchimiya, 2016). As a result, this process will improve the thermochemical properties of biomass, and will produce a hydrophobic solid product with an increased energy density. For instance, biochar has a high C/H ratio, which is indicative of its lack of volatile hydrocarbons, such as acetic acid and phenol. Thus, biochar is a useful product in the pyrolysis process of biomass. In addition, the biochar can be further activated by the chemical or physical activation to make more pores in the activated charcoal or activated carbon, which is more accessible for adsorption (Marsh and Rodriguez-Reinoso, 2006).

Depending on the heating rate and process temperature, pyrolysis can be divided into torrefaction, slow pyrolysis (or conventional carbonization), and fast pyrolysis (Manya, 2012; Meyer et al., 2011). Torrefaction, a mild pyrolysis, is currently being considered for improving the thermochemical properties of biomass, and producing a hydrophobic solid product with an increased energy density (Basu, 2013). In this process, the biomass is generally heated from 200 to 300°C in the absence of oxygen to make it lose its fibrous nature. As a consequence, the torrefied biomass will be easily grindable, and it can be directly pressed to form densified pellets without adding binders. However, other temperature ranges (i.e., 300–400°C) have been suggested (Tsai and Liu, 2013). Torrefied biomass products have been proposed as available feedstocks for direct combustion, gasification, coking in blast furnace (for reduction in carbon footprint of making steel), and coal cocombustion due to its high energy content, good grindability, and hydrophobic properties. By contrast, pyrolysis may be roughly classified as slow and fast processes based on the heating rate. In the slow pyrolysis, the residence time of released vapor in the pyrolysis zone is in the order of minutes or longer. Also, the biomass is heated slowly (e.g., heating rate is less than 50°C/min) in the absence of oxygen to a relatively low temperature (e.g., 400–600°C). This process is used primarily for char production. In the fast pyrolysis, the vapor residence time in the process zone is on the order of seconds or shorter (less than 0.1 s). More significantly, the biomass is heated rapidly (e.g., heating rate is larger than 100°C/min, even 10,000°C/min) in the absence of oxygen to a relatively high temperature (e.g., 450–700°C). This type of pyrolysis is to produce more biooil and noncondensed gas (Basu, 2013). In contrast to considerable studies to evaluate the benefits of biochar when applied to metal-contaminated water and soil, there is little work available in published literature on coffee-derived biochar as a soil amendment and as an adsorbent (Kim et al., 2014; Meyer et al., 2011). In the study by Kim et al. (2014), the biochar was applied to acid mine drainage, the heavy metal concentrations were decreased and the pH was increased.

10.6 THERMOCHEMICAL PROPERTIES OF EXHAUSTED COFFEE RESIDUE AND ITS RESULTING BIOCHAR

Coffee processing by-products, including coffee pulp, coffee husk, and coffee mucilage, often pose serious environmental problems. For this reason, efforts have been made to develop methods for its utilization as a raw material for the production of feeds, caffeine, pectin, and so on. In view of reusing exhausted coffee residue as a biomass precursor for the production of pyrolysis products, several research works have been performed on this subject. However, these focused on the preparation of activated carbon using chemical and physical activation. For example, by using water vapor and potassium carbonate (K_2CO_3) as activating agents, Castro et al. (2011) studied the preparation of activated carbons from SCGs and their pore properties and applications for phenol adsorption. As noted by another research (Jutakridsada et al., 2016), the ground coffee residue was used as a precursor for the production of activated carbon by using zinc chloride ($ZnCl_2$). The optimal resulting activated carbon was applied for adsorption of copper ion due to its SSA of 831 m^2/g. Although several researches focused on the reuse of exhausted coffee residue as a biosorbent for the removals of pollutants from the aqueous solutions (Kyzas, 2012; Liu et al., 2015), very little work on its pyrolysis is available in the published literature (Tsai and Liu, 2013; Vardon et al., 2013). Thus, the data on the thermochemical properties of resulting biochars are also limited.

Based on its chemical and physical properties, biochar can be used for a range of applications: as an agent for soil improvement, improved resource use efficiency, remediation and/or protection against particular environmental pollution, and as an avenue for greenhouse gas (GHG) mitigation." As mentioned earlier, biochar is produced through the process of pyrolysis as a solid by-product. Its characteristics, including SSA (pore property), calorific value, and carbon and inorganic contents generally depend on how biochar is produced by processes and feed parameters, such as temperature, residence time, and type of biomass (Basu, 2013). Among them, pyrolysis temperature may be the most important parameter of influencing the properties of biochar. In the pyrolysis study by Tsai et al. (2012), the calorific values of biochar from exhausted coffee residue (30.4–31.9 MJ/kg) were relatively high as compared to that (i.e., 28.0–32.0 MJ/kg) of fossil coal. Furthermore, the pyrolysis temperature of 400°C was found to be optimal for the higher calorific values. Interestingly, there was no obvious change to the calorific value of resulting biochar at the higher temperature ranging from 500 to 700°C. As compared to the calorific value of exhausted coffee residue (23.5 MJ/kg, listed in Table 10.2) an increase of 36% was obtained for the resulting biochar. In the study by Vardon et al. (2013), a slow pyrolysis was conducted at the following process conditions: heated to 450°C at a rate of 50°C/min,

Table 10.3 Elemental Analysis of Biochars-Derived Waste Coffee Grounds Reported in the Literature

References	EA (wt.%, db)					HCV (MJ/kg)	Comments
	C	H	N	S	O		
Tsai et al. (2012)	55.10[a]	6.36	2.25	ND	36.29[b]	—	SCGs: biochar[a]
Tsai et al. (2012)	56.63	6.22	2.45	ND	34.70[b]	—	SCGs: biochar[c]
Tsai et al. (2012)	58.82	6.01	2.64	ND	32.53[b]	—	SCGs: biochar[d]
Tsai et al. (2012)	66.77	5.46	3.04	ND	24.73[b]	—	SCGs: biochar[e]
Tsai et al. (2012)	67.03	4.95	2.74	ND	25.28[b]	—	SCGs: biochar[f]
Vardon et al. (2013)	76.2	5.6	3.9	0.05	—[g]	31.0	SCGs: biochar[h]
Vardon et al. (2013)	72.6	5.0	4.3	0.10	—	28.5	Defatted coffee grounds: biochar[h]

ND, Not detectable.
[a]Torrefaction conditions: directly torrefied at 230°C, and then maintained for time of 10 min.
[b]By difference.
[c]Torrefaction conditions: directly torrefied at 260°C, and then maintained for time of 10 min.
[d]Torrefaction conditions: directly torrefied at 290°C, and then maintained for time of 10 min.
[e]Torrefaction conditions: directly torrefied at 320°C, and then maintained for time of 10 min.
[f]Torrefaction conditions: directly torrefied at 350°C, and then maintained for time of 10 min.
[g]Not determined.
[h]Pyrolysis conditions: heated to 450°C at a rate of 50°C/min, and then maintain for retention time of 2 h.

and then maintained for 2 h (i.e., retention time) with a nitrogen sweep gas. As compared to the calorific values of the feedstocks (20.1–23.4 MJ/kg, seen in Table 10.2), coffee-derived biochars displayed higher energy densities (28.3–31.0 MJ/kg, seen in Table 10.3). On the other hand, the coffee-based biochar is very similar to the commercial coal-based char on the carbon content on a dry basis (72.6–76.2 wt.%, seen in Table 10.3). However, the fuel-bound nitrogen will contribute to nitrogen oxides (NO_x) emission from the combustion or cofiring of the coffee-based biochar. As in common biomass fuels, there is less than a detectable amount of sulfur in the biochar, suggesting that sulfur oxides (SO_x) would not be emitted in a large extent.

In the torrefaction study by Tsai and Liu (2013), the biochar torrefied at 320–350°C contained about 67 wt.% of carbon (Table 10.3), which is obviously higher than that (i.e., 52.5 wt.%) of the raw precursor (i.e., exhausted coffee residue) as listed in Table 10.2. More significantly, the carbon content of torrefied biomass increased mostly within the temperature ranging from 290 to 320°C, which should be the temperature window of torrefaction for coffee residue. As also shown in Table 10.3, the torrefied products contained high concentrations of O and H, but the data significantly decreased with increasing temperature. As a result, both the molar ratios of O/C and H/C of torrefied biomass decreased from 0.494 and 1.385, respectively, to 0.283 and 0.886, respectively.

10.7 THE POTENTIAL FOR AGRICULTURAL AND FABRIC APPLICATIONS OF BIOCHAR DERIVED FROM EXHAUSTED COFFEE RESIDUE

As mentioned earlier, biochar is enriched in carbon (C), and even more in soil nutrients, such as phosphorus (P), calcium (Ca), magnesium (Mg), potassium (K), and nitrogen (N). Therefore, biochar can enhance soil fertility (Vai and Chang, 2016). Furthermore, biochar has an important role in mitigating GHGs, especially in methane (CH_4) and carbon dioxide (CO_2). It could be attributed closely to the fact that its application can significantly increase the amount of carbon retained in the soil with a stable form. Alternatively, the carbon in agricultural and forestry residues will be released to the atmosphere over a short time as gaseous form (i.e., CH_4 and CO_2) due to the microbial decomposition. According to the report by the Intergovernmental Panel on Climate Change (IPCC, 2013), the main sources, including agricultural activities, waste management, energy use, and biomass burning, contribute to CH_4 emissions, which account for 16% of anthropogenic GHG based on global emissions from 2010.

Converting agricultural waste or food processing by-product into a porous biochar that holds carbon makes soils more fertile. As a result, people can boost food security, discourage deforestation, and preserve cropland diversity (Lehmann and Joseph, 2015). In brief, soil improvement that uses biochar may be obtained by the following benefits:

- Reduced leaching of nitrogen and other nutrients into groundwater due to the applications of chemical (nitrogen) fertilizers.
- Possible reduced emissions of potent GHG (i.e., nitrous oxide and methane) due to the reduced use of nitrogen fertilizers and its chemically stable form.
- Increased CEC resulting in improved soil fertility due to its high SSA and the presence of surface hydroxyl groups.
- Moderation of soil acidity as a result of acid rain and chemical fertilizer application (leaching loss of alkali metals and alkaline earth metals).
- Increased water retention and aeration due to the porous structure of biochar used as a sponge-like medium.
- Increased number of beneficial soil microbes due to the porous texture of biochar used as a good growth environment.
- Decreased use of chemical fertilizer (especially in nitrogen fertilizers, made from ammonia), thus mitigating nitrous gas (N_2O, one of potent GHGs) emission.
- Improved water quality by reducing contaminated runoff and nutrient loss due to its porosity of providing adsorption capacity.
- Increased interactions between plant and microbe due to the porous structure and rich minerals in the biochar.

Table 10.4 Inorganic Contents of Waste Coffee Grounds and Their Resulting Biochars Reported in the Literature

References	Major Inorganics (wt.%, db)				Minor Inorganics (ppm, db)						Comments
	P	K	Ca	Mg	Na	Zn	Fe	Mn	Cu	B	
Tsai et al. (2012)	0.112	0.403	0.103	0.143	ND	ND	ND	—[a]	ND	—	SCGs
Vardon et al. (2013)	0.18	0.81	0.20	0.20	51.8	6.3	38.8	2.8	0.17	—	SCGs[b]
Vardon et al. (2013)	0.17	0.74	0.17	0.22						—	Defatted coffee grounds[c]
Vardon et al. (2013)	0.48	1.94	0.56	0.60						—	SCGs: biochar[d]
Vardon et al. (2013)	0.25	1.08	0.54	0.36						—	Defatted coffee grounds: biochar[d]

ND, Not detectable.
[a]Not determined.
[b]SCGs derived from commercial coffee beverage production.
[c]Defatted coffee grounds produced from SCGs after lipid extraction.
[d]Pyrolysis conditions: heated to 450°C at a rate of 50°C/min, and then maintain for retention time of 2 h.

As shown in Table 10.4 (Tsai et al., 2012; Vardon et al., 2013), the data on the major inorganics (i.e., P, K, Ca, and Mg) and minor inorganics (i.e., Cu, Zn, Na, Fe, Mn, and B) of exhausted coffee residue and its resulting biochar revealed that that they have substantially smaller concentrations of Al, As, B, Ba, Cd, Cr, Cu, Fe, Na, Ni, Pb, Se, Si, Ti, and Zn. Moreover, the inorganic compositions between SCGs and defatted coffee grounds are similar. As can be expected, the macronutrients Ca, K, Mg, and P are significantly high in the resulting biochar. Overall, pyrolysis of exhausted coffee residue enriched the relative weight fraction of major nutrients in the resulting biochars, which may account for the soil productivity increased when they are used as soil amendments (Vardon et al., 2013).

To upgrade the additional value from coffee residue and its resulting carbon materials, some coffee-related products have been developed in Taiwan. For example, coffee charcoal yarn is made by coffee-derived charcoal and polyester. First, the phenols, esters, and oils must be further removed from the coffee residues, leaving them odorless. Then, the exhausted coffee feedstock is pyrolyzed (or carbonized) to make charcoal, which is again ground to nanosize powder. Thereafter, this fine powder is mixed with melt polymer to make polyester fiber. It should be noted that the polyester is derived from recycled plastic PET bottles. Due to its functions of absorbing water (fast drying) and odor, and blocking harmful UV light, the fabric made from coffee charcoal yarn has great applications in a great variety of ecofriendly products, including coffee yarns, coffee fabrics, clothing, underwear, bedding, and shoes. Therefore, there may be an interesting slogan "Drink your coffee and wear it, too," or "Drink it and wear it."

10.8 CONCLUSIONS

Biochar is the carbon-rich material that is produced from biomass (such as wood or crop residues) under the pyrolysis process. It can be directly used as a solid fuel, or a soil amendment to improve agriculture due to its persistence in soil and nutrient-retention properties. More significantly, its carbon sequestration can be used to actively remove carbon dioxide from the atmosphere, with the environmental implication for mitigation of climate change. In this chapter, SCGs were discussed to explore the possibilities of utilizing them as an excellent precursor for the production of biochar via pyrolysis. In addition to their thermal conversion and characterization described here, their potential for agricultural and energy applications were also addressed to make coffee grounds an ecofriendly material in several ways: soil improvement, mitigation of climate change and nutrient water pollution, waste management, and bioenergy production. On the other hand, the coffee-derived charcoal material can be processed to get a very fine powder, which can be used for the production of value-added products (e.g., functional textiles). Also, the biochar can serve as a good precursor for producing activated carbons or modified biochars that enable high adsorption affinity and selectivity for removals of pollutant in wastewater.

REFERENCES

Acchar, W., Dultra, E.J.V., 2015. Ceramic Materials from Coffee Bagasse. Springer, Heidelberg.

Al-Hamamre, Z., et al., 2012. Oil extracted from spent coffee grounds as a renewable source for fatty acid methyl ester manufacturing. Fuel 96, 70–76.

Basu, P., 2013. Biomass Gasification, second ed. Elsevier, London.

Binod, P., Pandey, A., 2015. Introduction. In: Pandey, A., Negi, S., Bindo, P., Larroche, C. (Eds.), Pretreatment of Biomass: Processes and Technologies. Elsevier, Amsterdam.

Bok, J.P., et al., 2012. Fast pyrolysis of coffee grounds: characteristics of product yields and biocrude oil quality. Energy 47, 17–24.

Bonilla-Hermosa, V.A., Duarte, W.F., Schwan, R.F., 2014. Utilization of coffee by-products obtained from semi-washed process for production of value-added compounds. Bioresour. Technol. 166, 142–150.

Bressani, R., 1979. The by-products of coffee berries. In: Braham, J.E., Bressani, R. (Eds.), Coffee Pulp: Composition, Technology, and Utilization. International Development Research Centre, Ottawa.

Caetano, N.S., et al., 2014. Spent coffee grounds for biodiesel production and other applications. Clean Technol. Environ. Policy 16, 1423–1430.

Campos-Vega, R., et al., 2015. Spent coffee grounds: a review on current research and future prospects. Trends Food Sci. Technol. 45, 24–36.

Cao, Y., et al., 2008. Mercury emissions during cofiring of sub-bituminous coal and biomass (chicken waste, wood, coffee residue, and tobacco stack) in a laboratory-scale fluidized bed combustor. Environ. Sci. Technol. 42, 9378–9384.

Castro, C.S., et al., 2011. Phenol adsorption by activated carbon produced from spent coffee grounds. Water Sci. Technol. 64, 2059–2065.

Chen, W.H., 2015. Torrefaction. In: Pandey, A., Negi, S., Bindo, P., Larroche, C. (Eds.), Pretreatment of Biomass: Processes and Technologies. Elsevier, Amsterdam.

Chen, W.H., Lu, K.M., Tsai, C.M., 2012. An experimental analysis on property and structure variations of agricultural wastes undergoing torrefaction. Appl. Energy 100, 318–325.

Chia, C.H., et al., 2015. Characteristics of biochar: physical and structural properties. In: Lehmann, J., Joseph, S. (Eds.), Biochar for Environmental Management: Science, Technology and Implementation. second ed. Routledge, New York, NY.

Cho, D.W., et al., 2015. Carbon dioxide assisted sustainability enhancement of pyrolysis of waste biomass: a case study with spent coffee ground. Bioressour. Technol. 189, 1–6.

Choi, I.S., et al., 2012. Conversion of coffee residue waste into bioethanol with using popping pretreatment. Bioresour. Technol. 125, 132–137.

Demirbas, A., 2005. Biomass co-firing for boilers associated with environmental impacts. Energy Sources A 27, 1385–1396.

Esquivel, P., Jimenez, V.M., 2012. Functional properties of coffee and coffee by-products. Food Res. Int. 46, 488–495.

Fischer, A., et al., 2015. The effect of temperature, heating rate, and ZSM-5 catalyst on the product selectively of the fast pyrolysis of spent coffee grounds. RSC Adv. 5, 29252–29261.

Franca, A.S., Oliveira, L.O., 2009. Coffee processing solid wastes: current uses and future perspectives. In: Ashworth, G.S., Azevedo, P. (Eds.), Agricultural Issues and Policies. Agricultural Wastes, Nova.

Intergovernmental Panel on Climate Change (IPCC), 2013. Climate Change 2013: The Physical Science Basis. IPCC, Geneva, Switzerland.

Ippolito, et al., 2015. Biochar elemental composition and factors influencing nutrient retention. In: Lehmann, J., Joseph, S. (Eds.), Biochar for Environmental Management: Science, Technology and Implementation. second ed. Routledge, New York, NY.

Jeguirim, M., Limousy, L., Dutournie, P., 2014. Pyrolysis kinetics and physicochemical properties of agropellets produced from spent ground coffee blended with conventional biomass. Chem. Eng. Res. Des. 92, 1876–1882.

Jutakridsada, P., et al., 2016. Adsorption characteristics of activated carbon prepared from spent ground coffee. Clean Technol. Environ. Policy 18, 639–645.

Kim, M.S., et al., 2014. The effectiveness of spent coffee grounds and its biochar on the amelioration of heave metals-contaminated water and soil using chemical and biological assessments. J. Environ. Eng. 146, 124–130.

Klass, D.J., 1998. Biomass for Renewable Energy, Fuels, and Chemicals. Academic Press, San Diego, CA.

Kondamudi, N., Mohapatra, S.K., Misra, M., 2008. Spent coffee grounds as a versatile source of green energy. J. Agric. Food Chem. 56, 11757–11760.

Kwon, E.E., Yi, H., Jeon, Y.J., 2013. Sequential co-production of biodiesel and bioethanol with spent coffee grounds. Bioresour. Technol. 136, 475–480.

Kyzas, G.Z., 2012. Commercial coffee wastes as materials for adsorption of heavy metals from aqueous solutions. Materials 5, 1826–1840.

Lee, J.W., et al., 2010. Characterization of biochars produced from corn stovers for soil amendment. Environ. Sci. Technol. 44, 7970–7974.

Lehmann, J., Joseph, S., 2015. Biochar for environmental management: an introduction. In: Lehmann, J., Joseph, S. (Eds.), Biochar for Environmental Management: Science, Technology and Implementation. second ed. Routledge, New York, NY.

Li, X., Strezov, V., Kan, T., 2014. Energy recovery potential analysis of spent coffee grounds pyrolysis products. J. Analyt. Appl. Pyrolysis 110, 79–867.

Limousy, L., et al., 2013. Gaseous products and particulate matter emissions of biomass residential boiler fired with spent coffee grounds pellets. Fuel 107, 323–329.

Limousy, L., et al., 2015. Performance and emissions characteristics of compressed spent coffee ground/wood chip logs in a residential stove. Energy Sustain. Dev. 28, 52–59.

Liu, C., et al., 2015. The role of exhausted coffee compounds on metal ions sorption. Water Air Soil Pollut. 226, 289.

Lowell, S., et al., 2006. Characterization of Porous Solids and Powders: Surface Area, Pore Size and Density. Springer, Dordrecht.

Manya, J.J., 2012. Pyrolysis for biochar purposes: a review to establish current knowledge gaps and research needs. Environ. Sci. Technol. 46, 7939–7954.

Marsh, H., Rodriguez-Reinoso, F., 2006. Activated Carbon. Elsevier, Amsterdam.

Masek, O., et al., 2008. A study on pyrolytic gasification of coffee grounds and implications to allothermal gasification. Biomass Bioenergy 32, 78–89.

Meyer, S., Glaser, B., Quicker, P., 2011. Technical, economical, and climate-related aspects of biochar production technologies: a literature review. Environ. Sci. Technol. 45, 9473–9483.

Mukome, F.N.D., Parikh, S.J., 2016. Chemical, physical, and surface characterization of biochar. In: Ok, Y.S. et al., (Ed.), Biochar: Production, Characterization, and Applications. CRC Press, Boca Raton, FL.

Murthy, P.S., Naidu, M.M., 2012. Sustainable management of coffee industry by-products and value addition—a review. Res. Conserv. Recyc. 66, 45–58.

Narita, Y., Inouye, K., 2014. Review on utilization and composition of coffee silverskin. Food Res. Int. 61, 16–22.

Neves, L., Oliveira, R., Alves, M.M., 2006. Anaerobic co-digestion of coffee waste and sewage sludge. Waste Manag. 26, 176–181.

Oliveira, T.L., et al., 2015. Study of biomass applied to a cogeneration system: a steelmaking industry case. Appl. Therm. Eng. 80, 269–278.

Pimchuai, A., Dutta, A., Basu, P., 2010. Torrefaction of agriculture residues to enhance combustible properties. Energy Fuels 24, 4638–4645.

Ponte, R., et al., 2014. Ultrasound-assisted production of biodiesel and ethanol from spent coffee grounds. Bioresour. Technol. 167, 343–348.

Sheng, C., Azevedo, J.L.T., 2005. Estimating the higher heating value of biomass fuels from basic analysis data. Biomass Bioenergy 28, 499–507.

Silva, M.A., 1998. The use of biomass residues in the Brazilian soluble coffee industry. Biomass Bioenergy 14, 457–467.

Skreibert, A., et al., 2011. TGA and macro-TGA characterisation of biomass fuels and fuel mixtures. Fuel 90, 2182–2197.

Smith, J.M., 1981. Chemical Engineering Kinetics, third ed. McGraw-Hill, New York, NY.

Taylor, L., Antonio, J., 2006. Coffee as a Functional Beverage. In: Wildman, R.E.C. (Ed.), Handbook of Nutraceuticals and Functional Foods. second ed. CRC Press, Boca Raton, FL.

Tsai, W.T., Liu, S.C., 2013. Effect of temperature on thermochemical property and true density of torrefied coffee residue. J.Analyt. Appl. Pyrolysis 102, 47–52.

Tsai, W.T., Liu, S.C., Hsieh, C.H., 2012. Preparation and fuel properties of biochars from the pyrolysis of exhausted coffee residue. J. Analyt. Appl. Pyrolysis 93, 63–67.

Uchimiya, S.M., 2016. Biochar Production Technology. In: Ok, Y.S. et al., (Ed.), Biochar: Production, Characterization, and Applications. CRC Press, Boca Raton, FL.

Vai, Y., Chang, S.X., 2016. Biochar Effects on Soil Fertility and Nutrient Cycling. In: Ok, Y.S. et al., (Ed.), Biochar: Production, Characterization, and Applications. CRC Press, Boca Raton, FL.

Van Loo, S., Koppejan, J., 2008. The Handbook of Biomass Combustion and Co-Firing. Earthscan, London.

Vardon, D.R., et al., 2013. Complete utilization of spent coffee grounds to produce biodiesels, bio-oil, and biochar. ACS Sustain. Chem. Eng. 1, 1286–1294.

Yang, S.I., et al., 2014. Application of biomass fast pyrolysis part II: the effects that bio-pyrolysis oil has on the performance of diesel engines. Energy 66, 172–180.

Zuorro, A., Lavecchia, R., 2012. Spent coffee grounds as a valuable source of phenolic compounds and bioenergy. J. Cleaner Prod. 34, 49–56.

FURTHER READING

Kumarathilaka, P., 2016. Biochar. In: Ok, Y.S. et al., (Ed.), Biochar: Production, Characterization, and Applications. CRC Press, Boca Raton, FL.

Suzuki, M., 1990. Adsorption Engineering. Elsevier, Amsterdam.

Energy applications of coffee processing by-products

11

Lionel Limousy*, Mejdi Jeguirim*, Madona Labaki**

*Institute of Materials Science of Mulhouse, Mulhouse, France;
**Lebanese University, Fanar, Jdeidet, Lebanon

ABSTRACT

This chapter explores the possibilities of using coffee processing by-products (e.g., coffee husks and spent coffee grounds [SCG]) for energy applications (e.g., biofuels, biodiesel, bioethanol). In particular, the recovery of energy from biomass through thermochemical processes (pyrolysis, gasification, combustion, hydrothermal treatment, etc.) and biochemical processes is presented. The energy recovery from biomass is an ecological route to produce energy from renewable sources, reduce waste, produce cleaner-burning fuels, protect the environment, reduce fossil fuels consumption and dependence, decrease fuel costs, lower greenhouse gas (GHG) emissions, and find a solution for the limited availability of fossil fuels. Moreover, it has a good impact on economic, social, and agricultural development and ensures a regular supply of energy.

Keywords: combustion; torrefaction; pyrolysis; gasification; bioethanol; biodiesel

11.1 INTRODUCTION

Coffee processing by-products are considered as biomass since they are composed of organic molecules and could be converted through thermochemical or biochemical processes into biogas, biofuel, biodiesel, or bioethanol, or could be directly subjected to combustion. A pretreatment technique, such as drying and torrefaction could be applied in the case of high humidity content in order to remove water for the coffee processing by-products. In fact, moisture harms the performance of the thermochemical process and influences the quality of gas produced (de Oliveira et al., 2013; Gómez-de la Cruz et al., 2015). Removing moisture increases the energy value of the coffee byproduct (Gómez-de la Cruz et al., 2015).

The main objective of this chapter is to present all the conversion processes that have already been developed industrially and projects that have high probabilities of commercialization due to their sustainable character, concerning energy applications of coffee processing by-products. It provides a detailed presentation of all works that were published during the last decades in the field of coffee by-products valorization for energy purposes.

Before describing the production processes for biofuels, combustion and thermogravimetric analysis of coffee by-products are presented. Indeed, to better understand the thermochemical processes, it is crucial to study the thermal behavior under different atmospheres of coffee wastes.

11.1.1 COMBUSTION

This part will be devoted to the valorization of solid coffee by-products corresponding to coffee husks and spent coffee grounds. The use of coffee husks to prepare fuels was mainly developed in South America while researches corresponding to the direct valorization of spent coffee grounds are mainly concentrated in Europe. This particularity can be explained by the fact that South America is one of the most important coffee suppliers in the world. It is also important to mention that the use of spent coffee grounds as alternative fuel was encouraged in Europe by the ECS (European Committee for Standardization), which has led to several research and development projects (Alakangas et al., 2006). As mentioned previously, coffee husks and spent coffee grounds can be directly valorized through different combustion devices for domestic applications. In both cases, raw materials need to be densified in order to improve storage (reduction of volume) and transportation (cost) conditions and also to obtain good combustion efficiencies. These by-products can be used directly (raw material) or blended with other biomasses in order to optimize the densification process and adjust the quality of the solid fuel (formulation of agro-fuels) and/or the combustion efficiency. Among the different available biomasses to formulate solid fuels, wood sawdust was mainly used to prepare densified solid fuels (pellets, briquettes, and logs). It can be used to ensure the cohesion of the biomass particles to obtain good mechanical properties, but also to limit the slagging phenomenon when mineral content needs to be reduced (boilers, fluidized bed). Then, the percentage of sawdust can range from a low amount (5–10 wt.%) to 80–90 wt.% depending on the combustion process (furnace, stove, boiler).

11.1.1.1 Densification of coffee processing by-products

The low bulk density of coffee husks and spent coffee ground limits their direct use for combustion to the local market. In fact, the cost of coffee husks and spent coffee ground transportation and storage is too high to make profits in the case of exportation. Then, the only possibility to valorize these by-products is to carry out densification after drying when it is necessary (humidity higher than 15 wt.% before densification). Different technologies can be used to obtain high-density fuels from coffee husks and spent coffee ground, such as pelletization, briquetting, or logging.

11.1.1.1.1 Pellets production

Few investigations have been reported on pellets production from coffee husks. Particularly, pellets have been produced with different systems: a Kahl pelletizer (Germany, model 38-780) operating at a production capacity of 164 kg/h and at a temperature of 125°C was used by Cubero-Abarca et al. (2014), a Eng-Maq pelletizer (Brazil, model 0200V) operating at a production capacity of 110 kg/h and with a temperature range between 80 and 95°C was used by Souza Faria et al. (2015) and a laboratory device (TDP benchtop press) operating at room temperature by Gil et al. (2010). Before densification, biomasses present very low densities ranging between 118 kg/m³ for silverskin to 172 kg/m³ for the outer skin. They reached respectively apparent density values of 756 and 709 kg/m³ after the pellets production, which enable and facilitate their transport and their supply to conventional combustion systems. Coffee by-product pellets have also been manufactured using coffee pulp (Cubero-Abarca et al., 2014), and they presented an average apparent density of 600 kg/m³ (Souza Faria et al., 2015). The densification process can increase the density of the raw biomass by a factor of 4 to 6, depending on the coffee husk part that is used and also on the densification conditions (temperature, compression pressure).

Spent coffee ground pellets were produced by Limousy et al. (2013) with a Kahl 14/175 pelletizer at 70°C and a frequency of 50 Hz. Spent coffee ground powder was provided by a coffee supplier and presented an average particle size of 0.6 mm. Pellets present a bulk density of 1211 kg/m³, a diameter of 6 mm, and an average length of 20 mm. The properties of the pellets described previously are given in Table 11.1.

In these works, pellets were prepared with the objective of having marketable products. Different standards can be targeted for these fuels, such as: the DIN-EN-14961-6 standard (European norm) or the NF agro-pellet standard (French norm). It is complicated to obtain formulations that respect these standards by using pure coffee by-products. A way to succeed is to prepare blends containing

Table 11.1 Mechanical and Physical Characteristics of Pure Coffee Husks and Spent Coffee Grounds (SCG) Pellets

	Coffee Pulp (Cubero-Abarca et al., 2014)	Coffee Hulls (Souza Faria et al., 2015)	Silverskin (Souza Faria et al., 2015)	SCG (Limousy et al., 2013)
Length (mm)	20.3	16.1	14.9	15–25
Diameter (mm)	6.12	6.1	6.1	6.0
Moisture (%)[a]	10.1	10.7	11.2	11.8
Friability	0.95	—	—	—
Durability (%)	75.24	94.5	97.6	—
Compression stress (kg/cm²)	26.86	—	—	—
LHV[b] (kJ/kg)	11,591	17,810	18,310	17,520

[a]Wet basis.
[b]Lower Heating Value.

adapted proportions of wood sawdust depending on the ultimate analysis of coffee by-products. Many formulations were developed and tested these last years; nevertheless, only a few reached a high combustion quality.

Another important aspect concerns the characteristics of the pellets, which are produced. Several parameters can be estimated to quantify the quality of pellets such:

- Friability: (number of pellet pieces/number of initial pellets) after a drop of 1 m height
- Compression stress (maximum stress: kg/cm^2): [2 × maximum load (kg)/3.14 × diameter (cm) × length (cm)]
- Mechanical durability: pellet weigh before the test/pellet weigh after the test
- Abrasion index: mass of particles lower than 2 mm relative to the initial mass of pellets after 3000 revolutions in a rotary drum

These measurements are more or less empirical, but they are representative of the mechanical properties and therefore the pellets quality.

11.1.1.1.2 Briquettes and logs production

Briquette is another form of fuel that can be produced from coffee by-products, which is more devoted to household boiler or stove markets. Coffee husk briquettes were mainly developed in South American countries (Brazil, Costa Rica). The production of briquettes was tested both in a laboratory setup (hydraulic press, applied pressure of 10 tons), and also with sawmill machines (mechanical and hydraulic piston press) (Fonseca Felfli et al., 2011; Suarez and Luengo, 2013). Coffee husks can be easily transformed into briquettes since this biomass presents a homogeneous particle size and low moisture content (10–15 wt.%). The production of coffee husk briquettes is mainly limited to the local market in Brazil (Fonseca Felfli et al., 2011). The sailing distance is limited to 300 km around the manufacturing plants; beyond this distance it becomes economically unviable to deliver. Another problem with coffee husk briquettes is caused by their high friability, which may generate excessive quantities of dust during their transport. A potential way to avoid this drawback is the introduction of 10–20 wt.% of sawdust in the coffee husk before briquetting. Despite this constraint, four Brazilian companies produce high-quality briquettes and export their production in the United States and in Europe. The briquettes price ranges from 50 to 150 €/t depending on the transport distance. Ciesielczuk et al. (2015) reported on another work on the preparation of briquettes containing 10 or 25 wt.% of SCG blended with beechwood sawdust. The developed process consists in a cold briquetting using a Hocker Polytechnik press, which operates at 120 bars. In the present case, the targeted market corresponds to Poland.

Spent coffee grounds have been used to produce compressed fuel logs in the west of France since 2012 (Limousy et al., 2015). Different suppliers distribute currently these logs to the French domestic market. The RID Solution Company uses a log maker (Fig. 11.1) equipped with a maintenance rail and circular saw to produce logs with 300 mm length and 90 mm of diameter. It operates at a temperature of 120°C and a pressure of 2 tons/cm^2, and has a production capacity of 1.5 tons/h. Different formulations

FIGURE 11.1 Log Maker (NIELSEN) Used by the RID Solution Company (France) to Produce Compressed Logs Containing 10 to 20 wt.% of SCG

Reprinted with permission from Limousy, L., Jeguirim, M., Labbe S., Balay, F., Fossard, E., 2015. Performance and emissions characteristics of compressed spent coffee ground/wood chip logs in a residential stove. Energy Sustain. Develop. 28, 52–59. Copyright Elsevier 2016.

were developed these last years, containing from 20 wt.% of spent coffee ground blended with pine sawdust until 100 wt.%. Only formulation containing 20 wt.% SCG was industrially developed because it met the needs of the "NF bois de chauffage" labeling (Fig. 11.2). In the present industrial process, it is necessary to dry the raw SCG due to its initial moisture ranging between 40 and 55 wt.%. Hence, a drier operating temperature of 400°C at the inlet and of 70°C at the outlet is used for the moisture removal. The SCG leaves leave the drier with residual moisture lower than 10 wt.%.

11.1.1.2 Combustion of coffee husks

Coffee husk represents about 20 wt.% of the coffee cherry weight (Fonseca Felfli et al., 2011). Depending on the processing technique used to extract the coffee beans (wet or dry process), this by-product can present different compositions. The dry process leads to the recovering of all the material corresponding to the shell present all around the beans (outer skin, pulp, parchment, and silverskin) (Fig. 11.3), which has moisture content close to 10 wt.%. In Kenya, a specific term is used for this material: *Mbuni husks*. For the wet process, the solid waste contains only the parchment and the silverskin due to the removal of the outer skin and the pulp by mechanical

FIGURE 11.2 Fuel Logs Containing SCG Produced by the RID Solution Company (France) for Domestic Use

Reprinted with permission from Limousy, L., Jeguirim, M., Labbe S., Balay, F., Fossard, E., 2015. Performance and emissions characteristics of compressed spent coffee ground/wood chip logs in a residential stove. Energy Sustain. Develop. 28, 52–59. Copyright Elsevier 2016.

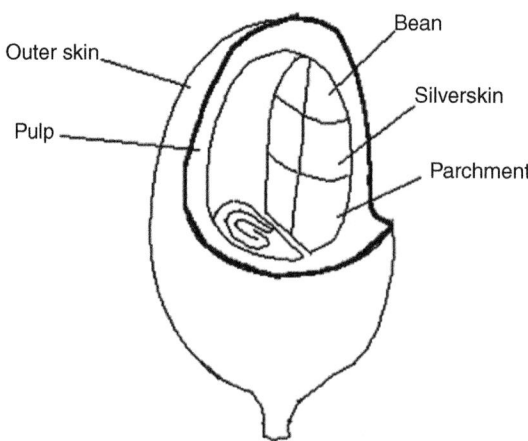

FIGURE 11.3 Description of the Containing of a Coffee Cherry

Reprinted with permission from Saenger, M., Hartage, E.-U., Werther, J., Ogada, T., Siagi, Z., 2001. Combustion of coffee husks. Renew. Energy 23, 103–121. Copyright Elsevier 2016.

Table 11.2 Chemical and Physical Characteristics of Different Coffee Husks Transformed to Fuels

	Mbuni Husks (Saenger et al., 2001)	Parchment Husks (Saenger et al., 2001)	Coffee Husk (Fonseca Felfli et al., 2011)	Coffee Husk (Suarez and Luengo, 2013)	Coffee Husk (Gil et al., 2010)
Proximate analysis (% wet basis or % dry basis)					
Moisture	11.4	10.1	13.1	10.0	6.7
Volatiles	64.6	72.0	73.2[a]	78.5[a]	79.4[a]
Fixed Carbon	20.0	17.0	23.1[a]	19.1[a]	16.1[a]
Ash	4.1	0.9	3.7[a]	2.4[a]	4.5[a]
Ultimate analysis (wt.%, dry basis)					
C	43.9	46.8	47.5	47.5	43.2
H	4.8	4.9	6.4	6.4	6.3
O	49.6	47.1	43.7	43.7	43.2
N	1.6	0.6	—	—	2.6
S	0.1	0.6	—	—	0.2
Thermochemical and physical properties					
Bulk density (kg/m^3)	184	301	—	196	—
Higher heating value (MJ/kg)	16.1	18.2	—	—	20.1
Lower heating value (MJ/kg)	—	—	18.4	18.4	—

[a]dry basis.

and fermentation processes, respectively. This byproduct is termed *parchment husks* (Saenger et al., 2001).

The chemical and physical properties of the different used coffee husks are given in Table 11.2.

A first observation is to attribute the minerals coming from the outer skin of the coffee cherry while sulfur is originated from the parchment and the silverskin. Both minerals and sulfur contents have negative impacts on the combustion process. Minerals induce the formation of slag but also the emission of small particles (PM1) while sulfur leads to sulfur dioxide ($SO_{2(g)}$) and sulfuric anhydride ($SO_{3(g)}$) presence in the combustion gas. Second, the presence of nonnegligible amounts of nitrogen in Mbuni husks will lead to NO_x emission during combustion.

Suarez and Luengo (2013) have carried out an interesting study on the combustion of coffee husk briquettes in Cuba. The objective of this study was to compare the use of coffee briquettes with firewood in a bakery furnace. They have produced briquettes with 80 mm of diameter and 140 mm length with a density of 840 kg/m^3. During the briquette production, the migration of lignin was observed at the surface of briquettes due to the applied pressure by the densification system. This phenomenon protects the fuel from enzymatic degradation.

The combustion characteristics of coffee husk briquettes correspond to: 5 min of lighting time, 10 min of flaming time, 25 cm of flame length, and 30 min for combustion of char. The combustion parameters of coffee husk briquettes are similar than the ones of firewood: thermal efficiency of 64%, fuel feed rate 20 kg/h, specific production of bread 16.2 kg/kg of fuel. This result indicates that coffee husk briquettes can replace firewood in bakery or pizzeria furnaces. The only limit corresponds to the cost necessary to produce the briquettes and the final selling price for customers.

The combustion of coffee husks has also been investigated by Saenger et al. (2001), their study focuses on the behavior of "Mbuni and parchment husks" during the combustion process, from the feed system to the composition of the exhaust gas. The combustion tests were carried out in a laboratory-scale furnace (single-particle combustion) and with a fluidized bed combustor (FBC). The main information that can be extracted from this work corresponds to the combustion behavior of coffee husks due to their composition. The high volatile fraction of both coffee by-products induces an easy ignition (rapid devolatilization) and a short combustion time. This parameter has to be considered to adapt the fuel feeding to the combustion system (heat release, cooling of the feeding system) and to avoid blockages due to pyrolysis and slag formation. As mentioned, the devolatilization step of coffee husks is a key parameter for the design and the operating mode (air distribution) of the combustion systems.

The composition of ash coming from both Mbuni and parchment husks indicates a high concentration of K_2O (37 wt.%) and SiO_2 (15 wt.%). The main problem comes from the presence of high content of K_2O, which tends to lower the melting point of ash and to be at the origin of slagging. The addition of Fe_2O_3 or kaolin is a good way to prevent this drawback.

Gil et al. (2010) studied the combustion behavior of different biomasses blended with coal or pine sawdust. Among these biomasses, coffee husks (5 wt.%) were blended with pine and chestnut sawdust (95 wt.%). The results of combustion test, obtained with a TG instrument, showed that coffee husks have a different behavior from other biomasses. The volatile matter burned at a lower rate than other biomasses, while char combustion happened at a higher rate and for a lower maximal temperature. This phenomenon is attributed to the chemical composition of coffee husks (Rhen et al., 2007).

11.1.1.3 Combustion of spent coffee grounds

Only a few works were carried out on the combustion of spent coffee grounds (SCG), especially on domestic systems (boiler and stove). Limousy et al. (2013, 2015) have performed developments in order to both produce fuels from spent coffee grounds and to adapt these fuels to conventional combustion systems. Wei et al. (2016) have studied the oxy-fuel cocombustion of spent coffee grounds with coal in a thermogravimetric setup to evaluate the potential synergistic effect of the fuels.

The use of spent coffee grounds in conventional systems of combustion confronts a major problem. As shown previously, the volatile fraction is high in spent coffee grounds, which is an advantage for combustion ignition. Nevertheless, the densification of pure SCG leads to the formation of dense pellets or logs. Then, combustion performances and gaseous emissions observed with pure densified SCG fuels are not satisfying enough to reach French or European standards. Conventional densified wood logs and pellets obtained from sawdust have a specific behavior in the combustion chamber. They tend to expand when combustion takes place, generating surfaces available to oxygen diffusion and then to combustion propagation. Using densified SCG (pellets, briquettes, or logs) does not lead to this phenomenon. This is due to the high cohesion of SCG particles that leads to a different behavior for SCG (it consumes more than it burns). Better cohesion of biomass can occur if SCG is blended with another biomass, such as wood sawdust: it promotes an increase of combustion time, as well as heat release while fuel remains dense and does not expand (a limitation of the combustion kinetic).

11.1.1.3.1 Combustion of pellets containing SCG

Different pellets formulation containing SCG were tested by Limousy et al. (2013) and Jeguirim et al. (2016) in order to compare the combustion performances of SCG to DIN+ pellets, which correspond to the best quality of wood pellets for boilers (German standard). They studied the combustion of pure and blended SCG (50/50 wt.%) with pine sawdust. The physicochemical properties of the different pellets are presented in Table 11.3. The experimental setup used for this study consists in a domestic pellet boiler (Okofen, 12 kW) equipped with a gas and a particle analyzer (Fig. 11.4).

In their study, Limousy et al. (2013) estimated the combustion and the boiler efficiencies in order to distinguish the quality of the fuel and the compatibility of the different fuels with the boiler. They observed that the combustion efficiency (estimated via the NF EN 12809 standard) of pure SCG pellets was high (86.3%) and very close to efficiencies obtained for pine and blended pellets (90.8% and 91.9% respectively). This result encourages the use of SCG in boilers. An interesting result corresponds to the combustion efficiency obtained with the blended pellets (50/50 wt.%), which is higher than the one obtained with the DIN+ pellets. A potential reason for this result comes from the synergistic effect between SCG and wood that has also been highlighted by Wei et al. (2016) during their investigations in coal/SCG blend

Table 11.3 Combustion Characteristics of Coffee Husk Briquette (Suarez and Luengo, 2013)

Characteristics	Value[a]
Lighting time (min)	5
Flaming time (min)	10
Length of the flame (cm)	25
Time for combustion for char (min)	30

[a]Average of 3 pieces.

FIGURE 11.4 Experimental Setup Used for the Combustion Tests Carried Out With a Domestic Boiler (Okofen 12 kW)

Reprinted with permission from Limousy, L., Jeguirim, M., Dutournie, P., Kraeim, N., Lajili, M., Said, R., 2013. Gaseous products and particulate matter emissions of biomass residential boiler fired with spent coffee ground pellets. Fuel. 107, 323–329. Copyright Elsevier 2016.

oxy-combustion tests. The presence of high K content in SCG is suspected to be the origin of such behavior, since potassium may act as a catalyst for biomass oxidation. Concerning boiler efficiency, it can be observed that the combustion of pure SCG pellets leads to a poor recovery of energy (yield of 64.1%), in comparison with the blended and the DIN+ pellets (83.5% and 84.3%, respectively). The high volatile matter content in SCG leads to a rapid ignition of the pellets and to a fast energy release inside the combustion furnace. The boiler heat exchanger and the combustion furnace of the pellet boiler are designed for a specific biomass. The heat release must be close to the one of wood pellets in order to reach high boiler efficiency. This difference of combustion is also observable through the ash residue obtained with the pure SCG pellets. It presents the same shape as the raw pellets, which confirms the fast devolatilization of the organic matter from the pellets. However, the boiler efficiency obtained with the blended pellets is comparable to the one of DIN+ pellets. It confirms that the use of adapted quantities of SCG in blends allows obtaining identical performances compared with DIN+ pellets. This result is remarkable in that it validates the possibility of using SCG to produce high performance agro-pellets containing SCG, which represents several advantages:

- Valorization of a worldwide available byproduct [more than 6 million tons each year (Jeguirim et al., 2014a)].
- To reduce the pressure on wood sawdust.
- To moderate the price of domestic pellets.
- To create a new circular economy (low distance market).

From an environmental point of view, the combustion of SCG leads to gas and particulate emissions different from those of DIN+ pellets. Limousy et al. (2013) observed a degradation of the exhaust gas composition when increasing the amount of SCG in the pellets. Emissions of CO were extremely affected with pure SCG pellets with an average value around 3000 mg/Nm³, whereas they were limited at 600 and 260 mg/Nm³ for the blended and DIN+ pellets (all the data are given at 10 vol.% O_2). NO_x concentration was also affected but in the present case it corresponds to higher nitrogen content in the pellets when SCG was used (2.9, 1.5, and 0.1 wt.% for pure SCG, the blend and DIN+ pellets respectively). NO_2 was absent when DIN+ combustion was carried out, whereas concentrations of 20 and 80 mg/Nm³ were observed with blended and pure SCG pellets respectively. The behavior of VOC's emissions was more indicative of the boiler efficiency (Fig. 11.5), because they were unstable for the pure SCG pellets whereas they presented constant values for the other formulations. This result was attributed to the fast release of the volatile matter from the SCG pellets when the feeding system was operating.

Particle matter (PM) emissions were monitored during the combustion tests to estimate the impact of SCG on the concentration and on the size of particles (from

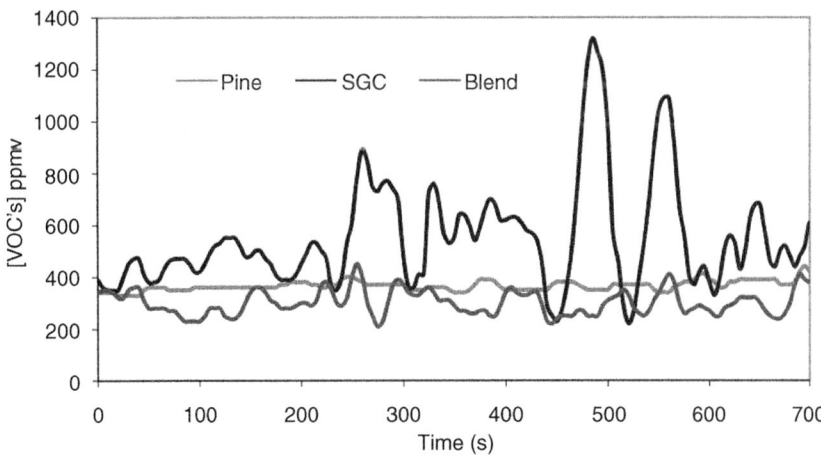

FIGURE 11.5 Profile of VOC's Emissions Given at 13%vol of O_2 (ppmv) Obtained During the Combustion of Different Pellets Containing SCG

Reprinted with permission from Limousy, L., Jeguirim, M., Dutournie, P., Kraeim, N., Lajili, M., Said, R., 2013. Gaseous products and particulate matter emissions of biomass residential boiler fired with spent coffee ground pellets. Fuel. 107, 323–329. Copyright Elsevier 2016.

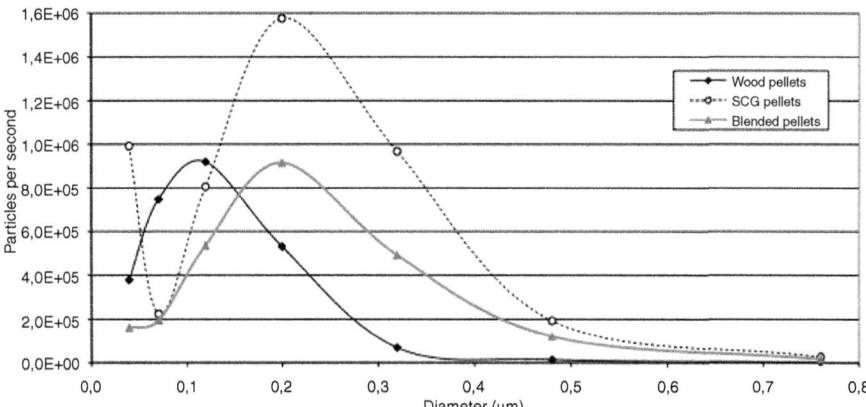

FIGURE 11.6 PM1 Size Distribution Observed During the Combustion of Different Pellets Containing SCG

Reprinted with permission from Limousy, L., Jeguirim, M., Dutournie, P., Kraeim, N., Lajili, M., Said, R., 2013. Gaseous products and particulate matter emissions of biomass residential boiler fired with spent coffee ground pellets. Fuel. 107, 323–329. Copyright Elsevier 2016.

29 nm to 10 μm). The concentration of PM increased with SCG content reached 1500 mg/Nm^3 for pure SCG pellets, 400 mg/Nm^3 for the blend, and 140 mg/Nm^3 for the DIN+ pellets (pine). The particle size distribution was also modified in the presence of SCG, but in a good way. In fact, the presence of SCG in the pellets tends to increase the particle size (Fig. 11.6), with the smallest being the more dangerous for health (especially PM1).

11.1.1.3.2 Combustion of densified logs containing SCG

Recently, Limousy et al. (2015) presented the first work performed on compressed logs. The originality and the complexity of this work resides in the development of new solid fuels combined with the wood stoves' performances. In this study, different fuels were compared with each other through the NF EN 13229 standard devoted to domestic wood stoves, and through the *green flame* French label. This label has been set up by the French Environmental and Energy Control Agency (ADEME) in 2000 to promote manufacturers developing very efficient stoves from an environmental and combustion performances. The *green flame* label allows purchasers to recover a part of their investment in the stove through tax credit benefits. This label can be obtained for a stove respecting two criteria: average CO emissions lower than 0.3 vol.% and yield of combustion higher than 75% (five stars label). These criteria evolve year after year to incite stove manufacturers to develop new stove technologies with increasing performances and efficiencies (reduction of the CO emissions to 0.15 and 0.12 vol.% and PM to 50 and 40 mg/Nm^3 for 6- and 7-star stoves, respectively, since 2015).

In this study, combustion tests were carried out in an 8 to 12 kW cast-iron stove (XP 68, Lorflam, France). The experimental setup was installed in the laboratory of Lorflam Company (Lanester, France) (Fig. 11.7). The solid fuel was supplied by

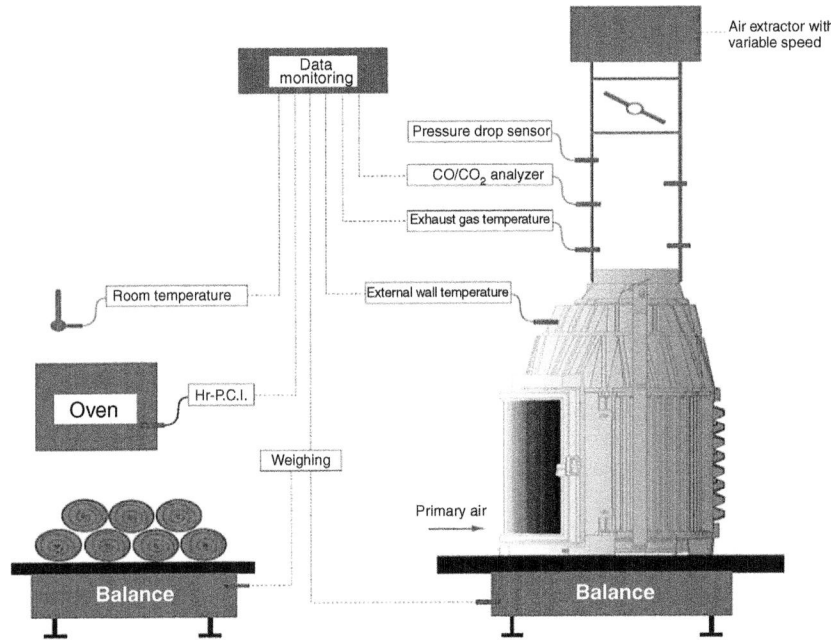

FIGURE 11.7 Experimental Setup (Synoptic and Laboratory Photo) Used During the Combustion of Fuel Logs Dontaining SCG

Reprinted with permission from Limousy, L., Jeguirim, M., Labbe S., Balay, F., Fossard, E., 2015. Performance and emissions characteristics of compressed spent coffee ground/wood chip logs in a residential stove. Energy Sustain. Develop. 28, 52–59. Copyright Elsevier 2016.

Table 11.4 Physicochemical Properties of the Different Pellet Formulations Containing SCG Used by Limousy et al. (2013)

Pellets	Moisture (%)	VM[a] (%)	FC[a] (%)	Ash[a] (%)	LHV[e] (kJ/kg)
Pure SCG	11.78	68.94	17.46	1.82	17.52
Blend	6.65	78.23	13.94	1.18	17.91
DIN+	7.90	77.61	14.19	0.30	17.80

[a]On wet basis.

RID Solution Company (Theix, France). With the experience of pellets containing SCG, RID Solution has developed an adapted variety of solid fuel logs containing 20 wt.% SCG blended with pine sawdust (Fig. 11.2, Obiflam variety). The combustion tests were carried out with 3 different fuel loads (Table 11.4). The main interest in comparing these different tests was to validate the adaptability of compressed logs to woodstoves.

In fact, compressed wood logs have been available in the French market for many years. The main problem with this solid fuel is that the energy release is very fast. This kind of fuel is more adapted to stoves that can store energy (Finnish mass stoves, pizzeria ovens, kachelöfens, etc.), than to classical stoves (made of cast iron or steel), because it generates overheating. The fuel developed by RID Solution integrates 20 wt.% of SCG in order to limit this problem by modifying the behavior of the compressed logs during combustion: no expansion and opening of the logs. Then the residential market can be investigated with new solid fuel with the benefit of recycling a well-known by-product.

The combustion tests were carried out for 45 min with these logs in specific conditions described in the NF EN 13 229 standard. The main difference with other combustion systems corresponds to the draught (12 Pa of pressure drop), which is automatically driven by the combustion rate. CO_2, CO, O_2 emissions were followed during the tests in order to estimate the combustion quality and the yield of combustion. Surprisingly, CO_2 and O_2 profiles were very different when comparing the tests carried out with the SCG compressed logs and the other tests (Fig. 11.8A–B).

The ignition started immediately with the beech logs (maximal concentration of CO_2 observed after 3 min), whereas it was observed only after more than 10 min for the compressed logs containing 20 wt.% SCG. The delay modifies the heat release in the stove, generating a faster combustion of the compressed logs compared to conventional wood logs. This parameter can be managed by a modification of the airflow entry associated to the design of the air distribution system. The combustion behavior observed with the intermediary system (melt of compressed and conventional logs) is close to the one obtained with conventional logs. The yield of combustion was higher for the compressed logs (20 wt.% of SCG) than for the other loads (82.25% compared to 82.05% for the mixed load and 81.75% for beech logs).

(A)

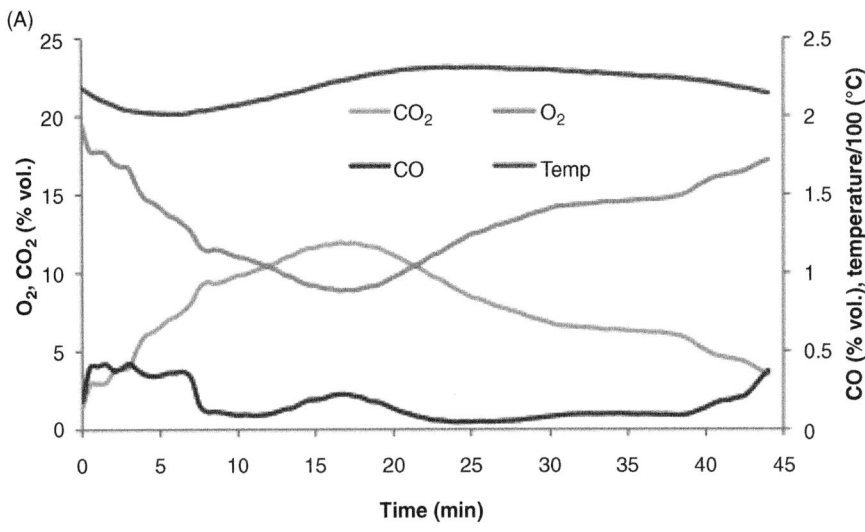

(B)

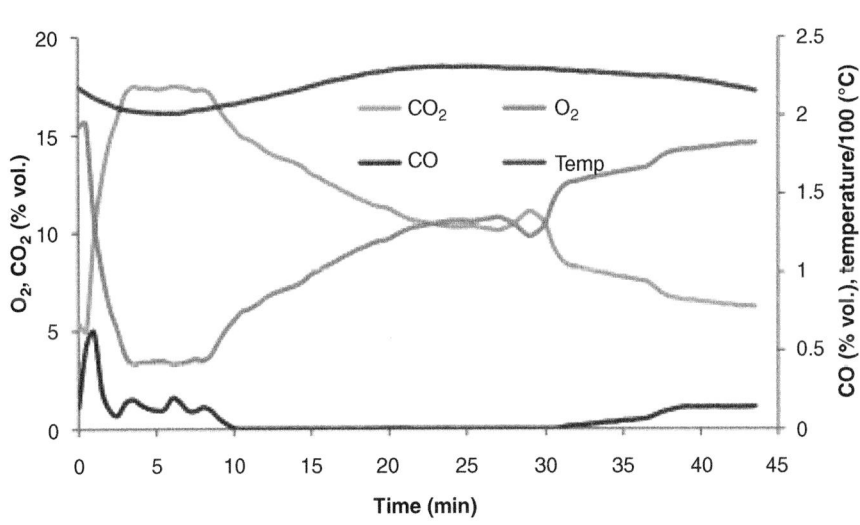

FIGURE 11.8

Concentrations of CO_2, CO, and O_2 (% vol) observed during the combustion tests performed with (A) compressed logs (20 wt.% SCG + 80 wt.% pine sawdust) and (B) conventional beech logs.

Reprinted with permission from Limousy, L., Jeguirim, M., Labbe S., Balay, F., Fossard, E., 2015.
Performance and emissions characteristics of compressed spent coffee ground/wood chip logs in a
residential stove. Energy Sustain. Develop. 28, 52–59. Copyright Elsevier 2016.

The stove used for this study has a 5-star green flame label, which was verified by the combustion test. Calculations performed with the emissions values collected during the different tests showed that this label was also obtained with the compressed logs containing 20 wt.% of SCG.

It means that the compressed logs containing 20 wt.% spent coffee grounds blended with sawdust are totally adapted to wood stove devices, and that some minor adaptations have to be done in order to improve the heat release obtained with such solid fuels.

11.1.2 PYROLYSIS, TORREFACTION, GASIFICATION, AND HYDROTHERMAL CONVERSION PROCESSES

Before describing the different biochemical and thermochemical processes, the thermogravimetric analysis of coffee by-products will be presented. Indeed, to better understand the thermochemical processes, it is crucial to study the thermal behavior under different atmospheres of coffee wastes.

11.1.2.1 Thermogravimetric analysis

Li et al. (2014) studied the thermal behavior under inert atmosphere of spent coffee grounds (SCG), using two different heating rates, 10 and 60°C/min, in order to simulate the slow and fast pyrolysis. Fig. 11.9 presents the curves of weight losses versus temperature (TG curves) and their derivatives (DTG). The weight loss occurred in four steps:

1. The first maximum obtained at 75°C for 10°C/min is due to moisture release
2. The second maximum obtained at 300°C for 10°C/min is due to hemicellulose decomposition
3. The third maximum obtained at 335°C for 10°C/min is due to cellulose decomposition
4. The fourth maximum obtained at 390°C for 10°C/min is due to the lignin structure decomposition. This stage is the main contributor to the final mass of char.

Increasing the heating rate from 10 to 60°C/min^{-1}, leads to the classical shift of the maxima to higher temperatures (Kan et al., 2014; Li et al., 2014). Steps 2 and 3 are characterized by a high weight loss rate and are qualified by devolatilization steps. Step 4, also called char formation stage, is characterized by a low weight loss rate (Jeguirim et al., 2014a).

Similar thermal behaviors were obtained by Jeguirim et al. (2014b), who studied coffee husk with different other biomasses. Similar results were also obtained during the thermal analysis of coffee residue performed by Chen et al. (2012) and Tsai and Liu (2013). However, the three latter studies obtained lower values for the maximum temperature of hemicellulose decomposition.

Comparing coffee husk and corn cob, Jeguirim et al. (2014b) noted lower temperatures for hemicellulose and cellulose decomposition for coffee husk. This

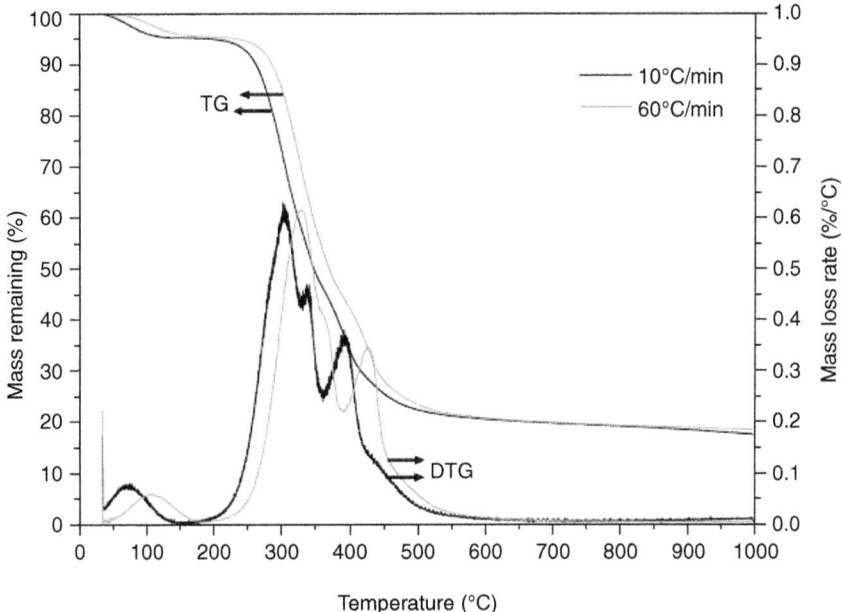

FIGURE 11.9 Thermal Gravimetric (TG) and Derivatives of the Thermal Gravimetric (DTG) Curves, Under Nitrogen Atmosphere, of Spent Coffee Grounds

Heating rates: 10 and 60°C/min, total gas flow: 20 NmL/min.

Reprinted with permission from Li, X., Strezov, V., Kan, T., 2014. Energy recovery potential analysis of spent coffee grounds pyrolysis products. J. Anal. Appl. Pyrol. 110, 79–87. Copyright Elsevier 2016.

observation was ascribed to the higher ash content in coffee husk and then to the catalytic effect of some minerals, such as potassium on the thermal degradation behavior.

11.1.2.2 Pyrolysis

Pyrolysis consists of thermal degradation of materials in absence of oxygen. The range of temperature is usually 400–600°C, but higher or lower temperatures could be used depending on the further use of the pyrolysis products. In fact, pyrolysis produces gas (syngas), liquid (biocrude or bio-oil), and solid fractions (biochar) (Hughes et al., 2014). The percentage of each fraction depends on the pyrolysis type, operating conditions, biomass nature, and so forth. Typical pyrolysis processes require water content lower than 40 wt.% to convert suitably the biomass into bio-oil. Indeed, the presence of water increases the energy necessary to the degradation of biomass by increasing the heat of vaporization of the feedstock.

Pyrolysis could be divided into three main categories: slow, fast, and flash pyrolysis. These categories differ according to operating conditions: process temperature, heating rate, solid residence time, biomass particle size, and so forth. Relative distribution of the solid, liquid, and gas fractions varies accordingly with the category

Table 11.5 Characteristics of the Different Fuel Loads Used to Estimate the Combustion Performances of the Densified Logs Containing 20 wt.% of SCG (Limousy et al., 2015)

Fuel Loads	Humidity (%)	Ash Content (%)	LHV (kJ/kg)
SCG/Pine sawdust (20/80 wt.%): 1.8 kg	10	0.62	17,386
Beech wood logs: 2 kg	17	0.49	14,557
50% of each fuel: 1.8 kg	13.5	0.56	15,971

of pyrolysis. Jahirul et al. (2012) summarized the different types of pyrolysis in Table 11.5.

Pyrolysis is nowadays considered for the production of transportation fuels and other products (Hughes et al., 2014). Fast pyrolysis is considered as a new technology and has been extensively studied during the past decade (Jahirul et al., 2012). The fast pyrolysis is more time and energy saving compared to slow pyrolysis. Moreover, it leads to higher amounts of bio-oils, with more adequate properties for its application as liquid fuels (Hughes et al., 2014; Jahirul et al., 2012).

11.1.2.2.1 Slow pyrolysis of defatted coffee grounds

Vardon et al. (2013) investigated the effect of defatting on the pyrolysis products of spent coffee grounds (SCG). The lipids extracted were converted into biodiesel. The yields of the different fractions—solid, liquid, and gaseous—issued from the slow pyrolysis of the defatted and nondefatted SCG were determined and summarized in Fig. 11.10. Particularly, due to lipid extraction, the yield of bio-oil decreased from 27.2% for SCG to 13.7% for the defatted grounds (Vardon et al., 2013). The higher yield of aqueous-phase obtained for defatted SCG is ascribed to the elevated content of cellulose and hemicellulose, due to the lipid extraction. Indeed, cellulose and hemicellulose can degrade into water-soluble organics. The biochar (27%–28%) and gas (21%–24%) yields did not change significantly by the effect of defatting. They remained comparable to the known values of SCG.

Slow pyrolysis of defatted coffee grounds generated valuable by-products, bio-oil, and biochar (Vardon et al., 2013). The characteristics of the bio-oil issued from the pyrolysis of coffee are discussed in a further section (see section titled "Characteristics of crude oil").

11.1.2.2.2 Fast pyrolysis of upgraded coffee grounds

Residues from industrial production of liquid coffee (upgraded coffee grounds or UCG) were studied by pyrolytic gasification (fast pyrolysis by steam) by Mašek et al. (2008). In three different laboratory-scale reactors, the UCG pyrolysis under steam was carried out. The process was performed in two-stages: (1) the fast pyrolysis of UCG by steam and (2) the combustion of the as-obtained solid

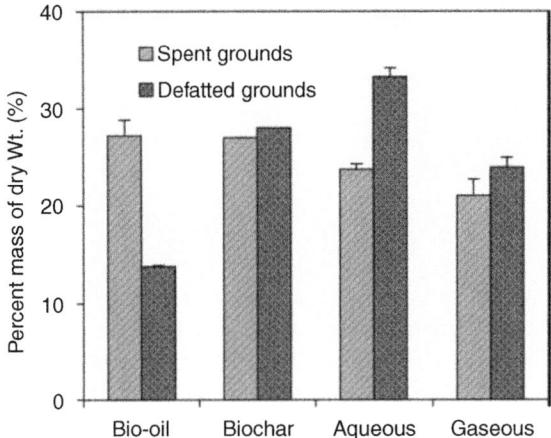

FIGURE 11.10 Mass Balance Yield (wt.%) for the Product Phases Derived From the Slow Pyrolysis of Spent Coffee Grounds and Defatted Coffee Grounds

Reprinted with permission from Vardon, D.R., Moser, B.R., Zheng, W., Witkin, K., Evangelista, R.L., Strathmann, T.J., Rajagopalan, K., Sharma, B.K., 2013. Complete utilization of spent coffee grounds to produce biodiesel, bio-oil, and biochar. ACS Sustain. Chem. Eng. 1, 1286–1294. Copyright American Chemical Society 2016.

residue with air. 88% of the carbon contained in UCG was converted into volatiles and gases by fast pyrolysis at 800°C. Furthermore, the degree of conversion was not influenced by the steam presence or concentration. This fact is an advantage for allothermal (heat by external source) gasification, since the amount of steam needed as carrier gas could be minimized without a negative effect on conversion. Moreover, the high degree of conversion suggests that the amount of char produced is equivalent to the one required in a combustor to provide heat for endothermic reactions in the pyrolyser. Along with the gas, a high amount of tar was produced. Tar is difficult to eliminate by steam under 900°C. It could be combusted by air to feed the reactor in energy or could be catalytically cracked into noncondensable gases. The HHV of the product gas was 14.9 MJ/Nm^3 (Mašek et al., 2008).

The authors concluded that SCG is a valuable source of energy with efficient factors: high degree of conversion, elimination of residue material, and particle sizes in the order of millimeters (reducing the need for fuel preprocessing).

11.1.2.2.3 Fast pyrolysis of spent coffee grounds

Li et al. (2014) studied the pyrolysis of spent coffee grounds (SCG) provided from a cafeteria in the library building of Macquarie University, Sydney, Australia. The heating rate did not significantly affect the fraction of each type of product issued from pyrolysis. The maximum yield for liquid was 66 wt.%, reached at around 630°C. The pyrolysis efficiency of SCG was estimated at 77%–85%, depending on the moisture content.

The combustion of the pyrolytic products could result in more than 22,000 MJ/ton. It was evaluated to 6% of the annual heating energy of the library building. It was estimated that about 1450 kg of CO_2 equivalent were saved by using coffee grounds as renewable energy source.

Bok et al. (2012) compared the fast pyrolysis of coffee grounds and wood (mallee), in the temperature range of 400 to 600°C. They concluded that the reaction temperature was the most important factor in fast pyrolysis of coffee grounds. The highest yield of biocrude (64.85%) was obtained at 550°C. The high heating value (HHV) of coffee grounds (22.74 MJ/kg) was higher than wood (17.32 MJ/kg). The higher HHV was linked to the higher amount of carbon and a lesser amount of oxygen present in coffee grounds compared to wood.

To thermally decompose lipid contained in coffee grounds and thus to obtain better quality of biocrude oil, pyrolysis temperatures higher than 400°C are required. Indeed, lipids are actively decomposed at 450°C (Bok et al., 2012). Ngo et al. (2015) pyrolyzed spent coffee waste at 550–750°C. They observed that the gas fraction increased with increasing the pyrolysis temperature, whereas the liquid yields decreased. The char yield varied in the range of 15–23 wt.%. At 550°C with a flow rate of 20 mL/min of N_2, the highest yield of bio-oil was obtained, 58.5 wt.%. Increasing the flow rate of the gas from 20 to 500 mL/min results in an increase of the liquid and char yields and a decrease of the gas yield. The highest gas and liquid yield was 63.4 wt.% for coffee waste.

Kelkar et al. (2015) studied the fast pyrolysis of spent coffee grounds at 429–550°C, using a compact, transportable, screw conveyor reactor. It was shown that increasing the temperature, up to 505°C, leads to an increase of the liquid (bio-oil) yield. Conversely, the decrease of residence time results in an increase of the liquid yield. The highest bio-oil yield (61.8 wt.%) was obtained at 500°C whereas the highest char one, 20.6 wt.%, at 429°C. Using a screw conveyor reactor instead of a fluidized bed reactor increased the bio-oil yield and decreased the char yield. In addition, increasing the screw speed results in an increase of the bio-oil yields.

Yang et al. (2012) pyrolyzed coffee beans to produce oil and then studied mixtures of the pyrolysis oil with petroleum diesel prepared by emulsification methods. The mixture (5 vol.% of oil), considered as low-add ratio, resulted in:

- A lower overall calorific value of the fuel.
- An increase of combustion efficiency.
- An earlier diesel engine ignition time.
- A relatively low indicated specific fuel consumption (ISFC), at low rpm (revolutions per minute), and significant higher ISFC at higher rpm.
- An improvement of the total heat release of pure diesel at moderate and high rpm.
- A higher performance during high-rpm operations compared to lower rpm.
- Lower pollutants emissions.

To improve the thermochemical properties of exhausted coffee residue, Tsai et al. (2012) carried out pyrolysis at 400–700°C with a heating rate of 10°C/min.

11.1.2.2.4 Microwave fast pyrolysis of coffee hulls

Fernández and Menéndez (2011) compared the conventional and the microwave pyrolysis of coffee hulls at 500–1000°C. Microwave system saves energy and time. In addition, microwave heating produces higher gas yield along with elevated syngas content compared to the conventional method. It seems that the hot-spot phenomenon, that occurs only under microwave heating, is responsible for the latter enhancement.

11.1.2.2.5 Copyrolysis

The coffee waste was copyrolyzed with other biomasses or wastes. The copyrolysis of different fractions of coffee wastes and polypropylene (PP) was studied by Zanella et al. (2013) at 360–420°C. The increase of the fraction of PP reduces the amount of light liquid products (with carbon atom number lower than 6, <C6) and increases the fraction of heavy condensate products. A decrease of the amount of light liquid products and a stabilization of those of medium- (C6–C16) and high-molecular-weight products (>C16) were obtained with the increase of temperature.

On the other hand, Soysa et al. (2015, 2016) copyrolyzed, at 400–600°C, in bubbling fluidized bed reactor, coffee ground with woody biomass (Douglas fir), with a ratio of weights of 1:1. The highest biocrude oil yield (54 wt.%) and the highest energy yield (55 wt.%) were obtained at 550°C. The authors demonstrated the beneficial effect of mixing coffee waste and woody biomass on the enhancement of the properties of the bio-oil obtained and the decrease of the energy activation (135 kJ/mol).

11.1.2.2.6 Catalytic pyrolysis

The catalytic pyrolysis involves the use of a catalyst during the process. The catalyst could be either directly mixed with biomass or placed in a fixed bed immediately downstream of the pyrolysis reactor (Hughes et al., 2014). Zeolite is the most-used catalyst in such processes till now. The gas products issued from pyrolysis could be adsorbed in the pores of the zeolite where cracking and rearrangement reactions produce olefins and aromatic compounds. It is an interesting feature since oxygenated compounds are converted into hydrocarbons without the use of hydrogen. However, coke is susceptible to be formed on the catalyst, especially when the biomass contains high amounts of lignin. Fig. 11.11 gives the aromatic yields obtained by Hughes et al. (2014) during the catalytic pyrolysis of coffee bean waste, which was previously subjected to biochemical processing. The catalyst used was HZSM-5 zeolite. The aromatic yields are in the range of 5–9 wt.%. These values were considered lower than those obtained from carbohydrate and protein-rich feedstocks, about 30 wt.%, because after the biochemical processing, the coffee was mainly composed of lignin or recalcitrant polysaccharides (Hughes et al., 2014).

Fischer et al. (2015) found similar results with ZSM-5 during the fast pyrolysis of spent coffee grounds; selectivity towards deoxygenated olefins and aromatic compounds: -benzene, xylene, and toluene increased in presence of ZSM-5. Indeed, ZSM-5 was found to catalyze oligomerization, cyclization, and aromatization

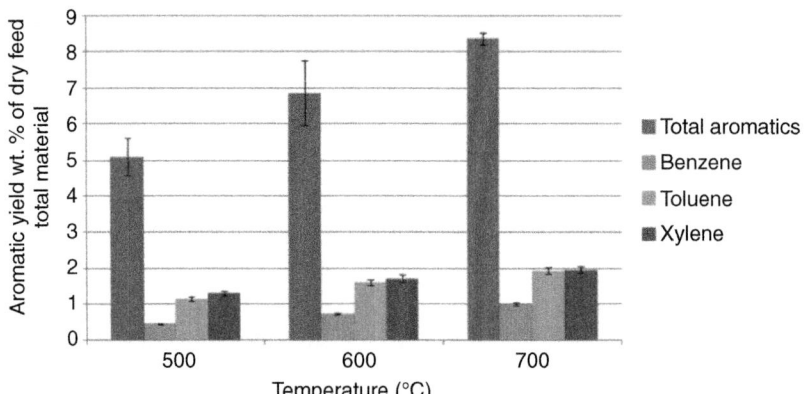

FIGURE 11.11 Production of Aromatics by Catalytic Pyrolysis of the Residue From Coffee Waste Biochemical Conversions

HZSM-5 zeolite was used as catalyst.

Reprinted with permission from Hughes, S.R., Lopez-Nuñez, J.C., Jones, M.A., Moser, B.R., Cox, E.J., Lindquist, M., Galindo-Leva, L.A., Riaño-Herrera, N.M., Rodriguez-Valencia, N., Gast, F., Cedeño, D.L., Tasaki, K., Brown, R.C., Darzins, A., Brunner, L., 2014. Sustainable conversion of coffee and other crop wastes to biofuels and bioproducts using coupled biochemical and thermochemical processes in a multi-stage biorefinery concept. Appl. Microbiol. Biotechnol. 98, 8413–8431. Copyright Springer 2016.

reactions of straight parafins and olefins to benzene and xylene. It also increases the CO yield in the gas fraction, via decarbonylation reactions. Fatty acids pyrolyzed mainly through decarboxylation and decarbonylation reactions.

Kan et al. (2014) used a series of NiCu impregnated on γ-Al$_2$O$_3$ (NiCu/γ-Al$_2$O$_3$) as catalysts during the pyrolysis of coffee grounds. Nickel content was 14 wt.% and copper content was varied between 0 and 6 wt.%. The catalyst presence and the increase of copper content increased the concentrations of CO and CO$_2$ in the gas. This observation was explained by the favoring of the decarbonylation and the decarboxylation of the oxygenated compounds present in the coffee grounds. Furthermore, in presence of the catalyst, a more homogeneous distribution of the organic compounds in the bio-oil was obtained. However, the amount of bio-oil produced was lower than in absence of catalyst. In addition, the catalyst enhanced char gasification, leading to a lower amount of char at the end of the pyrolysis process (17.5–19.6 wt.% with catalyst versus 22.2 wt.% without catalyst). The increase of the gas fraction in the presence of catalyst was ascribed by the authors to the catalytic conversion of the intermediates from pyrolysis on the Ni and Cu active sites.

11.1.2.2.7 Characteristics of crude oil

Romeiro et al. (2012) devoted a study to the characterization of the bio-oil issued from the pyrolysis at 380°C of soluble coffee grounds. The fraction of the pyrolysis oil was good, 50 wt.%. The other fractions were: char: 29 wt.%, gas: 15 wt.%, and acids: 6 wt.%.

The obtained oil possess the following properties:

- chemical composition: 0.4 wt.% water, 0.04wt.% sulfur, high contents in hydrocarbons,
- kinematic viscosity: 266 mm^2/s,
- density at 20°C: 0.960 kg/m^3, and
- calorific value: 34.3 MJ/kg[1]

Only the viscosity of the oil was higher than the required value for the use as a fuel. Blending the oil with additives, such as diesel, could resolve this point.

The characteristics of the bio-oil obtained by the slow pyrolysis of spent and defatted coffee grounds (Vardon et al., 2013) are listed in Table 11.6. Defatting decreases the higher heating value (HHV), also called energy density. However, the SCG kept values of HHV not far from those of petroleum crude oils (41–48 MJ/kg). Furthermore, defatting leads, after slow pyrolysis, to bio-oil with lower aliphatic functionality and a higher number of low-boiling oxygenates (Vardon et al., 2013).

Li et al. (2014) pyrolyzed coffee grounds, with two different heating rates, 10 and 60°C/min. The bio-oil obtained contained caffeine and lipids as in the raw material. Moreover, palmitic and linoleic acids were the major compounds, giving a great potential for biodiesel production. Indeed, 21 chemical compounds were identified in the bio-oil and their relative contents were determined by Li et al. (2014). A higher heating rate improved the quality of the bio-oil by improving its elemental composition. Bok et al. (2012) and Ngo et al. (2015), like Li et al. (2014), performed a gas chromatrography coupled to mass spectrometry (GC-MS) analysis of the bio-oil derived from coffee grounds. Among the compounds identified by the three groups of researchers were: caffeine, phenol, 4-methylphenol, catechol, acetic acid, butanoic acid, dodecanoic acid, hexadecanoic acid, palmitic acid, linoleic acid, 1-eicosanoic acid, 2-ethoxyethyl acetate, furan-2-ylmethanol, 2-propanone 1-hydroxy-, 2-propanone 1-(acetyloxy)-, cycloheptanone, di(ter-buty) silyl ester, hydroquinone, 3,4-dihydropyran, 1,2-cyclopentanediene, 2-furanmethanol, 1,2-benzenediol, 1,2-cyclopentadione 3-methyl-, 2(5H)-furanone, cyclopropyl carbinol, 1,2-benzenediol, 2-butanone 3-hydroxy-, 1,3-cyclopentanedione, 1,2-ethanediol, pyridine, 1-hydroxy-2-butanone, indolizine,

Table 11.6 Typical Operating Parameters and Products Yields for Pyrolysis Process (Jahirul et al., 2012)

Pyrolysis Process	Solid Residence Time (s)	Heating Rate (°C/s)	Particle Size (mm)	Temperature (°C)	Product Yield (%)		
					Bio-oil	Biochar	Biogas
Slow	450–550	0.1–1	5–50	275–675	30	35	35
Fast	0.5–10	10–200	<1	575–975	50	20	30
Flash	<0.5	>1000	<0.2	775–1030	75	12	13

9,12-octadecadienoic acid (Z,Z)-, 2-cyclopenten-1-one 2-hydroxy-3-methyl, and so forth.

The major compounds of the bio-oil are phenol derivatves, aldehydes, ketones, acids, and alcohols (Ngo et al., 2015). In particular, a noticeable amount of fatty acids (dodecanoic and hexadcanoic; C12 and C16) was noted (Ngo et al., 2015), revealing a thermal stability of such bio-oil; the thermal stability of a molecule increasing with its molecular weight. However, the quality of bio-oil is reduced by the presence of oxygen atoms in the organic compounds. Some process should be applied to this bio-oil to make it suitable for use as a fuel.

Spent coffee grounds led to bio-oil richer in hydrocarbons and lipids than bio-oils deriving from other feeds (woody or herbaceous biomass) (Kelkar et al., 2015). In addition to oxygenated organic compounds, typical of bio-oils, the bio-oil was composed of hydrophobic compounds, such as fatty acids, fatty acid esters, medium-chain parafins, olefins, and caffeine.

During the copyrolysis of coffee hulls and polypropylene (PP), Zanella et al. (2013) noted the presence of pyridine, phenol, C12, and C13, sometimes with hydroxyl groups. In addition, hexadecanoic acid, caffeine, and hydrocarbons up to C30 were detected in the high-molecular-weight products. These authors obtained during the copyrolysis temperatures higher than 400°C a liquid comparable to fossil fuel oil in molceular weight range and composition.

Soysa et al. (2015, 2016) also evidenced the presence of high amounts of oxygenated compounds, alcohols and ethers, in the bio-oil issued from the copyrolysis of woody (Douglas fir) biomass and waste coffee. The high contents of alcohols and ethers improves the homogeneity, decreases the viscosity and the density of the bio-oil and therefore produces better properties of bio-oil. At 550°C, the H/C atomic ratio was 1.92, and the O/C atomic ratio was 0.43 (significantly low). Furthermore, combining coffee waste and wood increases the HHV (21.4 MJ/kg at 550°C) of the bio-oil and the contents of high mass yield in it and decrease water content (Kelkar et al., 2015; Soysa et al., 2015, 2016). The pyrolysis of a mixture of coffee waste and wood enhanced the bio-oil properties.

Fischer et al. (2015) studied the pyrolysis of spent coffee grounds and identified in the bio-oil the presence of, mainly, fatty acids, linear hydrocarbons, furans, pyridines, phenols, ketones, and aromatic hydrocarbons. Increasing the pyrolysis temperature decreases the selectivity to fatty acids and increases the selectivity toward aromatic and linear hydrocarbons. The addition of a ZSM-5 catalyst led to the increase of deoxygenated olefins and aromatic compounds and the decrease of linear hydrocarbons and furans.

Biocrude oil derived from coffee grounds showed higher HHV than wood (mallee, 17.9 MJ/kg) (Bok et al., 2012). In fact, HHV of biocrude oil in the former case decreased from 23.19 MJ/kg at a pyrolysis temperature of 450°C to 20.03 MJ/kg at 600°C. However, at 400°C, it was of 12.04 MJ/kg. The HHV of biocrude oil issued from coffee grounds was found to be higher than many biomasses (wood, etc.) (Bok et al., 2012). Thus, it could be used in burners, gas turbines, diesel engines, and so forth, for combustion purposes. Oxygenated products lower the value of HHV.

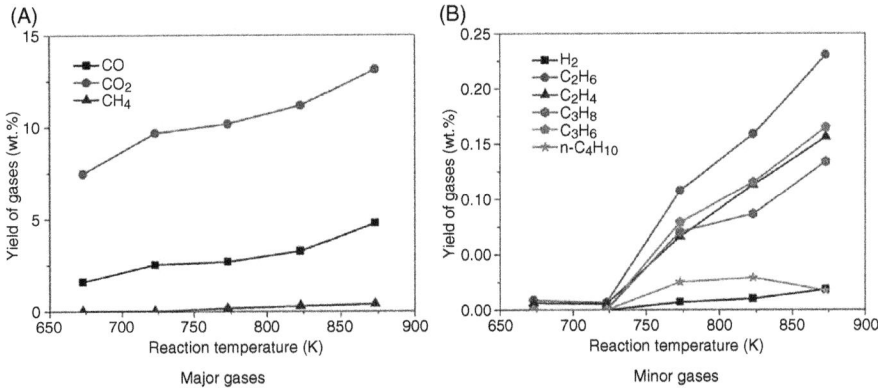

FIGURE 11.12 Yields of Noncondensable Gases Issued From the Fast Pyrolysis of Coffee Grounds

(A) Major gases, (B) minor gases.

Reprinted with permission from Bok, J.P., Choi, H.S., Choi, Y.S., Park, H.C., Kim, S.J., 2012. Fast pyrolysis of coffee grounds: characteristics of product yields and biocrude oil quality. Energy 47, 17–24. Copyright Elsevier 2016.

Copyrolysis at high temperature could lead to the decrease of the oxygenate amounts with the parallel increase of CO, CO_2, and H_2O.

11.1.2.2.8 Characteristics of the gaseous products

The pyrolysis of coffee grounds leads to three main gases, CO, CO_2, and CH_4 (Bok et al., 2012; Li et al., 2014; Ngo et al., 2015). The concentration of CO and CO_2 could be increased by the use of a catalyst, such as $NiCu/\gamma\text{-}Al_2O_3$ (Kan et al., 2014).

The variation of the gas yields with pyrolysis temperature was investigated by different authors (Bok et al., 2012; Ngo et al., 2015). The gas was mainly composed of H_2, CO, CO_2, CH_4, C_2H_6, C_2H_4, C_3H_8, C_3H_6, and $n\text{-}C_4H_{10}$. Among them, CO_2 has the highest content, followed by CO. An example of the evolution of the yield of each gas versus the pyrolysis reaction temperature is presented in Fig. 11.12 (Bok et al., 2012). Generally, the yield of a gas increased with increasing the pyrolysis temperature, except for $n\text{-}C_4H_{10}$, whose content decreased for temperatures 550°C (823 K). The amount of H_2 was very low (Bok et al., 2012; Ngo et al., 2015). The reactions that produce hydrogen are not favored at pyrolysis temperatures of 400–750°C (Ngo et al., 2015).

The use of ZSM-5 as catalyst was found to increase the CO yield (Fischer et al., 2015), CO was formed by decarbonylation reactions.

11.1.2.3 Torrefaction: application to coffee residue and coffee husk

Torrefaction is a thermochemical process that aims to decrease the water and volatiles contents from the biomass, thus improving some of its fuel properties: higher energy density, hydrophobic behavior, elimination of biological activity, easier grindability, more homogeneous composition and so forth. Torrefaction is sometimes called mild pyrolysis since it operates, like pyrolysis, under inert or reducing atmosphere, that is, in the absence of oxygen, but at temperatures of 200–350°C, which are lower than those of pyrolysis. The operating pressure is generally the atmospheric one. During torrefaction, the lignocellulosic compounds—hemicellulose, cellulose, and lignin—are subjected to degradation, leading generally to the departure of CO_2 and H_2O, and hence a decrease in H and O contents and an increase in C contents are detected in the remaining torrefied residue. Among lignocellulosic compounds, hemicellulose is the one most subjected to degradation during torrefaction.

Chen et al. (2012) and Tsai and Liu (2013) investigated the torrefaction of coffee residues (CR). Dhungana et al. (2012) studied the torrefaction of coffee husk (CH).

It was found that coffee by-products are more sensitive to torrefaction than many other biomasses (sawdust, rice husk), mainly because of their higher contents in lignocellulosic compounds, especially hemicellulose (Chen et al., 2012). Fig. 11.13 compares the ability to be torrefied of CR, sawdust, and rice husk. The higher percentages of weight losses due to torrefaction were obtained for CR compared to sawdust and rice husk.

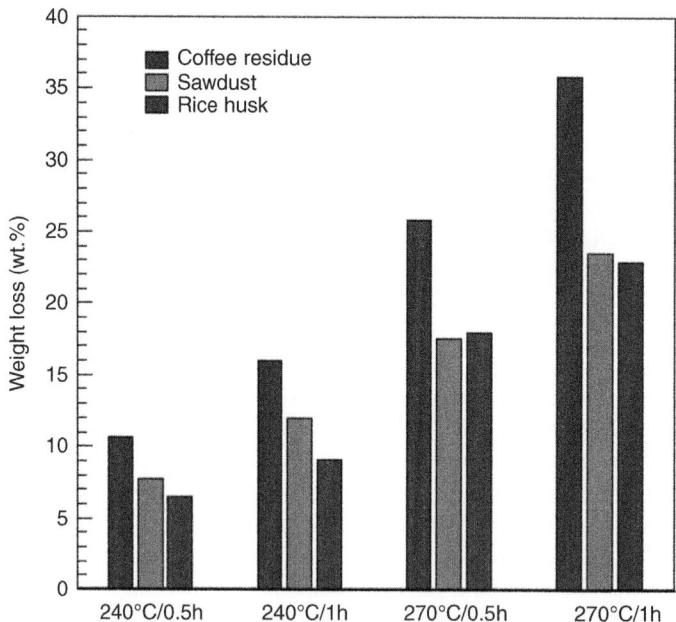

FIGURE 11.13 Percentages of weight losses obtained for coffee residue, sawdust, and rice husk at various torrefaction temperatures (240 and 250°C) and durations (0.5 and 1 h)

Many parameters influencing the torrefaction process were studied, especially the temperature and the duration (residence time). The effect of temperature was more noticeable than of the effect of residence time. Indeed, Fig. 11.13 (Chen et al., 2012) clearly demonstrate that, for a fixed residence time, the weight loss increased with temperature. However, for a fixed temperature, the weight loss increased with residence time but to a lower extent than with temperature. Tsai and Liu (2013) studied a larger range of temperature (230–350°C) than Chen et al. (2012) (240 and 270°C), and demonstrated that even if the weight loss percentage, also called torrefaction yield, increased with temperature, there is a temperature where the maximum of weight loss occurs. This temperature was of 290°C in the work of Tsai and Liu (2013). The latter value was in agreement with the results of thermogravimetric analysis where the depolymerization of hemicellulose occurred, resulting in a partial carbonization and volatilizatiion of the lignocellulosic compounds contained in the coffee residue.

Usually, torrefaction leads to the increase of HHV as reported in Table 11.7. HHV of torrefied CR and CH remain comparable to that of fossil coal (28–32 MJ/kg). Table 11.7 indicates also that increasing torrefaction time leads to the increase of HHV but to the decrease of energy yield. In all cases, the energy yield remains in the range of 88%–99%. Indeed, the weight loss will decrease the energy yield whereas the increase of HHV increases the energy yield. Since the impact of weight loss is greater than HHV impact, the energy yield will decrease with increasing temperature and duration.

The increase of carbon content and the decrease of H and O contents were also detected when the torrefaction temperature increased. Tsai and Liu (2013) also proposed the temperature range 290–320°C as a temperature window for coffee residue torrefaction, because in this range the carbon content of the torrefied CR increased the most. Examination by Fourier transform infrared (FTIR) (Chen et al., 2012) of the bands relative to hemicellulose, cellulose, and lignin, reveals that the increase of torrefaction temperature results in the consumption of more hemicellulose and cellulose in the torrefied product.

Some physical properties are also influenced by torrefaction, such as the true density (or bulk density) and the friability. It was reported that the surface becomes

Table 11.7 Elemental Analysis and Higher Heating Value (HHV) of Slow Pyrolysis Bio-oils Derived from Spent and Defatted Coffee Grounds

Bio-Oil Feedstock	C (%)	H (%)	N (%)	O (%)	S (%)	HHV (MJ/kg)
Spent grounds	74.0 (0.9)	9.8 (0.2)	2.6 (0.1)	13.4 (0.9)	0.17	32.3
Defatted grounds	70.9 (0.5)	8.0 (0.0)	4.3 (0.0)	16.4 (0.5)	0.39	27.0

Values in parentheses indicate standard deviations from triplicate measurements (Vardon et al., 2013).

smoother after torrefaction because of the degradation of hemicellulose; the torrefied product is more friable or brittle than the raw one (Chen et al., 2012; Tsai and Liu, 2013). Furthermore, the true density of CR after torrefaction increased with temperature for temperatures greater than 290°C. This bulk density exhibits a reverse trend at 230–290°C, where it diminishes with increasing temperature (Tsai and Liu, 2013).

Finally, it could be noted that torrefaction enhances the thermochemical properties of coffee wastes, rendering them suitable to be used as a cofuel with coal (Dhungana et al., 2012; Chen et al., 2012; Tsai and Liu, 2013).

11.1.2.4 Gasification

Gasification is a process where the biomass or organic waste is heated, generally between 500 and 1000°C, in the presence of a partially oxidative atmosphere, using an agent (steam, air, steam-oxygen, steam-air, carbon dioxide, etc.) to lead to chemical products, mainly composed of hydrocarbons, hydrogen, carbon monoxide, and carbon dioxide, called synthetic gas or syngas. Gasification process leads also to the production of small amounts of tar, char, and ash. The syngas, after some purification, will be directly burnt to give energy or power, or will be converted into other compounds that will be used as high-quality fuel or chemical products, such as hydrogen, methane, methanol, and ethanol. To find the amount of gasifying agent necessary to an efficient gasification of biomass, it is necessary to determine the amount of moisture and also the chemical composition of the biomass studied (Bouraoui et al., 2015, 2016).

11.1.2.4.1 Application to coffee wastes

Several authors (de Oliveira et al., 2013) reported that gasification is, among the conversion technologies, the most promising for generating heat, hydrogen, ethanol, and electricity. They have also stated that moisture content lower than 30 wt.% is suitable for gasification process. Moreover, ash content higher than 5 wt.% could cause damage to the equipment of gasification process (turbines, motors, cleaning systems). In this context, the analysis carried out by de Oliveira et al. (2013) indicated 0.94 wt.% of ash in the coffee wood and 1.71 wt.% in the coffee husk.

A high density of biomass is more suitable for the economic benefit of gasification process. Indeed, biomass with lower density is consumed faster in the reactor; hence a large volume of biomass would be necessary to generate the same amount of heat as biomass with higher densities. Furthermore, the cost of energy produced will be higher because of the transport and storage fees of lower density biomass. Regarding this factor, de Oliveira et al. (2013) determined the densities of coffee husk and coffee wood, respectively 138.8 and 416.7 kg/m^3, and deduced that coffee husk requires large volumes to produce the same energy as coffee wood.

The estimation of the energy that could be produced by coffee husk and coffee wood in Brazil is respectively of 11.3 and 49.5 PJ (de Oliveira et al., 2013). Therefore, these wastes could be used as input in biomass conversion into energy. Then, biomass conversion into energy is technically and economically feasible, even in remote locations where no access to electrical distribution is available.

Coffee grounds, where the moisture was reduced to 10 wt.%, were gasified by steam to produce fuel gas in a pilot dual fluidized bed gasification (DFBG) (Xu et al., 2006). At 800°C, a high carbon conversion (more than 70%) into gaseous products with high HHV (higher than 14.6 MJ/Nm^3) were obtained. Nevertheless, tar load was up to 50 g/Nm^3. The tar yield was decreased, to a certain limit, by the increase of steam/waste (coffee grounds) mass ratio and the decrease of the waste particle size. Introducing some amount of air into steam reduces also (about 60%) the tar yield. Nonetheless, O_2/C molar ratios should be kept below 0.1 in order to keep the HHV value of the product gas higher than 12.5 MJ/Nm^3. Increasing the residence time of the coffee waste in the reactor results in a slight and limited decrease of the tar yield (Xu et al., 2006). The tar obtained was rich in sulfur and nitrogen evidencing that during steam gasification, sulfur and nitrogen present in the coffee grounds are transferred to the tar. Lower amount of carboxylic groups is found in the tar compared to the coffee grounds (Xu et al., 2006). The product gas was composed of H_2, CO, CO_2, CH_4, C_2H_4, C_2H_6, C_3H_6, and C_3H_8. The composition of the product gas depends mainly on the gasification temperature and the air feed. Gasifying the coffee grounds by pure steam at 800°C (with a steam/waste mass of about 1.0) produces a syngas with a H_2/CO ratio of 0.55–0.65 and a normalized CO concentration of 37–38 mol% (Xu et al., 2006).

Wilson et al. (2010) gasified coffee husks at 700°C, 800°C, and 900°C. The activation energy of the process was about 161 kJ/mol. Increasing the reaction temperature from 700 to 900°C led to:

- An increase of the gasification process.
- A decrease of the amount of char formed.
- An increase of CO concentration in the syngas, whatever the gasification agent is (pure N_2, 2, 3, and 4 vol.% O_2 diluted in N_2).

At 900°C with 2 vol.% oxygen concentrations, the low heating value (LHV) of the syngas was 14.47 MJ/Nm^3. For 4 vol.% oxygen concentration, the gasification rate was higher (96%) than for lower O_2 concentrations. In addition, at 800°C less than 4 vol.% O_2, the highest CO/CO_2, 2.17, was obtained.

Couto et al. (2013) performed an experimental and numerical study on the gasification of coffee husks in a pilot plant, with fluidized bed technology. The operating temperature was 800°C and the feed rate was about 70 kg/h. The produced syngas was rich in hydrogen and carbon monoxide and also contained some methane; all the latter gases confer heat capacity to the syngas. Because of the partial combustion process, carbon dioxide was present in high amounts in the syngas. High quantity of nitrogen was also detected. Overall, the gasification of the coffee husks in such conditions gave: 70% of gas fraction, composed mainly of nitrogen, carbon dioxide, carbon monoxide, hydrogen, and methane, and 30% of tar and ashes (chars).

Another study (Brito et al., 2012), carried out by the same research group, dealt with the thermal gasification of different agro-industrial residues, among them coffee husks, in the same conditions. The net heat value was improved by the increase of gasification temperature and feed rate of residue. An economic study was also done. A comparison between the cost of natural gas and the cost of the recovery of energy from agro-industrial wastes by gasification, in the Alto-Alentejo region industries in Portugal, shows clear economic potentialities for the latter process.

11.1.2.4.2 Catalytic gasification of coffee wastes

Chaiklangmuang et al. (2015) also carried out the catalytic gasification at 500–650°C, under a steam pressure of 30 kP_a, in a quartz-fixed bed, of coffee residues. The catalyst used was nickel added to a coal matrix (Loy Yang brown coal, LY), LY-Ni. The total gas yield increased significantly with increasing temperature from 500 to 600°C and decreases with further increase of temperature. Thus, the optimum temperature to get high gas yield is 600°C.

Furthermore, at a temperature of 600°C, more gas products, H_2, CO, CO_2, and CH_4, are produced by the catalytic steam gasification of coffee residue, with hydrogen being the dominant gas. The CO_2 gas concentration is higher than CO in the temperature range 500–600°C. At 650°C, CO concentration becomes higher than that of CO_2. Then, LY-Ni catalyst was efficient since it converted almost all the volatile biomass to light fuel gases, such as H_2 and CO. An appropriate catalyst addition led to high gas fraction and also to high light fuel yield in the produced gas.

11.1.2.4.3 Cogasification of coffee wastes

Velez et al. (2009) investigated the cogasification under steam and air, in a fluidized bed, of Colombian coal with 6–15 wt.% of different biomasses, sawdust, rice, and coffee husk. The presence of biomass reduced the CO_2 emissions and increased the hydrogen content, even if some decrease of energy efficiency was detected. The syngas obtained was mainly composed of hydrogen, carbon monoxide, methane, and hydrocarbons. Tar deposition and ash agglomeration were also observed. The presence of the latter compounds is considered as a disadvantage of the gasification process. The authors concluded that cogasification is an alternative to produce fuel gas to heat and generate electricity and also to convert refuse coal to gas with low heating value.

11.1.2.5 Hydrothermal conversion: application to spent coffee grounds

Hydrothermal liquefaction (HTL) "is the thermochemical conversion of biomass into liquid fuels by processing in a hot, pressurized water environment for sufficient time to break down the solid biopolymeric structure to mainly liquid components" (Elliott et al., 2015). Typical hydrothermal processing conditions are: temperatures of 250–550°C, pressures of 1–25 MPa, durations of 20–60 min, and absence or presence of catalyst. At such temperatures and pressures, sub-/supercritical water medium is occurring. HTL is a way to treat wet materials without the need to evaporate water or to dry since water acts as reactant, solvent, and/or catalyst. The high pressure maintains the water in the liquid phase. During HTL, heteroatoms are converted into N_2 and inorganic acids that could be neutralized with bases. Hence, HTL is considered as a clean process because no undesirable toxic compounds are obtained, such as SO_x, NO_x, or ammonia. HTL could be described as a thermal depolymerization where long carbon chains are thermally cracked to shorter ones, with the simultaneous remove of oxygen under H_2O (dehydration) or CO_2 (decarboxylation) forms and production of bio-oil with high H/C ratio and low viscosity, more suitable for applications as biofuel (Hughes et al., 2014). The principal reactions that occur during hydrotreatment are hydrodeoxygenation, hydrodesulfurization, hydroformylation, and hydrogenation (Hughes et al., 2014).

Different conditions influence the properties of the products resulting from HTL: the composition of feedstock, the temperature and heating rate, the pressure, the solvent, the residence time, and the catalyst. A homogeneous or heterogeneous catalyst could be used to improve the quality and yields of the desired products.

HTL was recently applied to the valorization of spent coffee grounds (SCG) (Yang et al., 2016). Indeed, SCG contain high amounts of moisture (50–60 wt.%), hence it is not energetically and technically feasible to be pyrolyzed. Moreover, the resulting oil is rich in oxygenated compounds (35–60 wt.%) and has lower HHV (17–23 MJ/kg) compared to petroleum fuel. Furthermore, the average oil content extracted from SCG is relatively low (10–15 wt.%) (Yang et al., 2016). Therefore, to optimize the treatment of SCG, recently, Yang et al. (2016) liquefied SCG in hot-compressed water to bio-oil, under N_2 atmosphere. Furthermore, the size of SCG, 200–300 µm, makes it suitable for HTL reactor, with no need to previous grinding. Grinding process is energy consuming and thus lowers the energy balance. Compared to lignocellulose biomass, SCG have higher contents of lipid ($\approx$15%) and protein ($\approx$17%) with lower content of lignin ($\approx$24%).

Yang et al. (2016) examined the effects of operating conditions on the yield and properties of the bio-oil obtained. Their results could be resumed in Table 11.8.

The authors reached the maximum bio-oil yield, 47.3 wt.%, for a retention time of 10 min, at 275°C, under 2.0 MPa, and water/feedstock mass ratio of 20:1. In these optimal conditions, Yang et al. (2016) examined the heating value and the chemical composition of the as-obtained bio-oil:

- The HHV was 31.0 MJ/kg, higher than the HHV (20.2 MJ/kg) obtained for the oil issued from the fast pyrolysis of the SCG.
- Characterization by Fourier transform infrared (FT-IR) and gas chromatography coupled to mass spectrometry (GC-MS) evidenced that the long chain aliphatic acids and esters are the main volatile compounds of such a bio-oil.

More investigations in the application of HTL to coffee by-products valorization should be carried out. For example, the use of solvents other than water or the use of cosolvents, and the use of catalysts, as well as the examination of their effects on the yields and properties of the resulting bio-oil (Yang et al., 2016).

Table 11.8 Heating Value (MJ/kg) (HHV) and Energy Yield (%) Obtained at Diffferent Durations and Temperatures of Torrefaction of Coffee Residue and Coffee Husk

Product Temperature	Before Torrefaction	Residence Time (min)					Residence Time (min)				References
		10	15	30	60	10	15	30	60	
			Heating value (MJ/kg)				Energy yield (%)			
Coffee husk	20.16									Dhungana et al. (2012)
250°C		—	20.20	21.24	22.06		98.8	95.4	94.1	
280°C		—	20.38	24.43	24.95		98.8	93.0	91.2	
Coffee residue	20.6									Chen et al. (2012)
240°C		—	—	22.0	22.9	—	—	95.7	94.4	
270°C		—	—	25.2	28.2	—	—	92.2	88.4	
Coffee residue	23.5									Tsai and Liu (2013)
260°C		22.9	—	—	—	n.d.	—	—	—	
290°C		24.8	—	—	—	n.d.	—	—	—	
320°C		27.3	—	—	—	n.d.	—	—	—	
350°C		29	—	—	—	n.d.	—	—	—	

HHV before torrefaction is also given for each case.

11.1.3 BIOETHANOL, BIODIESEL, AND BIOGAS PRODUCTION

11.1.3.1 Production of Bioethanol from Coffee Process Byproducts as Feedstock

Alternative liquid fuels, such as bioethanol are essential to reduce the dependence on crude oil and to sustain economic development with environment. In particular, bioethanol can be produced from renewable biomass and adapted to existing fuel supply systems leading to cleaner combustion. Bioethanol is usually produced from food crops, such as corn and sugarcane through pretreatment, hydrolysis, and fermentation stages. However, the excessive use of these feedstocks increased food prices leading to economic concern on the competition between fuel and food crops production. Therefore, cheap and abundant nonfood lignocellulosic biomass sources, such as agriculture and agrifood processing residues are required to reduce the process cost of the biomass conversion to bioethanol (Sukumara et al., 2015).

Coffee processing residues (CPR) represent a promising potential for supplying considerable low cost and sustainable feedstock for bioethanol production. However, a higher degree of complexity, related to the characteristics and the composition variability of coffee processing residues, represents a limiting factor. In fact, CPR contains cellulose and hemicellulose fraction ranging from 31% for silverskin to 65% for coffee pulp (Franca et al., 2009; Mussatto et al., 2011). In addition, it contains lignin fraction ranging between 0.05% for spent coffee and 17.5% for coffee pulp (Franca et al., 2009). The presence of lignin limits the conversion of cellulose and hemicellulose into simpler forms and their removal from the lignocellulosic biomass is required.

11.1.3.1.1 Pretreatment techniques

Various pretreatments including concentrated acid hydrolysis (Mussatto et al., 2011), steam explosion (Choi et al., 2012) and alkali treatment (Menezes et al., 2014) have been used to remove lignin and to increase the production rate and the total yield of fermentable sugars during the hydrolysis step. Choi et al. (2012) used popping equipment for the pretreatment of coffee residue wastes (CRW). Different reactor pressure was applied ranging between 0.49 and 1.96 MPa during 10 min reaction time. After that, the sample was rapidly submitted to atmosphere pressure. The obtained results indicated that higher pretreatment pressure is essential to degrade the CRW cell wall. This degradation increases the CRW surface area and therefore improves the hydrolytic enzymes accessibility and the subsequent fermentation. The higher fermentable sugars concentration was 3.44 mg/mL obtained at a popping pressure of 1.47 MPa (Choi et al., 2012).

Menezes et al., 2014 have compared sodium hydroxide (NaOH) and calcium hydroxide [Ca(OH)$_2$] for the removal of lignin from the coffee pulp and for improving the accessibility of cellulose to enzymatic hydrolysis in order to provide fermentable sugars for bioethanol. The alkaline substances were used in an autoclave reactor at a temperature of 121°C. Authors found that NaOH alone was more suitable than Ca(OH)$_2$ with a good conversion of cellulose to glucose (60%) obtained for 4% NaOH and a time of approximately 25 min.

11.1.3.1.2 Hydrolysis step, saccharification

During the hydrolysis step, acid and enzymatic methods have been mainly used for coffee processing residues with varying efficiencies depending on treatment conditions and the properties of the hydrolytic agents. The dilute acid hydrolysis process is the oldest and simplest technology for producing bioethanol from biomass. Mussatto et al. (2011) have submitted spent coffee ground to hydrolysis reactions under various conditions of H_2SO_4 concentration, liquid-to-solid ratio, temperature, and reaction time. Authors found that these experimental conditions were adapted only for the hemicellulose hydrolysis, but not for cellulose hydrolysis. The chromatogram profile revealed the presence of glucose, arabinose, mannose, and galactose (from hemicellulose) in the SCG hydrolysate, while cellobiose and glucose (from cellulose) were detected in low amounts. The higher hydrolysis efficiency of 89.5% was obtained at temperature of 160°C; H_2SO_4 concentration of 100 mg/g dry matter; liquid-to-solid ratio of 10 g/g; and reaction time of 60 min.

The technical difficulties in recovering sugar from the acid motivate the use of the enzymatic method. This technology is more efficient and is accomplished at room temperature without hazardous waste generation. Enzymatic hydrolysis of coffee processing residues has been achieved essentially by the cellulases' actions. Kwon et al. (2003) have selected a cellulolytic enzyme complex during the saccharification process of spent coffee ground and the lipids extracted from the SCG. The enzyme hydrolyses were accomplished using a 2% (v/v) cellulase enzyme mixture at 50°C, with shaking at 200 rpm for 24 h. Authors found that the lipids extracted from the grounds showed a higher glucose conversion (18 g/L) in comparison with the spent coffee grounds (15 g/L). Authors concluded that the extracted lipids are a promising feedstock for bioethanol production.

Menezes et al., 2014 have selected the commercially enzyme Celluclast 1.5 L (Novozymes, Brazil) for coffee pulp hydrolysis at 50°C for 72 h. They found a glucose conversion ranged between 9.62 and 27.02 g/L depending on the alkaline substance concentration during the pretreatment step. The higher glucose conversion was obtained for the following pretreatment conditions: 4% (w/v) sodium hydroxide solution, with no calcium hydroxide, and 25 min treatment time.

11.1.3.1.3 Fermentation step and bioethanol production

Bacteria, fungi, and yeast microorganisms are generally used during the fermentation step of biomass feedstocks. Specific yeast *Saccharomyces cerevisiae*, also known as baker's yeast, has commonly been used to ferment the glucose derived from coffee processing residues to bioethanol. Gouvea et al. (2009) have used *S. cerevisiae* for the ethanol production from coffee husk (without pretreatment and hydrolysis steps). Authors found that fermentation yield decreased with yeast concentration increase. The highest ethanol production was 8.49 ± 0.29 g/100g dry basis (13.6 ± 0.5 g ethanol/L) obtained for 3 g yeast/L substrate at 30°C.

Mussatto et al. (2011) have compared three different yeast microorganisms namely, *Pichia stipitis*, *S. cerevisiae*, and *Kluyveromyces fragilis* for bioethanol production from spent coffee grounds (SCG) and coffee silverskin (CS) hydrolysates.

The latter was obtained by sulfuric acid hydrolysis. Authors found that *S. cerevisiae* consumed faster the sugar compounds from SCG hydrolysate comparing to the other strains, with almost total depletion after 24 h fermentation at 30°C. In addition, *S. cerevisiae* delivered the highest ethanol production from SCG hydrolysate (11.7 g/L, 50.2% efficiency). However, insignificant ethanol production (<1 g/L) was obtained from CS hydrolysate for all the evaluated yeast strains. Such results are probably due to the low sugars concentration present in this medium (approximately 22 g/L).

Similar investigations have used *S. cerevisiae* for the fermentation of the coffee waste hydrolysates (Choi et al., 2012; Kwon et al., 2003; Menezes et al., 2014). The main difference between these investigations consists of the pretreatment and hydrolysis techniques and the fermentation conditions. Choi et al. (2012) have used different popping treatments for coffee processing residue (CRW) followed by enzymatic hydrolysis and fermentation for 96 h at 37°C. Optimal condition for pretreatment was obtained at pressure of 1.47 MPa while the one for enzymatic hydrolysis was established at 18.3 mg cellulose/g CRW. This optimal condition leads to a hydrolysis efficiency of fermentable sugars equal to 85.6% and an ethanol production yield of 15.3 g/L. Menezes et al. (2014) have examined the fermentation of coffee pulp hydrolysate in an incubator for 48 h at 30°C, without pH control and agitation. The coffee pulp was submitted previously to an alkali pretreatment to make the cellulose accessible to enzymatic hydrolysis for providing fermentable sugars for bioethanol. The optimal conditions for pretreatment lead to the production of 13.66 g/L of ethanol with a yield of 0.4 g ethanol/g glucose. Kwon et al. (2003) have used spent coffee grounds and lipid extracted coffee grounds hydrolysates for fermentation, using S. cerevisiae at 30°C for 40 h. Approximately 22 g/L of ethanol was produced from both hydrolysates containing 50 and 58 g/L total sugar for coffee grounds and lipid extracted, respectively.

11.1.3.1.4 Case study

Gurram et al. (2015) have combined experimental and simulation investigations to perform techno-economical analysis of bioethanol production from coffee pulp and compared the obtained characteristics to other bioethanol feedstock. During the experimental investigations, authors have characterized the chemical composition of coffee pulp. The sugar characterization indicated the arabinose, galactose, glucose, xylose, and mannose contents as follows: 5.8%, 5.2%, 20.2%, 4.2%, and 4.7%, respectively.

Based on the coffee pulp characterization, Gurram et al. (2016) have carried out simulation using AspenPlus software to evaluate bioethanol production. The simulation was performed for a processing 10,000 tons/day of coffee pulp. The results indicated sugar and bioethanol yields of 2,100 and 1,050 tons/day, respectively. Such bioethanol production leads to an annual profit of $0.13 million. The simulation valued the capital investment cost to about $2 million and the maximum operation cost for process profitability to $1.87 million annually. The LCA indicated that the net value of energy of the whole process is economically viable, and the balance of CO_2 emission/reduction was on the environmental favor.

11.1.3.2 Production of biodiesel from coffee processing residues

Biodiesel is an alternative liquid fuel that has gained worldwide popularity in order to reduce the dependence on fossil fuels in the transport sector and to increase the renewable energy utilization. During the past decade, biodiesel production from vegetable oils has been increasingly performed due to their ecofriendly nature, as well as their easiness handling and storage. However, the biodiesel production cost is currently higher than fossil fuels, which restricts their spread to various applications. The major cause is attributed to biodiesel feedstock cost, which encourages the exploration of cheap and suitable raw materials. Therefore, several industries exploit animal fats and from waste vegetable oil to provide low-cost biodiesel. Moreover, several investigations have examined the potential of certain crops and agroindustrial residues with high oil content for the biodiesel production.

Few attempts have evaluated the potential of coffee processing residues as a source for biodiesel production (Caetano et al., 2012, 2014; Kondamudi et al., 2008). These investigations examined essentially the oil extraction from spent coffee grounds and their conversion to biodiesel through transesterification reaction. The other coffee processing residues did not receive particular attention due their low extracted lipids content.

11.1.3.2.1 Oil extraction and characterization

The lipid extraction from the spent coffee ground was performed with the Soxhlet system. The performance of various solvents on the quality of extracted oil was compared. Kondamudi et al. (2008) have carried out the extraction of oil from dried SCG using different solvents, such as dichloromethane, ether, and n-hexane under reflux conditions during 1h. The reaction mixture was analyzed each 5 min using HPLC technique. Authors have observed an increase in peak intensity for triglycerides with reflux time until the saturation point. A slight amount of free fatty acids (FFA), diglycerides (DG), and monoglycerides (MG) was also detected in the oil. Comparison between the different solvents indicated that more polar solvents extracted greater amounts of FFA and therefore more crude oil yields. The extracted oil yields were 13.4%, 14.6%, and 15.2% for hexane, diethyl ether, and dichloromethane, respectively. However, Kondamudi et al. (2008) noted that the pH value of the extracted oil decreased with the increase of the FFA amounts. In particular, the extracted oil obtained during hexane extraction is more neutral pH (6.8) compared to diethyl ether (4.7) and dichloromethane (4.5). Therefore, hexane was designated as an appropriate solvent for the process of oil extraction.

Caetano et al. (2014) have evaluated different solvents (hexane, ethanol, n-heptane, n-octane) and some mixtures (hexane/isopropanol) for lipids extraction from SCG using a pilot-scale Soxhlet. Authors have also determined characteristics of the extracted lipids, such as iodine number, acid value, water content, kinematic viscosity, density, and the high heating value. Comparison between solvents' performance showed that octane was the solvent allowing the higher crude oil yield. However, the extraction, using octane took a longer time and its recovery rate by distillation was very low. In contrast, hexane was the solvent causing lower oil yield, with the shortest extraction time and the highest recovery rate. Isopropanol exhibited a very good compromise between the oil extraction capacity and the solvent recovery, followed by the hexane and isopropanol mixture (50/50% vol./vol.). The latter allowed a higher extracted oil fraction but with a lower solvent recovery. Based on the solvents cost and the energy consumption required for crude oil extraction, the mixture of hexane and isopropanol 50:50 (%vol./vol.) was selected to accomplish further and pure hexane/isopropanol mixture (50/50 %vol./vol.). The HHV of the extracted lipids for both solvents were in same order of magnitude. The obtained values were 40.8 MJ/kg and 39.4 MJ/kg for pure hexane and the hexane/isopropanol mixture, respectively. The higher values indicate that the extracted oil could be used for direct combustion. A similar trend was observed for the density with 912 and 927 kg/m^3 for pure hexane and the hexane/isopropanol mixture. In contrast, a significant difference was observed for the viscosity and the acidity. In fact, authors obtained a lower viscosity for the oil extracted with pure hexane (14.9 mm^2/s comparing to the mixture (33.0 mm^2/s) while the acidity is higher for pure hexane (9.9 mg KOH/g SCG) comparing to the mixture (3.9 mg KOH/g SCG). The extracted oil acidity for both solvents is too high to be directly transesterified and previous esterification pretreatment step is required for biodiesel production.

11.1.3.2.2 Biodiesel production
Similar procedures were applied in the literature for biodiesel production from the crude oil extracted from spent coffee ground. Kondamudi et al. (2008) first heated the coffee oil at 100°C to remove the traces of water present, and then mixed the oil with 40 vol.% methanol and 1.5 wt.% catalyst (KOH). The reaction mixture was refluxed at ~70°C for the biodiesel production and the reaction time was defined by the disappearance of the oil (mostly triglycerides, TG) peaks in the HPLC. These operating conditions were previously optimized and during these conditions 100% oil-to-biodiesel transformation took place. Caetano et al. (2014) have produced biodiesel from the lipids extracted by pure hexane and pure hexane/isopropanol mixture by transesterification. The transesterification reaction was performed by adding 40 wt.% of methanol with previously dissolved 1.4 wt.% of NaOH catalyst for 3 h at 60°C.

11.1.3.2.3 Biodiesel characterization
A large amount of parameters and corresponding procedures for biodiesel characterization are imposed by the ASTM and EN standards. Several parameters were evaluated using different analytical techniques. The main results found in literature are shown in Table 11.9.

Kondamudi et al. (2008) found that biodiesel characteristics could reach the ASTM standards. Authors found the biodiesel consists essentially of methyl esters and 99% of the total composition includes palmitic acid (51.4%), linoleic acid (40.3%), and stearic acid (8.3%) Table 11.10.

In contrast, Caetano et al. (2012) have found that although the acid value and the viscosity have been considerably decreased, the produced biodiesel does not fulfill

Table 11.9 The Operating Parameters and Their Effects During Hydrothermal Liquefaction (HTL) of Spent Coffee Grounds

Parameter	Range Studied	Effect
Retention times	5–25 min	Maximum yield (31.63%) of bio-oil is reached for 10 min
Reaction temperature	250–300°C	• The effect of temperature is more significant than that of retention time. • 275°C produced the highest bio-oil yield (35.29%)
Water/feedstock mass ratio	5:1–20:1	The highest bio-oil yield (47.28%) was reached for a ratio of 20:1.
Initial pressure of process gas (N_2)	0.5–2.0 MPa	No significant effect on the distributions of the products.

From Yang, L., Nazari, L., Yuan, Z., Corscadden, K., Xu, C., He, Q., 2016. Hydrothermal liquefaction of spent coffee grounds in water medium for bio-oil production. Biomass Bioenergy 86, 191–198.

Table 11.10 Characterization of Biodiesel

Characteristics and Standards			Results	
Parameter	Test Method	Limit	Kondamudi et al. (2008)	Caetano et al. (2014)
Acid value [(mg KOH)/(g fuel)]	ASTM D 664	<0.5	0.35	2.14
Iodine value [(g I_2)/(100 g fuel)]	EN 14214	120	n.m	46.5
Kinematic viscosity at 40°C (mm^2/s)	EN 14214	3.5–5.0	5.84	12.88
Density at 15°C (kg/m)	ASTM D 445	1.0–6.0	—	—
	EN 14214	860–900	n.m	911

the EN 14214 standard specifications. The methyl ester yield and composition indicate transesterification reaction was incomplete. The authors attributed such behavior to the high water content and the high acid value in the oil, which caused soap formation, consuming catalyst and hindering the reaction.

The comparison between both studies indicate that oil drying, the neutralization of the excess acid before transesterification, and the use of KOH catalyst are crucial for obtaining a high biodiesel quality.

11.1.3.3 Production of biogas from coffee production waste

Biogas is an alternative gaseous fuel that has attracted significant attention with the increasing interest on renewable and sustainable energy technology. Biogas refers generally to a mixture of different gases, including methane (CH_4), carbon dioxide (CO_2), hydrogen (H_2), carbon oxide (CO), and and smaller amounts of hydrogen sulfide (H_2S), emitted during the breakdown of organic matter in the absence of oxygen. The biogas system includes different biological processes and energy conversion stages. This system is separated into four steps: feedstock supply, biogas production, digestate use, and biogas exploitation.

Generally, biogas can be produced from various feedstocks, such as agricultural waste, municipal waste, green waste, manure, and sewage sludge, using an anaerobic digester. The produced gases, such as methane, hydrogen, and carbon monoxide (CO) can be oxidized. The energy release can be used for any heating purpose, such as drying and cooking. The produced biogas can also be used in a gas engine to generate electricity.

Few investigations have examined the biogas production from coffee production waste. The main works deal with the hydrogen and/or methane production from coffee manufacturing wastewater containing coffee processing residues under mesophilic and thermophylic conditions (Fia et al., 2012; Jung et al., 2010, 2012; Neves et al., 2006). These investigations deal essentially with the feedstock characterization and the evaluation of the biogas production at laboratory scale.

11.1.3.3.1 Feedstock supply and characterization

Various coffee wastes were used as feedstocks for biogas production. Neves et al. (2006) have evaluated five different coffee solid wastes from instant coffee substitute production. These wastes included different blends of barley, rye, malted barley, chicory, and coffee with different chemical properties. These different wastes presented pH values between 4.5 and 5.0, total solids (TS) between 131 and 217 g/kg waste, volatile solids (VS) between 127 and 215 g/kg waste, and chemical oxygen demand (COD) between 109 and 208 g/kg waste.

Jung et al. (2010, 2012) evaluated the potential of coffee drink manufacturing wastewater (CDMW) for biogas production. This CDMW is composed from the gatherings of returned goods and the wastewater generated during the mixing process of crude coffee, starch, lactose, and sugar. This feedstock was a high-strength organic wastewater (184 g COD/L), mostly comprised of carbohydrates (71%). The soluble portion (SCOD/TCOD) was 78%, and total kjeldahl nitrogen (TKN) concentration was 3046 mgN/L.

Dinsdale et al. (1996) have examined the anaerobic digestion of coffee waste containing coffee grounds. Two waste streams were collected at two different periods (2 months). The waste streams present presented pH values between 4.3 and 4.6, total solids (TS) between 23.9 and 27.5 g/L, volatile solids (VS) between 23.1 and 27.1 g/L waste, total chemical oxygen demand (COD) between 35.9 and 40 g/L waste and Kjeldahl nitrogen (TKN) concentration between 80 and 120 mgN/L.

11.1.3.3.2 Biogas production

Different reactor designs have been used for the biogas production for coffee process wastewater including completely stirred tank reactor (CSTR) upflow anaerobic sludge blanket (UASB), upflow anaerobic fixed bed reactors (UAFD) and anaerobic membrane bioreactor (AnMBR). Jung et al. (2012) have examined the anaerobic digestion of coffee drink manufacturing wastewater (CDMW) using a continuous two-stage UASB reactor system. In a first-stage thermophilic (55°C) dark fermentative H_2 production was performed while in a second-stage mesophilic conditions (35°C) were applied to produce CH_4. A maximum CH_4 yield of 325 L CH_4/kg COD was achieved with an abatement of 93% of the COD under an organic load rate (OLR) of 3.5 g COD/L/d. The developed two-stage UASB reactor system achieved a biogas conversion of 88.2% (H_2: 15.2%; CH_4: 73%) and COD removal by 98%.

Fia et al. (2012) evaluated the conversion performance of three UAFB reactors during the treatment of coffee bean processing wastewater. Various support materials for biomass immobilization were used in the tested reactors, including blast furnace cinders, polyurethane foam, and crushed stone. Authors showed that the support materials characteristics influence strongly the degradation of the organic material. In particular, they demonstrated that the porous structure of the polyurethane exhibited a better performance comparing to the support materials used traditionally for fixed bed reactors (blast furnace cinders, porous ceramics). The reactor presented an average efficiency of removal of 80% (nonfiltered samples) when operated under the OLR of 4.41 kg/m^3 d and a HRT of 1.06 days.

Recently, Battista et al. (2016) have investigated the influence of basic and acid pretreatments on biogas production from coffee waste. Chemical pretreatments were performed in batch mode on a mixture prepared from coffee waste with a total solid concentration of 10% w/w. During these experiments, the biogas generated after acid and basic pretreatments was compared with biogas generated in absence of pretreatment stage. Authors demonstrated that the basic pretreatment has a beneficial effect on the lignin and cellulose hydrolyses. This basic pretreatment allowed permitted a biogas production of about 18 NL/L with methane content of almost 80% v/v.

11.2 CONCLUSIONS

The thermochemical valorization of coffee by-products in the presence of oxygen can be considered with some restrictive conditions. Densification seems to be a key parameter to produce marketable fuels with coffee by-products, improve their combustion efficiency, but also obtain transportable solid fuels. The best valorization pathway is obtained by blending coffee by-products with wood sawdust. It allows minimizing of the mineral content of the fuel (avoiding slagging and corrosion of the combustion systems) and also the production of specific solid fuels for bakery furnace or stove applications. SCG can be used both in oxy-combustion systems and in compressed logs. The energetic potential of these by-products is worldwide and must be organized locally to limit the final cost of the developed fuels and to present a good alternative to conventional solid fuels.

Wastes from coffee operations are complex materials. It was shown that coffee wastes could contribute to decrease CO_2 emissions by fossil fuels used in Brazil for steam and heat production (Protásio et al., 2013) more than other biomass, such as rice husk, sugarcane bagasse, maize harvesting wastes, and bamboo cellulose pulp. Indeed, coffee wastes possess a higher equivalent in fossil fuel volume than the other biomass with the highest heating value.

Coffee husks are low-cost residue and environmentally friendly. They could serve as an alternative for fuel. In fact, they are able to generate heat, steam, and electric power (de Oliveira et al., 2013). Biomass pyrolysis is an option to valorize waste that could help to reduce landfilling or incineration. The properties of spent coffee grounds, such as low ash, high elemental carbon, and high lipid content, render this waste promising for bioenergy production. Bio-oil with high combustion energy, low water content, and low carbonyl content is desired to reduce corrosion and reactivity.

The abundance of spent coffee grounds and the quality of its bio-oil make this waste a potential valuable bioenergy feedstock. It is advised to replace the conventional pyrolysis by the microwave pyrolysis to increase the gas yield and syngas production, with reducing energy consumption and time. Copyrolysis of coffee waste with another biomasses is encouraged since it could lead to bio-oil with better properties. Copyrolysis or pyrolysis at high temperatures led to a greater amount of gas and to the decrease of oxygenated compounds. The use of an appropriate catalyst enhances the selectivity towards the desired reactions and then the desired products or compositions of bio-oil and biogas. Formation of tar and ash can still be a problem for gasification.

To treat coffee waste, Hughes et al. (2014) proposed, for application in Colombia, the integration of multiple technologies: biochemical and thermochemical. Between coffee harvests, the biomass used will be other agricultural wastes, present in Colombia, such as sugarcane bagasse, palm rachis, cocoa beans, borra (spent coffee grounds), and so forth. These authors stated that such a technical system is economically feasible. The residue of biochemical conversions of coffee waste could be pyrolysed to produce several chemical compounds in addition to gasoline and fuel. Pyrolysis and hydrotreatment will lead to compounds of industrial interest, as for example, propylene to butyraldehyde, 1-hexene to heptaldehyde, 1-octene to nonaldehyde, decene to C11 aldehyde (target being detergent alcohols), benzene from pyrocrude to produce cyclohexane, ethylbenzene to make styrene, cumene to make phenol, and acetone to make methacrylates and methyl isobutyl ketone (Hughes et al., 2014).

The biochemical conversion of coffee processing by-products into bioethanol, biodiesel, and biogas was investigated in this chapter. The direct conversion of coffee residues to bioethanol was not found to be a desirable option due to the higher lignin contents and the presence of triglycerides and free fatty acids (FFAs). Therefore, pretreatment techniques, such as concentrated acid hydrolysis, steam explosion, and alkali treatment have been used to increase the production rate and the total yield of fermentable sugars during the saccharification step. The lipids extraction from the coffee residues seems also crucial to increase bioethanol production. These extracted lipids could be converted to bioethanol following the same steps as the defatted coffee residues. It can be also converted into fatty acid methyl ester (FAME) and fatty acid ethyl ester (FAEE) via the biodiesel transesterification reaction. The biogas production from the wastewater of coffee industries was examined, too. A combined procedure in a continuous up-flow anaerobic sludge blanket (UASB) including a first-stage thermophilic (55°C) dark fermentative for H_2 production followed by second-stage mesophilic conditions (35°C) for CH_4 production seems to be a promising strategy. A basic pretreatment could also increase the yield of biogas production.

REFERENCES

Alakangas, E., Valtanen, J., Levlin, J.E., 2006. CEN technical specification for solid biofuels: fuel specification and classes. Biomass Bioenergy 30, 8–914.

Battista, F., Fino, D., Mancini, G., 2016. Optimization of biogas production from coffee production waste. Biores. Technol. 200, 884–890.

Bok, J.P., Choi, H.S., Choi, Y.S., Park, H.C., Kim, S.J., 2012. Fast pyrolysis of coffee grounds: characteristics of product yields and biocrude oil quality. Energy 47, 17–24.

Bouraoui, Z., Jeguirim, M., Guizani, C., Limousy, L., Dupont, C., Gadiou, R., 2015. Thermogravimetric study on the influence of structural, textural, and chemical properties of biomass chars on CO_2 gasification reactivity. Energy 88, 703–710.

Bouraoui, Z., Dupont, C., Jeguirim, M., Limousy, L., Gadiou, R., 2016. CO_2 gasification of woody biomass chars: the influence of K and Si on char reactivity. Comptes Rendus Chimie 19, 457–465.

Brito, P.S.D., Rodrigues, L.F., Calado, L., Oliveira, A.S., 2012. Thermal gasification of agro-industrial residues. WIT Trans. Ecology Environ. 163, 95–102.

Caetano, N.S., Silva, V.F.M., Melo, A.C., 2012. Valorization of coffee grounds for biodiesel production. Chem. Eng. Trans. 26, 267–272.

Caetano, N.S., Silva, V.F.M., Melo, A.C., Martins, A.A., Mata, T.M., 2014. Potential of spent coffee grounds for biodiesel production and other applications. Clean Technol. Environ. Pol. 16 (7), 1423–1430.

Chaiklangmuang, S., Kurosawa, K., Li, L., Morishita, K., Takarada, T., 2015. Thermal degradation behavior of coffee residue in comparison with biomasses and its product yields from gasification. J. Energ. Inst. 88, 323–331.

Chen, W.-H., Lu, K.-M., Tsai, C.-M., 2012. An experimental analysis on property and structure variations of agricultural wastes undergoing torrefaction. Appl. Energy 100, 318–325.

Choi, I.S., Wi, S.G., Kim, S.B., Bae, H.J., 2012. Conversion of coffee residue waste into bioethanol with using popping pretreatment. Biores. Technol. 125, 132–137.

Ciesielczuk, T., Karwaczynska, U., Sporek, M., 2015. The possibility of disposing of spent coffee ground with energy recycling. J. Ecol. Eng. 16, 133–138.

Couto, N., Silva, V., Monteiro, E., Brito, P.S.D., Rouboa, A., 2013. Experimental and numerical analysis of coffee husks biomass gasification in a fluidized bed reactor. Energy Proc. 36, 591–595.

Cubero-Abarca, R., Moya, R., Valaret, J., Tomazello Filho, M., 2014. Use of coffee (Coffee Arabica) pulp for the production of briquettes and pellets for heat generation. Ciencia Agrotecnica 38 (5), 461–470.

DE Oliveira, J.L., DA Silva, J.N., Pereira, E.G., Oliveira Filho, D., Carvalho, D.R., 2013. Characterization and mapping of waste from coffee and eucalyptus production in Brazil for thermochemical conversion of energy via gasification. Renew. Sust. Energy Rev. 21, 52–58.

Dhungana, A., Dutta, A., Prabir, B., 2012. Torrefaction of non-lignocellulose biomass waste. Can. J. Chem. Eng. 90, 186–195.

Dinsdale, D., Hawkes, F.R., Hawkes, D.U., 1996. The mespholic and thermophylic anaerobic digestion of coffee waste containing coffee grounds. Water Res. 30, 371–377.

Elliott, D.C., Biller, P., Ross, A.B., Schmidt, A.J., Jones, S.B., 2015. Hydrothermal liquefaction of biomass: developments from batch to continuous process. Biores. Technol. 178, 147–156.

Fernández, Y., Menéndez, J.A., 2011. Influence of feed characteristics on the microwave-assisted pyrolysis used to produce syngas from biomass wastes. J. Anal. Appl. Pyrol. 91, 316–322.

Fia, F.R., Matos, A.T., Fia, R., Cecon, P.R., 2012. Treatment of wastewater from coffee bean processing in an aerobic fixed bed reactors with different support materials: performance and kinetic modeling. J. Environ. Manag. 108, 14–21.

Fischer, A., Du, S., Valla, J.A., Bollas, G.M., 2015. The effect of temperature, heating rate, and ZSM-5 catalyst on the product selectivity of the fast pyrolysis of spent coffee grounds. RSC Adv. 5, 29252–29261.

Fonseca Felfli, F., Mesa, P.J.M., Dilcio Rocha, J., Filippetto, D., Luengo, C.A., Pippo, W.A., 2011. Biomass briquetting and its perspectives in Brazil. Biomass Bioenergy 35, 236–242.

Franca, A.S., Oliveira, L.S., Ferreira, M.E., 2009. Kinetics and equilibrium studies of methylene blue adsorption by spent coffee grounds. Desalination 249, 267–272.

Gil, M.V., Oulego, P., Casal, M.D., Pevida, C., Pis, J.J., Rubiera, F., 2010. Mechanical durability and combustion characteristics of pellets from biomass blends. Biores. Technol. 101, 8859–8867.

Gómez-de la Cruz, F.J., Cruz-Peragon, F., Casanova-Pelaez, P.J., Palomar-Carnicero, J.M., 2015. A vital stage in the large-scale production of biofuels from spent coffee grounds: the drying kinetics. Fuel Proc. Technol. 130, 188–196.

Gouvea, B.M., Torres, C., Franca, A.S., Oliveira, L.S., Oliveira, E.S., 2009. Feasibility of ethanol production from coffee husks. Biotechnol. Lett. 31, 1315–1319.

Gurram, R., Al-Shannag, M., Knapp, S., Das, T., Singsaas, E., Alkasrawi, M., 2016. Technical possibilities of bioethanol production from coffee pulp: a renewable feedstock. Clean Technol. Environ. Pol. 18 (1), 269–278.

Hughes, S.R., Lopez-Nuñez, J.C., Jones, M.A., Moser, B.R., Cox, E.J., Lindquist, M., Galindo-Leva, L.A., Riaño-Herrera, N.M., Rodriguez-Valencia, N., Gast, F., Cedeño, D.L., Tasaki, K., Brown, R.C., Darzins, A., Brunner, L., 2014. Sustainable conversion of coffee and other crop wastes to biofuels and bioproducts using coupled biochemical and thermochemical processes in a multistage biorefinery concept. Appl. Microbiol. Biotechnol. 98, 8413–8431.

Jahirul, M.I., Rasul, M.G., Chowdhury, A.A., Ashwath, N., 2012. Biofuels production through biomass pyrolysis: a technological review. Energies 5, 4952–5001.

Jeguirim, M., Limousy, L., Dutournie, P., 2014a. Pyrolysis kinetics and physicochemical properties of agropellets produced from spent coffee ground blended with conventional biomass. Chem. Eng. Res. Devel. 130, 188–196.

Jeguirim, M., Bikai, J., Elmay, Y., Limousy, L., Njeugna, E., 2014b. Thermal characterization and pyrolysis kinetics of tropical biomass feedstocks for energy recovery. Energ. Sustain. Devel. 23, 188–193.

Jeguirim, M., Limousy, L., Fossard, E., 2016. Characterization of coffee residues pellets and their performance in residential combustor. Int. J. Green Energ. 13, 608–615.

Jung, K.W., Kim, D.H., Shin, H.S., 2010. Continuous fermentative hydrogen production from coffee drink manufacturing wastewater by applying UASB reactor. Int. J. Hydro. Energ. 35, 13370–13378.

Jung, K.W., Kim, D.H., Lee, M.Y., Shin, H.S., 2012. Two-stage UASB reactor converting coffee drink manufacturing wastewater to hydrogen and methane. Int. J. Hydro. Energ. 37, 7473–7481.

Kan, T., Strezov, V., Evans, T., 2014. Catalytic pyrolysis of coffee grounds using NiCu-impregnated catalysts. Energ. Fuels 28, 228–235.

Kelkar, S., Saffron, C.M., Chai, L., Bovee, J., Stuecken, T.R., Garedew, M., Li, Z., Kriegel, R.M., 2015. Pyrolysis of spent coffee grounds using a screw-conveyor reactor. Fuel Proc. Technol. 137, 170–178.

Kondamudi, N., Mohapatra, S.K., Misra, M., 2008. Spent coffee grounds as a versatile source of green energy. J. Agric. Food Chem. 56, 11757–11760.

Kwon, E.E., Yi, H., Jeon, Y.J., 2003. Sequential co-production of biodiesel and bioethanol with spent coffee grounds. Biores. Technol. 136, 475–480.

Li, X., Strezov, V., Kan, T., 2014. Energy recovery potential analysis of spent coffee grounds pyrolysis products. J. Anal. Appl. Pyrol. 110, 79–87.

Limousy, L., Jeguirim, M., Dutournie, P., Kraeim, N., Lajili, M., Said, R., 2013. Gaseous products and particulate matter emissions of biomass residential boiler fired with spent coffee ground pellets. Fuel 107, 323–329.

Limousy, L., Jeguirim, M., Labbe, S., Balay, F., Fossard, E., 2015. Performance and emissions characteristics of compressed spent coffee ground/wood chip logs in a residential stove. Energ. Sustain. Devel. 28, 52–59.

Mašek, O., Konno, M., Hosokai, S., Sonoyama, N., Norinaga, K., Hayashi, J.-I., 2008. A study on pyrolytic gasification of coffee grounds and implications to allothermal gasification. Biomass Bioenergy 32, 78–89.

Menezes, E.G.T., Do Carmo, J.R., Alves, J.G.L.F., Menezes, A.G.T., Guimaraes, I.C., Queiroz, F., Pimenta, C.J., 2014. Optimization of alkaline pretreatment of coffee pulp for production of bioethanol. Biotechnol. Prog. 30 (2), 451–462.

Mussatto, S.I., Machado, E.M.S., Martins, S., Teixeira, A., 2011. Production, composition, and application of coffee and its industrial residues. Food Bioproc. Technol. 4, 661–672.

Neves, L., Oliveira, R., Alves, M.M., 2006. Anaerobic co-digestion of coffee waste and sewage sludge. Waste Manag. 26, 176–181.

Ngo, T.-A., Kim, J., Kim, S.-S., 2015. Fast pyrolysis of spent coffee waste and oak wood chips in a micro-tubular reactor. Energy Sources A 37, 1186–1194.

Protásio, T.P., Bufalino, L., Tonoli, G.H.D., Guimarães Junior, M., Trugilho, P.F., Mendes, L.M., 2013. Brazilian lignocellulosic wastes for bioenergy production: characterization and comparison with fossil fuels. BioResources 8 (1), 1166–1185.

Rhen, C., Ohman, M., Wasterlund, L., 2007. Effect of raw material composition in woody biomass pellets on combustion characteristics. Biomass Bioenergy 31, 66–72.

Romeiro, G.A., Salgado, E.C., Silva, R.V.S., Figueiredo, M.K.-K., Pinto, P.A., Damasceno, R.N., 2012. A study of pyrolysis oil from soluble coffee ground using low temperature conversion (LTC) process. J. Anal. Appl. Pyrol. 93, 47–51.

Saenger, M., Hartage, E.-U., Werther, J., Ogada, T., Siagi, Z., 2001. Combustion of coffee husks. Renew. Energ. 23, 103–121.

Souza Faria, W., De Paula Protásio, T., Trugilho, P.F., Corradi Pereira, B.L., De Cassia Oliveira Carneiro, A., Rogerio Andrade, C., Benedito Guimaraes Junior, J., 2015. Transformation of lignocellulosic waste of coffee into pellets for thermal power generation. Coffee Sci. 11 (1), 137–147.

Soysa, R., Choi, S.K., Jeong, Y.W., Kim, S.J., Choi, Y.S., 2015. Pyrolysis of Douglas fir and coffee ground and product biocrude-oil characteristics. J. Anal. Appl Pyrol. 115, 51–56.

Soysa, R., Choi, Y.S., Choi, S.K., Kim, S.J., Han, S.Y., 2016. Synergetic effect of biomass mixture on pyrolysis kinetics and biocrude-oil characteristics. Kor. J. Chem. Eng. 33, 603–609.

Suarez, J.A., Luengo, C.A., 2013. Coffee husk briquettes: a new renewable energy source. Energy Sources 25 (10), 961–967.

Sukumara, S., Amundson, J., Badurdeen, F., Seay, J., 2015. A comprehensive techno-economic analysis tool to validate long-term viability of emerging biorefining processes. Clean Technol. Environ. Pol., 1–14.

Tsai, W.-T., Liu, S.-C., 2013. Effect of temperature on thermochemical property and true density of torrefied coffee residue. J. Anal. Appl. Pyrol. 102, 47–52.

Tsai, W.-T., Liu, S.-C., Hsieh, C.-H., 2012. Preparation and fuel properties of biochars from the pyrolysis of exhausted coffee residue. J. Anal. Appl. Pyrol. 93, 63–67.

Vardon, D.R., Moser, B.R., Zheng, W., Witkin, K., Evangelista, R.L., Strathmann, T.J., Rajagopalan, K., Sharma, B.K., 2013. Complete utilization of spent coffee grounds to produce biodiesel, bio-oil, and biochar. ACS Sustain. Chem. Eng. 1, 1286–1294.

Velez, J.F., Chejne, F., Valdes, C.F., Emery, E.J., Londoño, C.A., 2009. Co-gasification of Colombian coal and biomass in fluidized bed: An experimental study. Fuel 88, 424–430.

Wei, Y., Chen, M., Niu, S., Xue, F., 2016. Experimental investigation on the oxy-fuel co-combustion behavior of anthracite coal and spent coffee grounds. J. Therm. Anal. Calor. 124, 1651–1660.

Wilson, L., John, G.R., Mhilu, C.F., Yang, W., Wlodzimierz, B., 2010. Coffee husks gasification using high temperature air/steam agent. Fuel Proc. Technol. 91, 1330–1337.

Xu, G., Murakami, T., Suda, T., Matsuzawa, Y., Tani, H., 2006. Gasification of coffee grounds in dual fluidized bed: performance evaluation and parametric investigation. Energ. Fuels 20, 2695–2704.

Yang, S.Y., Wu, M.S., Wu, C.Y., Chen, K.H., Wu, T.M., Hsu, Y.-L., Lin, P.H., Ku, Y.Y., 2012. The performance of a diesel engine blended with coffee bean residue pyrolysis oil. Adv. Mater. Res. 591–593, 325–332.

Yang, L., Nazari, L., Yuan, Z., Corscadden, K., Xu, C., He, Q., 2016. Hydrothermal liquefaction of spent coffee grounds in water medium for bio-oil production. Biomass Bioenergy 86, 191–198.

Zanella, E., Zassa, M.D., Navarini, L., Canu, P., 2013. Low-temperature co-pyrolysis of polypropylene and coffee wastes to fuels. Energy Fuels 27, 1357–1364.

FURTHER READING

Hernandez, M.A., Susa, M.R., Andres, Y., 2014. Use of coffee mucilage as a new substrate for hydrogen production in anaerobic codigestion with swine manure. Biores. Technol. 168, 112–118.

Qiao, W., Takayanagi, K., Shofie, M., Niu, Q., Yu, H.Q., Li, Y.Y., 2013. Thermophilic anaerobic digestion of coffee grounds with and without waste activated sludge as co-substrate using a submerged AnMBR: system amendments and membrane performance. Biores. Technol. 150, 249–258.

Quintero, J.A., Rincon, L.E., Cardona, C.A., 2011. Production of bioethanol from agroindustrial residues as feedstocks. In: Pandey, A., Larroche, C., Rocke, S.C., Dussap, C.-G., Gnansounou, E. (Eds.), Biofuels: Alternative Feedstocks and Conversion Processes. Kidlington. Academic Press/Elsevier, Kidlington/Oxford/San Diego, CA.

Vermicompost derived from spent coffee grounds: assessing the potential for enzymatic bioremediation

12

Juan C. Sanchez-Hernandez*, Jorge Domínguez**

*University of Castilla-La Mancha, Toledo, Spain; **University of Vigo, Vigo, Spain

ABSTRACT

Vermicomposting provides a low-cost, rapid, and environmentally safe solution for the disposal of the large amounts of spent coffee grounds (SCGs) generated annually worldwide. In the last decade, there has been a huge increase in the number of studies concerning the revalorization of SCGs as a biodegradable material of interest in the pharmaceutical, cosmetic, food, and energy industries. However, little attention has been paid to this organic waste as a potential biostimulatory amendment for bioremediating polluted soils. This chapter provides an overview of the main current applications of SCGs and discusses the results obtained in the processing of SCGs through vermicomposting on a pilot scale. In the very short term, vermicomposting reduced substantially the residue mass, yielding a nutrient-rich and enzymatically active vermicompost. The levels and reactivity of carboxylesterases (CbEs) in the vermicompost obtained are evaluated. Although these esterases have a well-recognized function as pesticide-detoxifying enzymes in animals, their role as extracellular detoxifying enzymes in vermicompost is unknown. Thus, degradation and detoxification of chlorpyrifos (a model organophosphorus pesticide) was examined in the presence of both liquid and solid vermicomposts derived from SCGs. Degradation of the pesticide in liquid vermicompost followed typical first-order kinetics ($t_{1/2}$ = 4.74 day^{-1}). Furthermore, CbE activity was strongly inhibited, suggesting a bioscavenging role for these esterases in response to chlorpyrifos contamination. The high levels of CbE activity, and other enzymes of interest regarding contaminant transformation as laccases, in the vermicompost derived from SCGs compared to levels naturally occurring in soils, suggest that this revalorized product has a great potential for the bioremediation of pesticide-contaminated agricultural soils.
Keywords: esterases; organophosphorus pesticides; liquid vermicompost; enzyme fingerprinting; bioremediation; urban agriculture

Handbook of Coffee Processing By-Products. http://dx.doi.org/10.1016/B978-0-12-811290-8.00012-8

12.1 INTRODUCTION

The amount of caffeine (1,3,7-trimethylxanthine) in coffee beans is in the range of 0.8%–4.0% (w/w) depending upon the variety of coffee, that is, Arabica or Robusta (Murthy and Naidu, 2012). The stimulatory properties of this purine alkaloid on brain function and behavior may explain why coffee is one the most commonly consumed beverages in the world. Coffee beans also contain a complex mixture of different substances, including sugars, lipids, proteins, vitamins, minerals, alkaloids, and phenolic compounds, some of which may help reduce the risk of several chronic diseases, such as type 2 diabetes, Parkinson's disease, cardiovascular diseases, and some types of cancer (Doepker et al., 2016; Higdon and Frei, 2006; Ludwig et al., 2014). Worldwide consumption of coffee continues to increase, and large amounts of coffee by-products are generated during the manufacture of coffee powder (instant coffee). Depending upon the method (dry or wet) used to produce coffee powder, four types of residues are generated: coffee pulp, coffee husks, silverskin, and spent coffee grounds (SCGs) (Murthy and Naidu, 2012). Globally these residues represent a serious environmental hazard, and their revalorization should be a matter of significant concern to regulatory and environmental agencies. In the last decade, efforts have been made to develop biotechnological processes that use coffee waste to produce value-added products of environmental significance and of interest to the pharmaceutical, cosmetics, and food industries.

Among the coffee by-products generated during the production and consumption of coffee, SCGs represent the most important from an environmental viewpoint. SCGs comprise a powdered organic residue with high levels of humidity (80%–85%) and acidity (pH around 5.0). The residue is obtained during the production of instant coffee powder by treatment of raw coffee powder with hot water or steam under pressure (Mussatto et al., 2011a). It is estimated that about 2-kg wet mass of SCGs are generated from each kilogram of instant coffee produced, representing an annual global production of around 6 million tons (Mussatto et al., 2011a). Surprisingly, despite the large volume of SCGs produced, this organic residue has only recently received particular attention, unlike other coffee by-products, such as coffee pulp, husks, and silverskin (Cruz et al., 2015; Murthy and Naidu, 2012). A literature survey conducted using two common bibliographic search engines (Scopus and ISI Web of Knowledge) revealed the scarce interest that SCGs have aroused in the past two decades (Fig. 12.1). However, since 2010 there has been a significant increase in the number of scientific studies concerning the use of SCGs as a feedstock to produce value-added products of great interest in many industrial sectors (Campos-Vega et al., 2015). The purpose of this chapter is, first, to review recent findings on the industrial applications of SCGs, and subsequently, to discuss the use of vermicomposting to process SCGs and highlight the potential capacity of SCG-derived vermicompost to enzymatically degrade organophosphorus (OP) pesticides.

This chapter is divided into four parts. Section 12.2 presents a brief overview of the most common applications of SCGs in the production of biofuel and

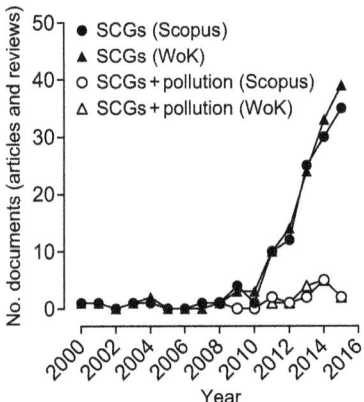

FIGURE 12.1 Global Scientific Production (Limited to Articles and Reviews) Dealing With Spent Coffee Grounds (SCGs) and Their Applications or Effects in the Environment (SCG + Pollution)

Survey was performed September 24, 2016 using two common bibliographic searching engines (ISI Web of Knowledge, or WoK, and Scopus).

other revalorized materials of interest in the pharmaceutical, food, and cosmetics industries. Section 12.3 focusses on the potential agricultural applications of SCGs, whereas Section 12.4 considers vermicomposting as a complementary low-cost and environmentally safe solution to the disposal of SCGs, making SCG-derived vermicompost a suitable material for alleviating pesticide pollution. The results of a case study are discussed in the context of pesticide detoxification by extracellular enzymes present in vermicompost derived from SCGs.

12.2 MULTIPLE APPLICATIONS OF SPENT COFFEE GROUNDS: AN OVERVIEW

SCGs comprise a suitable feedstock for producing liquid biofuel because of the high lipid content of the waste material; 10%–28% dry mass according to Campos-Vega et al. (2015). Research has mainly focused on developing new methods, or improving already-established methods, of obtaining this oil fraction with satisfactory yields. Thus, Soxhlet extraction using organic solvents, such as *n*-hexane (Caetano et al., 2014), hydrothermal liquefaction (Yang et al., 2016), catalytic hydrodeoxygenation (Döhlert et al., 2016), fast pyrolysis (Kelkar et al., 2015), and ultrasonic-assisted extraction (Rocha et al., 2014) are examples of some of the methods proposed for extracting oil from SCGs to produce liquid biofuel. Furthermore, SCGs have proved to be a viable solid fuel (pellets and biochar) for heating purposes (Jeguirim et al., 2014; Tsai et al., 2012; Zuorro and Lavecchia, 2012).

In the food industry, SCGs have been used in animal feed (Seo et al., 2015) and also as an ingredient in bakery products destined for human consumption (Martinez-Saez et al., 2017), as well as a source of dietary fiber with human health benefits (López-Barrera et al., 2016). SCGs have even been used in fermentative processes because of their high content of sugars (Mussatto et al., 2011b). From a pharmaceutical viewpoint, SCGs represent a source of a vast variety of chemicals with bioactive properties. Accordingly, many studies have demonstrated that SCGs contain readily extractable antioxidant substances, such as polyphenols, caffeine, and Maillard-reaction products, which may be exploited for pharmaceutical and nutritional purposes (Bravo et al., 2012). Furthermore, oils isolated from SCG by supercritical fluid extraction have been used in the formulation of topical creams, such as moisturizers and antiseborrheics (Ribeiro et al., 2013), and also as sunscreens that provide high levels of protection against UVB/A radiation (Marto et al., 2016).

The potential use of SCGs in the field of new ecofriendly materials has also been explored recently. For example, oil-free SCGs were used satisfactorily as a filler to reinforce polypropylene composite (Wu et al., 2016). Similarly, SCGs have been combined with other waste by-products derived from steel production and coal combustion to produce a geopolymer material with suitable physical features for use in pavement construction (Kua et al., 2016). By combining SCGs and pure clays, Sena da Fonseca et al. (2013) succeeded in developing a ceramic material that exhibited ideal physical and mechanical performance for construction purposes.

The presence of microplastic debris in the oceans represents, nowadays, a serious environmental hazard that must be tackled urgently (Eerkes-Medrano et al., 2015; Lambert et al., 2013). The challenge is now to develop biodegradable plastics that can be produced cheaply and that also have desirable physical and biodegradable properties. In this way, SCGs have proved suitable for elaborating biodegradable plastic composites (Baek et al., 2013; García-García et al., 2015; Wu, 2015), thus opening up various possibilities for manufacturing ecofriendly plastic materials to alleviate the environmental pressure caused by the more recalcitrant and toxic microplastics.

12.3 ENVIRONMENTAL APPLICATIONS OF SPENT COFFEE GROUNDS

12.3.1 USE OF SPENT COFFEE GROUNDS IN AGRICULTURE

Landfill disposal has traditionally been the main method used to deal with the high volume of SCGs generated yearly on a global scale. However, some studies have revealed that this practice is not free of environmental risk because of the presence of potentially toxic substances, such as caffeine, polyphenols, and tannins in fresh SCG (Mussatto et al., 2011b). However, some agricultural applications provide encouraging solutions to the disposal of SCGs. Although this organic residue is frequently used in domestic agriculture as a soil amendment, the beneficial effects on soil quality and plant health have only recently been investigated, and there is some controversy regarding the effects of SCGs in agriculture. For example, some studies using lettuce as a plant model have demonstrated beneficial effects on plant health when SCGs are applied at doses between 5% and 30% (v/v), including enhanced photosynthetic pigments, vitamins, minerals, and antioxidant capacity (Cruz et al., 2014a,b, 2015); however, the effects were highly dependent on the state of the SCGs (fresh or composted) and the application rate. Use of composted SCGs was the preferred option because the concentrations of caffeine and polyphenols present in fresh SCGs were greatly reduced by composting. Moreover, these studies also demonstrated that the use of fresh SCGs as a soil amendment may reduce micronutrient bioavailability due to the presence of metal-chelating substances in the fresh waste (Cruz et al., 2014b). By contrast, a recent study by Hardgrove and Livesley (2016) showed that direct application of noncomposted SCGs to different types of soil significantly reduced the growth of several horticultural plants (broccoli, leeks, radishes, sunflowers, and violas) at an application rate as low as 2.5% v/v. Although bioactive compounds in the SCGs may be responsible for this toxic effect, the exact mechanism leading to growth inhibition was not elucidated. Similarly, Yamane et al. (2015) examined the effect of SCGs (application rate, 10 kg/m^2) on green manure crops in a 2-year-long field study. Although SCGs caused significant inhibition of plant growth in the first cropping season, the effect decreased in successive seasons. The authors of the study suggested that caffeine, polyphenols, and tannins caused inhibition of plant growth. One interesting result of this study was the pesticide-like effect of fresh SCGs in controlling weed cover, a phenomenon also reported by Hardgrove and Livesley (2016). Despite the increasing interest in the use of SCGs in agriculture, further research is required to develop scientifically based guidelines for use of this waste material as a soil amendment.

12.3.2 BIOREMEDIATION WITH SPENT COFFEE GROUNDS

Most bioremediation studies concerning SCGs have focused on the capacity of the waste product to remove dyes and metals from water. For example, Lavecchia et al. (2016) used noncomposted and water-washed SCGs to remove Pb from water. The kinetics of adsorption of this metal to SCGs fitted a Langmuir model, with a maximum adsorption of 2.46 mg/g. Franca et al. (2009) also used fresh SCGs to remove the cationic dye methylene blue and the maximum adsorption capacity achieved was 18.7 mg/g. However, better results are achieved with activated SCGs. For example, SCG-derived carbonaceous materials have been used to remove dyes from water after a prior activation step in which the SCGs are impregnated with either H_3PO_4 (Reffas et al., 2010) or a mixture of H_3PO_4 and $ZnCl_2$ (Namane et al., 2005). Similarly, Jung et al. (2016) transformed SCGs into an activated carbonaceous material with excellent properties for removing dyes from aqueous media. Less laborious treatment of SCGs with solar energy ($\sim$200°C for 5 days) produced a more efficient degreased adsorbent for removal of Ni and Cd from water than water-washed SCGs (Yen and Huang, 2015; Yen and Lin, 2015). However, scarce attention has been given to the isolation of SCGs after sorbed dyes or metals from water. In this sense, the study by Safarik et al. (2011) is an original example in which SCGs were pretreated with a methanolic ferrofluid solution (1:5, w/v) to produce magnetically modified SCGs able to remove up to 73.4 mg of acridine orange per gram of dried SCGs by simple magnetic separation techniques. Although the aforementioned studies have demonstrated the potential usefulness of fresh and activated SCGs (or the pyrolytic-derived product) in bioremediation of contaminated waters, these laboratory-scale results have not yet been corroborated at a larger scale.

The high organic matter content of SCGs causes proliferation of soil microbes when the material is applied to soil as an amendment; however, little attention has been paid to the potential use of SCGs in the bioremediation of polluted soils. In situ bioremediation is the preferred approach for recovery of contaminated soils because it is cost effective and environmentally friendly. This cleanup technology is based on the capacity of microorganisms and plants to immobilize, accumulate, and metabolize environmental contaminants in soil. However, in situ bioremediation has important drawbacks related to the biochemical, physiological, and ecological characteristics of microorganisms and plants. For example, effective microbially induced degradation of soil contaminants is often a lengthy process that is highly dependent on the microbial growth rate and bioavailability of target contaminants. Stimulation of native microbial communities by addition of organic matter, in a procedure referred to as *biostimulation*, is often the most suitable solution to long-term maintenance of in situ microorganism-assisted bioremediation. SCGs comprise an attractive material because of their high nutrient content and water-holding capacity, although prior treatment (e.g., composting) is required to reduce the phytotoxicity of some of the components. Fenoll et al. (2011, 2014a,b, 2015) have studied the sorption capacity of SCGs in response to pesticide input. To examine pesticide leaching and persistence, these researchers added degreased SCGs (washed with hot water, air dried, refluxed with *n*-hexane, and finally sieved at <3 mm) to disturbed soil columns under laboratory conditions. They reported that application of the degreased SCGs to a soil with a low organic matter content reduced the leaching of fungicides, insecticides, and herbicides, probably as a result of the presence of a greater number of organic ligands available for pesticide binding. Moreover, these studies have also shown that treatment of soil with SCGs may lead to degradation of pesticides via a biostimulatory effect (Fenoll et al., 2014b). Overall, the findings of the studies by Fenoll et al. (2011, 2014a,b, 2015) suggest that SCGs have qualities that make the product suitable as an organic amendment for bioremediation purposes, to increase pesticide adsorption, reduce the leaching capacity of pesticides, and increase biodegradation of pesticides. However, biodegradation is not always achieved because of the low bioaccesibility of some pesticides (e.g., phenylurea herbicides), which are strongly sorbed to the organic matter of SCGs (Fenoll et al., 2015).

12.4 VERMICOMPOSTED SPENT COFFEE GROUNDS: AN ENVIRONMENTALLY SAFE VALUE-ADDED PRODUCT

Vermicomposting is defined as a biooxidative process in which dead organic matter is transformed into a fine and porous peat-like material called vermicompost, which has a high water-holding capacity and nutrient burden, and well-established microbial community; physicochemical and biological properties that undoubtedly improve plant health (Domínguez and Edwards, 2011a). During vermicomposting, detritivorous earthworms cooperate with microorganisms, thus accelerating the stabilization of organic matter and significantly modifying its physical, chemical, and biological properties (Domínguez, 2004; Domínguez and Edwards, 2011a,b; Domínguez and Gómez-Brandón, 2012, 2013). Unlike composting, there is no typical thermophilic phase (45–65°C) in vermicomposting, and enzymes and other chemicals of concern for plant growth and development are therefore not destroyed by thermal denaturing (Datta et al., 2016). Biochemical degradation of organic matter during vermicomposting is mainly due to microbial enzymes, although earthworms are the key drivers of the process. Aerobic conditions are favored in organic waste as a result of the feeding, casting (deposition of feces), and tunneling activities of earthworms. Earthworms also stimulate microbial proliferation (by increasing the surface area available for microbial colonization after fragmentation and ingestion of fresh organic matter), modify microbial biomass and activity (via dispersion of microorganisms in casts), and interact closely with other biological components (e.g., protozoa and springtails) of the vermicomposting system, thereby affecting the microbiome and microfaunal community structure (Domínguez et al., 2003; Lores et al., 2006). A more detailed description of the vermicomposting process is outside the scope of this chapter, and a comprehensive description of vermicomposting (covering suitable earthworm species and raw organic feedstock, impact of vermicompost on soil quality and plant fitness, and the most common domestic- and industrial-scale vermicomposting systems) is provided in the reference book by Edwards et al. (2011).

SCGs contain significant amounts of bioactive compounds, such as polyphenols, caffeine, and tannins, which may represent an environmental hazard when the fresh waste material is applied to soils as an amendment. Vermicomposting of SCGs is a rapid solution to reduce the toxicity of these compounds, and yields revalorized SCG-derived products rich in nutrients, microorganisms, and extracellular enzymes with beneficial effects on soil quality and plant health. The excellent qualities of vermicompost as a soil amendment have been extensively investigated (Edwards et al., 2011). However, very few studies have examined the qualities of vermicompost in relation to removal of pesticide residues from soils. Nogales and coworkers have shown that vermicomposts produced from diverse types of organic waste (from the winery distillery and olive oil industries) significantly reduced the mobility of pesticides, such as imidacloprid and diuron (and their metabolites) in soils with low organic matter contents (Fernández-Bayo et al., 2007, 2008, 2009). Enhanced adsorption of pesticide in soil was suggested to be due to the high content of organic ligands (such as lignin) in vermicompost. Apart from this adsorptive capacity, the biostimulatory effect of vermicompost on soil has scarcely been investigated. In a laboratory study, Fernández-Gómez et al. (2011) used the community-level physiological profiling approach to compare the microbial functional response of four different types of vermicompost to pesticide exposure. These authors found that vermicomposts with high functional diversity were most suitable for bioremediation purposes. Hence the aforementioned studies demonstrate the capacity of vermicompost to reduce the impact of pesticide residues in soil via two main mechanisms: by increasing the number of binding sites for pesticides with a moderate to high sorption coefficient, and by increasing the biodegradation capability. However, the role of certain extracellular enzymes of vermicompost as catalysts for degrading or transforming pollutants is a topic of increasing concern. In this sense, some studies have examined how a group of esterase enzymes with known function in pesticide metabolism, that is, carboxylesterases (CbEs), affects pesticide detoxification in soil bioturbed by earthworms (Sanchez-Hernandez et al., 2014a,b, 2015). These studies postulate the intriguing idea of using earthworms as biological vectors in the enzymatic bioremediation of OP-contaminated soils. On the basis of the previous results, the question that arises is whether vermicompost, and particularly vermicompost derived from SCGs, provides a rich source of these pesticide-detoxifying CbEs.

In the present study, SCGs were obtained from a local supplier (the cafeteria in the Faculty of Biological Science, University of Vigo). The SCGs were processed in pilot-scale vermireactors (3 m^2) held in a greenhouse. The earthworm species *Eisenia andrei* (commonly known as redworm) was used in the vermireactors (Fig. 12.2). *E. andrei* (Oligochaeta, Lumbricidae), an epigeic earthworm with a worldwide distribution, is tolerant to a wide range of temperature and moisture conditions, and is the most common earthworm species used in vermicomposting facilities (Domínguez and Edwards, 2011b). SCGs were added to the vermireactor in successive layers, as they were processed by earthworms. Population density and biomass of earthworms were determined periodically, and samples of SCG-vermicompost were also collected to determine the chemical and biological properties.

The initial earthworm population density in the vermireactor was 3153 ± 184 mature earthworms/m^2 and 7987 ± 1271 juveniles/m^2 (mean ± SD, $n = 5$), corresponding to an earthworm biomass of 700 ± 156 g mature earthworms/m^2 and 740 ± 154 g juveniles/m^2, respectively (Fig. 12.2A). After the addition of 60 kg of SCGs, the density and biomass of earthworms increased continuously and significantly over 14–28 days, reaching a maximum earthworm density of 13634 ± 1578 individuals/m^2 for an earthworm biomass of 1732 g fresh mass/m^2. The cocoon density reached the maximum level after 4 weeks (Fig. 12.2A). To study the effects of the SCG diet on earthworm growth, a complementary experiment was conducted with 30 newly hatched (baby earthworms) (0.055 ± 0.002 g, initial mean weight) held individually in SCGs. The results of this trial showed an exponential growth rate within the first 6 weeks (Fig. 12.2B), demonstrating the viability of vermicomposting of SCGs despite the high content of caffeine and other bioactive ingredients. This growth curve clearly indicated a phase of adaptation of juveniles to SCGs, probably marked by physiological adjustment to nutrients supplied by SCGs and energy expenditure

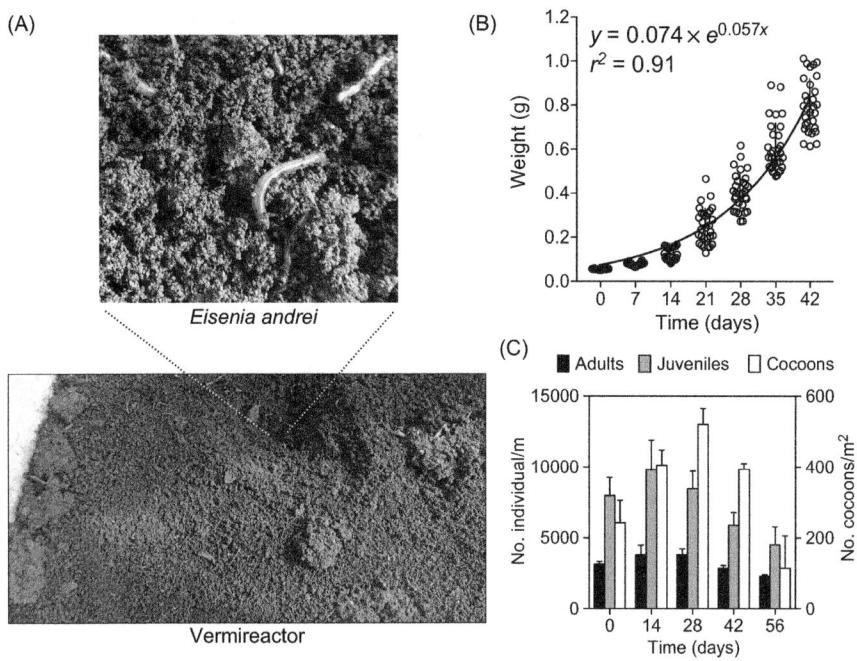

(A)

Eisenia andrei

Vermireactor

(B)

$y = 0.074 \times e^{0.057x}$
$r^2 = 0.91$

Weight (g)

Time (days)

(C)

■ Adults ▨ Juveniles ☐ Cocoons

No. individual/m

No. cocoons/m^2

Time (days)

FIGURE 12.2 Vermicomposting of Spent Coffee Grounds (SCGs) Using *E. andrei*

(A) Adult individuals (*E. andrei*) in the vermireactor. (B) Growth rate of juvenile ($n = 30$ individuals) *E. andrei* individually kept in Petri dishes containing SCGs. (C) Time-course dynamics of earthworm density (adults, juveniles, and cocoons) during the first 56 days of vermicomposting.

against exposure to potential toxic compounds (e.g., caffeine). Indeed, the typical growth curves of *E. andrei* in nontoxic substrates, such as cattle manure fit a hyperbolic-like model with a rapid growth rate in the first weeks of feeding (Elvira et al., 1996). Overall, the earthworm population in the vermireactor was similar to those reached in other types of organic residues, such as cow manure (8,000 individuals/m^2), and pig manure (14,600 individuals/m^2) (Monroy et al., 2006). The chemical components of fresh SCGs (caffeine, polyphenols, and tannins) may be toxic to earthworms and pose a serious threat to earthworm survival, growth, and reproduction. Nevertheless, the study findings indicate that vermicomposting of SCGs is feasible in terms of waste treatment, production of earthworm biomass, and yield of vermicompost with good characteristics. The chemical and biochemical properties and the microbial activity of fresh SCGs and the vermicompost derived from SCGs are summarized in Table 12.1. Concentrations of trace elements in the SCGs used in this pilot study were within the same range of variation as those reported by other authors. As a consequence of the rapid C mineralization during vermicomposting, there was a significant increase in the concentration of micro- and macronutrients in the SCG-derived vermicompost. Furthermore, the pH of the fresh SCGs was slightly acid and became neutral in the final product, and concentrations of cellulose and

Table 12.1 Mean (±SD, $n = 5$) Values of Chemical Composition and Microbial Activity of Spent Coffee Grounds (SCGs) as Raw Feedstock ($t = 0$ Days) and as Vermicompost ($t = 12$ Months)

	SCG Feed-stock ($t = 0$ Months)	SCG Ver-micompost ($t = 12$ Months)	SCGs[a]
pH	6.34 ± 0.10	6.94 ± 0.06	—
Electrical conductivity (µS/cm)	179 ± 25	118 ± 7.5	—
Moisture (%)	66.4 ± 1.1	79.7 ± 4.4	—
Organic matter (%)	98.7 ± 0.07	94.4 ± 1.13	90.5
Total carbon (%)	53.2 ± 0.76	46.7 ± 4.35	—
Total nitrogen (%)	2.34 ± 0.01	4.54 ± 0.28	2.3–2.79 ± 0.10
C/N ratio	22.7 ± 0.34	10.3 ± 0.46	16.91 ± 0.10
Phosphorus (g/kg d.w.)	1.10 ± 0.08	3.01 ± 0.32	1.47–1.8
Potassium (g/kg d.w.)	3.00 ± 0.42	6.52 ± 0.37	3.55–11.7
Calcium (g/kg d.w.)	1.23 ± 0.03	6.00 ± 0.52	0.77–1.2
Magnesium (g/kg d.w.)	1.40 ± 0.05	3.51 ± 0.14	1.29–1.9
Sulfur (g/kg d.w.)	0.43 ± 0.02	1.57 ± 0.67	1.6
Iron (g/kg d.w.)	0.067 ± 0.005	1.25 ± 0.27	0.052–0.11
Manganese (g/kg d.w.)	0.029 ± 0.001	0.24 ± 0.06	0.028–0.040
Zinc (g/kg d.w.)	0.022 ± 0.010	2.60 ± 1.43	0.008–0.015
Boron (mg/kg d.w.)	90.5 ± 8.3	66.1 ± 15.0	8.40 ± 1.10
Copper (mg/kg d.w.)	21.3 ± 0.44	60.0 ± 4.16	18.6–32.3
Basal respiration (mg CO_2/g d.w./h)	0.70 ± 0.18	0.30 ± 0.06	—
Lignin (%)	19.5 ± 0.96	57.7 ± 4.66	23.9 ± 1.7
Cellulose (%)	30.6 ± 2.13	17.2 ± 3.44	8.6 ± 1.8–12.4 ± 0.79
Hemicellulose (%)	28.1 ± 1.76	12.6 ± 4.36	36.7 ± 5.0–39.1 ± 1.94

[a]*Chemical composition of SCGs taken from Ballesteros et al. (2014), Mussatto et al. (2011a), and Murthy and Naidu (2012).*

hemicellulose were halved. All in all, values of most physicochemical properties of SCG-derived vermicompost matched with those considered to have the quality criteria of a good vermicompost (Edwards et al., 2011).

12.4.1 ENZYME ACTIVITIES IN VERMICOMPOST FROM SPENT COFFEE GROUNDS

The breakdown of macromolecules, such as cellulose, hemicellulose, lignin, and tannin during vermicomposting requires the action of a battery of extracellular enzymes, most of which are produced by microorganisms. Accordingly, changes in enzyme

activities during vermicomposting have been widely studied in an attempt to eluci-
date the biochemical interactions between microorganisms and earthworms during
organic matter decomposition, as well as to specify a biochemical fingerprint for mature vermicompost to be used as an indicator of vermicompost quality. The fraction of
organic compounds available during vermicomposting seems to be the driving force
of extracellular enzymes, such as β-glucosidase, phosphatase, urease, and protease
(Benitez et al., 1999). However, the increasing concentrations of humic substances
that appear as vermicomposting progresses provide chemical support for binding
extracellular enzymes, rendering them more stable and protecting them against proteases or adverse environmental conditions (desiccation or high temperatures). It is
therefore not surprising that the activity of extracellular enzymes remains high in
the mature vermicompost, marked by nutrient depletion and absence of earthworms
(Aira et al., 2007; Benitez et al., 2005; Castillo et al., 2013; Domínguez, 2004). In
fact, the activities of some extracellular enzymes are higher in the mature vermicompost than during the vermicomposting process (Castillo et al., 2013).

However, it is not clear whether the enzymatic burden of vermicompost has an
additive effect on soil biochemical performance, thereby accelerating the nutrient
decomposition in soil (Aira and Domínguez, 2011; Benitez et al., 2005). Likewise,
little is known about the effects of the extracellular enzymes present in vermicompost
on the persistence of environmental contaminants in soil. In an attempt to address
this question, some efforts have started characterizing the enzymatic profiles of vermicomposts derived from grape marc, cattle manure, or SCGs. The levels of several
extracellular enzyme activities in vermicompost produced from SCGs are presented.
Particular emphasis is given to laccase and CbE because of their potential role in
bioremediation of organic contaminants.

Enzyme activities in vermicompost were generally measured following established protocols in soil enzymology. Soil enzymes are diverse, as they include intracellular (living, resting, and dead microbial cells) and extracellular forms (free
in soil solution or associated with organo–mineral complexes of soil) (Nannipieri
et al., 2002). As vermicompost is a stabilized material with a high organic matter
content and a well-developed microbial community (Edwards et al., 2011), enzyme
activities are expected to be mainly intracellular and linked to organic complexes.
Most methods of determining soil enzymes measure catalytic activity in soil–water
suspensions, which are preferred as they integrate the activity of the multiple forms
in which enzymes are located in the soil matrix (German et al., 2011). Common
procedures for processing aqueous suspensions of soil generally involve continuous
magnetic stirring (Deng et al., 2013; Turner, 2010), homogenization in a blender
(Bell et al., 2013) or a Brinkmann Polytron (Saiya-Cork et al., 2002), or rotating
mixing (Jackson et al., 2013; Sanchez-Hernandez et al., 2015), before removal of ali-
quots to determine enzyme activity. However, these procedures do not yield efficient
(complete) homogenization of vermicompost. Thus, in a preliminary assay, CbE activity was compared in aqueous suspensions of vermicompost (3% w/v) prepared using two different procedures: by rotating mixing for 30 min, or by homogenization in
a glass–PTFE Potter–Elvehjem grinder (commonly used for tissue homogenization).

The activity of the esterase was 4 times higher in the vermicompost homogenates [142 ± 41 µmol/h/g dry mass for the hydrolysis of 1-naphthyl butyrate (1-NB) and 473 ± 110 µmol/h/g for the hydrolysis of 4-nitrophenyl butyrate (4-NPB)] than in the vermicompost: water suspensions obtained by rotating mixing (31.2 ± 4.0 µmol/h/g dry mass for 1-NB and 122 ± 16 µmol/h/g for 4-NPB). In view of these results, homogenates of vermicompost derived from SCGs were used, to determine the activity of 11 enzymes. Seven of these are representative of the C-, N-, P-, and S-cycles; two (laccase and peroxidase) are considered to measure the oxidative potential of soil (Bach et al., 2013), and the other two enzymes (dehydrogenase and catalase) are commonly used as indicators of microbial activity because of their intracellular location (Trasar-Cepeda et al., 1999; von Mersi and Schinner, 1991). Table 12.2 summarizes the assay conditions adapted to a microplate format for each enzyme activity.

The enzyme activities measured in the vermicompost derived from SCGs are summarized in Table 12.3. The enzyme activities of vermicomposts produced from horse manure and from grape marc are also included for comparison. The chemical nature of the feedstock and the presence of different substrates in the fresh organic waste processed during vermicomposting significantly influence the levels of enzyme activities in mature vermicomposts. Thus, higher levels of urease and protease activities were found in SCG-derived vermicompost than in the other two types of vermicompost as a result of a higher content of nitrogen compounds in SCGs. Likewise, the high oil content of SCGs also explains the higher CbE activity (1-NB) in the vermicompost produced from this organic waste than in the vermicompost obtained from horse manure. Similarly, SCGs and grape marc contain high concentrations of polyphenols, tannins, cellulose, and hemicellulose, thus explaining the higher activity of laccases and peroxidases in these vermicomposts than in vermicompost made from horse manure. As laccases and CbEs are important detoxifying enzymes in polluted soils, the high levels of these enzymes in vermicomposts made from SCGs and from grape marc suggest that vermicomposts are of potential use in the bioremediation of polluted soils.

Laccases have a long history of use in bioremediation. These copper-containing oxidases, which are mainly produced by white-rot fungi, plants, and some species of bacteria, catalyze the oxidation of various aromatic compounds, such as phenols and aromatic amines at the expense of molecular oxygen (Eichlerova et al., 2012). They can also transform or degrade environmental pollutants of current concern, such as chlorophenols, polycyclic aromatic hydrocarbons (PAHs), polychlorinated biphenyls (PCBs), trinitrotoluene, synthetic dyes, pesticides, and endocrine disrupters, such as nonylphenol and bisphenol A (Rao et al., 2014). Nevertheless, large amounts of crude preparations or purified laccases are needed to detoxify polluted soils, and this is one of the challenges in the biotechnological application of laccases in bioremediation. Induction of laccases in several wild fungal strains by specific chemical inducers or through genetically modified organisms has been the most common strategy used to obtain preparations of the enzyme (Gianfreda et al., 1999). However, this approach is not always the best option because of poor enzyme yields, high costs, and tedious purification procedures. Use of vermicompost may represent

Table 12.2 Enzyme Activities Measured in Vermicompost and Spectrophotometric Conditions for Microplate-Scale Assay

Enzymes (EC)	Assay Methods	Substrate (Final Concentrations)	Buffers	Wavelength (Product)	References
CbE (3.1.1.1)	Stopped assay (30 min at 20°C, orbital shaking incubator)	1-NB (2 mM); 4-NPE (2 mM)	Tris-HCl (pH = 7.4)	530 nm (1-naphthol), 405 nm (4-nitrophenol)	Sanchez-Hernandez et al. (2015)
Acid phosphatase (3.1.3.2)	Stopped assay (4 h at 20°C orbital shaking incubator)	4-Nitrophenyl phosphate (5 mM)	Modified universal buffer 20 mM (pH = 6.5)	405 nm (4-nitrophenol)	Popova and Deng (2010)
Alkaline phosphatase (3.1.3.1)	Stopped assay (4 h at 20°C orbital shaking incubator)	4-Nitrophenyl phosphate (5 mM)	Modified universal buffer 20 mM (pH = 11.0)	405 nm (4-nitrophenol)	Popova and Deng (2010)
β-Glucosidase (3.2.1.21)	Stopped assay (4 h at 20°C orbital shaking incubator)	4-Nitrophenyl-β-D-glucaropyranoside (5 mM)	Modified universal buffer 20 mM (pH = 7.4)	405 nm (4-nitrophenol)	Popova and Deng (2010)
Protease (3.4.21.92)	Stopped assay (4 h at 50°C orbital shaking incubator)	Casein (1% w/v)	Tris-HCl 0.1 M (pH = 8.1)	700 nm (tyrosine equivalents)	Schinner et al. (1996)
Urease (3.5.1.5)	Stopped assay (4 h at 20°C rotating mixer)	Urea (40 mM)	Unbuffered method	690 nm (ammonium)	Schinner et al. (1996)
Arylsulfatase (3.1.6.1)	Stopped assay (4 h at 20°C orbital shaking incubator)	—	Modified universal buffer 20 mM (pH = 7.4)	405 nm (4-nitrophenol)	Popova and Deng (2010)
Laccase (1.10.3.2)	Stopped assay (3 h at 20°C orbital shaking incubator)	Pyrogallol (5 mM) L-DOPA (25 mM)	Maleic acid 25 mM (pH = 7.4)	460 nm (oxidized substrates)	Bach et al. (2013)
Peroxidase (1.11.1.7)	Stopped assay (3 h at 20°C orbital shaking incubator)	Pyrogallol (5 mM)	Maleic acid 25 mM (pH = 7.4) + 0.3% H_2O_2	460 nm (oxidized substrates)	Bach et al. (2013)
Dehydrogenase (1.1)	Stopped assay (1 h at 40°C water bath shaking)	Iodotetrazolium chloride (0.14% w/v)	Tris-HCl 1 M (pH = 7.0)	464 nm (reduced iodonitrotetrazolium formazan)	von Mersi and Schinner (1991)
Catalase (1.11.1.6)	Stopped assay (5 min at 20°C, rotating mixer)	H_2O_2 (8.8 mM)	Unbuffered method	505 nm (colored product formed from oxidative coupling reactions between O_2, 4-aminoantipyrine, and phenol)	Trasar-Cepeda et al. (1999)

CbE, Carboxylesterase; L-DOPA, L-3,4-dihydroxyphenylalanine; 1-NB, 1-naphthyl butyrate; 4-NPB, 4-nitrophenyl butyrate.

Table 12.3 Mean (±SD, $n = 6$) Enzyme Activity in Vermicompost Obtained From Spent Coffee Grounds (SCGs), Horse Manure, and Grape Marc

Enzyme	SCGs	Horse Manure[a]	Grape Marc[a]
C-cycling enzymes			
CbE 4-NPB (µmol/h/g dry mass)	527 ± 57	172 ± 31	1079 ± 102
CbE 1-NA (µmol/h/g dry mass)	351 ± 61	155 ± 75	1180 ± 160
CbE 1-NB (µmol/h/g dry mass)	263 ± 68	35.2 ± 14.3	63.3 ± 30.1
β-Glucosidase (µmol/h/g dry mass)	12.3 ± 0.80	8.58 ± 1.17	29.4 ± 3.21
P-cycling enzymes			
Acid phosphatase (µmol/h/g dry mass)	5.00 ± 0.27	9.20 ± 0.16	16.3 ± 2.05
Alkaline phosphatase (µmol/h/g dry mass)	28.4 ± 1.00	8.05 ± 0.96	33.3 ± 3.63
N-cycling enzymes			
Urease (µmol NH_4^+–N/h/g dry mass)	387 ± 30	159 ± 5.30	78.5 ± 12.2
Protease (mg tyrosine equivalent/h/g dry mass)	21.7 ± 0.76	14.31 ± 0.66	13.10 ± 2.54
S-cycling enzymes			
Arylsulfatase (µmol/h/g dry mass)	4.82 ± 0.20	3.22 ± 0.29	0.67 ± 0.06
Oxidative enzymes			
Laccase (µmol pyrogallol/h/g dry mass)	12.7 ± 1.00	8.90 ± 0.93	30.0 ± 2.60
Laccase (µmol L-DOPA/h/g dry mass)	0.60 ± 0.13	0.35 ± 0.14	0.78 ± 0.24
Peroxidase (µmol pyrogallol/h/g dry mass)	7.45 ± 1.01	5.50 ± 0.34	5.60 ± 0.65
Microbial activity indicators			
Dehydrogenase (µmol INTF/h/g dry mass)	1.52 ± 0.21	0.92 ± 0.25	0.180 ± 0.02
Catalase (mmol H_2O_2 consumed/h/g dry mass)	180 ± 4.0	165 ± 7.2	103 ± 12.2

CbE, Carboxylesterase; L-DOPA, L-3,4-dihydroxyphenylalanine; INTF, iodonitrotetrazolium formazan; 1-NA, 1-naphthyl acetate; 4-NPB, 4-nitrophenyl butyrate.
[a]Data from Domínguez, J., Sanchez-Hernandez, J.C., Lores, M., 2017. Vermicomposting of winemaking by-products. In: Galanakis C. (Ed.), Handbook of Grape Processing By-Products. Academic Press, Elsevier, 55–78.

an alternative, cost-effective strategy for supplementing polluted soils with laccases. For instance, vermicomposts produced from polyphenol-rich feedstocks (SCGs or grape marc) had a higher laccase content than vermicompost made from horse manure, and even higher than in different types of soil (Bach et al., 2013). These findings encourage further research on the stability, catalytic efficiency, and reactivity of laccases after application of vermicompost (as solid or liquid preparations) to polluted soils.

Laccase-induced detoxification involves oxidation of contaminants to free radicals or quinones that subsequently undergo polymerization and may be incorporated into humic substance in formation (Riva, 2006). This detoxifying system leads to immobilization of the pollutant in naturally occurring polymeric compounds. CbE activity may provide a comparable detoxification system. The interest in these esterases for bioremediation purposes arises from their capacity to bind OP pesticides by covalent attachment between the phosphoryl moiety of the pesticide and the serine hydroxyl group of the active site of the enzyme (Jackson et al., 2011), as discussed in the next section.

In the present study, CbE activity was detected in vermicompost derived from SCGs by using three different substrates (Table 12.3). In animal tissues and soil, this enzyme is generally found as multiple isozymes with a markedly different substrate preference. In fact, our results showed that there was a significant difference between the hydrolysis rates of 1-NB and 4-NPB among the different types of vermicompost. Likewise, the enzyme activity was resistant to vermicompost desiccation and thermal denaturing (data not shown), which suggested an extracellular location linked to organic matter. To our knowledge, this is the first report of CbE activity in vermicompost derived from SCGs. The activity was one order of magnitude higher than in agricultural soils (Sanchez-Hernandez et al., 2015, 2017), measured using the same enzyme assay. The high CbE activity suggests that vemicompost derived from SCGs has an enzymatic bioremediation capacity, which would enhance its properties as revalorized product.

12.4.2 PESTICIDE DETOXIFYING ESTERASES IN VERMICOMPOST FROM SPENT COFFEE GROUNDS

This part of the study focused on the role of vermicompost-derived CbE activity in OP pesticide detoxification. The ubiquitous CbEs hydrolyze a vast range of carboxylic acid esters, thioesters, and amides (Yan, 2012). Their environmental interest has been addressed in a special issue published in the *Journal of Pesticide Science* in 2010 (Wheelock and Nakagawa, 2010). In pesticide toxicology, CbEs play a pivotal role in detoxifying anticholinesterase (organophosphate and methyl carbamate) and pyrethroid pesticides. These esterases can hydrolyze pyrethroids and methyl carbamates, as they contain a carboxylic ester in their chemical structure (Jackson et al., 2011). Detoxification of OP pesticides by CbEs, however, is achieved by an inhibition-based mechanism. The OP binds to the serine hydroxyl group of the active site of the enzyme to yield a stable "enzyme–inhibitor" complex. The enzyme cannot be reactivated in this phosphorylated state (Sogorb and Vilanova, 2002; Yan, 2012), and can undergo an "aging" process, whereby one alkyl group may leave the OP molecule in the presence of water (dealkylation), leading to permanent inactivation of the enzyme.

CbEs have multiple environmental applications. Thus, inhibition of CbE activity by OP pesticides is often used as a complementary indicator of the exposure of non-target to these types of agrochemicals (Wheelock et al., 2008). The esterases are not only important as simple biomarkers of pesticide exposure, but also because they contribute to developing insecticide resistance in some pest species (Farnsworth et al., 2010; Hemingway and Ranson, 2000). They have been satisfactorily included in the Toxicity Identification Evaluation scheme, whereby purified preparations of CbE are used in a logical weight-of-evidence approach to identify the cause of toxicity in environmental matrices, such as water or sediment (Phillips et al., 2010; Wheelock et al., 2008). More recently, the potential use of CbE activity in enzymatic bioremediation has been investigated in earthworm-bioturbed soil (Sanchez-Hernandez et al., 2014a,b, 2015). Although pending further research, the results of these studies suggest that soil CbE activity, induced by earthworm activity, irreversibly binds OP pesticides, thereby acting as biomolecular scavengers. In addition, earthworm casting was postulated as a significant source of stable and active CbE activity to soil, preventing saturation of the stoichiometric detoxification mechanism by CbEs caused by an excess of pesticides (Sanchez-Hernandez et al., 2014a,b).

In a similar approach, the sensitivity of CbE in vermicompost derived from SCGs to the insecticide chlorpyrifos (CPO) and its metabolite chlorpyrifos-oxon (CPoxon) was assessed. To this end, wet solid (80% w/v, moisture) and liquid vermicomposts were individually spiked with a low (1 ppm) and a high (10 ppm) concentration of both OP pesticides ($n = 4$ replicates per treatment), and incubated for 30 days at 20°C in darkness. CbE activity was measured 4 and 30 days after pesticide treatment. In the case of liquid vermicompost, the enzyme activity was determined directly in the test tubes according to the protocol in Table 12.2. For solid vermicompost, however, a subsample (1.5 g) was removed from each treatment and homogenized (3% w/v) in distilled water before determination of CbEs.

The results of this toxicity assay show that the response of the enzyme activity to the pesticides strongly dependent on the OP type, time of exposure, and the initial state of vermicompost (solid or liquid). In the liquid vermicompost, CPoxon significantly inhibited CbE activity after 4 days of exposure ($H_{(4)} = 17.8$, $P = 0.0013$, Kruskal–Wallis test), and the response was dose dependent (Fig. 12.3A). Although the CbE activity decreased in response to the CPO treatment, it was not statistically different from control values. This was expected because oxygen-analog metabolites of phosphorothioate pesticides display a higher affinity for the active sites of CbEs than the parent compounds (Chambers et al., 2010). Surprisingly, CbE activity in the solid vermicompost was not equally affected by OP treatments in short-term exposure (Fig. 12.3B).

For longer exposure times ($t = 30$ days), the CbE activity recovered fully in the liquid vermicomposts treated with CPoxon. However, spontaneous reactivation of the phosphorylated enzyme activity should not be ruled out. To investigate the possible mechanisms responsible for this recovery of enzyme activity, the catalase activity was measured as an indicator of microbial activity. There was a significant increase of this intracellular enzyme activity in all treatments (controls and pesticide-spiked vermicomposts) from 1.67 ± 0.03 mmol O_2 consumed per hour per milliliter vermicompost (mean $\pm$ standard deviation, $n = 20$) at $t = 4$ days to 4.82 ± 0.11 ($n = 20$) after 30 days of pesticide exposure. Therefore, enhanced microbial activity during the 30-day exposure time may contribute to reverse CbE activity in the liquid vermicomposts treated with CPoxon. Indeed, cometabolism of the pesticide by microorganisms may also contribute to restoring CbE activity. However, CbE activity was

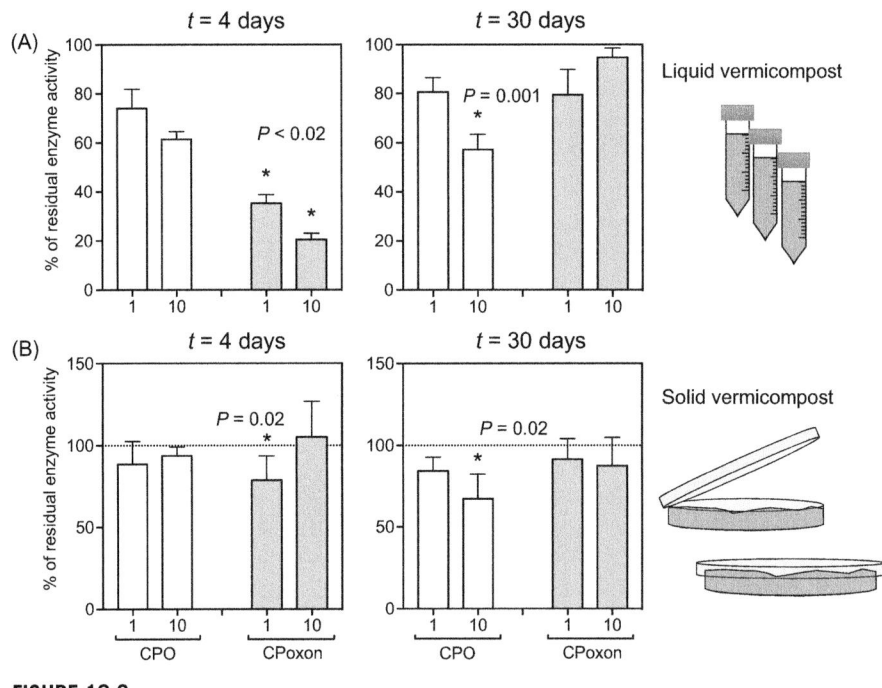

FIGURE 12.3

Effect of chlorpyrifos (CPO) and chlorpyrifos-oxon (CPoxon) on carboxylesterase (CbE) activity in both (A) liquid and (B) solid spent coffee ground (SCG)–derived vermicomposts treated with 1 and 10 ppm of pesticides. Data are the mean and standard deviation ($n = 4$ replicates). *Asterisk* indicates significant differences compared with controls (Wilcoxon rank-sum test).

significantly inhibited in liquid vermicomposts treated with the highest concentration of CPO relative to the activity in control samples (Fig. 12.3A). One plausible explanation for this is that the esterase enzyme was inhibited by formation of CPoxon during degradation of CPO, as freshly secreted CbE from microbial proliferation was inhibited by CPoxon on formation. Further experiments must be conducted to test this hypothesis and to provide solid evidence for the actual role of CbE activity in liquid vermicompost as a detoxifying mechanism. Nonetheless, the results obtained so far indicate the following: (1) CbEs acted as molecular scavengers for OP pesticides, displaying higher affinity for the toxic metabolite CPoxon than for the parent compound (CPO), and (2) the liquid vermicompost provided a better substrate for this CbE-induced detoxification system, probably because of higher bioaccessibility of the pesticides to the active site of the enzyme than in solid vermicompost.

Degradation of CPO in liquid vermicompost was confirmed by incubation of liquid vermicompost made from SCGs ($n = 4$ replicates) in the presence of 25 µg/mL CPO (20°C and dark). At this pesticide concentration, the degradation kinetics of the OP was monitored by liquid chromatography with diode-array detection

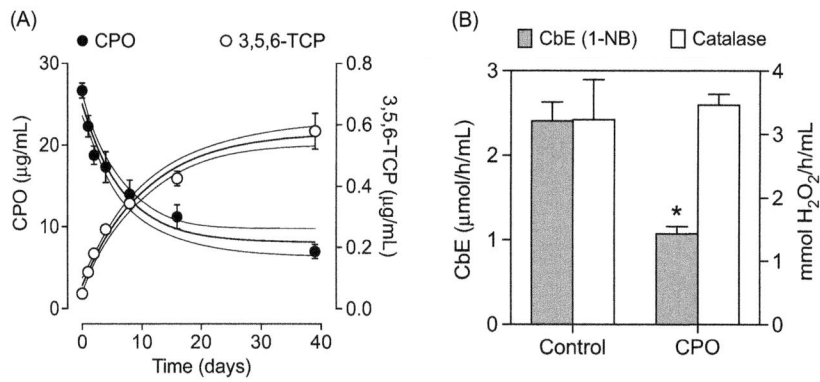

FIGURE 12.4

(A) Degradation kinetic of Chlorpyrifos (CPO) (25 µg/mL) in liquid vermicompost, and formation rate of the metabolite 3,5,6-trichloro-2-pyridinol (3,5,6-TCP). (B) Carboxylesterase (CbE) and catalase activities in control (pesticide free) and CPO-treated liquid vermicomposts after 40 days of pesticide treatment. The esterase activity was measured using the substrate 1-naphthyl butyrate (1-NB) as summarized in Table 12.1. Data are the mean and standard deviation ($n = 4$ replicates). *Asterisk* indicates Wilcoxon rank-sum test.

(Sanchez-Hernandez et al., 2015). Degradation of CPO fitted a first-order kinetics model with a half-life ($t_{1/2}$) of 4.74 day^{-1}, which was concomitant with a progressive increase in the concentration of the metabolite 3,5,7-trichloro-2-pyridinol (Fig. 12.4A). Moreover, at the end of the degradation trial, CbE activity was more strongly inhibited (55.3% of controls) in the liquid vermicompost spiked with CPO than in the controls (pesticide-free vermicompost), whereas the catalase activity did not change. The mean (±SD) catalase activity of uncontaminated liquid vermicompost was 3.23 ± 0.63 mmol O_2 consumed per hour per milliliter, whereas the mean activity of the enzyme was a 3.46 ± 0.17 mmol O_2 consumed per hour per milliliter in the CPO-spiked liquid vermicompost (Fig. 12.4B). The result of this assay confirmed that CbE activity in liquid vermicompost is involved in immobilizing CPoxon during degradation of CPO, reducing the potential toxicity of this metabolite to microorganisms (Singh and Walker, 2006).

Although it is still too early to draw firm conclusions, the data obtained in this study suggest that the role of CbEs as molecular scavengers of OP pesticides is highly dependent on the form of vermicompost that receives the pesticide. Homogenization of liquid vermicompost disperses extracellular enzymes and microorganisms in the aqueous medium and thus facilitates their contribution to pesticide degradation or transformation. In view of these findings, vermicompost could be potentially used for bioremediation of OP-contaminated soils following the conceptual scheme illustrated in Fig. 12.5. Binding of the pesticide to organic matter in vermicompost (step 1 in Fig. 12.5) represents the main route whereby vermicompost would reduce pesticide transport and toxicity in soil. Indeed, as discussed earlier, vermicomposts derived from SCGs and from grape marc provide a rich source of lignin that acts as an organic

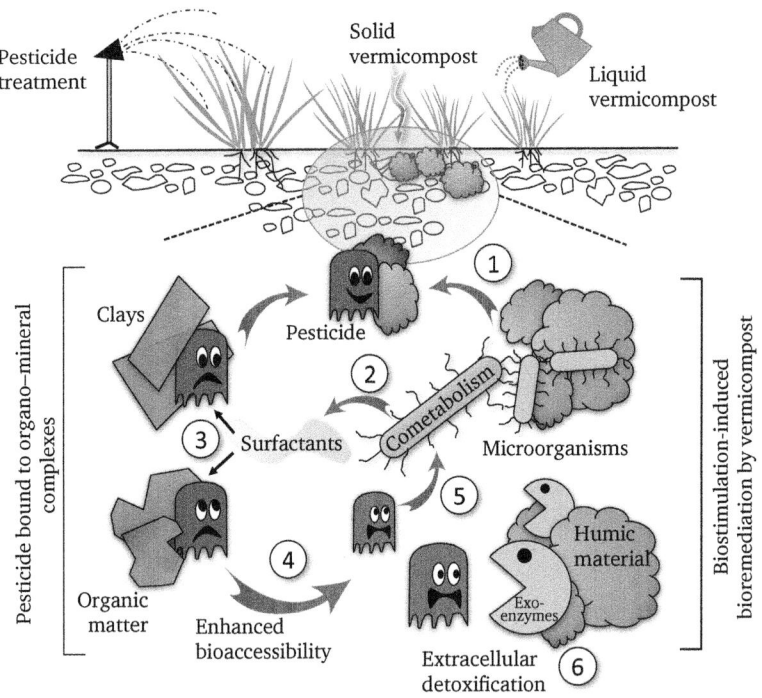

FIGURE 12.5 Conceptual Model of the Potential Role of Vermicompost (Liquid and Solid) on Pesticide Degradation and Transformation in Soil

ligand for pesticides. Furthermore, vermicompost can have a biostimulatory effect in soil when environmental conditions are favorable (step 2). The enhancement of microbial activity and biomass has two effects on bioavailability and biodegradation of pesticides. First, pesticides sorbed on clay and soil organic matter may be desorbed by the action of natural surfactants, such as humic acids and surfactants produced by microorganisms (e.g., glycolipids, phospholipids, lipopeptides, lipoproteins, and lipopolysaccharides) (Megharaj et al., 2011). As vermicompost is rich in humic substances and microorganisms, it is expected to exert a surfactant-like effect on sorbed pesticides to organo–mineral complexes of soil (step 3). Second, microbes may also directly attack sorbed pesticides through the catalytic action of exoenzymes (EE) that they release (Megharaj et al., 2011). Surfactants facilitate the bioavailability of pesticides for subsequent degradation by proliferating microorganisms (steps 4 and 5) and for transformation or breakdown by the action of extracellular detoxifying enzymes (CbEs, laccases, and peroxidases) associated with the organic matter in vermicompost (step 6). Pending further studies, the addition of liquid vermicompost to OP-contaminated soils or to agricultural soils that are periodically treated with OP will contribute to reducing the mobility and toxicity of these agrochemicals via the action of chemical (humic substances) and biochemical (e.g., CbEs) factors in the vermicompost.

12.4.3 POTENTIAL USE OF SCG-VERMICOMPOST FOR SAFE CULTIVATION IN URBAN AGRICULTURE

From an agronomic and environmental viewpoint, urban agriculture could benefit from SCG vermicompost. This kind of agriculture, defined as the production of crop and livestock within and around cities and towns, is a form of sustainable agriculture that has experienced an exponential growth worldwide in the last decade (Zezza and Tasciotti, 2010). Its popularity lies in its capacity to stimulate the local economy and social integration (Badami and Ramankutty, 2015), which is needed at a crucial time in human history: in 2007, for the first time, the world population became more urban than rural (Orsini et al., 2013). In this scenario, urban agriculture may be considered as a form of alleviating socioeconomic pressure caused mainly by two concomitant events: rural–urban migration, especially in developing countries, and the global financial crisis that initiated in 2007 (Hodson and Quaglia, 2009). Another bibliographic analysis of scientific studies involving the terms "urban agriculture" or "urban farming" shows an inflexion point in 2007 in the progress of urban agriculture studies since the 1980s (Fig. 12.6), which in fact coincides with these two aforementioned global events. These data could provide an idea of how important urban

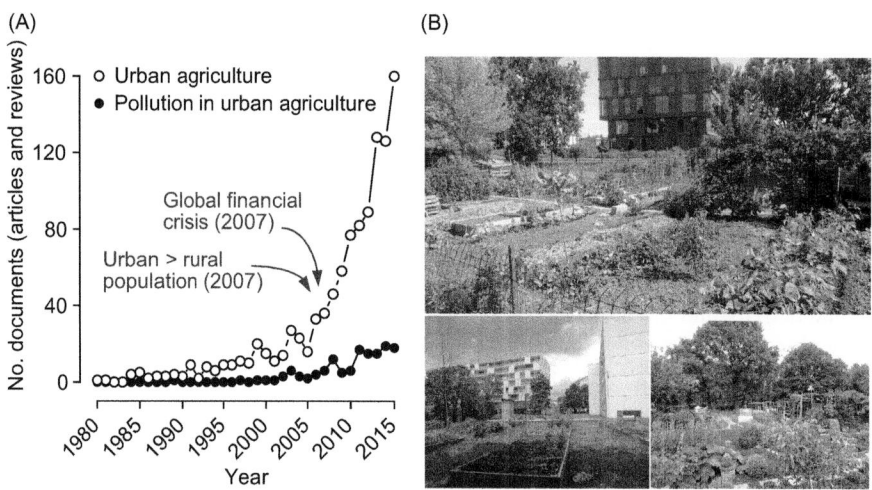

FIGURE 12.6

(A) Global scientific production (articles and reviews) dealing with urban agriculture and pollution. *Arrows* indicate two critical events in this century: the global financial crisis initiated in 2007 and, for the first time in human history, the urban population exceeding the rural, as of May 2007 (Orsini et al., 2013). Survey was performed using the bibliographic searching engine Scopus (October 2, 2016). (B) Photographs of several urban orchards in France.

Part B Photos by Camille Dumat, Available from: http://reseau-agriville.com

agriculture will be in cities around the world in the coming years, and how important the implementation of methodologies and monitoring programs for its sustainability and safety development will be.

One of the challenges of urban agriculture is the quality of urban soil to support safe crops. Urban soils are perhaps among the most contaminated soils in the world, and metals, such as Pb and Cd, as well as organic pollutants, such as PAHs and PCBs, are the most common environmental pollutants found in urban soils (Meuser, 2010). For example, a study performed in the city of Shanghai, China, that included the analysis of the 16 US-EPA priority PAHs in 57 soil samples collected in 2011, revealed concentrations of these aromatic hydrocarbons (ΣPAHs = 1970 μg/kg, mean value) that may pose a potential risk to human health. Similarly, a survey performed in east London, United Kingdom, in 2009 in which 76 soil samples were collected from an area of 19 km^2 (Vane et al., 2014) showed that total concentrations of the US-EPA PAHs were similar to other cities (ΣPAHs = 18 mg/kg, mean value), whereas concentrations of PCBs representing those congeners with a high chlorine content were higher (ΣPCBs = 120 μg/kg, mean value), which could suggest local sources for these PCBs, as opposed to a global atmospheric transport. Likewise, PAH concentrations were determined in soils collected from three European cities (Glasgow, Torino, and Ljubljana) during the spring of 2004 (Morillo et al., 2007), indicating that soils from Glasgow had the highest concentrations (1.48–51.8 mg/kg). In these latter two studies, a pyrogenic origin for PAHs was suggested, especially from motor vehicle exhaust. Moreover, climatic conditions and organic carbon content of soils were suggested as the main driving forces for PAH accumulation in soil.

Metal contamination is also very frequent in urban soils (Wortman and Lovell, 2013). High concentrations of Pb, for instance, are frequently detected in urban soils as a consequence of the legacy of Pb-based gasoline. Residues from traffic in cities (e.g., car tire and brake wear) and nearby industrial areas also contribute to increase metal concentrations in urban soils (Table 12.4). Accordingly, growing horticultural plants and fruits in urban environments has an elevated risk of metal accumulation from the soil (Antisari et al., 2015). However, metal pollution of the edible parts of vegetables occurs mainly by air deposition of metal-bound particles and simply washing the vegetables with water removes a significant fraction of metals (Nabulo et al., 2010; von Hoffen and Säumel, 2014). Nevertheless, although the high concentrations of metals in soil may pose a risk for plant uptake depending on plant species and soil physicochemical properties, it is important to consider that these pollutants may cause deterioration of soil quality, thus rendering the soil less productive from an agronomic viewpoint.

Vermicomposting provides a unique opportunity to revalorize SCGs in the cities that are, in turn, the main sources of this organic waste. Phosphate fertilization and compost amendment are two common practices in bioremediation of urban soils to reduce metal uptake by horticultural plants (Wortman and Lovell, 2013). Organic residues with a high organic matter content (>50%) and high concentrations of

Table 12.4 Mean and Range of Metal Concentrations (mg/kg) in Urban Soils From Several Cities Around the World

City/Country	Cd	Cu	Ni	Pb	Zn
New York (USA)	1.6	110	32	600	327
Bangkok (Thailand)	0.15	27	23	29	38
Hong Kong (China)	0.33	10	4	71	78
Berlin (Germany)	0.35	31	8	77	129
Oslo (Norway)	0.34	24	24	34	130
Connecticut (USA)	<0.5	40	12	176	163
Zagreb (Croatia)	0.5	18	49	23	70
Baltimore (USA)	0.56	17	2.8	100	92
Szczecin (Poland)	0.5–1.8	13.8–32.4	8.2–13.6	48–140	63–398
Sofia (Bulgaria)	0.11–0.33	32.6–65.7	—	30.9–34.3	37.9–81.0
Berger (Norway)	<0.1–1.5	4–2850	1–310	3–5780	8–998
Trondheim (Norway)	<0.1–11.3	1.7–706	6–231	9–976	7–3420
Seoul (South Korea)	1.0–4.4	11–471	—	93–1636	55–596
Beijing (China)	<0.01–0.97	2–282	2.8–169	5–117	22–400
Ljubljana (Slovenia)	—	14–135	14–45	10–387	56–581
Seville (Spain)	—	9–365	16–62	15–977	21–443
Torino (Italy)	—	15–430	77–830	14–1440	53–880

Data from Cheng, Z., Paltseva, A., Li, I., Morin, T., Huot, H., Egendorf, S., Su, Z., Yolanda, R., Singh, K., Lee, L., Grinshtein, M., Liu, Y., Green, K., Wai, W., Wazed, B., Shaw, R., 2015. Trace metal contamination in New York City garden soils. Soil Sci. 180, 167–174; Meuser, H., 2010. Contaminated Urban Soils. Springer, Dordrecht; Heidelberg; London (p. 318).

iron, phosphorus, and manganese are able to reduce Pb bioavailability (Wortman and Lovell, 2013). Hence, vermicompost derived from SCGs should be a suitable organic amendment because of the high concentrations of these elements and organic matter (Table 12.1). Additionally, vermicomposts are rich in humic substances, which favor binding of metals and organic pollutants. Another important issue of concern in urban agriculture is the high consumption of irrigation water, which can be aggravated in periods of water restriction in megacities. Vermicompost derived from SCGs has a very high water-holding capability that undoubtedly could alleviate water demand. Additionally, the innovative tendency of soilless cultures in urban agriculture to avoid the use of soil, obviously reducing the risk of plant contamination (Eigenbrod and Gruda, 2015), can benefit from liquid vermicompost, which is rich in nutrients and enzyme activities; these latter are essential for the mineralization of organic matter. Taken together, the physicochemical and biological properties of SCG-derived vermicompost make this revalorized product a promising strategic product for bioremediation (e.g., metal immobilization) and improvement (e.g., nutrient and extracellular enzyme addition) of urban agriculture soils, helping to reduce contamination of vegetable crops from these soils.

12.5 CONCLUSIONS

SCGs proved to be a suitable feedstock for vermicomposting, and a high density of earthworms was observed in the first few weeks of the decomposition process. Chemical and microbiological analyses revealed that mature and high-quality vermicomposts were obtained. Vermicomposting of SCGs is, therefore, a practical mean for the disposal of SCGs, yielding a value-added product with attractive properties for bioremediation purposes. In this way, highly polluted soils, such as those commonly used in urban agriculture could benefit from the physicochemical and biological properties of SCG-derived vermicompost. Measurement of selected extracellular enzyme activities involved in C-, N- S-, and P-cycling revealed higher activities of protease, urease, laccase, and CbE in vermicompost made from SCGs than in other types of vermicompost (e.g., vermicompost made from horse manure). The chemical composition of the primary feedstock influenced the levels of extracellular enzyme activity in vermicompost. Therefore, enzymatic fingerprinting can be considered for inclusion in the chemical and microbiological tests of vermicompost quality. Nevertheless, enzyme assays must first be standardized and accurate quality-control methods developed before biochemical fingerprinting can be used to define the quality of vermicompost.

The laccase and CbE activities were higher in the vermicompost made from SCGs than those from other vermicomposts produced from organic residues. This indicates the possibility of using this type of vermicompost in the enzymatic bioremediation of OP-contaminated soils. A survey of the scientific literature revealed that the use of vermicompost to decontaminate soil is relatively new and that further investigation must be conducted to reveal the real potential of the product in biostimulation-induced bioremediation. The CbE activity in vermicompost made from SCGs, particularly liquid vermicompost, was found to be sensitive to inhibition by CPoxon (toxic metabolite of CPO), leading to immobilization of this pesticide by irreversible inhibition of the enzyme activity. Although this detoxification mechanism is not catalytic, and is therefore susceptible to saturation due to phosphorylation of CbE activity, microbial proliferation may contribute to restoring the initial levels of extracellular CbE activity, thus contributing to the long-term maintenance of this enzymatic detoxification.

However, the toxicity assays were conducted with small sample sizes and the findings may not be transferable to field conditions. Thus, further research must be carried out to address some questions, such as the earthworm health during vermicomposting of SCGs, yields of vermicomposting of mixing feedstock rich in substrates that induce laccase and CbE activity, and the effect of application of liquid vermicompost on the biochemical performance of soil and its natural attenuation capacity.

ACKNOWLEDGMENTS

The referred study in this chapter was funded by the Spanish Ministerio de Economía y Competitividad (ref. CTM2014-53915-R and CTM2013-42540-R) and the Junta Castilla-La Mancha (ref. PEII-2014-001-P).

REFERENCES

Aira, M., Domínguez, J., 2011. Earthworm effects without earthworms: inoculation of raw organic matter with worm-worked substrates alters microbial community functioning. PLoS One 6, e16354.

Aira, M., Monroy, F., Domínguez, J., 2007. Earthworms strongly modify microbial biomass and activity triggering enzymatic activities during vermicomposting independently of the application rates of pig slurry. Sci. Total Environ. 385, 252–261.

Antisari, L.V., Orsini, F., Marchetti, L., Vianello, G., Gianquinto, G., 2015. Heavy metal accumulation in vegetables grown in urban gardens. Agron. Sustain. Dev. 35, 1139–1147.

Bach, C.E., Warnock, D.D., Van Horn, D.J., Weintraub, M.N., Sinsabaugh, R.L., Allison, S.D., German, D.P., 2013. Measuring phenol oxidase and peroxidase activities with pyrogallol, L-DOPA, and ABTS: effect of assay conditions and soil type. Soil Biol. Biochem. 67, 183–191.

Badami, M.G., Ramankutty, N., 2015. Urban agriculture and food security—a critique based on an assessment of urban land constraints. Glob. Food Sec. 4, 8–15.

Baek, B.-S., Park, J.-W., Lee, B.-H., Kim, H.-J., 2013. Development and application of green composites: using coffee ground and bamboo flour. J. Polym. Environ. 21, 702–709.

Ballesteros, L.F., Teixeira, J.A., Mussatto, S.I., 2014. Chemical, functional, and structural properties of spent coffee grounds and coffee silverskin. Food Bioprocess Technol. 7, 3493–3503.

Bell, C.W., Fricks, B.E., Rocca, J.D., Steinweg, J.M., McMahon, S.K., Wallenstein, M.D., 2013. High-throughput fluorometric measurement of potential soil extracellular enzyme activities. J. Vis. Ex. 81, pe50961.

Benitez, E., Nogales, R., Elvira, C., Masciandaro, G., 1999. Enzyme activities as indicators of the stabilization of sewage sludges composting with *Eisenia foetida*. Bioresour. Technol. 67, 297–303.

Benitez, E., Sainz, H., Nogales, R., 2005. Hydrolytic enzyme activities of extracted humic substances during the vermicomposting of a lignocellulosic olive waste. Bioresour. Technol. 96, 785–790.

Bravo, J., Juániz, I., Monente, C., Caemmerer, B., Kroh, L.W., De Peña, M.P., Cid, C., 2012. Evaluation of spent coffee obtained from the most common coffeemakers as a source of hydrophilic bioactive compounds. J. Agric. Food Chem. 60, 12565–12573.

Caetano, N.S., Silva, V.F., Melo, A.C., Martins, A.A., Mata, T.M., 2014. Spent coffee grounds for biodiesel production and other applications. Clean Technol. Environ. Policy 16, 1423–1430.

Campos-Vega, R., Loarca-Piña, G., Vergara-Castañeda, H.A., Oomah, B.D., 2015. Spent coffee grounds: a review on current research and future prospects. Trends Food Sci. Technol. 45, 24–36.

Castillo, J.M., Romero, E., Nogales, R., 2013. Dynamics of microbial communities related to biochemical parameters during vermicomposting and maturation of agroindustrial lignocellulose wastes. Bioresour. Technol. 146, 345–354.

Chambers, J.E., Meek, E.C., Ross, M.K., 2010. The metabolic activation and detoxification of anticholinesterase insecticides. In: Satoh, T., Gupta, R.C. (Eds.), Anticholinesterase Pesticides: Metabolism, Neurotoxicity, and Epidemiology. John Wiley & Sons, Hoboken, NJ, pp. 77–84.

Cruz, R., Gomes, T., Ferreira, A., Mendes, E., Baptista, P., Cunha, S., Pereira, J.A., Ramalhosa, E., Casal, S., 2014a. Antioxidant activity and bioactive compounds of lettuce improved by espresso coffee residues. Food Chem. 145, 95–101.

Cruz, R., Morais, S., Mendes, E., Pereira, J.A., Baptista, P., Casal, S., 2014b. Improvement of vegetables elemental quality by espresso coffee residues. Food Chem. 148, 294–299.

Cruz, R., Mendes, E., Torrinha, Á., Morais, S., 2015. Revalorization of spent coffee residues by a direct agronomic approach. Food Res. Int. 73, 190–196.

Datta, S., Singh, J., Singh, S., Singh, J., 2016. Earthworms, pesticides and sustainable agriculture: a review. Environ. Sci. Pollut. Res. Int. 23, 8227–8243.

Deng, S., Popova, I.E., Dick, L., Dick, R., 2013. Bench scale and microplate format assay of soil enzyme activities using spectroscopic and fluorometric approaches. Appl. Soil Ecol. 64, 84–90.

Doepker, C., Lieberman, H.R., Smith, A.P., Peck, J.D., El-Sohemy, A., Welsh, B.T., 2016. Caffeine: friend or foe? Annu. Rev. Food Sci. Technol. 7, 117–137.

Döhlert, P., Weidauer, M., Enthaler, S., 2016. Spent coffee ground as source for hydrocarbon fuels. J. Energy Chem. 25, 146–152.

Domínguez, J., 2004. State of the art and new perspectives on vermicomposting research. In: Edwards, C.A. (Ed.), Earthworm Ecology. CRC Press, Boca Raton, FL, pp. 401–424.

Domínguez, J., Edwards, C.A., 2011a. Relationships between composting and vermicomposting: relative values of the products. In: Edwards, C.A., Arancon, N.Q., Sherman, R.L. (Eds.), Vermiculture Technology: Earthworms, Organic Waste and Environmental Management. CRC Press, Boca Raton, FL, pp. 11–26.

Domínguez, J., Edwards, C.A., 2011b. Biology and ecology of earthworm species used for vermicomposting. In: Edwards, C.A., Arancon, N.Q., Sherman, R.L. (Eds.), Vermiculture Technology: Earthworms, Organic Waste and Environmental Management. CRC Press, Boca Raton, FL, pp. 27–40.

Domínguez, J., Gómez-Brandón, M., 2012. Vermicomposting: composting with earthworms to recycle organic wastes. In: Kumar, S., Bharti, A. (Eds.), Management of Organic Waste. InTech Open Science, Rijeka, Croatia, pp. 29–48.

Domínguez, J., Gómez-Brandón, M., 2013. The influence of earthworms on nutrient dynamics during the process of vermicomposting. Waste Manage. Res. 31, 859–868.

Domínguez, J., Parmelee, R.W., Edwards, C.A., 2003. Interactions between *Eisenia andrei* (Oligochaeta) and nematode populations during vermicomposting. Pedobiologia 47, 53–60.

Edwards, C.A., Subler, S., Arancon, N., 2011. Quality criteria for vermicomposts. In: Edwards, C.A. (Ed.), Vermiculture Technology: Earthworms, Organic Wastes, and Environmental Management,. Taylor & Francis Group, LLC, Boca Raton, FL, pp. 287–301.

Eerkes-Medrano, D., Thompson, R.C., Aldridge, D.C., 2015. Microplastics in freshwater systems: a review of the emerging threats, identification of knowledge gaps and prioritisation of research needs. Water Res. 75, 63–82.

Eichlerova, I., Šnajdr, J., Baldrian, P., 2012. Laccase activity in soils: considerations for the measurement of enzyme activity. Chemosphere 88, 1–7.

Eigenbrod, C., Gruda, N., 2015. Urban vegetable for food security in cities. A review. Agron. Sustain. Dev. 35, 483–498.

Elvira, C., Domínguez, J., Briones, M., 1996. Growth and reproduction of *Eisenia andrei* and *E. fetida* (Oligochaeta, Lumbricidae) in different organic residues. Pedobiologia 40, 377–384.

Farnsworth, C.A., Teese, M.G., Yuan, G., Li, Y., Scott, C., Zhang, X., Wu, Y., Russell, R.J., Oakeshott, J.G., 2010. Esterase-based metabolic resistance to insecticides in heliothine and spodopteran pests. J. Pestic. Sci. 35, 275–289.

Fenoll, J., Garrido, I., Hellín, P., Flores, P., Vela, N., Navarro, S., 2014a. Use of different organic wastes in reducing the potential leaching of propanil, isoxaben, cadusafos and pencycuron through the soil. J. Environ. Sci. Health B 49, 601–608, 49.

Fenoll, J., Garrido, I., Hellín, P., Flores, P., Vela, N., Navarro, S., 2015. Use of different organic wastes as strategy to mitigate the leaching potential of phenylurea herbicides through the soil. Environ. Sci. Pollut. Res. Int. 22, 4336–4349.

Fenoll, J., Ruiz, E., Flores, P., Vela, N., Hellín, P., Navarro, S., 2011. Use of farming and agro-industrial wastes as versatile barriers in reducing pesticide leaching through soil columns. J. Hazard. Mater. 187, 206–212.

Fenoll, J., Vela, N., Navarro, G., Pérez-Lucas, G., Navarro, S., 2014b. Assessment of agro-industrial and composted organic wastes for reducing the potential leaching of triazine herbicide residues through the soil. Sci. Total Environ. 493, 124–132.

Fernández-Bayo, J.D., Nogales, R., Romero, E., 2007. Improved retention of imidacloprid (Confidor®) in soils by adding vermicompost from spent grape marc. Sci. Total Environ. 378, 95–100.

Fernández-Bayo, J.D., Nogales, R., Romero, E., 2009. Assessment of three vermicomposts as organic amendments used to enhance diuron sorption in soils with low organic carbon content. Eur. J. Soil Sci. 60, 935–944.

Fernández-Bayo, J.D., Romero, E., Schnitzler, F., Burauel, P., 2008. Assessment of pesticide availability in soil fractions after the incorporation of winery-distillery vermicomposts. Environ. Pollut. 154, 330–337.

Fernández-Gómez, M.J., Nogales, R., Insam, H., Romero, E., Goberna, M., 2011. Role of vermicompost chemical composition, microbial functional diversity, and fungal community structure in their microbial respiratory response to three pesticides. Bioresour. Technol. 102, 9638–9645.

Franca, A.S., Oliveira, L.S., Ferreira, M.E., 2009. Kinetics and equilibrium studies of methylene blue adsorption by spent coffee grounds. Desalination 249, 267–272.

García-García, D., Carbonell, A., Samper, M.D., García-Sanoguera, D., Balart, R., 2015. Green composites based on polypropylene matrix and hydrophobized spend coffee ground (SCG) powder. Composites B 78, 256–265.

German, D.P., Weintraub, M.N., Grandy, A.S., Lauber, C.L., Rinkes, Z.L., Allison, S.D., 2011. Optimization of hydrolytic and oxidative enzyme methods for ecosystem studies. Soil Biol. Biochem. 43, 1387–1397.

Gianfreda, L., Xu, F., Bollag, J.M., 1999. Laccases: a useful group of oxidoreductive enzymes. Bioremed. J. 3, 1–26.

Hardgrove, S.J., Livesley, S.J., 2016. Applying spent coffee grounds directly to urban agriculture soils greatly reduces plant growth. Urb. Forest. Urb. Green. 18, 1–8.

Hemingway, J., Ranson, H., 2000. Insecticide resistance in insect vectors of human disease. Annu. Rev. Entomol. 45, 371–391.

Higdon, J.V., Frei, B., 2006. Coffee and health: a review of recent human research. Crit. Rev. Food Sci. Nutr. 46, 101–123.

Hodson, D., Quaglia, L., 2009. European perspectives on the global financial crisis: introduction. J. Common Mark. Stud. 47, 939–953.

Jackson, C.J., Oakeshott, J.G., Sanchez-Hernandez, J.C., Wheelock, C.E., Satoh, T., Gupta, R.C., 2011. Carboxylesterases in the metabolism and toxicity of pesticides. In: Satoh, T., Gupta, R.C. (Eds.), Anticholinesterase Pesticides: Metabolism, Neurotoxicity, and Epidemiology. John Wiley & Sons, Hoboken, NJ.

Jackson, C.R., Tyler, H.L., Millar, J.J., 2013. Determination of microbial extracellular enzyme activity in waters, soils, and sediments using high throughput microplate assays. J. Vis. Exp., e50399–e150399.

Jeguirim, M., Limousy, L., Fossard, E., 2014. Characterization of coffee residues pellets and their performance in a residential combustor. Int. J. Green Energy 13, 608–615.

Jung, K.-W., Choi, B.H., Hwang, M.-J., Jeong, T.-U., Ahn, K.-H., 2016. Fabrication of granular activated carbons derived from spent coffee grounds by entrapment in calcium alginate beads for adsorption of acid orange 7 and methylene blue. Bioresour. Technol. 219, 185–195.

Kelkar, S., Saffron, C.M., Chai, L., Bovee, J., Stuecken, T.R., Garedew, M., Li, Z., Kriegel, R.M., 2015. Pyrolysis of spent coffee grounds using a screw-conveyor reactor. Fuel Process. Technol. 137, 170–178.

Kua, T.-A., Arulrajah, A., Horpibulsuk, S., Du, Y.-J., Shen, S.-L., 2016. Strength assessment of spent coffee grounds-geopolymer cement utilizing slag and fly ash precursors. Constr. Build. Mater. 115, 565–575.

Lambert, S., Sinclair, C., Boxall, A., 2013. Occurrence, degradation, and effect of polymer-based materials in the environment. Rev. Environ. Contam. Toxicol. 227, 1–53.

Lavecchia, R., Medici, F., Patterer, M.S., 2016. Lead removal from water by adsorption on spent coffee grounds. Chem. Eng. Transact. 47, 295–300.

López-Barrera, D.M., Vázquez-Sánchez, K., Loarca-Piña, M.G.F., Campos-Vega, R., 2016. Spent coffee grounds, an innovative source of colonic fermentable compounds, inhibit inflammatory mediators in vitro. Food Chem. 212, 282–290.

Lores, M., Gómez-Brandón, M., Pérez-Díaz, D., Domínguez, J., 2006. Using FAME profiles for the characterization of animal wastes and vermicomposts. Soil Biol. Biochem. 38, 2993–2996.

Ludwig, I.A., Clifford, M.N., Lean, M.E.J., Ashihara, H., Crozier, A., 2014. Coffee: biochemistry and potential impact on health. Food Funct. 5, 1695–1717.

Martinez-Saez, N., García, A.T., Pérez, I.D., Rebollo-Hernanz, M., Mesías, M., Morales, F.J., Martín-Cabrejas, M.A., del Castillo, M.D., 2017. Use of spent coffee grounds as food ingredient in bakery products. Food Chem. 216, 114–122.

Marto, J., Gouveia, L.F., Chiari, B.G., Paiva, A., Isaac, V., Pinto, P., Simões, P., Almeida, A.J., Ribeiro, H.M., 2016. The green generation of sunscreens: using coffee industrial subproducts. Ind. Crops Prod. 80, 93–100.

Megharaj, M., Ramakrishnan, B., Venkateswarlu, K., Sethunathan, N., Naidu, R., 2011. Bioremediation approaches for organic pollutants: a critical perspective. Environ. Int. 37, 1362–1375.

Meuser, H., 2010. Contaminated Urban Soils. Springer, Dordrecht; Heidelberg; London, p. 318.

Monroy, F., Aira, M., Domínguez, J., Velando, A., 2006. Seasonal population dynamics of *Eisenia fetida* (Savigny, 1826) (Oligochaeta, Lumbricidae) in the field. C. R. Biol. 329, 912–915.

Morillo, E., Romero, A.S., Maqueda, C., Madrid, L., Ajmone-Marsan, F., Grcman, H., Davidson, C.M., Hursthouse, A.S., Villaverde, J., 2007. Soil pollution by PAHs in urban soils: a comparison of three European cities. J. Environ. Monit. 9, 1001–1008.

Murthy, P.S., Naidu, M.M., 2012. Sustainable management of coffee industry by-products and value addition—a review. Resour. Conserv. Recyc. 66, 45–58.

Mussatto, S.I., Ballesteros, L.F., Martins, S., Teixeira, J.A., 2011a. Extraction of antioxidant phenolic compounds from spent coffee grounds. Sep. Purif. Technol. 83, 173–179.

Mussatto, S.I., Machado, E.M.S., Martins, S., Teixeira, J.A., 2011b. Production, composition, and application of coffee and its industrial residues. Food Bioprocess. Technol. 4, 661–672.

Nabulo, G., Young, S.D., Black, C.R., 2010. Assessing risk to human health from tropical leafy vegetables grown on contaminated urban soils. Sci. Total Environ. 408, 5338–5351.

Namane, A., Mekarzia, A., Benrachedi, K., Belhaneche-Bensemra, N., Hellal, A., 2005. Determination of the adsorption capacity of activated carbon made from coffee grounds by chemical activation with $ZnCl_2$ and H_3PO_4. J. Hazard. Mater. 119, 189–194.

Nannipieri, P., Kandeler, E., Ruggiero, P., 2002. Enzyme activities and microbiological and biochemical processes in soil. In: Burns, R.G., Dick, R.P. (Eds.), Enzymes in the Environment: Activity, Ecology, and Applications. Marcel Dekker Inc., New York, NY, pp. 1–33.

Orsini, F., Kahane, R., Nono-Womdim, R., Gianquinto, G., 2013. Urban agriculture in the developing world: a review. Agron. Sustain. Dev. 33, 695–720.

Phillips, B.M., Anderson, B.S., Voorhees, J.P., Hunt, J.W., Holmes, R.W., Mekebri, A., Connor, V., Tjeerdema, R.S., 2010. The contribution of pyrethroid pesticides to sediment toxicity in four urban creeks in California, USA. J. Pestic. Sci. 35, 302–309.

Popova, I.E., Deng, S., 2010. A high-throughput microplate assay for simultaneous colorimetric quantification of multiple enzyme activities in soil. Appl. Soil Ecol. 45, 315–318.

Rao, M.A., Scelza, R., Acevedo, F., Diez, M.C., Gianfreda, L., 2014. Enzymes as useful tools for environmental purposes. Chemosphere 107, 145–162.

Reffas, A., Bernardet, V., David, B., Reinert, L., Lehocine, M.B., Dubois, M., Batisse, N., Duclaux, L., 2010. Carbons prepared from coffee grounds by H_3PO_4 activation: characterization and adsorption of methylene blue and Nylosan Red N-2RBL. J. Hazard. Mater. 175, 779–788.

Ribeiro, H., Marto, J., Raposo, S., Agapito, M., Isaac, V., Chiari, B.G., Lisboa, P.F., Paiva, A., Barreiros, S., Simões, P., 2013. From coffee industry waste materials to skin-friendly products with improved skin fat levels. Eur. J. Lipid Sci. Technol. 115, 330–336.

Riva, S., 2006. Laccases: blue enzymes for green chemistry. Trends Biotechnol. 24, 219–226.

Rocha, M.V.P., de Matos, L.J.B.L., de Lima, L.P., da Silva Figueiredo, P.M., Lucena, I.L., Fernandes, F.A.N., Gonçalves, L.R.B., 2014. Ultrasound-assisted production of biodiesel and ethanol from spent coffee grounds. Bioresour. Technol. 167, 1–28.

Safarik, I., Horska, K., Svobodova, B., Safarikova, M., 2011. Magnetically modified spent coffee grounds for dyes removal. Eur. Food Res. Technol. 234, 345–350.

Saiya-Cork, K.R., Sinsabaugh, R.L., Zak, D.R., 2002. The effects of long term nitrogen deposition on extracellular enzyme activity in an *Acer saccharum* forest soil. Soil Biol. Biochem. 34, 1309–1315.

Sanchez-Hernandez, J.C., Aira, M., Domínguez, J., 2014a. Extracellular pesticide detoxification in the gastrointestinal tract of the earthworm *Aporrectodea caliginosa*. Soil Biol. Biochem. 79, 1–4.

Sanchez-Hernandez, J.C., del Pino, J.N., Domínguez, J., 2015. Earthworm-induced carboxylesterase activity in soil: assessing the potential for detoxification and monitoring organophosphorus pesticides. Ecotoxicol. Environ. Saf. 122, 303–312.

Sanchez-Hernandez, J.C., Morcillo, S.M., del Pino, J.N., Ruiz, P., 2014b. Earthworm activity increases pesticide-sensitive esterases in soil. Soil Biol. Biochem. 75, 186–196.

Sanchez-Hernandez, J.C., Sandoval, M., Pierart, A., 2017. Short-term response of soil enzyme activities in a chlorpyrifos-treated mesocosm: use of enzyme-based indexes. Ecol. Indic. 73, 525–535.

Schinner, F., Kandeler, E., Ohlinger, R., Margesin, R., 1996. Methods in Soil Biology. Springer-Verlag, Berlin; Heidelberg; New York.

Sena da Fonseca, B., Vilão, A., Galhano, C., Simão, J.A.R., 2013. Reusing coffee waste in manufacture of ceramics for construction. Adv. Appl. Ceram. 113, 159–166.

Seo, J., Jung, J.K., Seo, S., 2015. Evaluation of nutritional and economic feed values of spent coffee grounds and *Artemisia princeps* residues as a ruminant feed using in vitro ruminal fermentation. PeerJ 3, e1343.

Singh, B.K., Walker, A., 2006. Microbial degradation of organophosphorus compounds. FEMS Microbiol. Rev. 30, 428–471.

Sogorb, M.A., Vilanova, E., 2002. Enzymes involved in the detoxification of organophosphorus, carbamate and pyrethroid insecticides through hydrolysis. Toxicol. Lett. 128, 215–228.

Trasar-Cepeda, C., Camiña, F., Leirós, C.M., Gil-Sotres, F., 1999. An improved method to measure catalase activity in soils. Soil Biol. Biochem. 31, 1–3.

Tsai, W.-T., Liu, S.-C., Hsieh, C.-H., 2012. Preparation and fuel properties of biochars from the pyrolysis of exhausted coffee residue. J. Anal. Appl. Pyrol. 93, 63–67.

Turner, B.L., 2010. Variation in pH optima of hydrolytic enzyme activities in tropical rain forest soils. Appl. Environ. Microbiol. 76, 6485–6493.

Vane, C.H., Kim, A.W., Beriro, D.J., Cave, M.R., Knights, K., Moss-Hayes, V., Nathanail, P.C., 2014. Polycyclic aromatic hydrocarbons (PAH) and polychlorinated biphenyls (PCB) in urban soils of Greater London, UK. Appl. Geochem. 51, 303–314.

von Hoffen, L.P., Säumel, I., 2014. Orchards for edible cities: cadmium and lead content in nuts, berries, pome and stone fruits harvested within the inner city neighbourhoods in Berlin, Germany. Ecotoxicol. Environ. Saf. 101, 233–239.

Mersi, W., Schinner, F., 1991. An improved and accurate method for determining the dehydrogenase activity of soils with iodonitrotetrazolium chloride. Biol. Fert. Soils 11, 216–220.

Wheelock, C.E., Nakagawa, Y., 2010. Carboxylesterases—from function to the field: an overview of carboxylesterase biochemistry, structure–activity relationship. J. Pestic. Sci. 35, 215–217.

Wheelock, C.E., Phillips, B.M., Anderson, B.S., Miller, J.L., Miller, M.J., Hammock, B.D., 2008. Applications of carboxylesterase activity in environmental monitoring and toxicity identification evaluations (TIEs). Rev. Environ. Contam. Toxicol. 195, 117–178.

Wortman, S.E., Lovell, S.T., 2013. Environmental challenges threatening the growth of urban agriculture in the United States. J. Environ. Qual. 42, 1283–1294.

Wu, C.-S., 2015. Renewable resource-based green composites of surface-treated spent coffee grounds and polylactide: characterisation and biodegradability. Polym. Degrad. Stabil. 121, 51–59.

Wu, H., Hu, W., Zhang, Y., Huang, L., Zhang, J., Tan, S., Cai, X., Liao, X., 2016. Effect of oil extraction on properties of spent coffee ground–plastic composites. J. Mater. Sci. 51, 10205–10214.

Yamane, K., Kono, M., Fukunaga, T., Iwai, K., Sekine, R., Watanabe, Y., Iijima, M., 2015. Field evaluation of coffee grounds application for crop growth enhancement, weed control, and soil improvement. Plant Prod. Sci. 17, 93–102.

Yan, B., 2012. Hydrolytic enzymes. In: Anzenbacher, P., Zanger, U.M. (Eds.), Metabolism of Drugs and Other Xenobiotics. John Wiley & Sons, Hoboken, NJ.

Yang, L., Nazari, L., Yuan, Z., Corscadden, K., Xu, C.C., He, Q.S., 2016. Hydrothermal liquefaction of spent coffee grounds in water medium for bio-oil production. Biomass and Bioenerg. 86, 191–198.

Yen, H.Y., Huang, S.L., 2015. Ni(II) removal from wastewater by solar energy-degreased spent coffee grounds. Desal. Water Treat. 57, 15049–15056.

Yen, H.Y., Lin, C.P., 2015. Adsorption of Cd(II) from wastewater using spent coffee grounds by Taguchi optimization. Desal. Water Treat. 57, 11154–11161.

Zezza, A., Tasciotti, L., 2010. Urban agriculture, poverty, and food security: empirical evidence from a sample of developing countries. Food Policy 35, 265–273.

Zuorro, A., Lavecchia, R., 2012. Spent coffee grounds as a valuable source of phenolic compounds and bioenergy. J. Clean Prod. 34, 49–56.

FURTHER READING

Edwards, C.A., 2011. Vermiculture Technology: Earthworms, Organic Wastes, and Environmental Management. Taylor & Francis Group, LLC, Boca Raton, FL.

Kim, M.-S., Min, H.-G., Koo, N., Park, J., Lee, S.-H., Bak, G.-I., Kim, J.-G., 2014. The effectiveness of spent coffee grounds and its biochar on the amelioration of heavy metals-contaminated water and soil using chemical and biological assessments. J. Environ. Manage. 146, 124–130.

Index

CPI Antony Rowe
Eastbourne, UK
September 14, 2019